THE BOLETES OF MICHIGAN

Victor Clare Potter
1920-1964

THE BOLETES OF MICHIGAN

by

Alexander H. Smith and
Harry D. Thiers

Ann Arbor

The University of Michigan Press

*Published with the assistance of a grant
from the Horace H. Rackham School of
Graduate Studies of the University of Michigan.*

DEDICATION

THIS work is dedicated to Victor Clare Potter, of Ithaca, Michigan; born May 14, 1920, died January 11, 1964. Victor Potter was a victim of arthritis, but this did not dampen his enthusiasm for natural history. He had special crutches made so that he could carry a collecting basket on one of them, and, so equipped, he made, in the short period of less than fifteen years, a collection of fleshy fungi of approximately 20,000 specimens, which is the most significant collection in existence from the central part of the Lower Peninsula. Many of them were boletes, including one of the most peculiar species of *Leccinum* yet described, which now bears his name.

He clearly showed by his own efforts that the amateur naturalist can still make significant contributions to biology. Indeed, the project on the boletes of Michigan has profited significantly from the activities of a number of amateurs in the state.

CONTENTS

INTRODUCTION

THE boletes, or fleshy pore fungi, are a conspicuous element of the summer and fall "mushroom" flora of Michigan. Many are very attractive because of their bright colors or curious ornamentation in the form of scales on the pileus or reticulation or other ornamentation on the stipe. In addition, many species stain blue immediately when the surface of the tubes or the context is injured, and this never fails to excite the interest of the collector.

The typical bolete, like the typical "mushroom," consists of a pileus, hymenophore, and stipe (cap, tubes, and stalk in nontechnical terms). The true mushrooms have gills on the underside of the pileus instead of tubes, but boletes are gathered as mushrooms by most collectors. Many of the species are among the best native edible fungi and are worth considerable study on this account. We hope our book will help to increase the amount of the crop that is used for food.

The terms applied to the bolete parallel those used for true mushrooms. Both the "bolete" and the "mushroom" consist of the same materials as the spawn and are the fruit bodies of the fungous plant. The fruit bodies produce the spores, and since the spores in both groups are produced on a cell called a *basidium* (pl. *basidia*), the fruit bodies which produce the basidia are termed basidiocarps—the term that will be used hereinafter.

The basidia (fig. 1) are produced in a layer lining the tubes; the ends of the cells point into the tube cavity. The basidium is the terminal cell of a filament, or branch of a filament, and these cells arrange themselves with their ends facing the center of the tube. The layer of basidia is technically known as the *hymenium,* and whatever part of the basidiocarp bears the hymenium is known as the *hymenophore.* Thus, the tube layer in a bolete is the hymenophore; in the true mushroom it is the gills; in a hydnum it is the teeth. In the coral fungi it is simply the subhymenium under the smooth surface of the basidiocarp.

In some boletes, as in many mushrooms, a layer of tissue covers the young hymenophore, and by the time the latter is mature this tissue breaks away, leaving the tubes exposed to the environment and allowing the spores, when shed, to be carried away by air currents. The layer of tissue is termed a veil, and when its remains are left as a more or less

1

distinct ring of tissue on the stipe the ring is called an annulus. Often, however, the remains of the veil adhere to the margin of the pileus, and the margin is then said to be appendiculate or the margin itself may project as a thin membrane which finally becomes divided into segments. The tissue of the veil may be soft and cottony, membranous, or slimy. If slimy, very often the pileus surface is slimy also, and when such a surface dries it frequently has the appearance of having been varnished.

It will be noted that the tubes of a bolete are nearly always arranged vertically even if the stipe must curve to accomplish this. The vertical alignment enables the spores, when they are shot off from the basidia (for a short distance), to fall the length of the tube without being caught by the basidia or spores below. The spores, or most of them, fall free of the tubes and are carried away by air currents to new habitats which, if favorable, will serve as new locations for the fungous plant. Many spores, of the millions produced, fall in unfavorable localities and either do not germinate or, if they do, the young fungous plant soon dies.

The spores (figs. 2-4) apparently may rest in the soil for a fairly long time without starting to grow. Growth begins, we presume, when a bubble of protoplasm exudes through an opening in the spore wall and, later, filaments grow out of the bubble. Spores of boletes are very difficult to germinate.

Spores have rather distinctive shapes and markings in many species (figs. 22, 47, 113), and since these features are constant for a species the differences are very important in bolete classification. Spores can be collected easily by cutting off the stipe of a basidiocarp at the level of the tubes, standing the pileus on a piece of white paper with the tubes down but supported so as to keep them from touching the paper, and covering the setup with a container of some sort to cut off air currents. If the basidia are producing and discharging spores, in an hour or two the accumulation of colored spore dust will be noted under the hymenophore. The color of this "spore deposit" is important in classifying boletes. Excess moisture should be allowed to evaporate before a final notation is made as to the color of the deposit.

With this much of an introduction to the bolete basidiocarp, we now consider the plant which produces it. Some of us still regard fungi as plants, albeit they are a rather special type. The bolete plant is a mass of threads (spawn), in the main much like the spawn of the mushroom. Technically, we designate both as *mycelium*. The threads making up the mycelium are branched and, depending on the species, vary from 2 to

20 μ thick. Partitions, termed *septa* (sing. *septum*), are present at irregular intervals to divide the thread into typically tubular cells (with their sides parallel). The threads of the mycelium are called *hyphae* (sing. *hypha*), and they grow outward from the point where the spore dropped, usually in all directions, unless hindered by some obstacle. If the food material is evenly distributed, as in the soil for instance, the plant (mycelium) grows out in all directions—producing a circle if one were to draw a line around the outer edge. The youngest most active hyphae are at the outer edge of the circle and the oldest are at or near the center, which is the point of inoculation.

The basidiocarps are produced from food stored by the living mycelium and usually originate as a compact knot of hyphae, or at a point on a compound strand of mycelium termed a *rhizomorph.* In other words, the basidiocarps are formed of mycelium—essentially like that comprising the vegetative stage, but of course many changes occur as the basidiocarps develop. When mature we find many modifications of hyphae, and these are often characteristic of particular species. Also, the hyphae may be aggregated into patterns which we designate as tissues, such as hymenophoral trama (the context of the hymenophore). Much of the information to be learned in order to identify boletes accurately is concerned with the spores, the types of hyphae in the basidiocarp and their arrangement, and the accessory tissues, such as veils which may be present or absent depending on the species.

In a work designed for the identification of species there is little use in tracing the history of our knowledge of boletes, interesting as it is. Instead, let us consider the present status of the taxonomic features touched on so far as these are used in organizing a modern classification. These are discussed in detail later.

We present herein a reasonably complete technical account of each species and give a thumbnail sketch of it for the mushroom hunter, with, wherever possible, an illustration. So rapidly has popular interest developed in fleshy fungi in recent years that it is no longer appropriate to try to write "for the amateur" or "for the specialist." It seems much more practical simply to give the essential details in terms that can be understood by anyone who applies himself to the subject.

The bolete flora of Michigan has been under study for years: first by C. H. Kauffman from about 1904 to 1928, by Smith since 1933, and by Thiers at intervals since 1950. Smith in particular is indebted to W. H. Snell for aid in the field in 1934 and for consultations on various problems over the years. The field work in Michigan has been variously financed; in part by the senior author personally, in part by the University of Michigan Herbarium field expedition funds, in part by grants from the Faculty Research Fund of the University of Michigan, and in part as incidental effort while collecting in other groups, such as *Pluteus* and *Rhizopogon,* on grants from the National Science Foundation.

A complete set of the specimens reported here has been deposited in the University of Michigan Herbarium. All color terms within quotation marks indicate that the plate of that name in R. Ridgway, *Color Standards and Color Nomenclature,* matched the fungus part designated. Ridgway terms not set off by quotation marks indicate an approximation of the Ridgway plate of that name. The term *Melzer's* is used as an abbreviation of Melzer's reagent, the well-known chloralhydrate-iodine solution now used universally in the study of the fleshy fungi.

FORMULAE USED IN BOLETE STUDIES

1. Potassium hydroxide (KOH): A 2.5 percent aqueous solution.
2. Ammonia: (NH_4OH): A 14 percent (approximately) aqueous solution.
3. Iron salts ($FeSO_4$ or Fe_2Cl_2): A 10 percent aqueous solution.
4. Melzer's reagent:

KI . . . . 1.5 gr.   Water . . . . 22.0 gr.

Iodine . . . . . 0.5 gr.   Chloral hydrate . . . . . 22.0 gr.

The drawings for figs. 20 through 123 were made with the aid of a camera lucida. The magnification for the spores as reproduced is about 1650 X ; for the hyphae and hyphal cells and cystidia it is about 650 X . The drawings were made from type specimens unless otherwise indicated, except for European species which have been illustrated from Michigan collections.

THE SPORES

The features of the spores used in identification are the color in mass as obtained from air-dried spore deposits, size, absence or presence of ornamentation and if ornamentation is present the type and pattern shown, thickness of the wall and whether there is a germ pore at the apex, and the color as observed under the microscope in mounts in Melzer's and in weak potassium hydroxide (KOH).

In the boletes the spore deposit is colored—usually a yellow to olive-brown, but olive, dark yellow-brown such as bister, or cinnamon-brown are not uncommon. In a few species the deposit is pale yellow or rich yellow to ocher-yellow, rarely mustard-yellow. In two groups it is vinaceous to vinaceous-brown or lilac-brown to chocolate-brown (dark or light), and in a few it is blackish brown. The colors are those found to a large extent in the gill fungi.

Unfortunately for those who wish to make field identifications, it does take a little time to get a spore deposit, but this must be done if one's work is to be accurate. The best way to standardize the spore-deposit color is to dry out the spores by placing the deposit on top of activated silica gel in a tightly closed dish for about 15 to 30 minutes, then recording the color, using a color chart if one is available or, if not, by identifying the color as closely as possible in one of the previously mentioned categories. The best way to keep the spore deposit is to fold it so the spores face to the inside of the fold and staple this to the card on which one has written the notes for that particular collection. The spores should not be "fixed" to the paper; some will be wanted for later study.

Spore size is usually the first feature ascertained with the aid of a microscope. One should always use spores from a deposit because, having been discharged from the basidia, they are considered mature. Bolete spores are mostly long and narrow. Most of the Michigan species have spores 9-25 μ in length and 4-9 μ in width. One should measure at least 10 spores selected to represent the smaller, larger, and the mean size as hundreds are scanned in a mount. The usual ranges in size run something like 7-9 $\times$ 2.5-3.5 μ in some species of *Suillus* to 16-24 $\times$ 5.5-8 μ in some *Leccinum* species. Rarely does one find basidiospores up to 35 μ long. In measuring spores one seeks to establish the typical size range for the species, not the absolute range which would include the smallest and largest spore found. It is one of the marvels of nature that each species has a characteristic size range with such a narrow span for spores—when one realizes that millions of spores are produced. However, accidents do occur, and as extremes one finds a few exceptionally small and a few exceptionally large spores in most mounts. If these are numerous enough we sometimes indicate their presence by using parentheses around the numbers indicating the extremes as follows: (6) 8-12 $\times$ (3) 4-5 (7) μ.

In *Tylopilus sordidus* exceptional spores, though relatively few, are characteristic of the species. These are much wider than spores falling within the normal range and have much thicker walls. A second "abnormality" which is not infrequently encountered is an extra long and narrower than usual spore. As yet we do not know the cause for either of these patterns of abnormality, but are inclined to wonder whether the nuclear cycle has anything to do with them. Large basidia, twice as large as the adjacent ones in the hymenium may produce larger spores than the normal range, but in the few instances in which we thought this situation occurred, the spores from the large basidia were normal in shape. We have seen only one or two instances of spores with an aborted side protuberance near the apex so as to produce a mitten-shaped spore—usually larger than normal. All of these types of abnormality appear among colored-spored fleshy fungi generally and are not peculiar to boletes.

Spore shape is important at the species level, and the descriptive terms used in this work are here defined and illustrated. As mentioned, most bolete spores are elongate (much longer than wide). If one observes a basidium bearing nearly mature spores (fig. 1), the surface of the spore facing the extended longitudinal axis of the basidium is termed the *face view* and the one at right angles to it is the *profile view.* The side opposite the face is the back view, but it is seldom used in descriptions. The spore is usually of one outline in face view and presents a quite different outline in profile view (figs. 3-4). For spores with the outline shown in profile view (fig. 4) the term inequilateral is used. If a transverse line is drawn across the spore about midway between the two extremities, it is noted that on the line which represents the face view (and which is referred to as the *ventral line*) the bulge in it is mostly distal to the transverse line that was drawn across the spore. On the dorsal line the widest part of the bulge is toward the proximal end of the spores (the end that was attached to the sterigmata). Thus, in profile view the optical section of the spore is not symmetrical. The concavity outlined by the depression of the ventral line just distal to the point of attachment to the sterigmata is termed the suprahilar depression (fig. 4). If this is very pronounced, as in fig. 104, we describe the spore as *inequilateral.* If the depression is scarcely evident, the shape in profile is described as *obscurely inequilateral* (fig. 120). A condition between these is referred to as *somewhat inequilateral* (fig. 37). Most spores fall in this last category. Spores which in face view resemble fig. 101 are termed *fusiform* and, if the ends are blunt, *subfusiform.* The term ovate is used if the spore is broadest toward the proximal end. If the side walls are about parallel and there is not much taper at either end, the outline is described as *oblong.* Spores will usually vary across 2 of these categories so frequently that in profile the spores are described as *obscurely* to *somewhat* inequilateral, and in face view *subfusiform* to *fusiform*. The term elliptic (fig. 109) is also used in its regular meaning, but in boletes most spores of this type fall into the category of being narrowly elliptic (fig. 70).

A feature which causes variability in spore size is the number of spores borne on a basidium. On some pilei some basidia may bear 2 spores and some 4. It has been worked out to show that the amount of protoplasm used in spore formation is about the same in both cases—so the 2 spores from 1 basidium have about the same total volume as the 4 from the other basidium, and hence are larger. If 2- and 4-spored basidia are present in about equal numbers, one gets a bimodal curve—with 2 peaks—when spores from a deposit are measured. We have genetically constant strains of gill fungi that are 2-spored and others that are 4-spored (which is the rule), but we also have some in which 1-, 2-, 3-, and 4-spored basidia occur on the same hymenophore, and in such a case a spore deposit will give a confusing size range. Hence it is important to preserve at least part of the pileus from which the deposit was taken—or check the number of spores on a basidium while it is still fresh and record the number in one's notes.

Spores deposited from the hymenophore should always be used for ascertaining spore size. It is true that in some species of boletes there are patches of hymenium on the upper part of the stipe, and the basidia there often produce spores. Unfortunately, these spores are often different in size, shape, and wall thickness from spores produced on the hymenophore proper. Why this happens, as far as we are aware, is still unknown. It is important to keep in mind, however, because often one finds spore dust over the stipe apex, and one is likely to assume that all the spores there have been dropped from the hymenium of the hymenophore proper whereas one may get a mixture. Spore size is most important at the level of distinguishing species.

Spore ornamentation, or the lack of it, is often used to group species into genera both in the gill fungi and in the boletes. In the boletes there are a number of patterns. J. Perreau-Bertrand has made excellent drawings of spores from species of many regions, but the main types are represented in the Michigan bolete flora.

First we have a pattern of longitudinal folds or wrinkles in the outer wall (fig. 113). These may be so well developed as to produce "wings," or they may be only very fine lines or creases. The genus *Boletellus,* with several species from Michigan, features this type. A second type used to help define a genus is warty to reticulate. This ornamentation involves warts or a mesh of ridges formed (possibly) by outer-wall material. In *Strobilomyces* this ornamentation correlates with the blackish to chocolate spore deposit and the globose to subglobose spore shape to define a clearly delimited genus at least as far as our flora is concerned.

In many species, as already intimated, there may be an outer and an inner wall, the latter usually being the thickest. The remains of the outer wall may not always be readily visible. The thickest of the walls, however, may at times show canals or lines extending through it and filled with a substance different from that of the wall proper. The ends of these canals will show as irregularly shaped spots on the spore surface under an oil-immersion lens. We do not regard this type of ornamentation as being as significant taxonomically as the 2 types already discussed, for in species such as *Tylopilus gracilis* one often finds part of the spores in a deposit unornamented and part ornamented. The electron microscope is now helping to solve the problem of the differentiation present in the spore wall in such species. We believe that spore ornamentation in fleshy fungi generally has arisen "de novo" many different times and that all boletes, for instance, with ornamented spores are not necessarily more closely related to each other than some in the group may be to species with unornamented spores. This is why we regard the family Strobilomycetaceae as defined by Singer as "artificial."

The wall in most bolete spores is around 0.2 μ thick, which means it is too thin to measure accurately by the light microscope. It is rather seldom that one encounters spores with walls much over 0.5 μ thick. Also it makes some difference at times if the spores are mounted in KOH—the wall swells more than it does in Melzer's. For these reasons the thickness of the wall does not figure prominently in our description of species in the boletes. Of greater significance is the presence of apical differentiation—in the form of a notched apex (fig. 113) caused by inflation of the wall around it, or by a discontinuity (germ pore) in an apical position. The latter is a rare occurrence in boletes, though a character of major taxonomic importance in the gill fungi.

Recently, it has been found that spores of boletes may give a bluish reaction in Melzer's, but this feature is not as clear as in the agarics because of certain complications, the first being the rather highly colored spore wall. This obscures the reaction. It has also been found that spores taken from dried hymenophore tissue are dingy blue mounted directly in Melzer's but soon change to reddish tawny (dextrinoid) in at least the distal half of the spore, the proximal half merely fading to yellowish. This reaction has been termed a "fleeting-amyloid" reaction. It is most evident on species in the section *Luridi* of *Boletus* and species of section *Subtomentosi*. Its significance still remains to be evaluated. In spores mounted in KOH the colors are usually yellower or more rusty brown than in water mounts.

THE HYMENIUM

The hymenium is composed mainly of the hyphal end-cells modified for spore production—the basidia. These (figs. 1, 6, 8) are arranged in a palisade and typically are clavate in shape. There is considerable variation in size—those from the part of the hymenophore nearest the pileus usually being larger (both longer and wider) than those near the tube mouths (pores). Most basidia of the hymenophore in Michigan boletes are 4-spored. However, it is not uncommon in mounts to see a few 2-spored basidia, and hence there is always the problem of whether all the sterigmata develop simultaneously. Not many secondary characters of the basidium are important in the recognition of genera and species in the boletes. Measurements of basidia are not particularly useful for the reasons stated.

One distinguishes the basidia in the hymenium from the sterile elements by observing that sterigmata have developed on them (fig. 1). Similar-appearing cells in the hymenium but lacking sterigmata are termed *basidioles* (figs. 6, 8)—meaning young basidia. If sterile elements of different shape and more random distribution are present, they are termed *cystidia; pleurocystidia* if on the hymenophoral surface and *cheilocystidia* if on the edge (the pores). Those found on the stipe are termed *caulocystidia*. These designations merely indicate position. As pleurocystidia are commonly present in boletes, their presence versus their absence has never become a major taxonomic feature per se.

Pleurocystidia in the Boletaceae are found in 4 simple shapes with intergradations between types in some instances: In *Boletus* and *Leccinum* is found mostly the .fusoid-ventricose type (fig. 5), which is to be regarded as a very simple and rather universal type in the fleshy fungi. Two variations from it are important, first to a type with rounded apex (fig. 9), termed utriform. If the cell is merely clavate but has a short apical protuberance, it is termed clavate-mucronate (fig. 7). In *Tylopilus* the contents of the cystidium in many species may be dextrinoid (dark reddish brown) or a dark yellow-brown in Melzer's. They are a type of gloeocystidia termed *chrysocystidia*. In *Suillus* (fig. 25) the cylindric to narrowly clavate type is often present in bundles, a very unusual feature for pleurocystidia in the fleshy fungi, and when revived in KOH there is considerable dark brown amorphous pigment around the bundle or adhering to the individual cells or the cells themselves have a dark brown content. In fresh material these same cystidia may stain vinaceous in KOH mounts. This type of cystidium, with its color reaction and occurrence in bundles, is an important generic character for *Suillus* though not every species in the genus has it.

The third important type is that commonly termed *pseudocystidia* (fig. 10). These cells are known in *Leccinum* and are used to distinguish species. Basically, they are relatively undifferentiated hyphal tips (filamentous cells often irregular in outline) with an oily, granular content as seen in various mounting media. Presumably, they connect to hyphae characterized by similar contents and are regarded as part of the *laticiferous system,* as for some cystidia in *Tylopilus;* these come under the heading of gloeocystidia. Thick brown-walled "setae" are known in some extralimital species of boletes.

The cheilocystidia, like the cheilobasidia, are often similar to those on the wall of the tube, but reduced in size. The caulocystidia, however, are often fairly distinctive, and their abundance in patches of caulohymenium, in addition to their tendency to develop dark pigment, constitutes the major feature of the genus *Leccinum.* They may be clavate, fusoid-ventricose, or nearly filamentose and are known to produce long hyphal-like necks in some. At times a secondary cross wall may form above the inflated part (fig. 85). They vary so much in size on one basidiocarp that size alone is not of major taxonomic significance.

The cystidia are to be regarded as specialized types of hyphal end-cells and possibly are concerned with secreting moisture to keep the humidity high in the tubes where the spores are developing. It is difficult to prove that cystidia are modified basidia since both originate as hyphal end-cells but serve different functions.

THE HYMENOPHORAL TRAMA

The hyphal arrangement in the tube trama is obscurely to distinctly bilateral. By this is meant that in a longitudinal section of a tube a narrow, central, downward-oriented strand of hyphae has on each side of it hyphae oriented downward and at the same time bending (diverging) toward the hymenium. As the latter layer is approached the cells usually become shorter and branching occurs to give rise to the hymenial elements. The diverging hyphae in most boletes are somewhat gelatinous, which causes the whole hymenophore, to some extent, to have a similar consistency. Because the texture is very pliant, good freehand sections of the tube layer are difficult to obtain. Also, the degree to which the individual hyphae are separated from each other, or the angle at which they diverge, is likely to be variable for a number of reasons. First, the more gelatinized the hyphae become the farther apart from each other they are separated in the trama because of the swelling effect the intervening slime exerts. Second, if sections are made from material killed and fixed in such solutions as FAA, one cannot be sure the picture one gets actually represents the arrangement in the living material even though the technique always gives the same result. Sections may be cut on a freezing microtome, but here again the freezing may very well effect the degree of gelatinization of the hyphae. Finally, the differences, even when shown to best advantage, are not sufficient to be of material aid in distinguishing genera or species in the family. The feature of bilateral hymenophoral trama is most valuable as a character to aid in distinguishing the Boletaceae as a family—especially from the true polypores.

In the trama of the basidiocarp, in addition to the hyphae which form the basic tissue (*matrical* hyphae), one often observes hyphae distinguished in some way by a content different from that of the matrical hyphae. The content is often in the form of globular material or a dense homogeneous material which refracts the light differently than do the other hyphae. Collectively, such hyphae are regarded as a *laticiferous system.* We regard them as *repository hyphae* containing, possibly, by-products of metabolism, and the system is best regarded as a *storage system* rather than a system of "vascular hyphae" (which implies the function of conduction). The degree of their specialization in the boletes varies. Smith has seen the ends of some actually forming basidia with spores attached. Rarely in the boletes do they actually contain a latex that is exuded on injured places of the basidiocarp or on the developing pore surface of the hymenophore.

THE STRUCTURE OF THE PILEUS

For taxonomic purposes the dermal layer of the pileus is the most important and will be treated first. The important characters are in the arrangement of the hyphae and, second, in their anatomical features.

Two arrangements are most commonly encountered; in one the hyphal elements are short and are in a more or less upright palisade. This we term a *trichodermium* (fig. 86). As the pileus expands the trichodermial elements become aggregated into tufts exposing the context of the pileus between them, and the pileus is said to have become areolate-rimose.

The second type is represented by a layer of appressed interwoven hyphae generally termed a cutis, or if the hyphae are gelatinous, a pellicle. The two types, upright elements contrasted to appressed elements, appear to be sharply distinct at first glance, but in many species the layer appears to have been trichodermial in origin but as the elements elongated they became tangled and appressed. Hence in working with individual species some difficulty in interpreting the dermal layer is to be expected. Whenever possible sections should be made from young pilei.

Though hyphal detail is very important taxonomically, one can expect some difficulties to arise in using the various characters. The simplest type of trichodermial element is an unmodified hypha 2-4 cells long with the terminal cell essentially tubular like the rest and lacking wall thickenings, incrustations on the wall, or distinctive content in the cells such as amyloid granules. One of the most important modifications is in the shape of the cells of the element. They, or some of them, may inflate and become nearly globose. In some species the distal 3-4 cells become nearly globose, and if they become so inflated that they are packed tightly together a cellular layer (fig. 111), or "*epithelium,*" is said to be present. In such cases the trichodermial origin of the layer may be all but obscured. Sometimes only the apical cell becomes greatly enlarged, and sometimes the basal cells become more inflated than the others. On the other hand most of the differentiation may take place in the terminal cell of the trichodermial element, and the cells beneath may be so reduced in number and length that the pileus surface appears to be mainly a tangled turf of terminal cells (fig. 90).

Incrustations on the hyphal walls may or may not be present, and if they are, they may be very conspicuous to almost invisible depending on the species. We also find the hyphal cells reacting in various ways to chemicals, such as weak KOH, by the production of blisters, or the outer layer of the wall separating into thin plates (fig. 15) as the inner wall swells (see *Leccinum insigne*). We also find the pigment which gives color to the pileus located in the dermal hyphae. Hence it is important to observe their color (which is usually dissolved in the cell content) in both KOH and in Melzer's. In some species mounted in Melzer's the pigment rounds up into globules and the hyphal cell resembles a tube full of colored beads (fig. 12). In some the same effect is produced by KOH. Viscidity of the pileus surface (whether it is sticky or slimy to the touch) for a given species may be caused by the breakdown of the hyphal walls or the actual secretion of slime by the hyphae—in the latter case the walls of the hypha remain clearly defined under the microscope.

The context of the pileus is typically composed of interwoven rather wide thin-walled hyaline hyphae, but in some, as mounted in Melzer's, the content is seen to stand out boldly and become bright yellowish orange to cinnabar-red or bright orange-brown. As yet we do not know the significance of this feature or the identity of the compound which colors when iodine is added; it may be glycogen.

Clamp connections (fig. 21a) are present in some boletes, such as *Suillus cavipes,* but are absent in many others. In some species they appear to be very rare, and the taxonomic importance of their presence or absence is to be doubted in such cases.

In a few species of boletes, such as *B. calopus,* some hyphal fragments or special walls such as transverse septa appear to be truly amyloid (blue in Melzer's), a feature characteristic of *Chroogomphus,* a group of gill fungi apparently derived from the boletes and which, possibly, should be included in this family.

HABITAT RELATIONSHIPS OF BOLETES

This topic is of great interest to all collectors regardless of their intent. The various facets of the subject will be taken up in the order of importance to the collector. First are associations with specific woody plants.

This subject has been rather well explored because it is important to foresters. It is well known that boletes are forest fungi and that a very large number are found only near certain kinds of trees. *Fuscoboletinus ochraceoroseus* is always near *Larix occidentalis* (western larch), *Leccinum atrostipitatum* has always been found with species of *Betula* (birch), and *Suillus americanus* with *Pinus strobus* (white pine), etc. The reason for this close association is that the mycelium of the fungus is living on the rootlets of the tree, where it forms small structures with the tree rootlet. The combination of fungous mycelium and tree rootlet is termed a *mycorrhiza.* It is thought that both the tree and the fungus benefit from this association. At least the fungus does not infect the whole root system of the tree—killing it as might happen, for instance, if *Armillaria mellea* (the honey mushroom) were the invader. As books have been written about mycorrhiza, the discussion here is merely intended to state that the relationship exists. In addition to most coniferous trees, boletes are known to form mycorrhiza especially with oak, aspen (popple), birch, and beech. The mushroom hunter will take his cue from this.

When hunting for boletes, often the easiest technique pays off. They have a habit of fruiting along old roads through the woods, along the edges of woodlots, around isolated trees in old fields or in pastures, and along fencerows where are trees of the genera mentioned. Boletes are often abundant in relatively thin oak woods on poor sandy soil, and, of course, conifer plantations are an ideal place to hunt them during the fall season. In the deep forest they are often around uprooted trees or in relatively open areas.

Boletes have two fruiting periods, and one finds different species during each. In hot wet summer weather many species of *Boletus* fruit in the hardwood forests, and in cooler wet weather, usually in the fall, there are tremendous fruitings of species of *Suillus* under conifers. Local conditions, however, are all-important in finding the places which support the most luxuriant fruitings. Also, one should not expect a heavy crop every year. The fruiting pattern is tied closely with the weather and as everyone knows that is usually unpredictable.

Some oddities in regard to habitat relationships have been encountered in Michigan. We (1967) discussed one situation in which a rare species, *Suillus albidipes,* fruited in a local pine plantation. It was estimated that 5,000 basidiocarps were found on a certain day, whereas in a second plantation less than 10 miles away hardly a *Suillus* basidiocarp was present on the same day. Generally, however, a fruiting will be found to be quite extensive (over a wide area) in the proper habitat.

Some genera are abundant in about 1 season in 10, and during such a season one is apt to experience considerable frustration in trying to identify all of them. The greatest variety of boletes fruits during hot moist summer weather.

In Michigan species of *Leccinum* have posed a very difficult taxonomic problem, and our treatment of this genus is definitely preliminary. The genus, as it is represented in Michigan, presents a picture of many undescribed variants, and it will be years before these have all been properly studied and assigned a place in the classification or simply regarded as unstable variants. Why does Michigan have such a population? We have theorized on this question and offer the following possible answers: First, the same variability may occur in other regions also, such as New England, when that flora is studied on the basis of hyphal characters. However, it is possible that in all areas cut over at the time the virgin timber was harvested, then burned over repeatedly, and that finally "came back" with a cover of such weed trees as aspen and birch that these areas (as far as *Leccinum* is concerned) furnished an invasion site offering many variants a chance to survive that they would not have had under other circumstances. These variants undoubtedly came from areas in which aspen and birch had escaped the devastation of American-type logging and subsequent treatment of the land, such as swamps and stream borders, and which in a few years produced enough spores to invade the burned-over sites as these were reoccupied by trees. But few fungi are homozygous; this means that the invaders were for the most part all slightly different genetically. These mycelia became established and are still producing basidiocarps, and since they are not all genetically alike, the basidiocarps present differences confusing to the taxonomist. We have located certain mycelia and find that each year the basidiocarps from them are constant in the characters we are using to distinguish them, but before we can recognize these variants as species we must demonstrate that there is a population of mycelia all of which produces similar basidiocarps, and there should be evidence from collections studied that this population occupies a considerable area. This may take years to accomplish and is the reason we make no claim that this publication is "complete" for the genus in the state. Obviously, new variants are arising continuously, and species from other areas always possess the potential of migrating into the state, especially if a temporary weather cycle of 5 to 10-year duration favors this process. Since on occasion mushroom books are published, or republished, and their publishers make the claim that theirs is the only *complete* one, it is worth mentioning that there is no *complete* technical monograph for any group of higher fungi in North America, and one is not likely for a long time, and none of the popular mushroom books was designed to be "complete."

GENERAL COMMENTS ON DISTRIBUTION

If the boletes of the state are reviewed from the standpoint of their occurrence in other areas, as one would expect, most are New England species described by Peck and by Frost. *Suillus* is a typical example (see *S. acidus, S. americanus, S. subaureus, S. subluteus, S. unicolor, S. brevipes,* etc.). European species in the aggregate are in the minority, and some of these we now believe may have been introduced in relatively recent times—*S. luteus* being an example. Few endemic species inhabit our region, but *S. sphaerosporus* may be cited as one of the most important. Others, such as *S. intermedius,* are segregates from a complex group of variants distributed over the northern United States from the Great Plains eastward to the Atlantic, but are most abundant in the Lake Superior area. *Fuscoboletinus* as a genus is one with its center of distribution in the Great Lakes region, especially in Canada (*F. glandulosus, F. spectabilis, F. paluster, F. sinuspaulianus, F. aeruginosus,* and *F. grisellus*). In the West are only *F. ochraceoroseus* and *F. aeruginascens.* However, as this genus is found generally in Canada where the associated forest trees occur, it is actually very likely that these species are more specifically associated with a particular northern type of forest. A few western species, such as *S. punctatipes,* occur as far east as Michigan. *S. tomentosus* is much more abundant in the western states than in the East, but as it was described from this region it is considered eastern even though the largest number of variants occurs west of the Great Plains. Accidents of discovery of this type are numerous. Species have often been known more from the localities in which some active mycologist worked than from the area they generally occupy.

In other large genera in the boletes, such as *Leccinum,* the pattern of species distribution is rather different. On the basis of present information the center of speciation is the Great Lakes region, since it contains the largest number of endemic, as well as many of the most unusual, species. Whether or not this, again, is due to adequacy of the sample studied from Michigan as compared to that for the other areas, such as New England or the Rocky Mountain states, remains to be seen. As compared with Europe, however, there can hardly be any doubt that our Great Lakes flora outnumbers the species known from that continent by more than 3 to 1, even allowing for different species concepts. In *Leccinum,* however, we find it interesting that a fairly large number of species in section *Leccinum* are endemic for western states but that the *L. holopus* group, for instance, is very poorly represented there even in areas where birch is found. In *Boletus* section *Mirabiles* we find *B. mirabilis* very rarely in Michigan but abundant in the Pacific Northwest. Its sister species, however, *B. projectellus* is considered southern, but during some years is very abundant, at least along the southern shore of Lake Superior. One might ask whether both species originated in the Great Lakes area and each migrated in a different direction.

At this stage our estimates as regards the place of origin of bolete species and their patterns of dispersion are extremely tentative because we are still at the "Alpha" level (beginning stage) in our understanding of species per se, and we really have little knowledge of their distributions based on an adequate sample for the areas considered. Past estimates are hopelessly inadequate. It is hoped that the present work will stimulate the detailed study of boletes in other states so that the degree of diversity in the group as a whole can be made evident and the peculiarities of distribution properly documented. For instance, why should *Tylopilus eximius* be frequent in New England and rare in Michigan? Most New England species are common here.

No work which brings together for the first time what is known of a large group of fungi for an area as diverse as the state of Michigan can hope to be "complete," as has been pointed out. The problems that remain involve field work in the western part of the Upper Peninsula and in the southwestern corner of the Lower Peninsula. More time should also be spent on the limestone areas of the state. However, because boletes fruit at the whim of the weather, one must follow the rain pattern each season, emphasizing different geological formations secondarily. There are still many complex groups in need of further study. For example, it is only in the past five years that we have finally realized the diversity which exists in *Leccinum*, and we are now busy collecting in old as well as new areas to obtain a representative sample of the genus.

GENERIC SEGREGATES FROM BOLETUS AND THEIR RELATIONSHIPS

When one raises to family rank the group of species comprising *Boletus* (sensu Fries) and includes in addition *Strobilomyces* and *Boletinus*, certain problems immediately confront the investigator. As is evident from the number of genera described in efforts to arrive at a natural classification of the boletes, there have always been—and for that matter still are—strongly diverging opinions as to how generic lines should be drawn for the group. This is not the place, in an account of a state flora, to review the long history of generic segregates from *Boletus*, for it started in 1821 with S. F. Gray, and January 1, 1821, is the starting date for the nomenclature of the group. Some justification, however, for the classification we are proposing seems appropriate.

Of the genera we recognize, there seems to be no significant difference of opinion among current investigators on *Strobilomyces*, *Gyroporus*, *Phylloporus*, or *Pulveroboletus* (in the sense of the type species).

If we consider the other genera in the order in which they appear in the text, *Suillus* should be considered first.

If any genus is to be judged on the basis of its combination of characters, *Suillus* is, next to *Boletus* itself, the most distinctive genus in the family. Perhaps it is pertinent to list some guidelines as to what should be required in the way of characters or combinations of them to justify erecting a genus, or recognizing one previously described. This approach, however, is rather mechanical. Our aim should be to arrive at a classification which is "natural" in the sense that the most recently evolved (related by descent from common ancestors) species should be placed together in a group. We have a number of categories in our system of classification: Order, family, genus, species, subspecies, variety, and form. Each of these above the rank of species may be divided into subgroups just as is indicated for species: Subgenus, section, subsection series, and stirps, for instance, under the heading of genus. The smallest group which we recognize above the rank of species is the *stirps* (pl. *stirpes*). In it we place a central species along with those obviously very similar to it. The stirps bears the name of the central species. It is understood, when the term stirps is used, that the author or authors regard the group as having common parentage. We often refer to a stirps as a central species with its satellites. The stirpes are then grouped into subsections, sections, and subgenera, depending on the degree of differences shown by the members of each group. Such a classification is by its plan of organization a natural one, at least in the eyes of those who propose it. But if we regard a classification from this point of view, it is at once evident that what is really important in establishing a genus is not "a bundle of characters," but rather what value the characters have in actually indicating the relationships of the fungi involved. This necessitates assigning different values to various characters since some are better indicators of relationship than others. One might define a genus, then, as a major segment of evolution (or "line") as this is measured by the populations that we find in nature and that exhibit the combinations of index characters we believe important in assessing degrees of relationship. This is why, generally, genera are so different as regards number of species: *Cortinarius* with nearly 800 species in North America, *Pulveroboletus* with 1—and it is at this point that sharp differences of opinion as to what constitutes a genus are encountered.

We have tried to arrive at a reasonably well-balanced approach to this problem, which means that we recognize as genera the more distinctive groups of obviously related species in the Boletaceae according to an evaluation of characters that can be applied on a broad base throughout the fleshy fungi.

Let us consider *Suillus* in the light of these comments. Its species for the most part are readily recognized by their aspect in the field, and this "thumbnail" characterization can be backed up by reliable morphological and anatomical features. In any natural genus we find a core of those species "typical" of the genus as well as some lacking 1 or 2 of the important features. In other words evolution as a process leaves in its wake certain nonconformist populations as well as those which conform to the genetic pattern of the group. The nonconformists often end up as genera containing only the one species—as does *Pulveroboletus.* It is a matter of judgment as to whether an odd species is distinct enough to deserve being placed in a separate genus, or whether it should be placed in a section or subgenus of a previously described genus. So it is with *Suillus.* We have a large core of clearly related species and then a certain number of "Anhang" (hangers on) species which do not have all the characters of the genus, but are "closer" to it than to any other genus. Any truly natural genus will show this pattern, since evolution as a process became highly refined long before man began to try to plot its course by studying the populations that it left in its wake. In order to make our assessment of the course of evolution as objective as possible, we must select some characters as a working basis for recognizing groups or our classification becomes "intuitive" to a dangerous degree (by intuitive is meant that an investigator insists that 2 groups are closely related though he cannot give good reasons for his opinion).

In *Suillus* some very concrete characters support the impression both the amateur and specialist readily detect in the field, namely, that two different fungi belong in the same genus. The features are:

1) A characteristic yellow coloration usually over the pileus tubes and stipe.
2) A viscid pileus.
3) A dull cinnamon to olive spore deposit.
4) Subcylindric to narrowly clavate pleurocystidia, often in bundles, which often undergo color changes with KOH or have incrusting material around the bundles when revived in KOH.
5) Caulocystidia in fascicles or patches which become colored as the basidiocarp develops, producing what are termed glandular dots.

The above characters also show a pattern of evolution within the genus. In addition to these features, certain others also show a progression of development within the genus, but their pattern of occurrence is such that meaningful generic separations cannot be made by using one or several in combination to subdivide *Suillus* further into smaller genera. These are: the presence of a veil and whether it leaves an annulus (ring) on the stipe. The veil in *Suillus* grows over the tube cavity from the margin of the pileus, and as we (1964) showed, it is present in some degree on species with wide pores as well as on species with minute pores and is present on specimens with a dry pileus as well as on those in which the pileus is viscid or slimy. Some species (*Suillus sibiricus,* for example) have a veil which part of the time leaves an annulus and the remainder of the time adheres to the margin of the pileus.

A second character involves clamp connections. For a time it was thought that this was a very important feature. Studies in pure culture, however, showed that for most species clamps could be found on the mycelium; but they are often rare, in some occasional, and in some readily detected, as in *S. cavipes.* Some authors have tried to maintain the genus *Boletinus* for species with clamps, a dry pileus, and a stipe which becomes hollow in the base. But *Suillus sibiricus,* a viscid species, also tends to have a hollow stipe base. *S. pictus* has a dry pileus and lacks clamps on the basidiocarp hyphae. In *S. lakei,* a western species, one can hunt an hour to find a clamp—but the clamps are present. One variety of this species has in the pileus such a poorly defined gelatinous layer that it often goes undetected. In the type variety, however, it is much more distinct. This interlocking pattern of characters among these fungi convinced us that *Boletinus* should not be ranked as a genus. We rebel at having "first-class" and "second-class" genera. Consequently, we maintain *Suillus* as we defined it in 1964. It has, in this concept, all the requirements for a truly natural genus, particularly since within it one can follow the evolution of a number of characters.

Let us now examine the important features of *Suillus* as a genus and see if evolutionary changes in certain of its characters can be recognized in various other genera. In Singer (1963) the Gomphidiaceae are recognized as a family of gill fungi considered related to the Boletaceae, particularly in the area of *Suillus.* We accept the idea of this relationship. But why do we believe that 2 such distinct groups of fungi are indeed related? Obviously, they have certain features in common: (1) elongate spores of the type known as *boletoid;* (2) greatly elongated pleurocystidia which in some react in KOH, giving characteristic color changes, and which are more like the cystidia of *Suillus* than of any other group of fleshy fungi; (3) caulocystidia are basically similar also and even occur in bundles as in typical species of *Suillus.*

The differences are that the spore deposit in the Gomphidiaceae is dark brown to blackish or olive-fuscous. The hymenophore is lamellate instead of poroid. There also are a number of secondary characters, such as the size of the cystidia and the size of the spores, with a tendency of the basidiocarps in *Gomphidius* to blacken when bruised, a feature not characteristic of *Suillus* but present in the genus to some extent. A major difference between *Suillus* and the Gomphidiaceae, presumably, is the color of the spore deposit. How can a black-spored group be related to one with cinnamon to cinnamon-buff or olive spores? There seems to be a sharp gap here which would be considered significant generally in the classification of the fleshy fungi. The answer to the question brings into the discussion *Fuscoboletinus,* a genus which has been the object of rather sharply diverging opinions. Its features are much like those of *Suillus,* and the controversy has centered around whether or not its species should be placed there. The strongly diverging character is the reddish to chocolate or purple-drab spore deposit. In this genus, also, the hymenophore varies from poroid to practically lamellate (see *F. paluster*). Also, in some species the hymenophore is pallid to gray, a feature of *Gomphidius.* Those who recognize *Fuscoboletinus* see in it a bridge between *Suillus* and the Gomphidiaceae. The bridging character is the progression of slightly different spore-deposit colors as one progresses from species to species—in other words there is a character- istic spectrum for the genus which connects plausibly to *Gomphidius* on the one hand and *Suillus* on the other. The key difference in the spectrum for *Suillus* and that of *Fuscoboletinus* is the development of red tones in the latter at the *Suillus* end of the line, and the dark chocolate colors mixed or shaded with violet at the opposite end. The characters associated with spore-deposit color in *Fuscoboletinus* are those mainly cited for *Suillus,* and we find them, but in a different degree of evolution, in the Gomphidiaceae—for instance, the spores are larger and with thicker walls, but the shape is essentially the same. The pleurocystidia and cheilocystidia are typically elongate and reactive in KOH, but in some the walls have thickened. In other words, though basic similarities in characters are still evident, the details in individual species have changed—this is why the group is distinct from the parent group. This is the essence of evolution: To us, in the fungi under discus- sion, the important feature is that each genus (*Suillus, Fuscoboletinus, Gomphidius*) marks a major step in the evolution of a very distinctive type of gill fungus from a poroid ancestor. This, as we have previously emphasized, is our justification for recognizing *Fuscoboletinus.*

Before making a final decision as to whether *Fuscoboletinus* is a "good" genus or not, one should not neglect the other genera of the Boletaceae, since parallelism in evolution is a most important consideration, especially in the fungi. One should ask: What is the overall color spectrum in the fleshy pore fungi? In the *Strobilomyces* group it is blackish brown, in *Boletellus* it is olive to olive-brown for the most part, in *Leccinum* it is dull cinnamon, olive, olive-brown, chocolate-gray, or olive-fuscous, in *Boletus* it is olive, olive-brown, yellow-brown, amber-brown, or more rarely cinnamon-brown; in *Tylopilus* it is vinaceous, vinaceous-cinnamon, cinnamon, wood-brown, or chocolate (reddish tones predominating). If we compare *Fuscoboletinus* in this series, the spores are lilac-drab, violaceous-brown, reddish brown, wood-brown, or avellaneous (gray-tinged pink). In its color spectrum, obviously, *Fuscoboletinus* is closest to *Tylopilus,* which has always been recognized as having reddish spores as a key character. Hence *Fuscoboletinus* as a genus fits logically into the general classification of the boletes.

The most ambiguous genus as regards spore-deposit color is *Leccinum.* Its spore-deposit color spectrum covers a wider gamut of variation than that of any other genus. However, *Leccinum* is a large and very distinct genus as defined on the basis of the darkening stipe ornamentation. It is so homogeneous that many investigators have assumed that there were less than a dozen species in it. This led to a neglect of the genus in favor of groups apparently more in need of critical study. Its position in our classification involves a relationship to *Tylopilus* on the one hand and possibly to the species of section *Subtomentosi* of *Boletus* on the other. As the caulocystidia are very unlike those of *Suillus,* a close relationship to that genus is not indicated, even though in a few species such as *L. subspadiceum* the stipe ornamentation might at first be described as glandular-punctate.

As far as we are concerned *Boletellus* is based on the feature of more or less longitudinally striate spores. In other respects the species seem to connect with the *Subtomentosi* of *Boletus*. Some of the species would be placed in the *B. chrysenteron* group were it not for this feature. However, on a world basis the longitudinally striate-spored species appear to form a core of species indicating a phylogenetic line and hence should have generic recognition. There has been considerable evolution of secondary characters such as stipe ornamentation in the group. *B. russellii* is an example. We refer those smooth-spored species Singer (1936) placed in *Boletellus* to the section *Subtomentosi* of *Boletus.* Even as thus restricted by the process of segregating groups with special features as genera, *Boletus* is still a large genus, but it is much more homogeneous than in its original sense. Of the groups recognized in it here we would single out for comment first of all subsection *Luridi* of section *Boletus*. It is based primarily on the color of the pores—orange to red to brown—but the core of the subsection is composed of species with very narrow (3-7 μ) hyphae composing the cuticle of the pileus. In addition to this core of related species we find, in groups separated by other characters, species with colored pores, as in *Leccinum*, which are not closely related to the *Luridi*. In other words, for the most part *Luridi* is a typical natural group even though the index character for it may seem superficial to some. It is true that the character of colored pores simply has its basis in pigment dissolved in the cell sap of the cheilocystidia. Such occurrence of pigment in special cells is well known throughout the fleshy fungi—it is almost routine. *Mycena* and *Lactarius* are relatively unrelated genera of gilled fungi which have species featuring the character.

Subsections *Calopodes* and *Boleti* of *Boletus* represent the core of section *Boletus*. We regard them as forming an end line in the evolution of those species with a *Boletus*-subtype arrangement of the hyphae of the hymenophore. We still need a truly critical account of the *edulis* group for North America—and for the world for that matter. It is in this group that new characters, such as iodine reactions on spores and context tissue, will aid in clarifying taxa. *B. calopus* is a good example, for in it the transverse septa of the hymenophoral hyphae have blue walls as mounted in Melzer's (see Watling & Miller, 1968). But as is evident from the tone of this whole discussion, these new characters need to be studied in all groups in the family to ascertain if a pattern of evolution can be established showing how they reached their present degree of development.

One genus left out of our system, and which each of us at one time or another has accepted, is *Xerocomus*. We have divided the species of this group on other features than the velvety to subtomentose pileus, wide tubes, and the arrangement of the hyphae of the hymenophoral trama—the *Phylloporus* subtype—by which the genus is currently defined. One reason for abandoning the genus is that other workers, Singer included, who have accepted it have not in their own work showed that they had a clear concept of what the genus should be. We have found it an impractical grouping as far as identification is concerned, and the relationships in section *Subtomentosi* do not follow well the guidelines laid down for recognizing the genus. In its makeup it seems comparable to subsection *Luridi*—a core of related species and others not related but with some of the stated diagnostic characters of the group. To us it is not a question of whether a slight difference in the arrangement of hyphae in the hymenophoral trama can be demonstrated, for we have done this, but rather what does the slight difference mean?

We regard the group starting with section *Piperati* to be the trunk, so to speak, of the evolutionary "tree" of the boletes.

As to the overall evolution of the boletes, we visualize *Gyroporus* with its very pale spores as perhaps being relatively primitive, but along with it we consider *Suillus* as being close to the point of origin. At least 3 lines can be followed in *Suillus.* One, as pointed out, leads to *Gomphidius,* another to section *Piperati* of *Boletus,* which presumably originated from the more unspecialized species of *Suillus,* a third to the highly evolved species of *Suillus,* such as *S. luteus,* in which one finds both a distinctly glandular dotted stipe and a characteristic annulus. From section *Piperati* it is easy to visualize a connection to the *B. pallidus* and *B. badius* types and these clearly lead into the more xerocomoid fungi of the section *Subtomentosi.* Throughout this progression, as in all major lines, we have "Anhang" species such as *B. parasiticus* which have made unusual adjustments—in the present case an ability (dependency) to parasitize species of *Scleroderma* (a gastromycete). At present *Phylloporus* is placed in *Paxillaceae,* though it does have some features of the *Subtomentosi.*

Gastroboletus, the gastromycete with *Boletus*-type gleba, is regarded as an artificial group in the sense that these species have evolved independently from various lines in the family Boletaceae and are more closely related to their parental types than to each other. The question of the origin of the boletes is an open one. Two possibilities suggest themselves: first, an origin from *Suillus*-like ancestors in the Gastromycetes, *Truncocolumella* for instance; and second, origin from *Gomphus*-like ancestors. As yet neither possibility has been explored sufficiently to give an unequivocal answer.

Lastly, one must always keep in mind that all taxonomy is tentative and is continuously being modified by more information on individual groups. As yet we lack biological studies on most of these fungi, and when these are made, they will certainly modify taxonomic concepts arrived at from a study of material collected in nature. This is not an idle generalization. The boletes as a group are so closely related that the problem of gene exchange across taxonomic categories, when we have reliable information, is bound to modify existing concepts. For instance, all of the boletes in section *Subtomentosi* are in a sense so closely related that one is inclined to interpret the various combinations of characters as presented by our present species concepts as representing a "hybrid swarm" situation, even though hybridization has not been proven in laboratory studies. It would be curious indeed if the higher fungi did not hybridize. Certainly, the field evidence in *Leccinum,* for instance, indicates this. As cytological and genetical studies progress we may even find the classification we have proposed here is too radical rather than too conservative.

The BOLETACEAE Chevalier

Flore Env. Paris p. 248. 1926 (as "ordre")

We recognize only 1 family for pore fungi with fleshy readily decaying basidiocarps typically centrally stipitate, and with the hymenophoral trama as seen in a longitudinal section having hyphae more or less in bilateral arrangement. We place the family in the Agaricales. We place *Phylloporus* in the *Paxillaceae,* following Watling (in press).

Type genus: *Boletus* Fries.

KEY TO GENERA

1. Spore deposit not obtainable—the tubes usually oriented in such a way that the spores cannot fall free of them; spores not discharged from the basidia in the normal manner . . . .*Gastroboletus*
1. Spore deposit readily obtained from basidiocarps in good condition and near maturity 2

    2. Pileus covered with coarse dry gray to blackish scales; spores globose to subglobose and their surface reticulate to verrucose; spore deposit blackish brown or nearly so . . . . *Strobilomyces*
    2. Pileus not as above; spores smooth or if elongated then either smooth or ornamented 3

3. Spores ornamented by longitudinal wings, folds, or striations . . . .*Boletellus*
3. Spores smooth or the ornamentation different than in above choice 4

    4. With any 2 or more of the following features in combination: (a) hymenophore boletinoid at maturity; (b) stipe glandular dotted; (c) veil leaving an annulus on stipe; (d) pileus glutinous to viscid; (e) pleurocystidia in bundles and as revived in KOH with incrusting brown pigment around or on the bundle 5
    4. Not with above combinations of features 6

5. Spore deposit grayish brown to wood-brown, vinaceous-brown, purplish brown, chocolate-brown to purple-drab . . . .*Fuscoboletinus*
5. Spore deposit dingy yellow to yellow-brown, pale cinnamon-tan, olive, olive-brown, or a greenish mustard-yellow . . . . *Suillus*

    6. Spore deposit pale yellow; spores more or less ellipsoid; stipe hollow at maturity . . . . *Gyroporus*
    6. Not as above 7

7. Veil dry and flocculent to almost powdery (in appearance) and typically leaving a zone or ring on the stipe when it breaks . . . .*Pulveroboletus*
7. Veil lacking or not as above if present 8

8. Spore deposit gray-brown, red-brown, vinaceous, vinaceous-brown, or purple-brown (if stipe bears dark points or squamules see *Leccinum*) .... *Tylopilus*
8. Spore deposit yellow-brown, rusty, yellow, olive, olive-brown, dark cinnamon-brown to pale cinnamon or amber-brown 9

9. Stipe scabrous-roughened with the ornamentation dark from the first or darkening as the basidiocarps age (often blackish finally) .... *Leccinum*
9. Stipe not ornamented as above: either naked, furfuraceous, pruinose, subsquamulose, reticulate, or lacerate-reticulate .... *Boletus*

GYROPORUS Quélet

Enchir. Fung. p. 161. 1886

The short, ellipsoid spores which are yellow in a deposit and the hollow stipe at maturity characterize this genus. A veil may or may not be present. The pores are small and nearly circular in outline. The cuticle of the pileus may be a trichodermium or a cutis of appressed fibrils. Type species: *Gyroporus cyanescens* Fries.

KEY TO SPECIES

1. Context of pileus instantly changing to blue to lilaceous to deep violet when cut . . . .*G. cyanescens* var. *violaceotinctus*
1. Context not changing to blue or violet readily 2

    2. Pileus mineral red to dark vinaceous-red . . . . *G. purpurinus*
    2. Pileus rusty to chestnut-brown, tawny or crust-brown . . . . . . *G. castaneus*

1. Gyroporus cyanescens (Fries) Quélet

Enchir. Fung. p. 161. 1886

var. **violaceotinctus** Watling

Notes from Royal Bot. Garden, Edinb. 29(1): 63. 1969

Illus. Pl. 1.

Pileus 4-12 cm broad, obtuse to convex, becoming broadly convex, sometimes nearly plane or shallowly depressed; margin thin, frequently splitting, incurved when young, entire; surface dry, uneven to pitted or wrinkled, appressed fibrillose, ground color pale straw color ("light buff") or occasionally near cinnamon-buff, entire basidiocarp often staining blue when old, fibrils and squamules occasionally cinnamon-buff to clay color. Context thick, (10-15 mm), brittle, pallid but instantly indigo-blue when cut or broken, odor and taste not distinctive.

Tubes pallid (white to "ivory-yellow") at first, gradually pale olive-yellow ("Naples-yellow" to "seafoam-yellow") with age (bright yellow becoming pallid), instantly indigo-blue where injured, deeply depressed around the stipe, 5-10 mm deep; pores small and round, about 2 per mm, instantly indigo blue when bruised.

Stipe 4-10 cm long, 1-2.5 cm thick at apex, usually enlarged downward and 2.5-3.5 cm thick at base, occasionally clavate, frequently

28

quite irregular in shape, hollow and easily broken in removing it from the substratum, surface dry, tomentose to appressed-fibrillose, often becoming glabrous toward the apex or with age, not reticulate, instantly blue when injured.

Spore deposit pale yellow ("colonial-buff" to "straw-yellow"); spores 8-10 X 5-6 μ, smooth, lacking an apical pore, elliptic in face view, subelliptic in profile or slightly reniform (the dorsal line in an optical section nearly straight), with an inconspicuous hyaline sheath, greenish hyaline in KOH, pale tan in Melzer's.

Basidia 24-30 X 9-12 μ, clavate, 4-spored. Pleurocystidia absent as far as observed. Cheilocystidia 33-56 X 7-10 μ typically abundant, rarely apparently absent, thin-walled, narrowly clavate to subcylindric, apex somewhat tapered and thickened, walls sometimes appearing pale yellow in KOH and nearly orange in Melzer's.

Tube trama of hyphae divergent from a distinct central strand, nonamyloid, often staining yellowish in KOH. Cuticle of pileus a thick tangled trichodermium of filaments which become appressed in age and resemble a cutis, the hyphae thin-walled, the walls picric yellow revived in KOH, the cells 8-15 μ or more in diameter and often sausage-shaped or variously somewhat inflated, smooth, nonamyloid, and content not distinctly colored in Melzer's. Context of floccose, interwoven hyphae, nearly hyaline to faintly yellowish in KOH, the walls nonamyloid, in Melzer's the content of many of the hyphae reddish orange and homogeneous. Surface of the stipe with hyphae similar to those of the pileus cuticle. Clamp connections regularly present.

Habit, habitat, and distribution.—Solitary to scattered or cespitose but also gregarious, along roads or railroad cuts on exposed soil or on humus in open hardwood stands. It is not uncommon, and most likely to be collected in quantity along sandy roads through hardwoods. It fruits during the summer and early fall. We have seen it most abundantly in Luce County.

Observations.—This is rated as an exceptionally good edible species in Europe, but in North America, where emphasis has been on avoiding species which turn blue, it has not come into much prominence as an esculent. Getting rid of the sand is a problem with us. It is one of the species which can be accurately identified in the field. The overall buff to olive-tinged coloration, the slight veil (observe buttons), hollow stipe, and above all else the quick change to blue when any part is touched, serve to characterize it. Var. *cyanescens* turns greenish yellow before progressing to blue. We have not found this variety in Michigan.

Material examined.–*Barry: Smith 51217. Cheboygan: Shaffer 1584; Smith 25924, 37630, 62976. Chippewa: Mains 32-521. Emmet: Kauffman 7-3-05; 7-31-05; Mains 32-750; Thiers 4255. Gratiot: Potter 3479, 3971, 5341, 6074, 8413, 8478, 9850, 10456, 11075, 12325, 12373, 13992. Houghton: Pennington. Jackson: Mazzer 4298; Smith 62843. Livingston: Smith 6808. Luce: Smith 39289, 39501, 42286; Thiers 3850, 4033. Mackinac: Smith 44109, 4411. Marquette: Bartelli 283; Kauffman 8-27-06. Ogemaw: Smith 67420. Ontonagon: Shaffer 3785. Schoolcraft: Smith 63056. Washtenaw: Shaffer 2586, 2621, 2637, 2771; Smith 6718, 18573, 62899, 64129, 72554; Thiers 4554, 4576, 4585; Wehmeyer 7-16-21; Zehner 85.

2. Gyroporus castaneus (Fries) Quélet

Enchir. Fung. p. 161. 1886

Boletus castaneus Fries, Syst. Myc. 1:392. 1821.
Boletus cyanescens fulvidus Fries, Syst. Myc. 1:395. 1821.
Boletus fulvidus Fries, Epicr. Syst. Myc. p. 426. 1838.
Boletus testaceus Persoon, Mycol. Eur. 2:137. 1825.
Suillus castaneus (Fries) Karsten, Bidr. Finl. Nat. Folk, 37:1. 1882.
Gyroporus castaneus var. *fulvidus* (Fries) Quélet, Enchir. Fung. p. 161. 1886.
Boletus rufocastaneus Ellis & Everhart, N. Amer. Fungi 2d Ser. no. 2302. 1890.
Coelopus castaneus (Fries) Bat., Bolets p. 12. 1908.

Illus. Figs. 21, 23; Pls. 2-4.

Pileus 3-7 (10) cm broad, obtuse to broadly convex, expanding to broadly convex to plane or shallowly depressed, the margin often split, margin entire and straight at first but often flaring at maturity; surface dry and unpolished to pruinose to almost pulverulent, sometimes appearing glabrous near the margin, sometimes appearing granulose-reticulate when old; color variable, chestnut-brown to biscuit color or tawny-orange to orange or reddish or cinnamon to bright yellow-brown ("cinnamon-brown" to "ochraceous-buff" or "ochraceous-orange" or "orange-rufous" to "Sanford's brown" to "Mars-yellow" to "pinkish cinnamon"), to fulvous to cinnamon or lateritious. Context up to 1 cm thick, white, unchanging when bruised or exposed, fragile, taste slightly acid to not distinctive, odor not distinctive, KOH—brownish, FeSO$_4$—brownish.

Tubes shallow (5-8 mm long), free or deeply depressed around the stipe, rarely adnate to subdecurrent, white when young, becoming yellowish ("cream color") or brighter yellow ("Naples-yellow" to

*The collections are arranged by counties and the latter alphabetically.

"barium-yellow" to "seafoam-yellow") at maturity, unchanging when bruised or merely changing to brownish; pores typically small (1-2 per mm), round to angular, in very old basidiocarps more angular and up to 1 mm diameter, not staining when bruised.

Stipe 3-7 (9) cm long, 6-8 (15) mm thick at the apex, about 2.3 cm near the base, often pinched off at the base, equal to slightly ventricose or subclavate, hollow in lower portion at least in age and very fragile, surface dry, unpolished, glabrous, colored evenly like the pileus or paler above ("cinnamon-buff" to "cinnamon"); veil absent.

Spore deposit a clear light yellow ("pale lemon-yellow"); spores 8-12.6 X 5-6 μ, smooth, lacking a germ pore, ellipsoid to occasionally ovoid, with a distinct hyaline sheath, pale tawny in Melzer's, nearly hyaline in KOH, relatively thin-walled.

Basidia 30-36 X 9-12 μ, clavate, 4-spored, yellowish hyaline in KOH. Cheilocystidia abundant, narrowly fusoid, 30-40 X 6-9 μ, hyaline in KOH, nonamyloid, thin-walled, smooth. Tube trama divergent and only subgelatinous, nonamyloid, hyphae about 7 μ broad.

Pileus cuticle a turf of pileocystidia often as terminal cells of short hyphae (cuticle basically a trichodermium), 50-72 X 8-16 μ, gradually tapered to an obtuse to subacute apex, hyaline in KOH, smooth, thin-walled, the penultimate cells about half the length of the terminal cell, the cells near the base of trichodermial element often greatly inflated; walls smooth and nonamyloid, content not colored in Melzer's. Context of floccose interwoven hyaline hyphae, walls smooth and nonamyloid and content as revived in Melzer's not colored. Stipe surface with hyphae having numerous hyaline sausage-shaped cystidioid end-cells somewhat appressed to projecting. Clamp connections present, but many pseudoclamps also present.

Habit, habitat, and distribution.—Single to scattered or gregarious in open oak and in mixed conifer and hardwood forests during late summer and early fall. It is fairly common in southern Michigan.

Observations.—Within the color range given in the description there may be several constant races occurring in our region. These should be restudied from fresh material in order to evaluate properly their status as taxa. In most herbarium material there are not sufficient notes with the specimens to contribute much toward such a study. We have, for instance, a large form in the state with a stipe 3 cm or more thick at the apex, but so far we have found only rather old or weathered specimens.

Materials examined.—Barry: Smith 72965, 73104. Cheboygan: Smith 57398, 58291, 61428, 63103; Thiers 3454. Emmet: Kauffman

9-12-05; Smith 63867. Gratiot: Potter 3484, 3523, 8015, 10455, 11074, 11457, 14473. Livingston: Homola 948; Hoseney 503, 540, 561A; Smith 1765, 1783, 6443, 64203. Luce: Smith 63037, 70289, 74212. Marquette: Smith 72683. Washtenaw: Kauffman 8-8-15; Shaffer 2587, 2615; Smith 1646, 6903, 62900, 64271, 72459, 72602, 72804; Thiers 4570, 4580.

3. Gyroporus purpurinus (Snell) Singer
Farlowia 2:236. 1945

Boletus castaneus f. *purpurinus* Snell, Mycologia 28:465. 1936.

Illus. Figs. 20, 22.

Pileus 1-5 (9) cm broad, convex, expanding to broadly convex or nearly plane, often with a central to eccentric depression; margin obtuse at first, thin in age, often recurved slightly; surface minutely velvety to subtomentose, dry, often uneven, in age slightly rimose at times; dark vinaceous-red ("neutral red," "mineral red"), this color retained in drying. Context white, fragile, unchanging when bruised, taste mild.

Tubes 5-8 mm deep, adnate-depressed becoming deeply depressed, convex, white, slowly becoming pale yellow, not staining appreciably when injured; pores 1-3 per mm (very small), white becoming yellow, unchanging on injury.

Stipe 3-6 cm long, 3-8 mm thick, equal or tapering upward, stuffed, soon with cavities in lower part and finally entirely hollow; surface somewhat roughened to velutinous, concolorous with the pileus or browner, apex often pallid. Veil absent.

Spores bright yellow in deposit, 8-11 X 5-6.5 μ, ovoid to ellipsoid, smooth, many suballantoid in profile view, yellowish in KOH, yellowish to pale rusty brown in Melzer's, wall only slightly thickened, no pore evident.

Basidia 23-30 X 9-12 μ, 4-spored, yellowish in Melzer's, soon becoming nearly hyaline as revived in KOH. Cheilocystidia 23-50 X 4-8 μ, narrowly fusoid-ventricose to lanceolate, hyaline in KOH or in Melzer's, thin-walled, smooth.

Tube trama bilateral, of hyaline subgelatinous hyphae. Pileus cuticle a collapsing trichodermium with narrowly fusoid-ventricose to obtusely lanceolate cells as the terminal cell of the individual elements, the cells 40-120 X (5) 8-20 μ, with a purplish pigment soon disappearing in KOH mounts, in Melzer's often with orange to brownish orange homogeneous content or some becoming hyaline, walls thin and nonamyloid, lower cells of trichodermal elements often inflated to 20 μ.

Context hyphae mostly hyaline as revived in KOH and with thin smooth walls, as revived in Melzer's lacking colored contents in the region under the cuticle (the tissue merely yellowish). Clamp connections present but not numerous.

Habit, habitat, and distribution.—In open hardwoods, rare in southern Michigan, but apparently more common southward, fruiting during summer and early fall. The best collection seen to date is that of Hoseney (1341) made August 5, 1969, in a woodlot in Ann Arbor.

Observations.—This is a very striking bolete in either fresh or dried condition, but the pigment is unstable in alkali or chloral hydrate. Kauffman collected it August 15, 1915, and his material was later sent to Snell. In the University of Michigan Herbarium we have a part of this collection (an isotype) and the collection by Mrs. Hoseney. Some color intergradation with *G. castaneus* is to be expected, particularly in herbarium specimens in contact with naphthalene. This compound apparently causes a gradual reddening of the dried cap in some of the color forms of this species. Shaffer 2638 appears to be such a specimen.

SUILLUS S. F. Gray
emended by Smith & Thiers

Contrib. toward a Monograph of N. Amer. Sp. of *Suillus*
p. 115, Pls. 46. 1964

Suillus S. F. Gray, Nat. Arr. Brit. Pl. 1:646. 1821.

Pileus fleshy and readily decaying; hymenophore tubulose and pores of various sizes and shapes; stipe central to eccentric; spore deposit after evaporation of excess moisture pale cinnamon to yellow-brown or dark brown to olive-brown to olive and in one an olive-mustard-yellow; pleurocystidia typically clavate to subcylindric or only slightly ventricose near the base, often coloring in KOH when fresh and when revived in KOH typically with masses of pigment around the base of the cystidium, or the cell content colored, the cystidia often in clusters; caulocystidia in many species in clusters on the upper part of the stipe and coloring like the pleurocystidia, darkening on aging naturally to produce the colored glandular dots or smears characteristic of so many of the species; pileus cuticle usually an ixotrichodermium; clamp connections present or absent; species mostly forming mycorrhiza with conifers.

Type species: *Suillus luteus.*

KEY TO SECTIONS

1. Spores subglobose, fresh deposits olive-mustard-yellow; veil thick and tough . . . . Sect. *Paragyrodon*
1. Spores in face view elliptic, oblong or subfusoid 2
   2. Pileus with a dry fibrillose to squamulose epicutis; or if not as above then the stipe both annulate and lacking distinct glandular dots or smears . . . . Sect. *Boletinus*
   2. Pileus viscid to slimy, glabrous or with agglutinated appressed squamules; stipe if annulate having glandular dots near the apex (if stipe not annulate then glandular dots may or may not be present) . . . . Sect. *Suillus*

Section PARAGYRODON (Singer) Smith & Thiers

Contrib. toward a Monograph of N. Amer. Sp. of *Suillus* p. 22. 1964

Paragyrodon Singer

Ann. Mycol. 40:25. 1942

The thick veil and subglobose spores are the distinguishing features of this section. The color of the fresh spore deposit is rather peculiar, but since it is a combination of olive and yellow this difference does not seem to be sufficiently distinct to justify a monotypic genus—at least not at the level at which we recognize genera in the boletes.

Type species: *Suillus sphaerosporus.*

4. Suillus sphaerosporus (Peck) Smith & Thiers

Contrib. toward a Monograph of N. Amer. Sp. of *Suillus* p. 22. 1964

Boletus sphaerosporus Peck, Bull. Torrey Club 12:33. 1885.
Paragyrodon sphaerosporus (Peck) Singer, Ann. Mycol. 40:25. 1942.

Illus. Pls. 5-7.

Pileus (4) 8-15 (20) cm broad, convex to plano-convex when young, expanding to plane or shallowly depressed in age or at times the margin uplifted and the shape becoming quite irregular; surface viscid to slimy, glabrous, spotted or streaked with patches of dried gluten; when young ocher to golden yellow ("ocher-yellow" to "buff-yellow" or "old gold"), becoming ochraceous-tawny and finally dark dingy yellow-brown, often with rusty stains in age, readily staining brown when bruised; margin incurved, entire, sterile. Context thick, up to 2 cm or more, whitish to yellowish, changing to vinaceous-brown ("vinaceous-tawny" to "testaceous") when exposed, odor and taste not distinctive or occasionally unpleasant.

Tubes shallow in relation to thickness of pileus (4-10 mm deep), adnate, becoming decurrent or with lines extending down the stipe, canary-yellow ("mustard-yellow"), becoming golden yellow and finally brown; pores large (about 1 mm diameter near maturity), angular, yellow when young, staining brown ("cinnamon" to "tawny") when bruised.

Stipe 4-10 cm long, 1-3 cm thick, tapered to the base, or equal or nearly so, solid, yellowish within but soon staining when cut or injured; surface pruinose to glabrous above the heavy tough gelatinous more or less median annulus, apex at times somewhat reticulate from decurrent tubes which are pale yellow (pale "mustard-yellow") but which discolor readily; veil thick, tough, membranous, with a pale buff exterior of unpolished appearance at first, the inner layer slowly gelatinizing, the lower edge adnate to stipe but tending to separate from it in places, staining dark brown when handled, breaking away from the pileus margin to leave a gelatinous annulus.

Spore deposit olive-mustard-yellow fresh, on standing (at least on old leaves and on the apex of the stipe) discoloring slowly to "snuff-brown" or darker (a dark yellow-brown). Spores 6-9 X 6-8 μ, globose to subglobose, smooth, yellowish hyaline in water mounts, ochraceous to pale bister in KOH, pale tawny to darker in Melzer's, with an inconspicuous hyaline outer sheath, unornamented, wall thickened slightly.

Basidia 4-spored, 18-22 X 9-11 μ, hyaline in KOH individually, nonamyloid. Pleurocystidia scattered to abundant, 20-32 X 8-12 μ, fusoid-ventricose with subacute apex, content dingy brown in KOH as revived, nonamyloid. Hymenium dark brown in KOH. Cheilocystidia abundant, similar to pleurocystidia, extending down the stipe as caulocystidia.

Tube trama gelatinous and consisting of hyphae diverging from a central strand which is brownish when revived in KOH, and hyphal walls and the cell content nonamyloid. Epicutis of pileus a thick (over 200 μ) layer of appressed narrow gelatinous nearly hyaline hyphae as revived in both KOH and Melzer's. Tramal body of brownish floccose tissue as revived in KOH, nonamyloid. Clamp connections present, abundant in some basidiocarps but difficult to find in others.

Habit, habitat, and distribution.—Solitary to gregarious under hardwoods, especially oak in Michigan, not uncommon in June and again in September.

Observations.—Although this is perhaps the fleshiest species of *Suillus* in Michigan, we have found no one who would recommend it for the table. It is one of the most frequently collected species in modern "suburbia" where original forest trees have been left as shade trees and the ground cover cleaned out but not planted to grass. It seems apparent that the fungus must have been living in the original forest, but did not fruit (or does not) as readily under those conditions as it does after the habitat has been changed.

Material examined.—Barry: Mazzer 4232. Jackson: Smith 18375, 18571, Zehner, 242. Livingston: Hoseney 348, 1032; Mazzer 4201; Smith 6067. Montcalm: Potter 13521. Washtenaw: Homola 1039, 1097; Hoseney 33, fall 1965; Kauffman 10-19-07, 7-21-17; Smith 9612, 15138, 15240, 18663, 18708, 62495, 62557, 64130, 74482, 64513, 66448, 72525, 72854, 72870; Thiers 4508.

Section BOLETINUS (Kalchbrenner) Smith & Thiers

Contrib. toward a Monograph of N. Amer. Sp. of *Suillus* p. 24. 1964

Boletinus Kalchbrenner
Bot. Zeit. 25:181. 1867

Stipe annulate by a fibrillose zone or a distinct membranous annulus or if an annulus or zone of fibrils is lacking then the pileus distinctly fibrillose-squamulose; stipe surface not glandular dotted though some encrusted caulocystidia may be present in some species; hymenophore sublamellate to poroid and often becoming decurrent on the stipe.

Type species: *Suillus cavipes.*

KEY TO SPECIES

1. Pileus dry and fibrillose-squamulose 2
1. Pileus viscid to slimy. 3
   2. Stipe hollow at base when mature; clamp connections regularly present on hyphae of basidiocarp. . . . . *S. cavipes*
   2. Stipe solid; clamps absent . . . . *S. pictus*
3. Stipe when sectioned staining green in lower part; spores 4-4.5 μ broad . . . . *S. proximus*
3. Stipe not staining as above; spores 2.8-3.5 μ broad . . . . *S. grevillei*

5. **Suillus cavipes** (Opat.) Smith & Thiers

Contrib. toward a Monograph of N. Amer. Sp. of *Suillus* p. 30. 1964

Boletus cavipes Opat., Comm. Bolet. p. 11. 1836.
Boletinus cavipes (Opat.) Kalchbr., Ic. Hymen. Hung. 52. 1877.
Boletus ampliporus Peck, Ann. Rept. N. Y. State Mus. 26:67 (351). 1874.

Illus. Pls. 8-9.

Pileus 3-10 (12) cm broad, obtuse to convex expanding to nearly plane or at times with an obtuse umbo; surface moist to dry but not viscid, densely tomentose to fibrillose-squamulose, dull yellow to reddish, reddish brown to dark reddish brown ("cinnamon-rufous," "tawny," "orange-cinnamon" to near "Vandyke-brown"), rarely ocher-yellow, the tips of the fibrils often pallid, margin involute and usually with adhering veil remnants. Context white to yellowish in age, soft, not changing to blue when bruised, odor and taste not distinctive.

Tubes 3-5 mm deep, decurrent, pale yellow to greenish yellow, more olive-ochraceous in age; pores 0.6-1 mm wide and 1-1.5 mm or more radially, angular, radiating from the stipe, simple to compound.

Stipe (3) 4-9 cm long, (5) 8-15 (20) mm thick, usually narrowed below and enlarged at apex, subequal at times, sometimes clavate, solid in upper part, soon hollow in base, usually concolorous with pileus or a paler shade, apex yellow ("amber-yellow") at times, lower portion decorated with veil remnants at first but these soon vanishing, at times with a slight annulus.

Spore deposit dark olive-brown when moist, "snuff-brown" to "Saccardo's umber") when air-dried; spores 7-10 X 3.5-4 μ, smooth, no apical pore evident, narrowly ovate to ventricose in face view, in profile narrowly and obscurely inequilateral, nearly hyaline in Melzer's, greenish hyaline in KOH.

Basidia 18-35 X 4-5 μ, 4-spored, hyaline in KOH and nearly so in Melzer's. Pleurocystidia scattered to numerous, subcylindric to fusoid-ventricose, hyaline, thin-walled 42-45 X 6-10 μ. Cheilocystidia similar to pleurocystidia or obventricose.

Tube trama subgelatinous and the hyphae somewhat divergent from a central strand or finally the central strand lacking and trama more or less interwoven. Pileus trichodermium of hyphae 8-15 μ wide, the cells more or less equal or constricted at the cross walls, hyaline in KOH, yellow in Melzer's. Contest hyphae hyaline, interwoven, nonamyloid. Clamp connections regularly present.

Habit, habitat, and distribution.—Gregarious to cespitose under larch (*Larix laricina*). It is common in the fall and fruits in quantity. It is edible and recommended by some as good. The greatest danger of poisoning from associating with this species, at least in southern Michigan, is of coming in contact with poison sumac while collecting the fungus.

Observations.—There is a great deal of color variation in this species and sometimes one is tempted to recognize color forms. However, if this were done, it should be on the basis that different pigments are involved. We have made no study of the chemistry of the pigments, but from field observations it appears that at least 2 are involved and that the color variants would have to be based on the varying amounts of each pigment present in each form. This could easily lead to an endless number of "forms."

Material examined.—Alger: Kauffman 9-6-27. Barry: Mazzer 2590. Cheboygan: Chmielewski 10-19-47; Shaffer 1992; Smith 1079, 36009, 50998. Chippewa: Smith 62179. Livingston: Baxter 10-8-27. Mackinac: Smith 43782; Marquette: Mains 33-595, 33-639; Smith 33-426. Washtenaw: Mains 31-791, 10-3-32; Kauffman 1591; Smith 31900, 38917, 64676, 64745, 66437, 10-2-29, October 1938.

6. Suillus pictus (Peck) Smith & Thiers

Contrib. toward a Monograph of N. Amer. Sp. of *Suillus* p. 31. 1964

Boletus pictus Peck, Ann. Rept. N. Y. State Cab. 23:128. 1872.
Boletinus pictus (Peck) Peck, Bull. N. Y. State Mus. 8:77. 1889.

Illus. Figs. 24-26; Pl. 10.

Pileus 3-12 cm broad, broadly conic to convex or hemispheric, the margin incurved, expanding to broadly conic-umbonate with a spreading or decurved margin, or nearly plane, occasionally shallowly depressed, margin typically appendiculate with veil fragments; surface dry, never truly viscid but at times tacky when wet (especially when young), coarsely fibrillose; cuticle at first red ("ocher-red"), fading on standing *in situ* and in extreme age, the fibrils often grayish, or tips of the squamules fuscous. Context up to 1.5 cm thick, yellow, changing to pinkish gray to reddish ("avellaneous" to "russet-vinaceous") upon exposure, floccose. Taste and odor mild.

Tubes not readily separable, adnate to decurrent, often extending down the stipe, about 5 mm deep, bright yellow; pores concolorous with sides, large at maturity (0.5-2 × up to 5 mm radially), smaller toward the pileus margin, angular, often compound, changing to reddish or brownish ("avellaneous") when injured.

Stipe 4-10 (12) cm long, 8-25 mm thick, equal or enlarged downward, occasionally with a clavate to subclavate base, solid, rarely hollow; context white in the interior or brownish at the base, cortex yellowish or occasionally concolorous with exterior, lower portion sheathed with remains of a fibrillose coating similar to the fibrils of the pileus and breaking into zones or patches, often terminating in a fibrillose annulus, with a whitish, delicate, more or less cortinate partial veil in addition, apex yellow and more or less reticulate from decurrent lines from the tubes.

Spore deposit clay color to tawny-olive as air-dried, olive-brown before moisture has escaped; spores 8-11 (12) × 3.5-5 μ, smooth, apical pore lacking, narrowly ovate in face view, narrowly inequilateral in profile view, olive yellowish in KOH mounts as revived, yellow to pale tawny in Melzer's.

Basidia 4-spored, 20-26 × 6-8 μ, hyaline in KOH, yellow in Melzer's. Pleurocystidia abundant, clavate, 30-40 × 9-12 μ, with cinnamon-colored content in KOH, much more numerous than the above are large prominently projecting cystidia 55-75 × 7-10 μ, subcylindric, flexuous, or straight and with a subacute to obtuse apex, hyaline and thin-walled but usually with bister incrusting material where surrounded by the hymenium. Cheilocystidia similar to the second type of pleurocystidia but frequently having yellowish content in KOH. Caulocystidia similar to the large incrusted pleurocystidia, occasionally in fascicles.

Tube trama gelatinous and divergent, nonamyloid. Pileus trichodermium of loosely tangled hyphae with short somewhat inflated cells near apex (9-15 μ in diameter), and with orange-buff content revived in KOH. Context hyphae pale bister in KOH and with incrusting material on the walls (in contrast to the hyphae of the trichodermium), the hyphae thin-walled, nonamyloid in all parts. Clamp connections absent.

Habit, habitat, and distribution.—Scattered to gregarious under white pine. It is regularly associated with this pine and fruits during the summer and fall. It is common where white pine grows naturally, but has, to date, not been found in plantations.

Observations.—The cystidial characters connect this species with the viscid group. The species is so distinctive in the field on the basis of the association with white pine and the red fibrils of the pileus that there is little room for error in identifying it. It is also rated as a good species for the table, and should be on the list of any hunter living or vacationing in white pine country.

Material examined.—Cheboygan: Charleton G-197; Chmielewski 66; Shaffer 1483; Smith 25982, 36858, 37880, 62937, 63960, 66941, 67227, 72403, 74708, 7-1-51, July 1951; Thiers 734, 3465, 3752, 4323. Chippewa: Pennington 8-12-06; Smith 38590, 71928; Thiers 3853, 4339. Crawford: Thiers 3390. Emmet: Kauffman 9-5-05; Smith 41902, 57397, 63767; Thiers 880, 3562, 3572. Gratiot: Potter 8429, 11191. Keweenaw: Povah FP 420, F1219. Mackinac: Smith 42903, 44113, 63909; Thiers 3620. Marquette: Bartelli 208. Montcalm: Potter 10094. Ogemaw: Cibula; Shaffer 2809; Smith 67506. Luce: N. J. Smith 140A; A. H. Smith 36811, 37090, 36963, 38192, 39499, 41818, 44116; Thiers 3401.

7. **Suillus proximus** Smith & Thiers

Contrib. toward a Monograph of N. Amer. Sp. of *Suillus* p. 42. 1964

Illus. Pl. 11.

Pileus 6-9 cm broad, obtuse, expanding to obtusely umbonate, convex or broadly convex; surface slimy-viscid, glabrous; with pale pinkish cinnamon streaks beneath the gluten and an overall yellow ground color, margin soon watery brownish where bruised. Context thick, yellowish to watery buff, in stipe apex with watery lemon-yellow streaks, typically slowly staining green when cut; odor acid-metallic, taste mild;

KOH on cuticle quickly olive, on context bluish gray; with $FeSO_4$ blue-gray on context and cuticle.

Tubes 5-6 mm deep, decurrent, dingy yellowish, soon dingy to dull pale cinnamon where cut; pores small but in age up to 1 mm broad, staining pale cinnamon brownish where bruised.

Stipe 8-9 cm long, 10-15 mm thick at apex, equal or nearly so, solid, yellow within at apex, dingy rusty brown below and turning green when cut in lower or midportion or throughout; surface at apex canary-yellow, vinaceous-cinnamon lower down, apex striate to subreticulate with decurrent lines from tubes, not glandular dotted; annulus superior, often with a gelatinous margin, remainder floccose and yellow.

Spore deposit "chestnut-brown" to "cinnamon-brown," moist but fading to pale cinnamon-brown as it air dries, finally becoming clay color; spores 7-10 X 4-4.5 μ, suboblong to obscurely inequilateral in profile, in face view narrowly elliptic to subcylindric, smooth, dingy pale ochraceous-brown in KOH, slightly paler in Melzer's.

Basidia 25-32 X 7-9 μ, pale dingy ochraceous in KOH (in the hymenium), about the same color in Melzer's, 4-spored; sterigmata very fine. Pleurocystidia in bundles or isolated, 35-60 X 6-8 μ, hyaline to bister revived in KOH, with incrusting bister pigment surrounding the base of the cluster, thin-walled. Cheilocystidia similar to pleurocystidia and very numerous, with more colored ones and more incrusting pigment in and around the bundles.

Tube trama of slightly divergent gelatinous hyphae 4-8 μ in diameter (sections revived in KOH), no incrusting pigments present, hyphae thin-walled; subhymenium of interwoven hyphae appearing cellular in sections from the cut ends, only subgelatinous in KOH.

Pileus cuticle a thick layer of interwoven gelatinous hyphae 3-7 μ in diameter (possibly a collapsed trichodermium), with thin walls yellowish in KOH, no incrusting pigment seen. Context of floccose nongelatinous interwoven hyphae 5-15 μ in diameter, with the cells often somewhat inflated, no incrusting pigment seen. Clamp connections absent.

Habit, habitat, and distribution.—Gregarious in swamps with *Larix* (larch) or often at the edge of such areas. To date it is known only from Oakland County. Smith 66436, 67759, 67806, 75160, and the type (Smith 64508).

Observations.—*S. grevillei* from the same locality had an olive-brown spore deposit. The difference in the color of the fresh spore deposits was very striking at first but on air-drying they both became about the same color. The green staining is the best field character for the recognition of this species.

8. Suillus grevillei (Klotzch) Singer

Farlowia 2:259. 1945

Boletus grevillei Klotzch, Linnaea 7:198. 1832.
Boletus elegans Fries, Epicr. Syst. Mycol. pp. 409-10. 1838.
Boletus clintonianus Peck, Ann. Rept. N. Y. Cab. 23:128. 1872.
Suillus grevillei var. *clintonianus* (Peck) Singer, Agaricales in Mod. Tax. p. 721. 1962.

Illus. Pl. 12.

Pileus 5-15 cm broad, hemispheric becoming broadly convex to nearly plane; surface glabrous, glutinous, "chestnut" on the disc, "empire-yellow" on the margin, at times finally bright yellow overall, pellicle separable, margin sterile. Context "straw-yellow" to "amber-yellow," soon rubescent ("salmon-buff" or "flesh color") on injury, rather thick (1-1.5 cm); taste mild to slightly astringent to bitterish, odor none to somewhat metallic.

Tubes adnate to depressed, becoming subdecurrent, 10-15 mm deep, "amber-yellow" (sulphur-yellow) but finally "olive-ocher" (olive-yellow), when bruised or cut becoming testaceous to vinaceous-brown; pores angular, 1-2 per mm, dissepiments rather thick and entire.

Stipe 4-10 cm long, 1-3 cm thick, equal to slightly clavate, solid, flavous to pale yellow within and over exterior, sometimes slightly sulphur greenish in base when cut but not truly green, surface soon developing chestnut variegations; annulate with a typically floccose annulus which may have a gelatinous outer layer in wet weather, distinctly reticulate above the annulus from extensions of the decurrent tubes, dingy in age, staining vinaceous-brown after handling.

Spore deposit olive-brown moist, "Sayal-brown" (dull cinnamon) air dried; spores 8-10 X 2.8-3.5 μ, more or less oblong in face view and obscurely inequilateral in profile view, smooth, with a faint hyaline outer sheath, pale olivaceous to ochraceous in KOH and Melzer's. Pleurocystidia numerous to scattered or rare, sometimes in fascicles, 40-60 X 6-8 μ, subcylindric to clavate, but some obscurely fusoid-ventricose, bister incrusting material at base only, content brownish to yellowish in KOH and Melzer's. Cheilocystidia in fascicles, incrusted with bister pigment at level of the hymenium, the fascicles scattered to numerous, the individual cystidia similar to the pleurocystidia. Caulocystidia solitary or in scattered small fascicles and with incrusting pigment as revived in KOH.

Tube trama gelatinous and of divergent hyphae. Pileus trichodermium soon collapsing, of long gelatinous hyphae 3-6 μ in diameter, the cells very long and narrow, end-cells not specialized in shape other than being tubular. Context of floccose hyaline to yellowish interwoven hyphae. Surface of stipe a layer of heavily incrusted hyphal tips. Clamp connections absent.

Habit, habitat, and distribution.—Cespitose to gregarious, often in arcs and always associated with *Larix* (larch) as a mycorrhiza-former. Common in the fall where tamarack (larch) grows.

Observations.—This is a common species with the pileus varying in color from rich reddish brown to pale yellow. The slime layer is very thick and causes a problem for any one attempting to collect this species for the table, but the species is edible.

Material examined.—Chippewa: Smith 38787, 72420. Emmet: Kauffman 9-1-05; Smith 38823, 42924, 43107. Kent: Kauffman 9-24-04. Livingston: Smith 6065. Marquette: Mains 33-599, 33-636. Oakland: Smith 10989, 18733, 64545, 67329. Ogemaw: Smith 50558, 50595, 67583, 67584, 67618. Washtenaw: Hintika 10-6-61, 11-3-61; Kauffman 9-24-04, 10-19-07, 8-14-25; Mains 31-743; Smith 62308, 64589.

Section SUILLUS

Typically, the stipe is glandular dotted (in *S. unicolor* and *S. subaureus* the glandules darken slowly and often not conspicuously, in *S. brevipes* they do not show in young basidiocarps), the spore print after moisture has escaped is typically dull cinnamon, and the pileus is somewhat slimy. A veil may either be well enough developed to leave an annulus, it may adhere entirely to the margin of the pileus, or it may be absent.

Type species: As for the genus.

KEY TO SPECIES

1. Context of pileus or stipe changing to blue at least slightly when cut . . . .*S. tomentosus*
1. Context not changing as in above choice 2

   2. Veil or false veil present (check young specimens). 3
   2. Veil absent, the pileus margin naked or nearly so when young 10

3. Veil or false veil adhering to pileus margin, no remains left on the stipe as a ring (annulus) 4
3. Veil leaving an annulus on the stipe 6

   4. Tubes and stipe ocher-yellow; stipe 3-9 mm thick; growing under or near *Pinus strobus* . . . . *S. americanus*
   4. Not as above 5

5. Pileus glabrous or with a slight patchy appearance along margin; pores 2-3 per mm . . . . *S. albidipes*
5. Pileus spotted with brown patches of veil remnants; pores 1-2 mm broad at maturity (if tube mouths are smaller and pileus has red spots or streaks or grayish patches of appressed tomentum see *S. subaureus* and *S. hirtellus*). . . . . *S. sibiricus*. . . . .*S. sibiricus*

    6. Stipe base conspicuously staining yellow when injured . . . . . *S. lutescens*
    6. Not staining as in above choice 7

7. Stipe 1-2.5 cm thick; annulus with a thin vinaceous-gray to purple-drab layer or zone on its outer side which may gelatinize in wet weather . . . *S. luteus*
7. Not as above 8

    8. Veil soft and baggy, flaring outward at lower edge before breaking (see Pl. 21) . . . . *S. subluteus*
    8. Veil not as above—much less conspicuous 9

9. Pileus white at first, taste of slime on pileus very acid . . . . *S. acidus*
9. Pileus distinctly colored when young 9a

    9a. Pileus ocher-yellow or darker; taste of slime on pileus very acid . . . .*S. intermedius*
    9a. Pileus pallid to dingy cinnamon; slime on pileus mild . . . *S. subalutaceus*

10. Pileus white at first; stipe having conspicuous pinkish brown smears. . *S. placidus*
10. Pileus more highly colored 11

    11. Pileus soon glabrous and deep dull yellow; odor sweetly fragrant (in large collections). . . . . *S. punctipes*
    11. Not as above 12

12. Pileus white in button states but becoming cinnamon to reddish cinnamon, glabrous but finally obscurely spotted from drying gluten . . .*S. granulatus*
12. Pileus glabrous or decorated with patches of fibrils but not as above 13

    13. Tubes boletinoid at maturity; pileus glabrous and pinkish brown to dull cinnamon or grayish brown; stipe conspicuously glandular dotted. . . . . *S. punctatipes*
    13. Not as above 14

14. Pileus when young distinctly spotted with patches or tufts of fibrils or tomentum. 15
14. Pileus glabrous and slimy when young. 16

    15. Stipe with a ferruginous zone in the base when sectioned . . . .*S. subaureus*
    15. Stipe with an olive-fuscous zone in the base when freshly sectioned . . . .*S. hirtellus*

16. Basidiocarp lemon-yellow overall . . . .*S. unicolor*
16. Not as above 17

    17. Pore surface with copious droplets of latex when young . . . . *S. lactifluus*
    17. Pore surface not showing droplets of latex at first or any other time. . . . .*S. brevipes*

9. Suillus tomentosus (Kauffman) Singer, Snell, & Dick

Mycologia 51:570. 1960

Boletus tomentosus Kauffman, Pap. Mich. Acad. Sci. Arts & Letters 1:117. 1921.
Boletus hirtellus var. *mutans* Peck, nom. nud.
Suillus hirtellus var. *mutans* (Peck) Snell, Lloydia 7:23. 1944.
Xerocomus lenticolor Snell & Dick, Mycologia 52:448. 1960.

Illus. Pl. 13.

Pileus 5-10 (15) cm broad, obtuse to convex, becoming broadly convex, the margin at first incurved somewhat and rather coarsely tomentose at first, gradually glabrescent; surface at first tomentose-floccose to squamulose overall but gradually glabrescent, the tomentum and squamules grayish, viscid beneath the tomentose-fibrillose layer; ground color evenly pale clear yellow to pale orange-yellow. Context thick (8-15 mm), pallid to yellow but paler than the pileus surface, changing slowly to blue or greenish blue when cut or bruised, odor none, taste slightly acid to mild.

Tubes 1-2 cm deep, pale dingy yellow becoming dingy olive-yellow and staining dingy greenish blue when injured, adnate to decurrent, rarely depressed around the stipe; pores bister to dark dingy cinnamon or vinaceous-brown when young but slowly becoming dingy yellow, small (about 2 per mm), staining bluish when injured, at least in age.

Stipe (3) 5-10 (12) cm long, 1-2 (3) cm thick at apex, equal or clavate, solid, yellow within but paler than pileus surface, staining greenish blue slightly when cut; surface concolorous with pileus or more orange, glandular dotted to glandular-punctate overall, no veil remnants adhering and no evidence seen that the tomentum of the pileus actually represents the remains of a veil.

Spore deposit dark olive-brown very slowly changing to dull cinnamon on drying; spores 7-10 (12) $\times$ 3-4 (5) μ, smooth, pale yellowish in KOH, in Melzer's yellowish to pale tan, narrowly inequilateral in profile view (some swaybacked), obscurely fusoid to elongate-ovate in a face view.

Basidia 4-spored, 26-34 $\times$ 5-8 μ, yellowish in KOH and in Melzer's, clavate. Pleurocystidia typically in bunches and at the level of the hymenium the cluster surrounded by brown incrusting material (when revived in KOH), when fresh hyaline to pale vinaceous-brown in KOH; individual cystidia 30-46 $\times$ 7-10 μ, subcylindric to narrowly clavate or narrowed to a blunt apex, often with incrusting debris, content hyaline to dingy yellow-brown when revived in KOH. Cheilocystidia similar to pleurocystidia, very numerous and tube dissepiments often with copious brown amorphous pigment. Caulocystidia abundant, similar to cheilo-cystidia.

Tube trama of gelatinous hyaline hyphae somewhat divergent from a floccose central narrow strand, the hyphae 5-9 μ in diameter. Pileus trichodermium with elements which become fused to form the squam-ules, these hyphae 8-12 μ broad, and with end-cells somewhat cystidi-oid and 10-16 μ broad in widest part. Clamp connections absent.

Habit, habitat, and distribution. — Scattered to gregarious under 2-needle pines, late August and early September, common during wet seasons in its typical habitat, especially so in the pine barrens of the Upper Peninsula as the work of Mrs. Bartelli has shown.

Observations. — In Michigan we have mostly the very pale yellow form, but though we have studied the matter, we have been unable to recognize any taxa that could be based on the intensity of the yellow pigment in the pileus. The species can be quickly distinguished from *S. hirtellus* and *S. subaureus* by the blue staining reaction of injured con-text. In pale forms one often finds the young pores yellow instead of brown as described, but again, we were unable to convince ourselves that there was a constant difference here. The species is even more variable in the Rocky Mountains. *S. variegatus* in Europe is very closely related. A critical comparative study of these two should be made.

Material examined. — Cheboygan: Smith 35834, 38368. Chippewa: Bartelli 2420; Smith 36806, 38609, 43952, 63980, 64019. Mackinac: Smith 8-21-53; Bartelli 2252, 251; Smith 72694, 72719. Ogemaw: Smith 67444. Luce: Smith 38322, 42227, 42297, 58013, 58178.

10. Suillus hirtellus (Peck) Kuntze

Rev. Gen. Pl. 3:535. 1898

Boletus hirtellus Peck, Bull. N. Y. State Mus. 2:94. 1889.
Rostkovites hirtellus (Peck) Murrill, Mycologia 1:14. 1909.
Ixocomus hirtellus (Peck) Singer, Ann. Mycol. 40:30. 1942.
Boletus subaureus var. *siccipes* Coker & Beers, Bol. North Carol. p. 83. 1943.

var. **hirtellus**

Illus. Pl. 14 (upper fig.).

Pileus 5-10 (15) cm broad, convex with an incurved margin, ex-panding to plane or slightly depressed; surface at first with appressed grayish fibrillose squamules and smaller reddish patches giving surface a spotted appearance, ground color bright yellow with a tendency to stain vinaceous-brown from handling, in age glabrous or nearly so, viscid when fresh; margin when young with cottony material along it like the tufts of fibrils on the pileus. Context pale yellow, not staining blue when bruised, odor and taste mild.

Tubes shallow, 3-8 mm deep, decurrent and with lines extending down the stipe a short distance, pale yellow to ochraceous; pores at maturity compound and elongated, about 1 mm wide, pale yellow when young, more orange-buff in age, staining slightly vinaceous-brown where bruised.

Stipe 3-8 cm long, 1-2 cm thick at apex, equal or enlarged downward, solid, yellowish within above but soon olive-fuscous in the base; surface glandular dotted, pale yellow overall including the dots and these blackening in age or from handling, base typically tapered to a point; veil absent (tomentum of pileus never connected to the stipe).

Spore deposit "Sayal-brown" (dull cinnamon); spores 7-9 X 3-3.5 μ, smooth, nearly oblong in face view, in profile somewhat inequilateral, apical and slightly turned up in many (as seen in profile), pale ochraceous in KOH, pale tawny in Melzer's reagent.

Basidia 4-spored, 17-23 X 4-5.5 μ, narrowly clavate, pale ochraceous in KOH. Pleurocystidia in prominent clusters, incrusted with brown amorphous material, individual cystidia (30) 40-70 (80) X 6-9 μ, clavate to subcylindric, hyaline to brown in KOH, often with refractive reticulate content. Cheilocystidia similar to pleurocystidia, much amorphous pigment present. Caulocystidia similar to pleurocystidia, much amorphous pigment present, occurring in clusters. Caulohymenium often showing sporulating basidia; subhymenium gelatinous.

Tube trama of gelatinous, divergent, hyaline hyphae as revived in KOH. Epicutis of pileus a pellicle of loosely interwoven yellowish to hyaline gelatinous hyphae 3-6 μ in diameter. Clamp connections absent.

Habit, habitat, and distribution. – Gregarious under pine and aspen throughout the aspen areas of the state, common in the summer and fall during wet weather.

Observations. – The stipe may not appear glandular dotted at first because the glandulae are the same color as the rest of the surface, but they quickly darken from handling. The species is distinguished with difficulty from *S. subaureus*, though the latter has less pileus ornamentation and the flecks it does have are more likely to become red by maturity. Material cited is not included for this species because of the difficulty in distinguishing the species on the basis of dried material, and because there has been much confusion previously in regard to it.

11. Suillus subaureus (Peck) Snell in Slipp & Snell

Lloydia 7:30. 1944

Boletus subaureus Peck, Ann. Rept. N. Y. State Mus. 39:42. 1886.
Rostkovites subaureus (Peck) Murrill, Mycologia 1:13. 1909.
Ixocomus subaureus (Peck) Singer, Rev. de Mycol. 3:45. 1938.

Illus. Pl. 15.

Pileus 3-12 (17) cm broad, convex with an incurved margin, expanding to broadly convex to nearly plane, often with a fluted or flared margin, margin often remaining inrolled for a long time, slightly downy but no veil or false veil is present on buttons; surface viscid beneath a covering of appressed tomentum which may be brownish at first but on many pilei becomes aggregated into patches or fibrillose scales which become scarlet to cinnabar-red and spotlike in appearance, surface rarely appearing glabrous with age, ground color yellow to apricot-yellow. Context up to 3 cm thick, yellow, unchanging or staining brownish to vinaceous-brown; taste mild to slightly acid, odor not distinctive.

Tubes subdecurrent when young becoming decurrent in age, (3) 4-10 mm deep, somewhat radially arranged, ochraceous to luteous becoming flavous, unchanging when bruised, somewhat separable from pileus context; pores about 2 per mm, angular and arranged so as to be inconspicuously lamellate, dingy yellow becoming dull ochraceous.

Stipe 4-8 cm long, 1-2 cm thick, equal to tapering downward to occasionally subclavate, solid, flavous within, then reddish tawny where cut, with a ferruginous to fulvous zone in the base as seen in a longitudinal section, yellow overall, glandular dots showing up in age when they have discolored, viscid at least when young, surface sordid brownish from handling, basal mycelium whitish.

Spore deposit olive-brown ("light brownish olive") fresh; spores 7-10 X 2.7-3.5 μ, narrowly elliptic to subfusiform in face view, in profile obscurely inequilateral, smooth, pale greenish yellow in KOH to nearly hyaline, scarcely darker in Melzer's.

Basidia 4-spored, 23-88 X 5-7 μ, clavate, hyaline to yellowish in KOH, yellowish in Melzer's. Pleurocystidia (25) 40-50 X 4-6 μ, clavate to cylindric, mostly in bunches and more or less covered with dark yellow-brown amorphous pigment as revived in KOH, in mounts of fresh material (in KOH) purple to lilac or rarely some of them hyaline, or with bister content, thin-walled. Cheilocystidia similar to pleurocystidia, very abundant, often crooked near the apex, the surrounding amorphous pigment very abundant. Caulocystidia abundant and scattered over the surface, 50-70 (100) X 3-5 μ, filamentous to cylindric (often flexuous), varying to narrowly clavate, subgelatinous, often in clusters, no pigment present but purplish in KOH when revived, some cystidia also with purplish content in KOH when fresh.

Tube trama of gelatinous somewhat divergent hyphae hyaline in KOH, the central strand of floccose yellowish hyphae. Pileus epicutis a pellicle of hyphae 3-6 μ broad, interwoven and gelatinous, no incrusting material present. Clamp connections absent.

Habit, habitat, and distribution.—Scattered to gregarious under aspen and scrub oak, summer and early fall, not uncommon.

Observations.—*S. subaureus* typically produces a heavy-set basidiocarp and these develop rather slowly so that the tubes remain shallow (3-5 mm deep) a long time and one might be led to regard this as a taxonomic feature. Its relative abundance in the state has not been clearly established as yet because of confusion with such related species as *S. hirtellus.*

12. Suillus punctipes (Peck) Singer

Farlowia 2:277. 1945

Boletus punctipes Peck, Ann Rept. N. Y. State Mus. 32:32. 1880.
Ixocomus punctipes (Peck) Singer, Ann. Mycol. 40:30. 1942.

Illus. Pl. 16.

Pileus 3-10 cm broad, convex becoming broadly convex or finally the margin spreading and wavy, when young covered by tufts of dull grayish brown tomentum, soon becoming glabrous and dull ochraceous-orange, then becoming pale dingy ochraceous and at times with obscure grayish tufts or patches of tomentum over marginal area, finally becoming pale yellow and entirely glabrous. Context thick and soft, pallid yellow and unchanging when cut, in KOH turning purplish to vinaceous; odor like that of *Hygrophorus agathosmus*; taste mild.

Tubes 4-6 mm deep (shorter than the thickness of the context at deepest point), adnate to short-decurrent, dull brownish young, becoming olivaceous to honey-yellow when mature; pores dark dingy brown ("bister") at first, slowly paler and and finally merely dingy yellow, small (about 2 per mm), round to angular.

Stipe 4-9 cm long, 9-14 mm at apex, equal to clavate, solid, dull ochraceous to dull ochraceous-orange within (color about as in *Chroogomphus rutilus*), base often cinnabar color, cut surface slowly staining dull brown; surface at first "bister" from a dense coating of viscid glandular dots, the flavous ground color showing through by maturity, when handled the fingers quickly becoming stained sordid ochraceous and the fragrant odor can be noted on them.

Spore deposit dark olive-brown fresh; spores 7.5-10 (12) × 3-3.5 μ, in face view elongate-elliptic to subfusiform, in profile narrowly inequilateral to swaybacked and with a broad suprahilar depression, smooth, pale dingy yellowish in Melzer's with a few becoming dull tawny-brown (faintly dextrinoid), dingy pale buff in KOH varying to pale olive yellowish.

Basidia 2- and 4-spored, 16-20 × 5.5-7 μ, sections of hymenium dingy yellowish in KOH, individual basidia nearly hyaline, yellowish hyaline in Melzer's. Pleurocystidia of 2 types: (1) as cystidioles 18-26 × 6-8 μ with hyaline content, with thin smooth walls and an obtuse to subacute apex; (2) in clusters with dark bister to dull rusty brown incrusting material around the base of the cluster, individual cystidia 36-55 × 7-11 μ and clavate to subfusoid, with hyaline to brown content in KOH and Melzer's. Cheilocystidia in masses and with brown content, incrusted around the base as revived in KOH, clavate to subfusoid and 50-70 × 7-14 μ. Caulocystidia similar in size and coloring to cheilocystidia and frequently forming an hymeniform layer over large areas of the upper part of the stipe.

Tube trama of subparallel elements in the central area and somewhat divergent toward the subhymenium, cells pale yellowish to hyaline in KOH and Melzer's; the hyphae 5-8 μ in diameter and subgelatinous. Epicutis of pileus a tangled trichodermium of gelatinous hyphae, the end-cells of which are narrowly clavate to tubular, 6-8 μ in diameter, the main filament 4.5-6 μ in diameter, some hyphae with dark brown content but most merely yellowish throughout, considerable debris held in the layer but incrustations were seen only near the base of the trichodermium and on the floccose hyphae beneath. No clamp connections present and no amyloid reactions present on any of the tissues as observed in Melzer's.

Habit, habitat, and distribution.—Scattered under spruce and balsam in a sphagnum bog, Burt Lake, Cheboygan County, late summer and early fall. Not uncommon in the one locality and probably widely distributed throughout the state in similar habitats.

Observations.—This is a well-defined species distinct from *S. hirtellus* by the conspicuous coating of glandulae on the stipe and more orange-buff colors. It differs from both *S. hirtellus* and *S. subaureus* in the more slender stature and more pulvinate pileus as well as in having a fragrant odor. To determine the presence of the latter one should have at least a half dozen basidiocarps.

Material examined.—Cheboygan: Smith 37856, 57483, 57892, 62947, 62979, 63340, 64042, 7-31-62; Thiers 3456, 3784, 3867, 4340. Chippewa: Smith 72836. Crawford: Smith 57498. Emmet: Thiers 3243, 3736. Luce: Smith 36952, 44115. Marquette: Bartelli 296.

13. Suillus luteus (Fries) S. F. Gray

Nat. Arr. Brit. Pl. 1:646. 1821

Boletus luteus Fries, Syst. Mycol. 1:385. 1821.
Cricunopus luteus (Fries) Karst. Rev. Mycol. 3:16. 1881.
Viscipellis luteus (Fries) Quélet, Enchir. Fung. p. 155. 1886.
Ixocomus luteus (Fries) Quélet, Fl. Mycol. Fr. p. 414. 1888.

Illus. Pl. 17.

Pileus 5-12 cm broad, hemispheric to obtuse at first, broadly convex to plane in age; surface glabrous and viscid but at times somewhat streaked beneath the slime, color dark reddish brown to dark yellow-brown, fading out to more or less ochraceous in age. Context white or tinted more or less pale yellow especially near the tubes and stipe apex; odor not distinctive, taste mild.

Tubes 3-7 mm deep, gradually shorter toward the pileus margin, adnate to subdecurrent, at first whitish to pale yellow, at length yellow to olive-yellow ("old gold" to "honey-yellow"), unchanging when cut; pores 3 per mm, in age 1-2 per mm, surface with a sheen and mouths yellow at first but becoming dark-dotted as clusters of cheilocystidia darken.

Stipe (3) 4-8 cm long, 1-2.5 cm thick, equal or attenuate at base, typically peronate by the whitish veil up to the membranous often reflexed rather persistent annulus which is purplish drab in a zone on the outer (or under) side and this material becoming gelatinous under conditions of high humidity; glandular dotted above and below the annulus, solid, white within at first, yellow in age.

Spore deposit dull cinnamon moist, pale cinnamon faded (air-dried); spores 7-9 X 2.5-3 μ, smooth, with a pronounced hyaline sheath, more or less oblong in face view, narrowly inequilateral in profile view, nearly hyaline in KOH, merely yellowish in Melzer's.

Basidia 4-spored, 14-18 X 4-5 μ, yellowish in Melzer's and in KOH. Pleurocystidia in scattered to rare bundles surrounded by bister incrusting material when revived in KOH (details of the bundle obscured by the covering of debris), when isolated the individual cystidia 20-35 X 5-7 μ and narrowly clavate, content usually bister as revived in KOH. Cheilocystidia similar, but tube edges often a mass of bister incrustation obscuring cystidial detail.

Tube trama gelatinous, of divergent hyphae. Pileus cuticle of narrow (2-4 μ) gelatinous filaments forming a collapsing trichodermium, hyaline in KOH and pale yellowish in Melzer's. Context of interwoven hyphae, hyaline nonamyloid and with cells somewhat inflated. Caulocystidia in numerous fascicles, the fascicles bister when revived in KOH. Clamp connections absent.

Habit, habitat, and distribution.—Scattered to gregarious under conifers, very abundant in conifer plantations in the southern part of the state where it is regularly collected for the table under the name "Slippery Jack."

Observations.—The well-formed annulus with the purplish drab to vinaceous-brown zone or thin layer of fibrils over the outside, and which usually gelatinizes, is the best field character for this species.

Material examined.—Livingston: Smith 66420. Oakland: Smith 64548. Washtenaw: Hintikka 10-6-61; Hoseney 10-22-65, 10-10-67; Kauffman 10-10-24, 10-1-24, 10-2-24, 8-16-25; Mains 32-825, 32-855; Rea 734; Smith 18832, 51169, 56142, 58352, 62363, 66400, 72864, 9-?-42.

14. Suillus lutescens Smith & Thiers

Contrib. toward a Monograph of N. Amer. Sp. of *Suillus* p. 71. 1964

Pileus 3-6 cm broad, convex to obtuse, expanding to nearly plane or remaining broadly umbonate, glabrous, glutinous, "buffy brown" to "pinkish buff" in age. Context yellowish pallid and slightly yellower when bruised, odor slightly fragrant, taste mild.

Tubes adnate, near "massicot-yellow" (pale yellow) young, adnate to decurrent, shallow (about 5-7 mm deep, but longer in very old basidiocarps). Pores minute, about 2-3 per mm when mature, pale "massicot-yellow" (clear yellow); no color change when bruised.

Stipe 5-12 cm long, 10-15 mm at apex, equal or nearly so, solid, yellowish within and near lemon-yellow in the cortex in upper part, interior watery mottled, surface pallid and overlaid with brownish glandular dots above and below the bandlike annulus, base conspicuously staining yellow where handled; annulus thick, soft, and readily collapsing, outer layer gelatinous, inner layer of soft floccose tissue.

Spore deposit dark cinnamon moist, or where smeared and dried, olive-brown; spores 7-10 (11) $\times$ 3-4 μ, smooth, subfusiform to narrowly ellipsoid in face view, obscurely inequilateral in profile, pale greenish yellow in KOH, pale tawny in Melzer's.

Basidia 4-spored, 20-26 $\times$ 5.5-7 μ, narrowly clavate, hyaline in KOH. Pleurocystidia scattered and at times with a slight amount of brown pigment around the base, 33-52 × 6-8 μ, content hyaline or coagulated and yellowish (when revived in KOH). Cheilocystidia in large clusters with copious brown amorphous pigment surrounding the base of the cluster; individual cystidia 50-90 × 6-10 μ, narrowly clavate, smooth, mostly with dark yellow-brown content, yellow-brown pigment often pervading throughout the cheilohymenium. Caulocystidia in large clusters, with amorphous brown pigment around the clusters; single cystidia 50-100 × 5-10 μ, narrowly clavate to narrowly fusoid-ventricose, some cylindric and flexuous, nearly all with dark brown content revived in KOH. Caulohymenium mostly with basidia paler than the cystidia but still colored and with amorphous pigment throughout the layer; subhymenium gelatinous.

Tube trama of somewhat divergent gelatinous hyaline hyphae. Epicutis of pileus a layer of widely spaced interwoven branched gelatinous hyphae 2-5 μ wide, yellowish in KOH, apparently trichodermial in origin; hypodermal area dark bister in KOH. Clamp connections absent.

Habit, habitat, and distribution.—Under conifers (spruce, fir, and pine), but with birch and aspen in the stand also, late summer and fall, apparently rare, Smith 72488 and the type (Smith 57893).

Observations.—This is a peculiar species because of the strong yellow staining reaction of the base of the stipe, the pale yellow tube mouths when young, the scattered pleurocystidia and gigantic cheilocystidia. In the pores it resembles *S. luteus*. In the tube trama near the pores the tramal hyphae of the central strand are often yellow-brown the same as the cheilocystidia.

15. Suillus acidus (Peck) Singer

Farlowia 2:271. 1945

Boletus acidus Peck, Bull. N. Y. State Mus. 105:15. 1906.

var. acidus

Illus. Pl. 18.

Pileus 3-10 cm broad, obtuse becoming convex, glutinous from a thick slime layer, glabrous, not streaked beneath the slime, white at first as in *S. placidus,* the slime soon yellowish and pileus becoming tinged greenish yellow, in age near "Naples-yellow" or finally "pale pinkish buff" (yellowish white), slowly developing brownish discolorations where bruised. Context pallid, odor slight, taste of pellicle decidedly acid, with KOH lilaceous, with $FeSO_4$ grayish olive, with Guaiac no color change.

Tubes pale lemon-yellow and beaded with drops when young (yellow in distinct contrast to pileus context), dingy ocher-yellow in age; pores dingy yellow, very minute (2-4 per mm), slowly brownish where bruised.

Stipe 5-12 cm long, 6-14 mm thick, equal, solid, surface sticky when fresh from exuded drops, with distinct glandular dots showing when mature, whitish at first, then pale pinkish buff, staining dingy olive to brownish over basal area when handled. Annulus thin and evanescent, cottony to the inside; interior of stipe orange-buff throughout, no olive line in the base as seen in a longitudinal section, with copious white mycelium around the base. .

Spore deposit dull cinnamon becoming pale cinnamon when air-dried; spores 7-9 $\times$ 2.8-3.2 μ, smooth, in face view suboblong to subfusoid, in profile view narrowly inequilateral, the suprahilar depression broad, yellowish hyaline in KOH, mostly pale tawny in Melzer's.

Basidia 4-spored, 18-23 $\times$ 5.5-7 μ, narrowly clavate, hyaline in KOH, yellowish in Melzer's. Pleurocystidia 40-60 $\times$ 5-8 μ, subcylindric to narrowly clavate, mostly with dark yellow-brown content revived in KOH and with amorphous incrusting material around the base of the cluster, in Melzer's the incrusting material partially dissolving to form pigment globules. Cheilocystidia in bunches and similar to pleurocystidia but more amorphous incrusting material present in the subhymenial tissue. Caulocystidia 25-38 $\times$ 5-7 μ, scattered to bunched in the caulohymenium which is heavily incrusted with dark brown amorphous material in KOH.

Tube trama of gelatinous-refractive hyphae diverging from a very poorly formed central strand. Pileus cutis a thick layer of interwoven gelatinous hyphae 3-8 μ wide, the cells of the hyphae tubular and hyaline in KOH, and in Melzer's yellowish. Context hyphae hyaline in KOH, in Melzer's with homogeneous orange-yellow content, 5-12 μ wide, thin-walled and lacking wall incrustations. Clamp connections absent.

Habit, habitat, and distribution.—Gregarious under red pine (*P. resinosa*), City Park, Midland, September 4, 1966. Smith 73307.

Observations.—Our account (1964) of this species under var. *acidus* applies to a pale yellow species with brown tube mouths when young. This was a mistake. The Midland collection clearly demonstrated that there is a white, annulate, strongly acid taxon, and it must be regarded as the type variety of Peck's species. Like *S. placidus,* the pilei become pale yellow in age. The species in its typical form was very abundant in the park at Midland on the date cited. The taxon described by Snell, Singer, & Dick (1959) under this name is very likely a composite description covering several closely related taxa.

15a. Suillus acidus (Peck) Singer, var. luteolus var. nov.

Pileus 4-10 cm latus, subumbonatus, viscosus, luteolus; sapor acidus; pori brunnei, demum sordide lutei; stipes 8-12 cm longus, 8-12 mm crassus, glandulose punctatus, annulatus; sporae 7-9 (10) × 3-3.5 μ. Specimen typicum in Herb. Univ. Mich. conservatum est; prope Burt Lake, Michigan, September 4, 1957, legit Smith 57938.

Pileus 4-10 cm broad, slightly umbonate to broadly convex, surface copiously glutinous; glabrous beneath the slime and over surface but in age more or less streaked with brownish fibrils beneath the slime; color at first pale yellow (a dull "warm buff"), in age nearer cinnamon-buff and more virgate, margin floccose-appendiculate at least when young. Context thin, firm, dull yellow, becoming dark yellow-brown in Melzer's, not changing when bruised, odor not distinctive, taste of gluten very acid.

Tubes 4-6 mm deep in large caps (rather shallow), adnate to short-decurrent, dull yellow; pores small (2-3 per mm), dull brown when young and instantly darker brown in KOH.

Stipe 4-12 cm long, 8-12 mm at apex, equal, solid, pallid to yellowish within, glandular dotted above and below the superior annulus, ground color pallid, often stained grayish around the base; glandular dots dingy brownish, blackish when dried; annulus of soft floccose material coated on outer side with a layer of slime.

Spore deposit dull cinnamon when air-dried; spores 7-9 (10) × 3-3.5 μ, smooth, olive-yellowish in KOH, in Melzer's pale tan, in face view osbcurely fusiform to suboblong, in profile narrowly inequilateral.

Basidia 4-spored, clavate, 16-20 × 5-6.5 μ, hyaline in KOH, yellowish in Melzer's. Pleurocystidia 26-40 × 6.5-10 μ, cylindric to clavate, hyaline or with yellow-brown content as revived in KOH, when fresh hyaline to dark lilac-brown in KOH, in fascicles and these with amorphous-brown pigment surrounding them at the level of the hymenium. Cheilocystidia abundant, 40-65 × 5-9 μ, cylindric to clavate, mostly with yellow to yellow-brown content; tube dissepiments dark yellow-brown from amorphous pigment. Caulohymenium with a subgelatinous subhymenium, a palisade of basidia and fascicles of caulocystidia resembling the cheilocystidia present and imbedded in the amorphous yellow-brown pigment, some clavate cells up to 14 μ wide at apex and some cells fusoid to obventricose also present.

Tube trama of divergent gelatinous hyphae from a narrow floccose central strand of hyaline to slightly brownish hyphae. Epicutis of pileus a collapsed tangled mass of gelatinous yellowish (in KOH) hyphae 3-7 μ wide, the hyphae probably originating as a trichodermium. Clamp connections absent.

Habit, habitat, and distribution.—Gregarious under pine, usually *P. resinosa*, September, not uncommon during wet seasons.

Observations.—This is the taxon that was described as the type variety in error by Smith & Thiers (1964). The white (type) variety, var. *luteolus*, and *S. subalutaceus* are so similar in the dried state that one cannot make an accurate identification of them without detailed notes on the basidiocarps when fresh. The distribution and the tree associates of the taxa in this group need further critical study.

16. **Suillus intermedius** (Smith & Thiers), stat. nov.

Suillus acidus var. *intermedius* Smith & Thiers, Contrib. toward a Monograph of N. Amer. Sp. of *Suillus* p. 74. 1964.

Illus. Pl. 19.

Pileus 5-10 (17) cm broad, convex to obtuse becoming plane or obscurely umbonate; glabrous, glutinous, pallid yellow when young, darkening to yellow-ocher and finally yellow-brown (near "clay color") to dingy pale cinnamon and then often streaked or spotted with drying gluten, margin often appendiculate with soft yellow patches of veil tissue. Context thick, yellowish white to pale ochraceous, unchanging when cut; odor none or slight, taste of slime acid to mild, of context mild.

Tubes 4-6 mm deep in large pileus, adnate to short-decurrent, pale yellow; pores small (about 2 per mm), pale yellow ("cartridge-buff") and beaded with hyaline drops when young, darker yellow in age, often spotted in age, unchanging when bruised or becoming slowly brownish.

Stipe 4-10 cm long, 5-12 mm thick, equal or nearly so, solid, cortex salmon-ochraceous in lower part; surface pallid yellowish with darker dingy glandular spots which are dingy brown and slowly change to fuscous by old age, surface darkening where handled and one's fingers soon stained black from handling it; annulate near the apex with a gelatinous soft, pale buff band of veil tissue (a floccose layer interior to a gelatinous one).

Spore deposit dull cinnamon ("Sayal-brown"); spores 8-11 × 3.5-5 μ, pallid dingy ochraceous in KOH and only slightly more tan in Melzer's, wall very slightly thickened; in face view obscurely fusiform, in profile inequilateral and with apex often slightly turned up.

Basidia clavate, 16-20 × 4-6 μ, 4-spored, hyaline to pale ochraceous in KOH. Pleurocystidia in bundles with the incrusting material around the base of the bundle dark bister in KOH, 38-55 × 8-12 μ, clavate to subfusoid, content hyaline or dark yellow-brown, walls thin, smooth or at times with pigment incrustations. Cheilocystidia similar to pleurocystidia but more numerous and more of them colored. Caulocystidia numerous and the entire caulohymenium becoming dark bister from amorphous material in age.

Tube trama somewhat divergent from an indistinct narrow parallel strand of floccose hyphae, subhymenial zone gelatinous, floccose hyphae colored to hyaline in KOH. Pileus cuticle a collapsed trichodermium of pale dingy ochraceous hyphae (in KOH) which have scarcely any granular content, little incrusting debris, and are gelatinous and 3-6.6 μ wide, their outline very distinct as revived in KOH. Clamp connections none.

Habit, habitat, and distribution.—Scattered to gregarious under pine, mostly *P. resinosa,* common in August and September after wet weather along the south shore of Lake Superior.

Observations.—During the summer of 1965 this species was collected repeatedly near Culhane Lake in Luce County. Its relationships are with *S. subolivaceus,* a western species, from which it differs in the color of the pileus and stipe. Both have the matted mycelium at the base, and both have the tubes copiously beaded with drops when young and fresh.

Material examined.—Cheboygan: Smith 72817. Chippewa: Smith 72860, 72871. Luce: Smith 58078. Mackinac: Smith 43783.

17. **Suillus subalutaceus** (Smith & Thiers), stat. nov.

Suillus acidus var. *subalutaceus* Smith & Thiers, Contrib. toward a Monograph of N. Amer. Sp. of *Suillus* p. 75. 1964.

Illus. Pl. 20.

Pileus 3-6 cm broad, obtuse to convex or in age with an obtuse umbo, surface thinly slimy-viscid, somewhat variegated, color when young "pinkish buff" to more yellowish but soon changing toward vinaceous-buff ("vinaceous-buff") especially on the variegations, often merely "light pinkish cinnamon," margin often appendiculate with remains of the veil. Context thick, pallid near the cutis, nearly lemon-yellow above the tubes, odor and taste mild; where bruised slowly becoming vinaceous brownish.

Tubes 4-6 mm deep (shorter than the depth of the pileus), decurrent, pale yellow and staining vinaceous-cinnamon where bruised or in contact with KOH; pores round, small, pale brownish, about 2 per mm.

Stipe 8-10 cm long, 10-15 mm thick, equal, pallid to yellowish, near apex the interior yellowish, solid; surface with numerous vinaceous-cinnamon glandular dots both above and below the annulus; annulus a broad band of floccose material the outer layer gelatinous.

Spore deposit tawny to "snuff-brown"; spores 8-11 X 3-3.5 μ, smooth, in face view subfusoid, in profile inequilateral, dingy pale yellowish in KOH and pale brownish in Melzer's.

Basidia 4-spored, clavate, yellowish in KOH and slightly more so in Melzer's. Pleurocystidia in clusters and more or less embedded in amorphous brown material; individual cystidia 30-48 X 5-7 μ, cylindric or nearly so, mostly hyaline but yellowish to brown in some. Cheilocystidia mostly in a palisade on the tube edge, hyaline to yellowish in KOH but amorphous pigment present around the basal part, 30-45 X 5-6 μ, subcylindric. Caulocystidia in large clusters almost obscured by amorphous material (as revived in KOH); individual cystidia with dark brown content, 40-65 X 5-8 μ, narrowly clavate to cylindric; amorphous pigment also present along the caulohymenium.

Tube trama of gelatinous somewhat divergent hyphae, central strand of parallel floccose to subgelatinous hyphae. Epicutis of pileus a gelatinous layer of semidecumbent to interwoven hyphae (originating as a trichodermium), air-dried 2.5-4 μ wide, the content pale dingy ochraceous in KOH. Clamp connections absent.

Habit, habitat, and distribution.—Gregarious in mixed stands of red and white pine, Luce County, and in the same habitat in Mackinac County, fairly common during wet seasons, August and September.

Observations.—In general appearance *S. lutescens* is very close to this species, but the latter shows a strong change to yellow in the base of the stipe when injured. It differs from *S. acidus* in the mild taste of the slime on the pileus, and the color of the pileus in addition.

18. **Suillus subluteus** (Peck) Snell ex Slipp & Snell

Lloydia 7:34. 1944

Boletus subluteus Peck, Bull. N. Y. State Mus. 1:62. 1887.

Illus. Pl. 21.

Pileus 3-8 cm broad, obtuse to convex, becoming obtusely umbonate or expanding to nearly plane; surface glabrous, glutinous; color "light ochraceous-salmon" to "light ochraceous-buff" (salmon-buff to pale buff) when young and fresh, gradually becoming darker dingy yellow-brown (often finally near "bister"), typically developing a fuliginous shade over an obscurely ochraceous-buff ground color, darkening somewhat on standing, taste (especially the gluten) acid at first but soon fading, odor not distinctive.

Tubes 6-10 mm deep (often deeper than depth of context), adnate to short-decurrent, pale olive yellowish young, becoming "light ochraceous-buff" (pale yellow) to orange-buff; pores round at first and 1-2 per mm, becoming somewhat angular in age, yellowish with brownish spots from clusters of cheilocystidia.

Stipe 4-7 cm long, 6-12 (16) mm thick at apex, equal or enlarged downward, solid, cortex ochraceous, central area paler, soon blackening around the wormholes; surface pinkish ochraceous at first and dotted above and below the baggy annulus with vinaceous dots which slowly blacken; annulus with a thick inner layer of soft floccose tissue and with a gelatinous layer over the exterior, in young specimens the lower margin flared outward in a rather characteristic manner, upper edge flared even more, in age collapsing and rather inconspicuous.

Spore deposit dull cinnamon after moisture escapes; spores 7-10 X 2.3-3.2 μ, smooth, pale greenish hyaline to yellowish in KOH, pale cinnamon to yellowish in Melzer's, narrowly oblong to narrowly elliptic in face view, inequilateral in profile, with the apical end slightly turned up dorsally.

Basidia 4-spored, 16-20 X 5-6.5 μ, clavate, nearly hyaline in KOH, yellowish in Melzer's. Pleurocystidia in fascicles with brown incrusting material surrounding the bundle at the level of the hymenium as revived in KOH, when fresh and mounted in 2.5 percent KOH dark lilac to lilac-brown; individual cystidia 30-45 X 7-11 μ, clavate, hyaline or with colored content as revived in KOH. Cheilocystidia similar to pleurocystidia or obfusoid, many with dark brown content as revived in KOH, colored like pleurocystidia in KOH when fresh. Caulohymenium with a broad gelatinous subhymenium, with numerous fascicles of caulocystidia surrounded by copious bister (as revived in KOH) amorphous pigment; also some areas of hymenium incrusted with similar pigment but no cystidia seen in them; caulobasidia in an interrupted hymenium with subgelatinous filamentose hyphal projections among them but these lacking apical differentiation into specialized types of cells.

Tube trama with a colored to hyaline narrow central strand of parallel floccose hyphae, from this gelatinous hyphae diverge to the subhymenium. Epicutis of pileus a tangled collapsing trichodermium of narrow (2-4.4 μ in diameter), branched, hyaline to yellowish hyphae, considerable debris present in the layer. Clamp connections none.

Habit, habitat, and distribution.—Gregarious under jack pine (*P. banksiana*) from late August into September, not common but often in quantity during wet seasons. We have found it most abundantly in jack pine bogs in deep moss.

Observations.—This is one of the most readily recognized species of *Suillus* in Michigan because of the baggy veil leaving an annulus that collapses and becomes inconspicuous in age, the ochraceous to salmon-ochraceous tones of the interior of the stipe, and the general darkening of the pileus as it ages. The most closely related species is the southern *Suillus cothurnatus* Singer. We now strongly suspect that *S. pinorigidus* Snell & Dick is the same as *S. subluteus.*

Material examined.—Chippewa: Shaffer 1947; Smith 43936, 50371, 58196. Emmet: Smith 42030, 42050. Luce: Smith 36961, 42475, 42290, 50235, 50403, 58019. Mackinac: Smith 43784. Marquette: Bartelli 2440. Ogemaw: Smith 67441, 67495, 67580.

19. Suillus albidipes (Peck) Singer

Farlowia 2:45. 1945

Boletus albidipes Peck, Bull. N. Y. State Mus. 57:22. 1912.
(Not *Boletus granulatus* var. *albidipes* Peck, Ann. Rept. N. Y. State Mus. 54:168. 1901.)
Suillus albidipes (Peck) Snell, Mycologia 37 (June): 378. 1945.
Suillus granulatus subsp. *albidipes* (Peck) Snell & Dick, Mycologia 53:232. 1961.

Illus. Pls. 22-23.

Pileus 4-10 cm broad, convex becoming broadly convex; surface glutinous to viscid, color white or pallid varying to near "vinaceous-buff" on young pilei, by maturity pale ochraceous or varying toward vinaceous-cinnamon (darkening as in *S. granulatus*), when old often spotted by the drying slime; margin at first decorated by a dry cottony roll of whitish to vinaceous-buff material representing a false veil which seldom comes in contact with the stipe, in age this tissue collapsing and evanescent but at times forming pallid patches which remain along the margin in mature specimens. Context white, slowly becoming yellow, odor and taste not distinctive, with KOH pink, then lilac-gray, with $FeSO_4$ olive-blue on context, slowly olive-gray on pileus surface.

Tubes about 5 mm deep where context is about 1 cm deep, pale dingy yellow, adnate becoming adnexed; pores round, minute (about 3 per mm), yellow, typically not staining much when bruised.

Stipe 3-6 cm long, 10-15 mm thick, equal, bulbous, or tapered at the base, solid, yellowish within but cortex of midportion reddish in age, lemon-chrome above, in age dingy brown in the base; surface white at first and not glandular dotted, but in age dark very fine glandular dots present lower down, slowly yellow above and reddish brown below.

Spore deposit dull cinnamon when moisture has escaped; spores 6.6-8.8 X 2.5-3 μ, thin-walled, greenish hyaline in KOH, yellowish in Melzer's, oblong in face view, oblong to obscurely inequilateral in profile, with 1-2 oil drops.

Basidia 15-20 X 4.4-6.6 μ, clavate, sterigmata about 2.5 μ long, 4-spored, content hyaline in KOH. Pleurocystidia 26-35 X 8-11 μ, clavate, hyaline in KOH or content dingy yellow-brown, rare, in fascicles mostly near tube edge and with dingy yellow-brown incrusting pigment surrounding the bundle at the level of the hymenium. Cheilocystidia similar to pleurocystidia but as dried having more yellow-brown content and more amorphous pigment present at the level of the basidia, or the tube edge entirely of short, clavate, colored to hyaline cells. Caulocystidia 30-50 X 7-10 μ, in fascicles, clavate to narrowly fusoid, some with bister content, some hyaline to pale ochraceous, with heavy deposits of dark rusty brown pigment around the bases of the bundles. Caulobasidia in a hymenium with a gelatinous subhymenium. Clamp connections absent.

Tube trama of a central strand of brownish to hyaline parallel hyphae 4-7 μ in diameter, with hyaline somewhat divergent and gelatinous hyphae leading to the hymenium, content hyaline to brownish in Melzer's. Pileus with a gelatinous epicutis of interwoven hyphae 2-6 μ in diameter, dingy ochraceous in KOH. Clamp connections absent.

Habit, habitat, and distribution.—Not uncommon during late summer and fall under pine, mostly white pine, in natural stands or in plantations where it is very abundant at times.

Observations.—*S. albidipes* is featured by the pale colors and the cottony pallid material making up the false veil. The spores are among the smallest in the genus. This species has frequently been confused with *S. granulatus,* but the false veil quickly distinguishes *S. albidipes* from the latter if young basidiocarps are available. In old ones the pileus of *S. granulatus* is typically a darker rusty cinnamon. The problem with *S. albidipes* centers around related western taxa with basically the same type of veil but with other distinguishing features, see Smith & Thiers (1964).

Material examined.—Cheboygan: Smith 67678. Livingston: Smith 64747, 75209, 75216. Luce: Smith 58179. Marquette: Bartelli 2442. Washtenaw: Kauffman 10-7-24, 10-8-24; Smith 66419; Thiers 4542.

20. Suillus sibiricus (Singer) Singer

Farlowia 2:260. 1945

Ixocomus sibiricus Singer, Rev. de Mycol. 3:46. 1938.
Boletus sibiricus (Singer) Smith, Mushrooms in Nat. Habitats p. 220. Sawyer's Inc., Portland, Ore. 1949.

Illus. Pl. 24.

Pileus 3-10 cm broad, convex to obtuse, expanding to plane or slightly umbonate; surface viscid to glutinous, spotted overall but especially toward the margin with "snuff-brown" to "cinnamon-brown" appressed squamules, ground color dingy olive yellowish to bright yellow; margin at first with a soft cottony roll of pale dingy yellow veil remnants, or the roll often breaking up into denticulations and finally the material evanescent. Context pale olive-yellow, slowly dull cinnamon when cut, taste acidulous, odor none or slight.

Tubes 1-1.5 cm deep, adnate becoming decurrent, dingy ochraceous to near honey-yellow; pores angular and often compound, 1-2 mm broad, dingy ocher-yellow, staining full cinnamon when bruised.

Stipe 5-10 cm long, 7-15 mm thick at apex, solid, dingy olive yellowish within, surface dingy ocher-yellow above, soon stained vinaceous at the base, glandular dotted overall; occasionally with an annulus but veil typically hanging on the edge of the pileus.

Spore deposit dull cinnamon; spores 8-11 × 3.8-4.2 μ, smooth, in face view narrowly elliptic, inequilateral in profile, pale dingy ochraceous in KOH, pale tawny to yellowish in Melzer's, smooth, a few reddish tawny.

Basidia 4-spored, 22-28 × 5-7 μ, clavate, nearly hyaline in KOH, yellowish in Melzer's. Pleurocystidia in fascicles surrounded by brown amorphous pigment, individual cystidia 40-70 × 6-9 μ, cylindric to narrowly clavate and often crooked, content of yellow to dark brown pigment as revived in KOH. Cheilocystidia similar to pleurocystidia, many somewhat ventricose, masses of amorphous pigment incrusted on the gill edge. Caulocystidia similar to cheilocystidia. Caulohymenium with a gelatinous subhymenium.

Tube trama of gelatinous divergent hyaline hyphae coming from a narrow parallel central strand. Epicutis of pileus a pellicle of appressed-interwoven hyphae 3-6 μ in diameter, gelatinous and dingy ochraceous in KOH. Clamp connections absent.

Habit, habitat, and distribution.—Gregarious under white pine, known in Michigan from the Yankee Springs Recreation Area in Barry County.

Observations.—This species is very close to the common *S. americanus,* but has a thicker stipe (commonly up to 15 mm), the stipes are often hollowed somewhat at the base and not infrequently furnished with an annulus. In the Michigan collections the colors are about the same as those of *S. americanus,* but see Smith, Thiers, & Miller (1966).

21. Suillus americanus (Peck) Snell ex Slipp & Snell

Lloydia 7:39. 1944

Boletus americanus Peck, Bull. N. Y. State Mus. 1:62. 1887.

Illus. Figs. 31, 34; Pl. 25.

Pileus 3-10 cm broad, obtuse with an incurved margin, expanding to broadly convex or with a low obtuse umbo; margin with soft cottony veil material appendiculate or continuous all along it, the veil yellowish but in age the collapsed patches brownish; surface bright yellow, viscid, with scattered appressed patches of fibrils buff to dingy cinnamon and variously distributed, at times streaked with reddish fibrils in places but veil never red, rarely with reddish spots. Context mustard-yellow, staining vinaceous-brown when cut, odor and taste not distinctive.

Tubes 4-6 mm deep, adnate to decurrent, dull yellow and staining vinaceous-brown where injured; pores large and often angular (up to 1.5-2 mm in diameter), mustard-yellow when young, duller ochraceous in age, drying dark yellow-brown.

Stipe 3-9 cm long, 4-10 mm thick, equal, often crooked, becoming hollow, cortex yellow, lacking salmon tints; surface lemon-yellow and covered with glandular dots which darken to "cinnamon-buff" or browner in age or from handling, surface generally vinaceous-brown where handled; annulus lacking, the material of the false veil seldom if ever touching the stipe.

Spore deposit dull cinnamon; spores 8-11 X 3-4 μ, narrowly and obscurely fusiform in face view, in profile narrowly inequilateral, dingy yellow to brownish in KOH, smooth, weakly dextrinoid (pale cinnamon in Melzer's).

Basidia 17-22 X 5-7 μ, 4-spored, clavate, yellowish to hyaline in KOH, yellowish in Melzer's. Pleurocystidia in fascicles surrounded by amorphous bister-brown pigment as revived in KOH; subcylindric to clavate or obventricose, 36-60 X 7-11 μ, content hyaline or dark brown. Cheilocystidia abundant along the dissepiments, hyaline to brown, 40-65 X 7-12 μ, cylindric, clavate or fusoid-ventricose, with a great deal of dark brown amorphous pigment at the level of the hymenium when sections are revived in KOH. Caulocystidia resembling the cheilocystidia, in massive clusters and dark brown amorphous material often obscuring the entire caulohymenium.

Tube trama of gelatinous hyaline hyphae diverging to the hymenium. Epicutis of pileus a pellicle of appressed gelatinous yellowish hyphae 2.5-5 μ in diameter. Clamp connections absent.

Habit, habitat, and distribution.—Common under white pine, solitary, gregarious, or cespitose, late summer and fall. It occurs throughout the state where white pine grows, but is not common in plantations until the trees have reached pole size or larger.

Observations.—The slender stipe, bright yellow color, yellow pores when young becoming wide in age, cottony tissue along the pileus margin at first, and the epicutis of the pileus present in the form of a pellicle rather than a trichodermium all aid in its accurate identification. *S. sibiricus* is the most closely related species, but has a thicker stipe and some basidiocarps in a group are usually annulate.

Material examined.—Cheboygan: Charlton G-113; Sharp 3819; Smith 42330, 50148, 62973, 63877, 72402; Thiers 3783. Chippewa: Kauffman 7-14-06; Smith 72419; Thiers 4030. Crawford: Smith 57499. Emmet: Shaffer 1892; Smith 37123, 42932, 43547, 58080; Thiers 1358, 3244, 3803. Gratiot: Potter 8705, 8946, 9024, 11208, 11546, 11856, 11947, 12115, 12875, 13244, 13492, 14371. Luce: Smith 39535, 50267. Mackinac: Smith 42896. Marquette: Bartelli 271. Montcalm: Potter 6561, 6657, 10110, 10251, 13179, 13662. Oakland: Smith 7221. Ogemaw: Smith 67544, 67587, 66924. Washtenaw: Homola 1070; Rea H-722; Smith 6912, 7167, 64596; Thiers 4569.

22. Suillus glandulosipes Smith & Thiers

Contrib. toward a Monograph of N. Amer. Sp. of *Suillus* p. 86. 1964

Pileus 2.3-6 (12) cm broad, convex to broadly convex, surface slimy-viscid when fresh, pale yellow becoming lightly pinkish tan, sometimes virgate beneath the slime, pellicle separable, margin decorated with remains of a false veil; the latter cottony and soft, soon evanescent. Context firm, ivory-yellow, up to 3 cm thick, with KOH lilac-gray, with FeSO$_4$ olive-green; odor and taste slight.

Tubes about 4 mm deep, depressed at the stipe or adnate to decurrent, yellow; pores small (about 1 mm wide), yellow.

Stipe 2.5-7 cm long, 4-12 (18) mm thick at apex, nearly equal or slightly enlarged downward, solid, interior white to "pale pinkish cinnamon," bright pale yellow in age; surface covered with pale brown glandular dots which become dark brown in age, in age with vinaceous stains over basal part where handled.

Spores 6-7.5 X 2.8-3.2 μ, pale yellow in KOH, smooth, oblong in face view, suboblong in profile. Basidia 4-spored. Pleurocystidia in fascicles, 34-50 X 6-9 μ, subcylindric to narrowly clavate, with incrustations around the bundle dark brown and the cystidia with content similarly colored as revived in KOH. Cuticle of pileus a thick layer of interwoven gelatinous hyphae 2-5 μ wide. No clamp connections observed.

Habit, habitat, and distribution.—Under jack pine (*P. banksiana*) Yellow Dog River pine barrens, Marquette County, June 25, 1968, N. J. Smith 1052.

Observations.—This species has the stipe almost as conspicuously dotted as *S. placidus* and hence cannot justifiably be classed with *S. albidipes.* Like other western species, however, it appears to be rare in the state. It should also be pointed out that the spore size is at the lower end of the range for typical *S. glandulosipes.*

23. Suillus placidus (Bonorden) Singer

Farlowia 2:42. 1945

Boletus placidus Bonorden, Mohl's Bot. Zeitung. 19:204. 1861.
Ixocomus placidus (Bonorden) Gilbert, Bolets p. 132. 1931.

Illus. Pls. 26-27.

Pileus 3-10 cm broad, broadly convex when young, remaining so, or becoming plane or finally with a wavy margin, margin naked; surface glabrous and viscid to slimy, often appearing varnished if weather is dry, soft to the touch, white to ivory-white at first, gradually becoming yellowish, finally pale lemon-yellow, dingy olive with age when water-soaked, slime often becoming grayish to blackish with age, surface often showing depressions as the soft context collapses. Context white but slowly lutescent, when cut often slowly becoming pale vinaceous, taste mild, odor mild to acidulous, with KOH dingy vinaceous, with ammonia red.

Tubes 3-8 mm deep (rather shallow), at first adnate to depressed around the stipe but soon decurrent, yellowish pallid when young, becoming yellow, unchanging when injured; pores pale yellow becoming yellow, small (1-2 per mm), often beaded with pinkish droplets of an exudate when young and fresh, often glandular dotted.

Stipe 4-12 cm long, (3) 5-12 (22) mm thick, equal, solid, becoming hollow, white within but soon lemon-yellow, surface white with "vinaceous-buff" to "vinaceous-fawn" smears and dots, ground color yellow in age, the glandular areas discoloring to gray or blackish on drying; no veil present at any stage.

Spore deposit dull cinnamon; spores 7-9 X 2.5-3.2 μ, nearly oblong in face view, somewhat inequilateral in profile; smooth, a hyaline sheath evident, nearly hyaline in KOH, pale yellow in Melzer's.

Basidia 24-28 X 6-7 μ, hyaline in KOH, yellowish in Melzer's, clavate, 4-spored. Pleurocystidia 49-60 X 6-9 μ, subcylindric to narrowly clavate, scattered, typically in fascicles, often with bister content and dark brown amorphous pigment masses incrusting the bundle at the level of the hymenium. Cheilocystidia in larger aggregations, more capitate and darker colored when revived in KOH but otherwise similar to the pleurocystidia. Caulocystidia numerous to abundant, often in fascicles, similar to the pleurocystidia or with more apical enlargement individually, abundant over large areas, content hyaline to bister when revived in KOH, mostly incrusted with amorphous material.

Tube trama of gelatinous divergent hyphae from a central floccose strand. Epicutis of pileus a layer of narrow hyphae (3-6 μ) with gelatinous walls and with bister content in KOH and in Melzer's. Hyphae of context floccose and hyaline. Clamp connections absent.

Habit, habitat, and distribution.—Scattered to abundant under white pine, late summer and early fall, not rare within the natural range of white pine but not found by us as yet in plantations.

Observations.—The white pileus, conspicuously smeared stipe, and lack of a veil in conjunction with proximity to white pine make this an easily recognized species.

Material examined.—Alger: Mains 32-198. Cheboygan: Chmielewski 15, 38; Smith 41893. Chippewa: Smith 71907, 72418. Crawford: Shaffer 2180; Smith 57500. Emmet: Smith 62974. Gratiot: Potter 3644, 4077, 5733, 8311. Luce: Smith 13304, 36918, 37006, 37150, 38049, 36930, 42228, 42629, 44114, 57950, 58117; Thiers 3860, 4334. Marquette: Bartelli 267, 372, 2526. Oakland: Smith 7230. Ogemaw: Smith 67556.

24. Suillus brevipes (Peck) Kuntze

Rev. Gen. Pl. 3:535. 1898

Boletus viscosus Frost, Bull. Buff. Soc. Nat. Sci. 2:101. 1874.
(Not *B. viscosus* Vent.)
Boletus brevipes Peck, Ann. Rept. N. Y. State Mus. 38:110. 1885.

Suillus brevipes var. *aestivalis* Singer, Farlowia 2:217. 1945.
Boletus plumbeotinctus Kauffman in Snell, Mycologia 28:472. 1936.

var. brevipes

Illus. Pl. 28.

Pileus 5-10 cm broad, hemispheric to broadly convex, in age broadly convex to nearly plane, the margin at times lobed slightly; surface glabrous and slimy, in wet weather the slime often 2 mm deep, color evenly dark vinaceous-brown to grayish brown, becoming paler to dull cinnamon and at times in age dingy yellow-ocher; margin in buttons naked to very faintly white-tomentose but veil lacking and no distinct roll of white cottony tissue present. Context white when young, becoming yellow in age at least in apex of stipe, soft, unchanging when bruised, odor and taste not distinctive.

Tubes 4-10 mm deep, adnate to decurrent, near "honey-yellow" (dingy yellow), darker and more olivaceous in age; pores small (1-2 per mm), round, not elongating radially appreciably, when young pale dingy yellow, in age spotted from the spores and the cheilocystidia.

Stipe 2-5 cm long, 1-2 (3) cm thick, short, solid, white within but finally yellow in the cortex or apex generally, surface white becoming pale yellow, unpolished to pruinose under a lens and when young lacking glandular dots, glandular dots at times visible in age but never well developed.

Spore deposit near "cinnamon"; spores 7-9 (10) X 2.8-3.2 μ, smooth, in face view narrowly elliptic to oblong, in profile obscurely inequilateral, pale yellowish in Melzer's and in KOH.

Basidia 4-spored, 18-24 X 5-6 μ, clavate, hyaline in KOH, yellowish in Melzer's. Pleurocystidia in bunches with amorphous brown pigment surrounding the base of the cluster, individual cystidia 35-50 X 6-9 μ, cylindric to clavate, content hyaline or either partly or entirely brown from coagulated pigment. Cheilocystidia similar to pleurocystidia or larger and more broadly clavate, with considerable amorphous pigment along the edge of the dissepiments. Caulocystidia like the cheilocystidia but the bundles not numerous.

Tube trama of divergent hyaline gelatinous hyphae from a narrow subgelatinous central strand. Epicutis of pileus a thick gelatinous pellicle of narrow (4-6 μ) hyphae, appressed-interwoven but possibly originating as a trichodermium. Clamp connections absent.

Habit, habitat, and distribution.—Scattered to cespitose under 2-needle and 3-needle pines, abundant in the fall especially in plantations.

Observations.—The very slimy glabrous dark vinaceous-brown pileus when young, the lack of a veil, pale yellow pores and white unspotted stipe at first, distinguish this species. It is often collected for food in pine plantations in the southern part of the state.

Material examined.—Chippewa: Shaffer 1948; Smith 43929. Huron: Mazzer 5217. Luce: Smith 38489, 58015. Oakland: Smith 75171. Washtenaw: Smith 66442, 67585.

24a. Suillus brevipes var. subgracilis Smith & Thiers

Contrib. toward a Monograph of N. Amer. Sp. of *Suillus* p. 99. 1964

Illus. Pl. 29.

Pileus 3-8 cm broad, obtuse to convex, in age broadly convex, glabrous, glutinous when wet, varnished when dry, pale pinkish tan when young, dingy ochraceous in age or on the disc about "cinnamon-buff"; margin curved in at first and naked at all times; in age the pileus surface appearing slightly virgate beneath the slime. Context white at first, slowly becoming pale lemon-yellow, olive-gray in $FeSO_4$, pinkish then vinaceous-drab in KOH, taste and odor not distinctive.

Tubes up to 1 cm deep, adnate to depressed around the stipe, lemon-yellow, pores up to 1 mm broad when mature, bright yellow, not staining either brownish or bluish when bruised.

Stipe 3-5 cm long, 7-12 mm at apex, solid equal or nearly so, cortex lemon-yellow with the pith paler, surface white and lacking glandular dots, gradually becoming bright lemon-yellow or apex lemon-chrome, some grayish brown discoloration of the surface fibrils occurs lower down.

Spore deposit pale cinnamon; spores 6.5-9 X 2.4-3 μ, smooth, thin-walled, obscurely inequilateral in profile, in face view nearly oblong, greenish hyaline in KOH, yellowish in Melzer's.

Basidia 16-20 X 4-5 μ, 4-spored, hyaline in KOH, clavate. Pleurocystidia in fascicles with rusty brown to vinaceous-brown incrusting pigment around the base of the bundle, individual cystidia with brown to vinaceous-brown content and either smooth or incrusted. Cheilocystidia abundant, tube edge often heteromorphous, also incrusted with reddish brown amorphous pigment, the details of the cystidia the same as for the pleurocystidia. Caulocystidia in patches or fascicles, individual cystidia 40-70 X 5-9 μ, cylindric to clavate, with vinaceous-brown content in KOH, bases surrounded by some amorphous incrusting pigment.

Tube trama bilateral and gelatinous but with conspicuous broad (up to 15 μ) laticifers with dark brown content as revived in KOH. Epicutis of pileus a thick layer of appressed-interwoven hyphae 2.5-5 μ in diameter and hyaline in KOH, but many of the hyphae (or segments of them) with a hyaline refractive content. Subcutis of floccose hyphae with pigment pockets which become vinaceous in KOH, or the entire region flushed vinaceous. Clamp connections absent.

Habit, habitat, and distribution.—Gregarious under 2-needle pines, in plantations in the southern part of the state, late summer and fall.

Observations.—This variety consistently has a thinner stipe, paler colors and more conspicuous laticiferous hyphae in the trama of the hymenophore. Also, the tubes are more brilliant yellow. It may be desirable eventually to recognize this as an autonomous species.

Material examined.—Oakland: Smith 66452. Ogemaw: Smith 67486. Washtenaw: Smith 72555.

25. **Suillus lactifluus** (Wither. ex S. F. Gray) Smith & Thiers

Mich. Bot. 7:16. 1968

Leccinum lactifluum S. F. Gray, Nat. Arr. Brit. Pls. 1:647. 1821.

Illus. Pl. 30.

Pileus 4-10 cm broad, obtuse to convex, expanding to plane; surface slimy, color pinkish buff to more or less pinkish cinnamon, in age darker reddish cinnamon, glabrous, in age streaked to some extent beneath the slime or finally distinctly virgate; margin sterile and overhanging the edge as a thin membrane 0.5-1.5 mm wide (as in species of *Leccinum* section *Leccinum*). Context watery pallid but with lemon-yellow areas especially near the tubes, dull lilac-blue with KOH, red in NH_4OH, olive with $FeSO_4$, odor slight, taste acid.

Tubes pale yellow ("baryta-yellow"), when young and just forming, duller in color and usually beaded copiously with droplets of a milk-like latex which evaporate by maturity, adnate to short decurrent or in intermediate stages slightly depressed, up to 8 mm deep; pores minute, 2-3 per mm at first but becoming 1-2 per mm at maturity, staining brownish where bruised ("cinnamon-buff").

Stipe 2-5 cm long, evenly enlarged downward or in age narrowed downward, virgate within with lemon-yellow streaks, staining cinnamon to vinaceous-cinnamon in the base; surface pale lemon-yellow overall at first, very inconspicuously punctate above but the ornamentation concolorous with remainder of the surface and not darkening by maturity in spite of the fact that the surface stains brown from being handled, colored dots show slightly on the dried basidiocarps.

Spore deposit cinnamon as air-dried; spores 7-9 × 3 μ, in profile suboblong, to slightly inequilateral, in face view oblong or nearly so, yellowish in KOH, in Melzer's dingy yellowish.

Basidia 4-spored. Pleurocystidia in clusters with vinaceous-brown incrusting material when viewed as revived in KOH and also many cystidia with colored content, hyaline when fresh but turning vinaceous with KOH; individual cystidia cylindric to narrowly clavate, 34-45 (50) × 6-9 μ. Caulocystidia in massive clusters, the individual cystidia 50-90 × 7-12 μ, when revived in KOH smooth and hyaline for the most part but with the content finely colloidal and in KOH red only in the content in the tip of the cell, smooth, pedicel flexuous, some surface hyphae with reddish content in KOH also, and with reddish debris between the cells; the red pigment in some clusters of cystidia slowly fading on standing in KOH. Epicutis of pileus of gelatinous hyphae 2-5 μ in diameter, hyaline to yellowish in KOH. Clamp connections absent.

Habit, habitat, and distribution.—Scattered to gregarious under white pine in a plantation at the University of Michigan Botanical Gardens in October. It is known in Michigan only from the one locality.

Observations.—For comments on this species see Smith and Thiers (1968). It is very easy to recognize by the sterile projecting margin of the pileus, the droplets of latex on the young tubes, the fact that the glandulae on the stipe do not darken as in *S. granulatus,* and that the stipe stains yellow-brown to cinnamon on handling.

26. Suillus unicolor (Frost in Peck) Kuntze

Rev. Gen. Pl. 3(2):536. 1898

Boletus unicolor Frost in Peck, Bull. N. Y. State Mus. 8:100. 1889.

Illus. Pl. 31.

Pileus 4-7 cm broad, plano-convex, expanding to nearly plane, thinly slimy viscid, perfectly glabrous or rarely with a few reddish dots near the margin, color "empire-yellow" to lemon and constant for the life of the basidiocarp. Context mild, odor not distinctive, when cut pale to bright lemon-yellow, with KOH giving a flush of red and finally bluish fuscous (the color change on the pileus surface similar).

Tubes decurrent, shallow (2-3 mm in young specimens, 5-6 mm in mature ones), gelatinous, bright yellow, readily separable from pileus; pores minute, bright yellow, not staining when bruised.

Stipe 4-6 cm long, 8-10 mm thick, solid, narrowed downward to a subradicating base, within lemon-yellow in the apex and olivaceous in the base, surface lemon-yellow but staining dull olivaceous where handled, near apex with raised areas of glandulae but the dots concolorous with remainder of the surface or in age finally brownish; copious basal mycelium pale yellow.

Spores 7-9 X 2.6-3 μ, smooth, in profile narrowly inequilateral, in face view elongate-subfusiform to nearly oblong, pale yellowish in KOH, pale tawny in Melzer's. Basidia 4-spored. Pleurocystidia in fascicles, when revived in KOH with vinaceous-brown content and vinaceous-brown incrusting debris around the cluster, individual cystidia cylindric to narrowly clavate, 34-47 X 5-9 μ. Pileus cutis of appressed, hyaline, narrow, tubular, gelatinous hyphae lacking clamp connections.

Habit, habitat, and distribution.—Scattered under aspen-birch-spruce stands near edge of swampy area, northwest corner of Montmorency County. July 26, 1967. Smith 74648.

Observations.—In the dried state the glandular dots are very conspicuous on the stipe. The yellow of the pileus is preserved in drying—an unusual feature in this group of species. Frost did not mention the stipe as staining olive for his *Boletus unicolor,* but we are assuming that this character was overlooked. There was no veil along the pileus margin and not any significant development of fibrils on the pileus, so it is not likely that our material represents a variant of *S. hirtellus. S. subaureus* is closest but the stipe in it has a ferruginous to orange zone in the base when cut, and the color changes on the surface are to brownish. Also, the basal mycelium is white, and the pileus has appressed tomentum over the surface to some extent as observed on young basidiocarps.

27. **Suillus granulatus** (Fries) Kuntze

Rev. Gen Pl. 32:535. 1898

Boletus granulatus Fries, Syst. Myc. 1:385. 1821.
Rostkovites granulatus (Fries) Karsten, Rev. Mycol. 3:16. 1881.
Viscipellis granulatus (Fries) Quélet, Enchir. Fung. p. 156. 1886.
Ixocomus granulatus (Fries) Quélet, Fl. Myc. Fr. p. 412. 1888.

Illus. Pl. 32.

Pileus 5-11 (15) cm broad, becoming broadly convex, viscid to glutinous when wet, glabrous or streaked or spotted with "cinnamon" on a pale buff ground color, in age more or less cinnamon overall or "orange-cinnamon," immature pilei often whitish for a long time, margin naked to minutely appressed-fibrillose; cutis in age often separating to form minute areolae. Context whitish when young but soon pale yellow, soft, unchanging, with a watery green line above the tubes, odor mild to slightly fragrant, taste of pellicle mild to acid, with $FeSO_4$ olive-gray; KOH on cutis olive-gray, on context "Natal-brown" (vinaceous-brown).

Tubes adnate-subdecurrent, about 1 cm deep, pallid at first, but pores small and pale yellow becoming dingy yellow when mature, not staining when bruised, when young a boletinoid configuration more or less evident, in age up to 1 mm broad and somewhat spotted from colored fascicles of cheilocystidia, staining dingy cinnamon in age when bruised, when very young often scantily beaded with droplets of a cloudy liquid.

Stipe 4-8 cm long, 1-2 (2.5) cm thick at apex, equal or narrowed to a point at the base, solid, white inside at first, but soon bright yellow in apical region, tinged cinnamon toward the base, surface whitish but soon bright yellow above, pallid downward but base becoming dingy cinnamon, covered overall by pinkish tan to vinaceous-brown glandular dots; veil none.

Spore deposit dingy cinnamon; spores 7-9 (10) X 2.5-3.5 μ, smooth, yellowish in KOH and Melzer's, oblong or tapered slightly to apex, in profile somewhat inequilateral.

Basidia 4-spored, 18-23 X 5-6 μ, clavate, hyaline in KOH, merely yellowish in Melzer's. Pleurocystidia in bundles with brown incrusted material around the base, individual cystidia 36-50 X 7-9 μ, clavate to subcylindric, hyaline or with yellowish to brown content. Cheilocystidia similar to pleurocystidia. Caulocystidia 40-70 X 7-10 μ, clavate, subfusiform to cylindric, mostly with colored content, bases of clusters surrounded by brown pigment, content of cystidia hyaline to colored and mostly coagulated; incrusting pigment present over much of the caulohymenium.

Tube trama of gelatinous divergent hyphae from a colored floccose central strand. Epicutis of pileus a gelatinous trichodermium of hyphae 4-7 μ broad, collapsing to form a pellicle in KOH, the content dingy ochraceous to bister, some incrusting material present in the layer. Clamp connections absent.

Habit, habitat, and distribution.—Scattered, gregarious or cespitose under various species of pine but abundant under white pines, especially in plantations. It fruits during the summer and fall and is one of the most common species in the state.

Observations.—This is one of the best species of *Suillus* for the table and the one most frequently collected. It is not as slimy as the Slippery Jack. The color pattern of the pileus is from white in buttons covered by the duff or those which have just emerged, to rather dark reddish cinnamon in old exposed pilei. It lacks a false vail and the stipe is covered by glandules which soon discolor to produce colored dots and smears. The pileus may appear obscurely spotted in age, but this is usually from the gluten which shrinks as it dries to produce the effect noted.

Material examined.—Barry: Mazzer 930-62. Cheboygan: Smith 39372, 42093, 58143, 1951 (no date); Thiers 3749, 3780, 4327. Chippewa: N. J. Smith 59; A. H. Smith 43937, 50387, 50782, 57216; Thiers 3140. Crawford: Smith 49574; Thiers 3741. Gratiot: Potter 11857, 11948, 12107, 12438, 12445, 12459, 12473, 12541, 12958, 13425, 13468, 13686, 13727, 14079, 14175. Livingston: Smith 66417, 75241. Luce: Smith 36172, 36805, 58017; Thiers 4058, 4105. Mackinac: Smith 42899; Thiers 3671. Marquette: Bartelli 257, 267, 370. Midland: Smith 72799. Montcalm: Potter 9026, 9033. Oakland: Nickell 42; Smith 7224, 66453. Ogemaw: Smith 50495, 67551. Roscommon: Smith 1202. Washtenaw: Kauffman 8-6-25; Smith 62361, 62369, 62507, 66399, 66421, 66442, 66476, 72802.

28. Suillus punctatipes (Snell & Dick)·Smith & Thiers

Contrib. toward a Monograph of N. Amer. Sp. of *Suillus* p. 94, 1964

Boletinus punctatipes Snell & Dick, Mycologia 33:36. 1941.

Illus. Pls. 33-34.

Pileus 8-16 (20) cm broad, convex or irregular from mutual pressure, broadly convex in age or margin spreading and wavy; margin naked when young, surface glabrous and glutinous but at times virgate beneath the slime; color pinkish cinnamon to "mikado-brown" (vinaceous orange-brown), or finally when old becoming grayish to violaceous-brown ("benzo-brown"). Context thick and white except for a yellow zone above the tubes and a vinaceous line under the cutis, odor and taste mild, KOH on white areas pinkish, on the vinaceous zone olive, on the cutis olive-gray.

Tubes up to 1 cm deep, when mature, mostly 4-7 mm deep, decurrent, pale yellow becoming ocher-yellow, not staining; pores boletinoid in arrangement from the first, in age conspicuously radial in arrangement and 1-3 (4) mm in diameter radially, cream color young, ocher-yellow mature and often grayish in age, not staining brown or blue when injured.

Stipe 6-10 cm long, 1-3 cm thick, solid, usually narrowed at the base, white within at first, staining vinaceous to vinaceous-brown where injured, at times entirely vinaceous-brown; surface white at first becoming yellow where not stained vinaceous-brown, glandular dotted or with masses of glandular cutis about 1 cm wide present, glandulae pinkish to vinaceous-brown (much as in *S. granulatus* only much larger).

Spore deposit pale "Sayal-brown" when moisture has escaped; spores 7.5-9 X 3-3.2 μ, oblong in face view, obscurely inequilateral in profile, smooth, hyaline in KOH and Melzer's, or slightly yellowish, some dark red-brown in KOH (from areas near the pleurocystidia ?).

Basidia 4-spored. Pleurocystidia clustered, 45-68 X 6-9 μ, cylindric to narrowly clavate, much vinaceous-red to vinaceous-brown amorphous pigment around the clusters in the hymenium but many cystidia with hyaline content at least in the projecting part (many basidia in vicinity of clusters also with red content and some with dull lilac-brown inclusions). Caulocystidia similar to pleurocystidia or smaller, the whole caulohymenium often dull vinaceous-red.

Pileus cutis a collapsed trichodermium of hyphae 4-10 μ in diameter and as revived in KOH with copious incrustation of fine granules, the layer gelatinous; laticiferous hyphae with dull brown content present, subcuticular layer with much vinaceous-brown pigment. Clamp connections none. All tissues nonamyloid.

Habit, habitat, and distribution. —Solitary in mixed conifers including *Pinus banksiana,* Tahquamenon Falls State Park.

Observations. —This species is known from a single collection in Michigan. It is likely to be regarded as a dark *S. granulatus* with boletinoid pore surface and broad patches of glandular surface on the stipe.

FUSCOBOLETINUS Pomerleau & Smith

Brittonia 14:156-72. 1962

Spore deposit reddish to vinaceous, vinaceous-brown to purplish brown to lilac-drab or grayish brown to pinkish brown after loss of moisture; hymenophore either yellow before maturity or white becoming grayish brown; remainder of features more or less as in *Suillus* sensu Smith & Thiers (1964).

Type species: *Fuscoboletinus sinuspaulianus.*

KEY TO SECTIONS

1. Tubes white to grayish, rarely yellowish in age. . . . . Sect. *Griselli*
1. Tubes yellow when young 2
    2. Stipe glandular-dotted . . . . Sect. *Pseudosuillus*
    2. Stipe not glandular-dotted 3
3. Pileus viscid or soon becoming so beneath the veil. . . . . . . Sect. *Fuscoboletinus*
3. Pileus dry and fibrillose . . . . Sect. *Palustres*

Section PALUSTRES (Singer) Smith

Mich. Bot. 4:30. 1965

Pileus dry and fibrillose to squamulose; hymenophore yellow, boletinoid to sublamellate.

Type species: *Fuscoboletinus paluster* (Peck) Pomerleau.

KEY TO SPECIES

1. Stipe 5-10 mm thick; cuticular elements dark red-brown in Melzer's; clamp connections present . . . . *F. paluster*
1. Stipe 1-3 cm thick; cuticular hyphae without red to red-brown content in Melzer's; clamp connections absent . . . . *F. ochraceoroseus*

75

29. **Fuscoboletinus paluster** (Peck) Pomerleau

Mycologia 56:708-9. Pl. 104. 1964

Boletus paluster Peck, Ann. Rept. N. Y. State Cab. 23:132. 1872.
Boletinus paluster (Peck) Peck, Bull. N. Y. State Mus. 8:78. 1889.
Boletinellus paluster (Peck) Murrill, Mycologia 1:8. 1909.

Illus. Pl. 35.

Pileus 2-7 cm broad, obtuse when young, expanding to convex-umbonate or margin plane to arched and disc often retaining a low umbo; surface deep red ("rose-doree" to "jasper-red") from the colored floccose-tomentose covering which breaks up into fibrillose squamules. Context yellowish white to golden or deeper yellow, unchanging when bruised, odor and taste (according to Snell) farinaceous or the taste slowly becoming persistently acid.

Tubes with large angular mouths more or less in radial arrangement or sublamellate with crossveins producing a merulioid effect, decurrent, yellow, not changing to blue or bluish green when injured, in age merely dingy ochraceous.

Stipe 3-6 cm long, 5-10 mm thick, equal or nearly so, solid, yellow to golden within; surface yellow above, somewhat roughened to furfuraceous and often with an evanescent annulus concolorous with squamules on pileus or the veil material variously distributed and finally evanescent, central part often reddish to bright red, yellow and tomentose at base.

Spore deposit dark purplish brown moist, fading to pinkish brown when air-dried; spores 7-9 X 3-3.5 μ, smooth, with a hyaline sheath, in face view somewhat boat-shaped, in profile inequilateral; greenish hyaline in KOH and barely yellow in Melzer's.

Basidia 20-24 X 5-6 μ, 4-spored, hyaline in KOH, yellowish in Melzer's. Pleurocystidia scattered, free of incrustations, 50-72 X 9-12 μ, subcylindric with obtuse apex, short-pedicellate below, hyaline in KOH, thin-walled and readily collapsing, nonamyloid. Cheilocystidia similar but content often smoky yellow in KOH and often shorter and more clavate.

Tube trama apparently interwoven and not particularly gelatinous, hyaline in KOH, nonamyloid. Pileus trichodermium of large (9-12 μ in diameter), nongelatinous hyphae orange to red or red-brown in Melzer's, the cells somewhat narrowed to the cross walls and greatly elongated. Context of floccose interwoven hyphae pale yellow in Melzer's and hyaline in KOH, smooth. Clamp connections present but often difficult to find.

Habit, habitat, and distribution.—Gregarious to cespitose in cold northern bogs, cedar swamps, and the like, often on very decayed conifer logs, apparently throughout the northern part of the state but it has been found as far south as the Yankee Springs area in Barry County.

Observations.—The reaction of the hyphae of the trichodermium in Melzer's, the almost lamellate hymenophore, and the bright red pileus distinguish this species.

Material examined.—Alger: D. Swartz 9-6-27. Barry: Mazzer 10-21-62. Chippewa: Smith 33-841. Marquette: Mains 33-596, 33-600, 33-601.

30. Fuscoboletinus ochraceoroseus (Snell) Pomerleau & Smith

Brittonia 14:157. 1962. Pl. 105

Boletinus ochraceoroseus Snell, Mycologia 33:35. 1941.

Illus. Pls. 36-37.

Pileus 8-15 (25) cm broad, convex to broadly convex and subumbonate, becoming plane or nearly so, or the margin slightly uplifted; surface dry and often uneven or pitted, fibrillose to fibrillose-squamulose, the scales sometimes more or less upright over the disc but usually appressed toward the margin, variable in color, often more or less bright lemon-yellow along the margin and pinkish beneath the fibrillose squamules on the disc, at times rose-pink to brick-red with little or no yellow visible or at times whitish from a dense fibrillose covering; margin incurved and usually appendiculate with fragments of the broken veil when young. Context thick, soft, pale bright yellow, often with a pinkish red zone under the fibrils, unchanging when bruised or showing a very slight change to bluish to greenish blue; odor acidulous, taste very slightly acrid, bitterish in cooked material.

Tubes shallow (about 5 mm deep), adnate to decurrent, boletinoid, bright straw-yellow to dull olive-ocher and finally becoming dingy brown; pores elongated to mostly angular (about 2-5 × 1-2 mm), radially arranged to sublamellate, compound.

Stipe 3-5 cm long, 1-3 cm thick, solid, subequal, the base often subbulbous and the apex frequently flared; nearly concolorous with the tubes and usually reticulate above from the decurrent tubes, more sordid and often reddish at base or at times brownish, unpolished or fibrillose below the annulus; veil thin and submembranous, pallid to yellowish, sometimes forming an evanescent annulus but usually adhering to the margin of the pileus.

Spore deposit "army-brown" (dark vinaceous-brown); spores 7.5-9.5 × 2.5-3.2 μ, smooth, subhyaline to yellowish in KOH, subcylindric to slightly inequilateral, a few dextrinoid.

Basidia 4-spored, 28-34 × 4.5-5.5 μ, narrowly clavate to subcylindric, hyaline to yellowish in KOH. Pleurocystidia scattered to abundant, 46-58 × 6-8 μ, subcylindric to subventricose, obtuse to abruptly acute, hyaline in KOH, thin-walled, often flexuous. Cheilocystidia similar to pleurocystidia (35-40 μ long) and at times in clusters.

Tube trama hyaline, gelatinous. Pileus trama of loosely interwoven hyphae 8-11 μ in diameter, hyaline in KOH, thin-walled. Pileus cuticle of more compactly interwoven hyphae not otherwise differentiated except for the red dissolved pigment which breaks down quickly in KOH. Clamp connections absent.

Habit, habitat, and distribution.—Scattered to gregarious during late summer and early fall under *Larix occidentalis* (western larch). Known only from the Pacific Northwest. It is included here because most other larch-associated species of boletes are not restricted to one tree-species, and hence this one is likely to be found in the Lake Superior region.

Observations.—*Suillus pictus,* at times, approaches this species in color but is associated with white pine, not larch, and the spore deposit colors of the two are quite different.

Section FUSCOBOLETINUS

For characters see key to sections.

KEY TO SPECIES

1. Stipe dry and fibrillose-squamulose below a zone left by the veil; veil not gelatinous . . . . *F. sinuspaulianus*
1. Veil gelatinous, often causing stipe to be viscid where its remains adhere. 2

    2. Pileus with floccose squamules over a gelatinous cutis; growing under larch in bogs . . . . *F. spectabilis*
    2. Pileus glabrous; under spruce and possibly other conifers but not larch . . . . *F. glandulosus*

31. Fuscoboletinus sinuspaulianus Pomerleau & Smith

Brittonia 14:165. 1962

Suillus sinuspaulianus (Pomerleau & Smith) Singer, Snell, & Dick, Mycologia 57:457. 1965.

Pileus 3-13 cm broad, at first subconic, then convex to broadly convex, becoming plane and sometimes subumbonate; surface smooth or somewhat wrinkled, obscurely innately fibrillose, glabrous, glutinous, deep ferruginous to chestnut-brown, sometimes fading to dull orange-brown, the separable veil appearing as patchlike scales from the drying gluten; margin at first incurved and appendiculate with veil remnants, and with a narrow sterile band. Context very soft and watery, dingy orange-buff to yellow; odor somewhat mealy; taste not distinctive.

Tubes broadly adnate to decurrent, separable from the context of the pileus, up to 1 cm thick near the stipe in large basidiocarps, becoming very soft and readily collapsing, dull yellow-brown ("buckthorn-brown"), almost "snuff-brown" in age, darkening when bruised; pores concolorous, angular, 1-2.5 mm broad, typically boletinoid and compound, with radiating lines, fimbriate when young, becoming densely spotted on the side with fine reddish or deep brown dots when old and dry.

Stipe 4-12 cm long, 1-3 cm thick, subequal and enlarged in the middle or below, central but sometimes a little eccentric, not glandular dotted anywhere, reticulate and dull yellowish brown above the annulus, fibrillose, floccose or squamulose, grayish brown to grayish red, lined or spotted with red below the annulus, solid, golden-yellow and watery within, reddish around wormholes and becoming slowly reddish toward the base when exposed.

Spore deposit deep purple-brown or dark chocolate-brown in mass when old; spores 8-10.5 (13.2) $\times$ 4-4.5 (4.8) μ, smooth, lacking an apical pore, in face view elliptical or narrowly ovate, inequilateral in profile, slightly depressed above the inconspicuous hilar appendage, mostly narrowed at the apex, greenish hyaline to pale fuscous in KOH, yellowish to dark tawny or in part dextrinoid, with slightly thickened wall.

Basidia 20-24 $\times$ 3.5-5 μ, hyaline in KOH, narrowly clavate, 4-spored. Pleurocystidia scattered or in clusters, 25-50 μ broad, subcylindric to very slightly fusoid-ventricose, 40-60 $\times$ 4-6 μ individually, often projecting 30-45 μ beyond the hymenium, with obtuse to subacute apices, typically thin-walled, hyaline to dingy brown either from coagulated content or brown incrustations in the basal area of only the clustered ones.

Tube trama of a narrow floccose central strand of interwoven hyphae flanked on either side by gelatinous subparallel to slightly divergent hyphae (as revived in KOH). Pileus with a cutis of appressed golden to lemon-yellow hyphae 8-12 μ broad, the cells readily collapsing, densely incrusted with pale yellow granules. Context hyphae floccose, hyaline to pale yellow in KOH and with fine yellowish incrustations, pale dull vinaceous-red in Melzer's. Clamp connections absent.

Chemical reactions: KOH on surface of the pileus brownish red, on context reddish, pinkish, or lilac, negative on tubes. NH_4OH on surface of pileus clear red or pink, context pinkish red, on tubes lilac. Melzer's on context deep yellowish brown with a greenish zone around the margin of the drop, on the stipe the same reaction.

Habit, habitat, and distribution.—Gregarious or scattered in conifer forests, under *Pinus, Abies,* and *Picea* in the fall. It has not been found yet in Michigan but it is to be expected along the south shore of Lake Superior.

Observations.—The dull ferruginous pileus, yellowish brown tubes which readily separate from the pileus in spite of the pores being boletinoid, the felty floccose veil and viscid pileus are the important field characters.

32. **Fuscoboletinus glandulosus** (Peck) Pomerleau & Smith

Brittonia 14:162. 1962

Boletinus glandulosus Peck, Bull. N. Y. State Mus. 131:34. 1909.
Suillus glandulosus (Peck) Singer, Lilloa 22:657. 1949.

Illus. Pls. 38-39.

Pileus 3-12 cm broad, convex, then flattened to shallowly depressed; surface glutinous to very slimy, the thick slime giving a spotted or reticulate appearance to the surface, with an ochraceous ground color showing off the netted pattern, red when young, becoming "mahogany-red" to chestnut or "burnt-sienna," unchanging or darker with age and the slime becoming blackish on old pilei; margin concolorous, incurved at first, often elevated or flared later, appendiculate with veil remnants, narrowly sterile. Context up to 12 mm thick, firm, watery yellow, flavous ("amber-yellow" to "pinard-yellow"), becoming paler and near lemon-yellow, reddish around larval tunnels, separated from the tube layer by a brilliant yellow line; taste mild; odor mild to slightly pungent and unpleasant.

Tubes subdecurrent to decurrent, separable from the context, 4-8 mm thick, brilliant yellow when young, duller yellow at maturity and often changing to olivaceous or "cinnamon-buff" in age; pores concolorous, angular, compound, and irregular in size and shape, up to 2 mm wide, typically boletinoid with radial lines, finely hairy when young under a lens, becoming spotted with small brownish black dots on the sides when old from discolored bunches of pleurocystidia.

Stipe 4-8 cm long, 8-15 mm thick, subcylindric and gradually tapered downward, without glandules, annulate, reticulate and bright yellow above and occasionally reddish punctate, smooth below and slimy yellow with reddish dots at first, becoming deep red to chestnut without flavous tones showing through the dingy colored sheath below the annulus; base yellow; context yellow and unchanging when exposed; annulus slimy and thick at first, collapsing and in age inconspicuous, deep red when young, becoming blackish brown in age; thin inner veil fibrillose and evident at times as patches which darken in age.

Spore deposit dull purplish drab or deep purple-brown; spores 7.5-11.5 X 3.5-5 μ, narrowly oblong in face view, inequilateral in profile, slightly depressed above the inconspicuous hilar appendage and narrowed toward the apex, smooth; wall slightly thickened, nearly hyaline in KOH, yellowish in Melzer's.

Basidia 26-30 X 7-9 μ, 4-spored. Pleurocystidia numerous in dense fascicles 30-50 μ in diameter, with bister incrusting material (as revived in KOH) surrounding the basal portion and forming purple-brown drops in Melzer's; individual cystidia subcylindric, 40-60 X 4-7 μ, thin-walled, the content hyaline to dull brown. Cheilocystidia similar to pleurocystidia.

Tube trama gelatinous, somewhat divergent from a floccose central strand, nonamyloid. Pileus context beneath the pellicle consisting of floccose interwoven hyphae with hyaline thin walls more or less incrusted with dingy to pale bister granules or particles, nonamyloid. Pileus cuticle a thick gelatinous layer of interwoven nonamyloid appressed and incrusted hyphae 4-8 μ in diameter. Stipe surface a layer of gelatinous hyphae similar to those of the pileus cuticle, no caulocystidia found. Clamp connections absent.

Chemical reactions: KOH on surface of the pileus deep red, on context pink. NH_4OH on surface of the pileus deeper and richer red than with KOH. $FeSO_4$ on surface of slime blue. Melzer's reagent on flesh red without a peripheral green zone.

Habit, habitat, and distribution.—Gregarious under conifers, usually hemlock or arborvitae in Michigan. It is rare in the state but is known from the Yankee Springs area in the southwest part, as well as from Tahquamenon Falls State Park, and near Marquette, in the Upper Peninsula.

Observations.—This species is at once distinguished from *F. spectabilis* by its glabrous pileus and in not being associated with larch. *F. sinuspaulianus* does not have a gelatinous veil.

Material examined.—Berrien: Rose October, 1965. Chippewa: Thiers 3847. Luce: Smith 36837, 37412, 39307, 57442, 50212. Marquette: Bartelli 2048, 2282.

33. Fuscoboletinus spectabilis (Peck) Pomerleau & Smith

Brittonia 14:161. 1962

Boletus spectabilis Peck, Ann. Rept. N. Y. State Cab. 23:123. 1872.
Boletinus spectabilis (Peck) Murrill, N. Amer. Fl. 9:160. 1910.

Illus. Pls. 40-41.

Pileus 4-10 cm broad, obtuse to convex, becoming flattened with or without an obtuse umbo; surface viscid to slimy beneath the remains of the veil which often form rather large grayish brown patches, scales, or streaks composed of soft tissue or tomentum; veil remnants finally rich red ("Morocco-red") throughout but finally fading to yellowish or brownish, color beneath the veil also soon some shade of red to rose. Context yellow, becoming deeper yellow by maturity, not changing to blue when exposed but slowly to pinkish and then brown, odor penetrating and disagreeable or somewhat aromatic, taste astringent.

Tubes 6-12 mm long, adnate to decurrent, pale yellow, usually changing to pinkish when injured, dull yellowish brown in age; pores concolorous with sides, angular, obscurely radiating, elongated, about 1 mm wide or more, often compound, dotted with small dark points when old and dry.

Stipe subcylindric, 4-10 cm long, 10-15 mm thick, solid, pale yellow throughout when young, in age olive in the base and chrome to ochraceous in the midportion, rarely becoming hollow, yellow above the annulus, sheathed with the remnants of the red veil and the veil leaving a gelatinous annulus; veil colored like the pileus, sheathing the stipe from base to annulus, soon fading to vinaceous-gray, yellowish beneath, duplex, floccose to the interior (towards the stipe) and gelatinous on outer (under) side, leaving patches on the stipe in age, these more or less evanescent in age.

Spore deposit purplish brown ("Vandyke-brown" to "warm sepia"); spores 9-13 (15.2) X 4-5.6 (6.5) μ, smooth, lacking an apical pore, in face view elliptic, in profile inequilateral with a pronounced suprahilar depression and beyond this narrowed to apex, revived in KOH pale yellow to ochraceous-tawny, in Melzer's strongly dextrinoid.

Basidia 4-spored, 30-36 X 5-7 μ, yellowish in KOH and in Melzer's reagent. Pleurocystidia in numerous clusters, surrounded at the base by material bister in KOH and reddish brown in Melzer's, individual cystidia 40-80 X 6-9 μ, cylindric and narrowed toward the apex, thin-walled. Cheilocystidia similar to pleurocystidia or the tube-edge often covered with amorphous bister material.

Tube trama hyaline and divergent from a yellowish (in KOH) central strand. Pileus cuticle a thick layer of gelatinous hyphae, with smoky yellow content in KOH, becoming hyaline, representing a matted trichodermium, hyphae of short cells 4-8 μ in diameter with no appreciable incrustation. Context of loosely interwoven floccose hyphae with colorless incrustations (sections bister in KOH if rather thick), nonamyloid. Stipe surface similar to pileus cuticle but not appearing gelatinous. Clamp connections absent.

Chemical reactions: KOH on pileus surface bleaching the color that is present, on the context dark green; NH_4OH no reaction on pileus and context; Melzer's deep greenish brown on pileus and stipe context.

Habit, habitat, and distribution.—Solitary to scattered under larch (*Larix laricina*), not uncommon in larch bogs during September in Michigan.

Observations.—This very characteristic species can hardly be mistaken for any other. The duplex veil which in button stages is broken into coarse squamules gives the pilei the appearance of a *Strobilomyces*. The spores are large for the genus, and the KOH reaction of the pileus context is most unusual in this genus or in *Suillus,* in fact we do not recall a species which shows it other than this one. The color of the spore deposit as it air dries is more of a vinaceous-brown, but still within the "spectrum" generally treated under the term purple-brown in the taxonomy of fleshy fungi, certain authors to the contrary notwithstanding.

Material examined.—Cheboygan: Smith 1094, 50589. Emmet: Kauffman 8-27-05; Smith 43820. Gratiot: Potter 6443. Oakland: Smith 38851, 67760. Ogemaw: Smith 67589. Washtenaw: Hintikka 10-6-61; Kauffman 186, 1594, 9-22-24; Smith 11016, 15313, 18795, 62208, 64655.

Section GRISELLI Smith

Mich. Bot. 4:30. 1965

Hymenophore whitish to gray and not becoming truly yellow at any stage of development, usually grayish brown at maturity.
Type species: *Fuscoboletinus aeruginascens.*

KEY TO SPECIES

1. Pileus slimy to viscid; context pallid but soon greenish blue when bruised or cut; pileus usually smoky gray to wood-brown, rarely with yellowish tones . . . . *F. aeruginascens*
1. Pileus slightly viscid to almost dry; context unchanging when cut or bruised; pileus usually pale dull olive to olive-gray. . . . . *F. grisellus*
1. Pileus coated with chocolate-colored slime . . . . *F. serotinus*

34. Fuscoboletinus aeruginascens (Secretan) Pomerleau & Smith

Brittonia 14:168. 1962

Boletus aeruginascens Secretan, Mycogr. Suisse 3:6. 1833.
Boletus elbensis Peck, Ann. Rept. N. Y. State Cab. 23:129. 1872.
Ixocomus viscidus (Fries) Quélet, Fl. Myc. Fr. p. 416. 1888.
Boletus solidipes Peck, N. Y. State Mus. Bull. 167:38. 1912.
Suillus aeruginascens (Secretan) Snell, Lloydia 7:25. 1944.

Illus. Pl. 42.

Pileus 3-12 cm broad, globose at first, then broadly convex to almost plane, sometimes subumbonate; surface slimy, innately fibrillose, fibrillose-squamulose, sometimes cracking when dry, smoky gray-brown to avellaneous, generally becoming paler to greenish gray, yellowish, or occasionally whitish in age, finally showing fuscous spots, margin at first inrolled and appendiculate with veil remnants, later with a thin sterile band 1-2 mm wide, pellicle separable. Context at first firm, later soft, white to yellowish white, turning bluish green when exposed; taste fruity or mild; odor not distinctive.

Tubes adnate or subdecurrent, separable from the context, 6-9 mm long, white, pale brownish or grayish to whitish at first, brownish gray to smoke-gray at maturity, staining bluish to olivaceous; pores concolorous, round at first, becoming irregular and angular in age and often up to 3 mm radially (boletinoid with radiating lines and the pores compound—several smaller ones in a slight depression bounded by a raised border of dissepiment).

Stipe 4-6 cm long, 8-12 mm thick, equal or subequal and tapering upward, sometimes attenuate below, solid, usually reticulate, pallid to greenish white above the annulus, somewhat viscid lower down, smoky gray, brownish gray or brownish below the annulus; context white at times becoming slightly yellowish; veil submembranous, grayish to yellowish, rarely bright yellow.

Spore deposit vinaceous-brown ("fawn-color") moist, nearer avel-. laneous when air-dried; spores 8.5-12 X 3.5-5 μ, smooth, lacking an apical pore, in face view elliptic to subfusiform, in profile obscurely inequilateral in a majority, nearly hyaline to dingy yellowish brown in KOH, in Melzer's slightly dextrinoid, wall thickened slightly.

Basidia 4-spored, 20-30 X 6-8 μ, short-clavate to elongate-clavate, hyaline to yellowish in KOH. Pleurocystidia mostly in clusters with yellow-brown incrustations concentrated around the base, with a hyaline to yellow-brown coagulated content, thin-walled for the most part, subcylindric to subcapitate. Cheilocystidia similar to pleurocystidia or more broadly clavate. Hymenophore with trama composed of a central strand of floccose tissue, more or less interwoven floccose hyphae hyaline in KOH, merely yellowish in Melzer's, without particles or incrustations. Clamp connections absent.

Chemical reactions: KOH on pileus no reaction, on context and stipe surface brown. NH_4OH on pileus no reaction, on context brown.

Habit, habitat, and distribution.—Scattered to gregarious during late summer and fall; common throughout the state under larch.

Observations.—This well-known species, associated with larch (*Larix*, also known as tamarack) is not exactly attractive, and for this reason, perhaps, is seldom collected for the table. The degree of yellow showing varies greatly from one region to another in the United States. In Michigan the variation Peck described as *Boletus elbensis* is the one usually found. It often occurs along with *F. spectabilis*, as one would expect.

Material examined.—Emmet: Smith 1244. Livingston: Smith 6066. Ogemaw: Smith 3190. Washtenaw: Smith 5085, 9-10-40, 10-?-50.

35. Fuscoboletinus serotinus (Frost) Smith & Thiers, comb. nov.

Boletus serotinus Frost, Bull. Buffalo Soc. Nat. Sc. 2:100. 1874.

Pileus 5-12 cm broad, obtuse to convex, expanding to plane with a low umbo or broadly convex, slimy-viscid when fresh with the slime chocolate color to date-brown ("Carob-brown" to "mummy-brown"), ground color pallid but obscured until most of the slime has vanished, decorated with pallid to grayish patches of floccose veil remnants along the margin at first, margin with an overhanging zone which is sterile and which soon disappears. Context pallid, when cut slowly becoming bluish then going to fuscous-gray but finally fading to merely dingy brownish, odor slightly pungent after basidiocarps have been sectioned, with $FeSO_4$ quickly green.

Tubes pallid grayish to pale drab-gray, when injured staining vinaceous-gray going to dingy brown (almost cinnamon-brown), decurrent, 8-15 mm deep; pores angular but usually isodiametric, pallid to grayish when young.

Stipe 5-9 cm long, 9-17 mm thick, equal or nearly so, solid, surface dry and grayish-fibrillose from floccose veil up to an evanescent annular zone, apex weakly reticulate, base pallid and with pallid mycelium, interior when cut pallid except for yellow in the base, after cutting slowly becoming dull blue throughout.

Spore deposit obscurely vinaceous-brown (near "warm sepia" as air-dried). Spores 8-11 X 4- 5 μ, smooth, somewhat inequilateral in profile, in face view subelliptic, in KOH brownish hyaline, in Melzer's dull brown (near "Prout's brown"), wall not appreciably thickened.

Basidia 4-spored, 26-32 X 6-8 μ, narrowly clavate, hyaline in KOH, yellow in Melzer's (from fresh material). Pleurocystidia in bunches and with cocoa-colored debris (in KOH) surrounding the bundle (use fresh material); individual cystidia hyaline in KOH or cocoa colored near tube edge, 52-80 X 9-14 μ, subcylindric or with flexuous walls, some slightly ventricose near base, wall thin and at times with debris adhering. Cheilocystidia in fascicles as are the pleurocystidia but often the content cocoa colored as mounted in KOH, the cells 30-58 X 8-11 μ, mostly cylindric to narrowly clavate.

Pileus epicutis of hyphae 3-6 μ wide, in a gelatinous matrix, having a pale vinaceous content when mounted in KOH, in Melzer's the content stringy and not distinctively colored, hyphal walls smooth, thin, and indistinct (from gelatinization ?), laticiferous elements rare.

Habit, habitat, and distribution.—Closely gregarious in a swamp near Cedar Lake, Washtenaw County, October 4, 1969, Hoseney.

Observations.—Our description is drawn from this material, but this variant has been found repeatedly in southern Michigan and identified as a variant of *F. aeruginascens.* We now regard the color change of the context to bluish and then dark bluish gray to fuscous as a sufficiently important feature to justify recognizing this variant as a species, and Frost, we believe, was the first to describe it. The color change reminds one of that of *Leccinum insigne*. On the pore edges the basidia (with sterigmata) as well as the cystidia may be colored. Frost mentioned the overhanging margin, the "sordid brown" color and the changing context.

36. Fuscoboletinus grisellus (Peck) Pomerleau & Smith

Brittonia 14:168. 1962

Boletinus grisellus (Peck), Mem. N. Y. State Mus. 3:169. 1900.

Illus. Pl. 43.

Pileus 3-8 cm broad, obtuse to broadly conic, then expanding to nearly plane; surface appressed-fibrillose to slightly hairy-squamulose, viscid beneath the fibrils when fresh, pale dull olive, becoming pallid to grayish olivaceous or occasionally tinted with yellow, avellaneous in age at times drying dull brown; margin inrolled when young, frequently decorated with remnants of the submembranous grayish pallid veil. Context thickish, cottony, whitish to faintly olivaceous; odor and taste slight not distinctive, white, unchanging when cut, bluish gray with $FeSO_4$.

Tubes 3-6 mm long, adnate to decurrent, scarcely separable, pallid to grayish or finally dull brownish; pores concolorous with sides, round to angular, or long and narrow near the stipe (sublamellate), typically boletinoid with radiating lines at maturity.

Stipe short, (2) 3-8 cm long, 0.5-1.5 cm, thick, equal, straight, solid, whitish within near apex, bright yellow in the base; surface whitish to pallid, becoming yellowish, rarely somewhat bluish where handled; somewhat reticulate at times above the annulus or veil zone from the tubes, somewhat squamulose to nearly glabrous below the annular zone; veil submembranous, seldom leaving a well-formed annulus.

Spore deposit gray-brown to chocolate-gray; spores 8-10 (11.5) × 4.3-4.8 (5) μ, smooth, in face view elliptic to narrowly ovate, in profile inequilateral, suprahilar depression shallow; wall somewhat thickened, in KOH hyaline to pale fuscous, in Melzer's yellowish to pale tawny.

Basidia pale fuscous brown in KOH, 24-27 × 5-7 μ, narrowly clavate. Pleurocystidia mostly in clusters, 40-60 × 5-7 μ, emerging up to 50 μ beyond the hymenium, subcylindric, hyaline or with dull brown content in KOH, incrusted around the base of the clusters with dark material brown to bister as seen in KOH, vinaceous in Melzer's. Cheilo-cystidia similar or narrowly clavate.

Tube trama consisting of a floccose central strand of interwoven hyphae flanked on either side by slightly divergent gelatinous hyphae; subhymenium of loosely interwoven much-branched gelatinous hyaline hyphae. Pileus cuticle composed of subparallel radially arranged some-what gelatinous to floccose hyphae 5-8 μ in diameter, many hyphae dull yellow-brown in KOH and some having uneven walls from the incrusted pigment. Context trama of very loosely interwoven, hyaline to sub-hyaline hyphae with fine incrustations, the cells often enlarged near the septa, yellowish to hyaline in Melzer's. Clamp connections absent.

Habit, habitat, and distribution.—Not frequent, but usually found in quantity associated with *Larix,* in bogs around the edges of hum-mocks, late summer and fall.

Observations.—This species is much less viscid than *F. aerugi-nascens,* and there is no characteristic color change when the context or surfaces are bruised. In addition to these features, the basidiocarps are much more characteristically conic-umbonate, and the fruiting pattern—around the edges of hummocks in bogs—is very different.

Material examined.—Cheboygan: Smith 1093, 62288. Livingston: Mains 32-865; Smith 6088. Marquette: Mains 33-603, 33-950, 33-983, 34-134. Oakland: Smith 34250, 62161, 66432. Ogemaw: Smith 31926. Roscommon: Smith 67816, 67817. Washtenaw: Kauffman 183, 10-10-07, 10-16-05; Smith 64652.

Section PSEUDOSUILLUS Smith

Mich. Bot. 4:30. 1965

The glandular dotted stipe is the distinguishing feature of this section and serves to emphasize the parallelism of species in this genus with those of *Suillus.*

Type species: *Fuscoboletinus weaverae.*

37. Fuscoboletinus weaverae Smith & Shaffer

Mich. Bot. 4:27-30. 1965

Pileus 3.5-7 cm broad, convex, expanding to broadly convex; sur-face viscid, glabrous except for marginal veil remnants; color when young near vinaceous-buff to avellaneous (grayish flushed pink) over disc and toward margin "orange-cinnamon" (pale reddish cinnamon), disc pinkish cinnamon in age and margin darker (vinaceous-brown–"Verona-brown"). Context 9-10 mm thick above edge of stipe, yellow or orange around larval tunnels, unchanging when cut, with a nondistinctive odor and a pleasant taste.

Tubes adnate to stipe, 1.5-4 mm wide, white at first becoming yellowish, unchanging when cut; pores angular and irregular in shape, sometimes compound, more or less radially arranged but not elongated radially, when young yellowish (paler than "light buff" to "cream color"), becoming more ochraceous in age, unchanging when injured, with prominent vinaceous-brown glandular dots.

Stipe 4-8.2 cm long, 9-15 mm thick at the apex, usually tapering downward (but at times enlarged to 25 mm thick) just above the middle; with "mikado-brown" (vinaceous-brown) glandular dots, tan but at times staining yellow at the base where bruised, with white to bright yellow mycelium at the base and in surrounding soil, sometimes with yellow context in the base. Veil fibrillose to cottony, white to yellow-ish, sheathing lower part of stipe when young and also covering the hymenophore, not forming an annulus and remains over lower part of stipe eventually evanescent.

Spore deposit moist "avellaneous" to near "cacao-brown," with loss of moisture "deep brownish drab" (purplish brown); spores $6.7\text{-}8 \times 2.7\text{-}3.2\,\mu$, in face view oblong to narrowly ovate, narrowly ovate in profile, smooth, in Melzer's moderately to strongly dextrinoid.

Basidia $20\text{-}29 \times 4\text{-}7\,\mu$, usually clavate, 4-spored, in KOH hyaline or near the pleurocystidia some with colored content. Pleurocystidia usually in clusters, individual cystidia $20\text{-}60 \times 3.3\text{-}9.6\,\mu$, the shorter ones not projecting beyond the basidia, usually clavate, at times sub-cylindric to subfusoid, hyaline at first (in water or KOH) as revived in KOH bister or with bister incrusting pigment, thin-walled. Cheilocystidia like the pleurocystidia.

Tube trama bilateral; subhymenium indistinct and prosenchymatous. Pileus cuticle $400\text{-}700\,\mu$ thick; an ixotrichoderm of gelatinous, septate, rarely branched hyphae $2.5\text{-}7.5\,\mu$ in diameter, densely incrusted with minute refractive particles. Caulocystidia resembling pleurocystidia but larger. Clamp connections absent.

Habit, habitat, and distribution.—Scattered to cespitose on humus in sandy soil of a mixed woods (*Quercus, Populus, Betula,* and *Pinus* species), known only from Minnesota, but included here because it is an obvious element of the western Great Lakes fungous flora.

Observations.—In view of the number of species which have hymenophoral cystidia in clusters and in which the clusters become incrusted with pigment and hence colored at least in age, it is not at all surprising to find a species in this genus in which these elements also occur in bunches on the stipe and hence produce the effect described as glandular dotted. It is interesting to note that the discovery of this species by Mrs. Weaver helps materially to complete the line of relationship to *Gomphidius,* where caulocystidia are also found in fascicles. Thus, we find a progression in the color of the spore deposit, a parallel pattern of caulocystidia as to manner of occurrence and chemical reactions with such reagents as KOH, and the development of gelatinizing veils. Certainly, this indicates a line of evolution, if such lines can be established on morphological and anatomical characters.

TYLOPILUS Karsten

Rev. Mycol. 3:16. 1881

Spore deposit vinaceous, vinaceous-brown, purplish brown or rusty ferruginous; veil absent in American representatives; tubes white to pallid when young or very rarely weakly yellowish just before maturity, colored by the spores when mature.

Type species: *Tylopilus felleus.*

Certain species of *Leccinum* have spore deposits that fit the above-described color-spectrum, but their stipes have darkening ornamentation. Species of stirps *Pseudoscaber* often have dark stipe ornamentation, but it is the same color as the stipe. Since no significant change in the color takes place, those species are retained in *Tylopilus*. These are the species formerly placed in the genus *Porphyrellus* (see Smith & Thiers, 1968).

KEY TO STIRPES

1. Stipe scabrous to roughened on a white to yellowish ground color but ornamentation not becoming dark brown to black . . . . Stirps *Chromapes*
1. Stipe not as above 2

    2. Some spores in a mount showing canals through the thicker inner wall (mount spores in Melzer's) . . . . Stirps *Gracilis*
    2. Spores consistently smooth 3

3. Pileus dark sooty brown, drab, olive-brown, violaceous-fuscous, or dark coffee color; spores pale to dark chocolate in deposit . . . . . Stirps *Pseudoscaber*
3. Spores some shade of pinkish to vinaceous in deposit; pileus violaceous to vinaceous-brown, tan or dull cinnamon 4

    4. Tubes soon staining black when bruised or very dark colored from the beginning . . . . Stirps *Alboater*
    4. Tubes pallid at first to grayish and distinctly vinaceous in age 5

5. Taste bitter, stipe reticulate. . . . . Stirps *Felleus*
5. Taste bitter; stipe not reticulate . . . . Stirps *Rubrobrunneus*
5. Taste mild; stipe reticulate . . . . Stirps *Indecisus*

Subgenus ROSEOSCABRA (Singer), comb. nov.

Leccinum section *Roseoscabra* Singer

Amer. Midl. Nat. 37:124. 1947

Stirps CHROMAPES

38. Tylopilus chromapes (Frost) Smith & Thiers, comb. nov.

Boletus chromapes Frost, Bull. Buff. Soc. Nat. Sci. 2:105. 1874.
Leccinum chromapes (Frost) Singer, Amer. Midl. Nat. 37:124. 1947.

Illus. Pls. 44-45.

Pileus (3) 5-11 (15) cm broad, convex to hemispheric, becoming broadly convex to plane or occasionally shallowly depressed in age, margin entire, fertile, often flared and irregular in age, in wet weather the surface dry, smooth to uneven, rarely somewhat pitted, subtomentose or in age with a felted-fibrous surface (matted down), occasionally appearing glabrous with age, somewhat tacky to the touch at times, beautiful pink ("chatenay-pink" to "light congo-pink" to "old rose," occasionally near "Pompeian-red" to "deep Corinthian-red" to "Prussian-red") overall at first but fading out to pallid in places and in age often alutaceous tinged with rose. Context 5-20 mm thick, floccose, white or faintly pink or rose tinted at first, soft, unchanging or rarely slowly becoming yellowish in spots; taste mild (pileus surface usually distinctly acid to the tongue but not so in some cases—Coker) (taste very slightly acid—Thiers); odor not distinctive.

Tubes 5-12 mm deep, depressed to nearly free, white at first, then yellowish ("light buff" to "pale ochraceous-buff") and finally flesh colored ("pale vinaceous-fawn"), in age more brownish ("wood-brown"), unchanging when bruised (at times rosy when bruised—Coker); pores small, 2-3 per mm, round to angular, white when young.

Stipe 4-13 (17) cm long, 1-2.5 cm thick at apex, equal to tapering toward either apex or base, usually pinched off at the base, solid, flesh firm, pallid above, deep chrome-yellow downward (mycelium chrome-yellow—Coker), surface dry, paler pink than pileus at first but soon fading to pallid, scabrous dotted above or nearly to base but ornamentation not darkening, unpolished and uneven over lower part.

Spore deposit vinaceous-fawn to near "wood-brown" (rosy brown or dull salmon—Coker & Beers, "chamois color"—Farlow, chestnut-brown—Smith); spores 11-17 X 4-5.5 μ, smooth, lacking an apical pore, shape in face view nearly oblong to narrowly ovate, in profile obscurely inequilateral, color nearly hyaline in KOH and yellowish to tawny occasionally in Melzer's, with a poorly defined gelatinous sheath.

Basidia 28-36 X 9-12 μ, hyaline in KOH, yellowish in Melzer's, 4-spored. Pleurocystidia 35-46 X 8-12 μ, scattered, fusoid-ventricose, hyaline in KOH and yellow in Melzer's, the content not distinctive, smooth, thin-walled, apex subacute. Cheilocystidia similar to pleurocystidia or more obventricose and obtuse.

Trama subparallel and subgelatinous in mature specimens. Cuticle of pileus a tangled turf of narrow hyphae 4-7 μ wide (about the same size as those of the context), yellow in KOH from dissolved pigment, almost all evidence of a trichodermium lost by the matting down of the elements. Context of hyaline floccose hyphae. Caulocystidia scattered or in fascicles, yellow in KOH, content not distinctive in Melzer's. Clamp connections not demonstrated for certain.

Habit, habitat, and distribution.—Solitary to scattered, usually under aspens in our region, fruiting most abundantly early in the summer, late June and early July near Pellston.

Observations.—This species is both beautiful and distinct in its pink pileus and yellow base of the stipe. It can be accurately identified in the field. McIlvaine stated that it was edible and delicious, but it is difficult to find specimens that are not riddled with insect larvae. During wet periods in early summer, however, it can be found in quantity. In Ammirati 3750, the mature pileus dried yellow.

The differences cited in the color of the spore deposit are no doubt partly due to the moisture content of the spore deposit when the color was noted, and in part to different strains of the fungus. The variations noted are nearly all clearly within the spore-color spectrum of the genus.

Material examined.—Cheboygan: Charlton G-207; Chmielewski 9; Kilburn 3607; Smith 36713, 36854, 37330, 37377, 37857, 58144, 62954, 63009, 64852; Thiers 2859, 2909, 3377, 3460, 3720, 3779, 3826. Emmet: Brask 3605. Gratiot: Potter 10040, 10143. Houghton: Kauffman 1906. Oakland: Smith 18749. Ogemaw: Smith 67422. Ontonagon: Ammirati 3750. Roscommon: Potter 10156. Luce: Smith 38275, 38525, 42603, 50236. Washtenaw: Smith 6700, 18549, 18699; Wehmeyer 7-17-59.

Subgenus PORPHYRELLUS (Gilbert) Smith & Thiers
comb. & stat. nov.

Porphyrellus Gilbert, Bolets, p. 99. 1931

Section GRACILES (Singer), comb. nov.

Porphyrellus section *Graciles* Singer, Farlowia 2:119. 1945

Stirps GRACILIS

39. Tylopilus gracilis (Peck) Henn., in Engler &.Prantl

Nat. Pflanzen Familien 11:190. 1900

Boletus gracilis Peck, Rept. N. Y. State Mus. 24:78. 1872.
Porphyrellus gracilis (Peck) Singer, Farlowia 2:121. 1945.

Illus. Pl. 46.

Pileus (3) 4-8 (10) cm broad, hemispheric to convex or finally broadly convex; surface dry and appearing granulose under a lens, becoming areolate, chestnut-brown to "tawny" or finally paler, often with more cinnamon, usually evenly colored except for the pale flesh showing in areolate pilei, margin fertile or only a very narrow sterile band present. Context up to 1 cm thick, white to pallid or tinged incarnate near the cuticle, unchanging when bruised or cut, floccose but soft to the touch, taste mild, odor not distinctive.

Tubes deeper than flesh of pileus are thick (up to 2 cm deep), often of unequal length, causing pore surface to be uneven, deeply depressed around the stipe, white to pallid becoming flesh colored to vinaceous-brown, unchanging when bruised; pores 1-2 per mm but often mostly 1 mm in age.

Stipe 6-15 cm long, 4-10 mm thick at apex, 1.5-2 cm (3 cm— Coker) at base, very long in relation to width of pileus, often curved, nearly concolorous with pileus in well-colored basidiocarps but often a paler cinnamon-tan, solid, white within, surface finely granular to pruinose and often longitudinally striate, base whitish.

Spore deposit dark vinaceous-brown to near reddish chocolate; spores 9.8-16.8 X 5-7.8 μ, narrowly ovoid to subelliptic in face view, inequilateral in profile view, punctate (under an oil-immersion lens) in many spores, moderately thick-walled, with an inconspicuous hyaline sheath, inner colored wall with minute canals ending in the pits at the surface, color in KOH pale buff to ocher, rusty brown in Melzer's (dextrinoid).

Basidia 2- and 4-spored, 24-36 X 8-12 μ, hyaline in KOH, yellowish in Melzer's. Pleurocystidia scattered, 35-50 X 6-11 μ, narrowly fusoid-ventricose with gradually tapered neck and subacute apex, thin-walled, hyaline in KOH and Melzer's. Cheilocystidia mostly narrowly clavate, basidioid, 40-60 X 9-12 μ, hyaline, thin-walled.

Tube trama gelatinous and divergent, hyphae nonamyloid. Pileus cuticle a trichodermium of hyphae 5-10 μ in diameter, the upper part with short cells but terminal cells often more or less cystidioid and longer (40-60 X 9-15 μ), hyaline in KOH or yellowish, nonamyloid. Surface of stipe a layer of interwoven to suberect hyphal tips some of which are contorted to cystidioid. Clamp connections absent.

Habit, habitat, and distribution.—Solitary to scattered, usually under aspen or hemlock in Michigan, fruiting most abundantly early in the summer, late June to early July locally.

Observations.—There is considerable variation in the color of the fresh spore deposit, but this stabilizes with loss of moisture. The species is amply distinct by its slender stature and pinkish cinnamon colors along with the spores, some of which show the markings as described, although in some pilei we have found it difficult to find ornamented spores. It is best to mount the spores in Melzer's to see the ornamentation.

Material examined.—Barry: Smith 73126. Cheboygan: Bartelli 160; Charlton G-214; Lange 1579; Smith 25771, 37335, 37555, 71927; Thiers 742, 1210, 3455, 3788. Emmet: Kauffman 1905; Smith 41904, 62970; Thiers 3530, 3535, 3745. Luce: Smith 37408, 39059, 39287, 39493, 42301, 72378; Thiers 3858, 3863. Marquette: Smith 72481. Montmorency: Smith 74739. Ontonagon: Peters 1147.

Section PORPHYRELLUS

Stirps PSEUDOSCABER

The species grouped here are all obviously closely related. They are united by: (1) the overall dark somber colors, (2) they stain waxed paper blue, (3) the spore deposit is in the chocolate-brown series, typically the hymenophore is white at first or merely pallid, and (4) the pileus is dry and unpolished to velvety. The type species is at once distinguished by the large spores and the context changing to blue readily when cut along with dark brown pores. *T. sordidus* is distinguished by its strongly inflated large cheilocystidia and a small percentage of very wide thick-walled spores in mounts from the hymenophore (see Smith & Thiers 1968). *T. cyaneotinctus* has a dull yellow stage in the development and maturation of the hymenophore as well as small spores. *T. porphyrosporus* has spores like *T. pseudoscaber,* but it does not stain blue readily (in our material) and the pores are merely pallid to gray at first. *T. umbrosus* has a somewhat radicating stipe and the context does not stain blue when injured. *T. fumosipes* is very close to *T. umbrosus,* but the latter can be distinguished in the field by the somewhat radicating stipe according to Atkinson's description.

We need more descriptive data from fresh material on most of these taxa since they have all been confused in a hopeless manner to date. The characters which seem to us most important are used in the key to species.

KEY

1. Typical range of spore size 13-16 (18) x 6-8 µ 2
1. Typical range 10-14 x 4-6 µ 3
    2. Pores coffee-brown when young; context staining blue then reddish when cut . . . . *T. pseudoscaber*
    2. Pores pallid to grayish young; context slowly becoming fuscous when cut . . . . *T. porphyrosporus*
3. Tubes pallid at first, becoming yellowish and finally dark brown in age . . . . *T. cyaneotinctus*
3. Tubes not passing through a yellow stage 4
    4. Cheilocystidia up to 15 µ or more wide in the inflated part, a small number of spores in mounts from hymenial tissue 7-11 µ wide and with walls 1-2 µ thick . . . . *T. sordidus*
    4. Spores and cheilocystidia not as above 5
5. Stipe base tapered into a short pseudorhiza . . . . *T. umbrosus*
5. Stipe base obtuse to rounded. . . . . *T. fumosipes*

40. Tylopilus pseudoscaber (Secretan), comb. nov.

Boletus pseudoscaber Secretan, Mycogr. Suisse 3:13. 1833.
Porphyrellus pseudoscaber (Secretan) Singer, Farlowia 2:115. 1945.

Illus. Figs. 37-39; Pl. 47.

Pileus 5.5-15 cm broad, convex becoming plano-convex to nearly plane when older; surface dry, dull subtomentose, appearing plushlike, often appearing innately fibrillose when older, occasionally conspicuously finely areolate and rarely fibrillose-scaly, colored very dark earth-brown to olive-brown or dark vinaceous-brown ("clove-brown," "bone-brown" to "mummy-brown"), margin entire, broadly decurved, fertile. Context 1-3 cm thick in the disc, tapering to the margin, soft and spongy, white, turning to bright blue when cut and then in a few minutes reddish brown, finally dull brown, staining white paper blue-green; odor sharp and rather pungent, strong; taste mild.

Tubes 1.5-2 cm long, subdecurrent to shallowly depressed when young, becoming broadly and deeply depressed with age, pale grayish brown ("olive-buff" to "light buff" to deep "olive-buff"), occasionally becoming near "avellaneous" (grayish) to "tilleul-buff" (pallid) when older, becoming greenish blue when bruised or exposed; pores small, round, "bister" (dark yellow-brown) when young, not stuffed.

Stipe 4-10 (12) cm long, 1-3 cm thick at apex, equal to subventricose to clavate (larger below), solid-stuffed, whitish within, not changing color readily, surface dry and dull, evenly and minutely furfuraceous and concolorous with pileus, occasionally appearing glabrous, often becoming black at the base with age, lacking a blue-green ring at apex or lower down, sometimes conspicuously ridged with long close lines with anastomoses, others only slightly reticulate and some not at all so.

Spore deposit distinctly reddish brown ("chestnut") moist, dark grayish brown as air-dried; spores (12) 14-18 X 6-7.5 μ, smooth, lacking an apical pore, shape in face view narrowly elliptic, in profile obscurely inequilateral with a shallow suprahilar depression, pale earth-brown in KOH, near "tawny" or more reddish in Melzer's (somewhat dextrinoid); occasional giant spores up to 27 μ long also present.

Basidia 4-spored, 45-50 X 9-15 μ, smoky brownish in KOH, yellowish in Melzer's. Pleurocystidia scattered, 50-65 X 10-16 μ, clavate-mucronate to fusoid-ventricose, usually with a dark brown content when fresh. Cheilocystidia similar to pleurocystidia or up to 30 μ broad in enlarged part, brown when fresh.

Tube trama divergent from a thin central strand, subhyaline in KOH, hyphae 7-9 μ broad, no amyloid walls or cell content seen. Pileus cuticle a collapsed trichodermium, the hyphae with elliptic to subclavate end-cells with the tips tapered to an acute to subacute apex, walls thin and nearly hyaline, the cells back from the terminal cell often curved, of equal diameter, merely yellowish in Melzer's, the content of the hyphae hyaline to yellowish in Melzer's, no amyloid particles present in most basidiocarps in the cuticular hyphae. Context of pileus of dingy to hyaline interwoven hyphae, all parts yellowish in Melzer's. Clamp connections absent.

Habit, habitat, and distribution.—Scattered to gregarious in hardwoods and mixed forests along old roads, etc. We have one collection from Luce County, Smith 37829.

Observations.—The Michigan collection was inadequate for a description so we have relied on material from the Pacific Northwest for our data. It may very well be that our concept is too broad, as some of the characters given are not those of the species as described by at least some European authors. However, Secretan (1833) described the species as having colored pores. Singer (1967), however, described the species with yellowish tubes and pores. We have as yet not found a large-spored taxon with truly yellowish tubes in North America. The colored pores clearly distinguish Secretan's species from *B. porphyrosporus* Fries. Fries stated: "Poris amplis, pentagonis, griseis, tactu umbrinus" (1874, p. 514). Hence a species with dark coffee-colored pores when young cannot legitimately be assigned here, and we do not accept the synonymy generally given in the literature. For further comment see *T. prophyrosporus*.

41. Tylopilus porphyrosporus (Fries), comb. nov.

Boletus porphyrosporus Fries, Bolet. Fung. Gen. N. 36, 1835.

Illus. Pls. 48-49.

Pileus 4-7 cm broad, broadly convex, the margin even, surface dry and subvelvety, the pile becoming matted down, very dark dull brown (olive-brown) slowly changing to drab but when properly dried with a distinct ochraceous undertone. Context pallid, slowly changing to fuscous on cut surface, odor acid-disagreeable, taste mild, with NH_4OH on cutis dull reddish to reddish brown, the same color with KOH; with $FeSO_4$ on cut surface of context distinctly olive.

Tubes nearly free, sharply depressed around the stipe, dark dull cocoa color, about 10 mm long, when cut the sides soon wood-brown; pores staining dark chocolate color when bruised (no blue stain developing), concolorous with the sides when young.

Stipe 4-7 cm long, 1 cm thick, solid, staining fuscous within; surface pruinose to punctate with the ornamentation colored like the pileus, pallid below but soon stained fuscous from handling.

Spore deposit chocolate-gray ("wood-brown"). Spores 13-17 X 6-8 μ, smooth, lacking an apical pore, as revived in KOH cinnamon-brown, paler yellowish brown in Melzer's, shape in face view oval to narrowly elliptic or subfusoid, in profile rather obscurely inequilateral, the suprahilar area mostly merely flattened.

Basidia 26-35 X 9-12 μ, clavate, yellowish in KOH but slowly becoming hyaline, merely hyaline in Melzer's. Pleurocystidia scattered, 38-64 X 10-16 μ, broadly ventricose with a narrow flexuous neck and subacute apex, thin-walled, smooth, content revived in KOH yellowish to nearly hyaline, when revived in Melzer's bister to dark sepia-brown and stringy to granular. Cheilocystidia mostly resembling basidioles.

Tube trama of the *Boletus* subtype. Pileus cutis a layer of tangled hyphae possibly representing a rudimentary trichodermium (but evidence not conclusive), hyphae curved, 4-8 μ wide, septa few, terminal cells long and tubular, apex obtuse to subacute, the walls hyaline, when revived in KOH with much fine hyaline incrusting or adhering debris present, content in KOH soon fading to yellowish hyaline, in Melzer's the content ocher-yellow to brownish ochraceous and homogeneous to stringy. Context hyphae 6-12 μ in diameter, many with colloidal content when revived in KOH, some brownish ochraceous in Melzer's and all smooth. Clamp connections absent.

Habit, habitat, and distribution.–Solitary along an old trail, Tahquamenon Falls State Park, Michigan, August 23, 1965, Smith 72394.

Observations.–The broad spores which are cinnamon-brown in KOH are a striking feature as is the ochraceous undertone of the dried pileus. The species was mistaken in the field for small specimens of *Tylopilus eximius*. We place this collection in Fries' species realizing there are some discrepancies—namely the lack of a change to blue in the cut context above the tubes. Since the situation in this group needs further clarification both in Europe and North America, we prefer not to erect on this difference a new taxon for the Michigan variant at this time. It does seem clear, however, that a transfer of Fries species to *Tylopilus* is justified. Smith 57600 from Tahquamenon Falls State Park has the stipe apex coarsely reticulate, and much amyloid debris adheres to the spores and some amyloid particles are found in them.

42. Tylopilus cyaneotinctus Smith & Thiers

Mycologia 60:952. 1968

Pileus 4.5-8.5 (13) cm broad, convex, finely tomentose, the tomentum sometimes fasciculate, scarcely at all viscid, dull becoming delicately areolated or more coarsely so near the margin; color light brownish drab when young, becoming darker brown and less drab at or after maturity, with a slight tint of blue-green near the margin. Context whitish (pallid), soft but not cottony, about 7-10 mm thick near the stipe, when cut showing tints of rose and also blue-green, and staining waxed paper deep blue-green; tasteless and with a slight nitrous odor.

Tubes small to medium, varying somewhat according to the size of the pileus, up to 1 cm deep, deeply depressed around the stipe, sometimes with lines from the tube dissepiments; pores round, nearly white when young, then faintly dingy yellowish progressing to avellaneous or in age finally darker, often dotted with brownish clusters of cystidia, when bruised turning deep blue-green then deep brownish red; not stuffed when young.

Stipe 4-6 cm long, 7-10 mm thick near center, crooked, nearly equal or tapering slightly upward, the base pinched to a point; surface finely dotted, often somewhat streaked, about the color of the pileus and tubes below, with a bluish to pea-green area near apex; context on cut surface becoming pinkish.

Spores 10-13.5 × 4-5.5 μ, smooth, lacking an apical pore, smoky brown in KOH and in Melzer's slightly more rusty but in a few spores amyloid particles may be present, in face view narrowly elliptic, in profile somewhat inequilateral.

Basidia 4-spored. Pleurocystidia fusoid-ventricose, 34-60 × 9-12 μ, neck often elongated, with dark yellow-brown content in Melzer's or (rarely) with amyloid granules filling the cell. Caulocystidia numerous, clavate, 28-56 × 9-16 μ, hyaline in KOH, with a hyaline (in KOH) refractive thickening in the apex.

Pileus cuticle a trichodermium, the cells of the elements with smooth walls and the cells inflated somewhat in the lower 3-5 cells of the filament (9-18 μ broad), the upper 1-3 cells more or less cylindric and 5-9 μ in diameter, the end-cell blunt and seldom inflated, content as revived in KOH weakly ochraceous, more ochraceous in Melzer's but remaining homogeneous. Hyphae of context next to the cuticle with an orange content in Melzer's. Clamp connections absent.

Habit, habitat, and distribution.—Mostly solitary on barren hillsides and in open woods or on roadbanks, etc., during warm humid summer weather, rare in Michigan and most likely to be found in the southern counties.

Observations.—The yellowish tubes, the thickening in the wall of the caulocystidia at the apex, and the narrow spores are distinctive. The pinched off stipe apparently to some extent resembles that of *T. umbrosus.*

Material examined.—Washtenaw: Kauffman 1915; Smith 1680, 18490, 64172, 64188, 64227, 64341.

43. **Tylopilus sordidus** (Frost) Smith & Thiers

Mycologia 60:950. 1968

Boletus sordidus Frost, Bull. Buff. Soc. Nat. Sci. 2:105. 1874.

Illus. Figs. 32-33, 35-36; Pl. 50.

Pileus (3) 6-12 cm broad, convex, slowly expanding to broadly convex and finally nearly plane, margin even and at times uplifted in age; surface dry and velvety, soon becoming conspicuously rimose-areolate, the pale context showing in the cracks; color olive-brown to dark dingy yellow-brown (olive-brown) to "Prout's brown," in age the context as exposed in the cracks pale watery green. Context thick and firm, yellowish white, slowly staining pale bluish green where cut, dingy vinaceous around the larval tunnels, often staining vinaceous gray slowly where cut (after the green has faded); odor pungent and peculiar but not strong, taste mild to slightly acidulous; KOH causing a vinaceous-gray color, and in $FeSO_4$ a greenish to bluish tint develops.

Tubes adnate to depressed or in age decurrent, 1-2 cm deep, grayish to olive-gray at first, chocolate color in age, staining blue and then darker brownish where injured; pores 1-2 per mm, isodiametric or nearly so, grayish at first, where injured becoming chocolate.

Stipe (2) 3-6 cm long, 10-15 mm at apex, equal or nearly so, solid, blue in the cortex in apical region when cut, reddish in base but slowly changing to brownish, in the central part slowly staining vinaceous when first cut; surface faintly brownish pruinose to unpolished, nearly concolorous with pileus but with a copper-blue zone at apex, whitish at base.

Spore print "benzo-brown" (purple-brown) in a good fresh deposit, nearer wood-brown when faded; spores 11-15 X 5-6.5 (7) μ, in profile inequilateral with a distinct suprahilar depression, in face view narrowly to broadly boat-shaped or narrowly ovate to nearly oblong but most often broadly ovate, in profile subelliptic to somewhat inequilateral, smooth, wall slightly thickened; dark dingy yellow (near bister) as revived in Melzer's, in KOH pale dingy yellow-brown.

Basidia 22-35 X 9-15 μ, 2- and 4-spored, broadly clavate. Pleurocystidia rare to scattered, 50-60 X 9-15 μ, fusoid-ventricose, smooth, thin-walled, content smoky brown in some as revived in KOH, as revived in Melzer's with a dextrinoid content or some showing a content of dark blue granules (amyloid). Cheilocystidia mostly basidiole-like and smoky ochraceous in KOH but finally greatly inflated.

Tube trama of long cylindric cells up to 15 μ in diameter, in a main strand and some diverging toward the subhymenium on either side as seen in a longitudinal section *(Boletus* subtype). Cuticle of pileus a trichodermium of septate hyphae 8-18 μ in diameter, the end-cells usually somewhat cystidioid, the lower cells often short and somewhat inflated, as revived in Melzer's the elements with dark yellow-brown pigment in the cells and some with incrusting pigment granules, paler and dull ochraceous in KOH. Clamp connections absent.

Habit, habitat, and distribution.—Solitary to gregarious or rarely in small clusters, on sandy open soil in thin oak woods, or on exposed soil in forests such as along woods roads or along streams on the exposed soil on the bank; summer and early fall, uncommon.

Observations.—Smith and Thiers (1968) speculated on this species being a hybrid between *T. pseudoscaber* and *T. umbrosus.* It is the commonest member of the stirps in Michigan and features a very irregular pattern of spore size. In its typical spore range it has the wide spores of *T. pseudoscaber* or *T. porphyrosporus* and the spore length of *T. umbrosus.* In Hoseney 538 from the George Reserve, Livingston County, July 12, 1967, it was found that some of the spores had distinct particles or rods of amyloid material within them, that the pleurocystidia frequently had the neck filled completely or in part with amyloid granules, and that the cheilocystidia were pedicellate-clavate to vesiculose and often were found with an amyloid cap of amorphous material or patches of such material adhered variously to the surface. This sort of amyloid reaction is rarely present on the collections here described as *T. pseudoscaber,* but further studies should be made on all members of this stirps. The reaction shows up to some degree, apparently, on most of the species.

A specimen sent by Frost to Peck has now been studied; it confirms the concept as published by Smith and Thiers (1968). The following is an account of it.

Spores 11-14 X 5-6.5 μ (12-16 X 7-11 μ), smooth, lacking an apical pore but a minute thin spot visible in some spores; shape subelliptic in face view, in profile subovoid to obscurely or somewhat inequilateral; color pale ochraceous in KOH, pale tan in Melzer's but with a faintly darker outer line (as if outer wall was weakly amyloid); wall 0.4-0.7 μ thick as revived in KOH, wall in the exceptionally wide spores 1-2 μ thick.

Basidia 4-spored, hyaline in KOH, clavate, 8-12 μ broad. Pleurocystidia fusoid-ventricose, 33-48 X 10-15 μ, thin-walled, content dull dextrinoid in Melzer's and granular to homogeneous, homogeneous and dull brown to hyaline in KOH. Cheilocystidia 26-35 X 12-18 μ, clavate to subglobose-pedicellate, hyaline to brownish in KOH but walls thin and soon gelatinizing.

Tube trama of tubular hyphae 6-10 μ in diameter, lacking any amyloid reactions, not or only slightly gelatinous except near the pores. Pileus cuticle a trichodermium of hyphae 10-20 μ broad, walls thin and not incrusted, content ochraceous in Melzer's, "colloidal" in KOH, cells short and somewhat inflated to long and tubular to slightly inflated, end-cells clavate to fusoid-ventricose. Hyphae of subcutis with highly colored (ochraceous) content in Melzer's. Clamp connections absent.

Observations.—We are not prepared at this time to assert that *B. fumosipes* is synonymous with Frost's species. The inflated cheilocystidia of the latter gelatinized, as is often true for the pore dissepiments in boletes, but must be regarded as an important feature of the species. Although amyloid material was not found in the Frost collection studied, under the present circumstances it cannot properly be assumed that the species will never show it.

Material examined.—Emmet: Smith 63888. Gratiot: Potter 13389. Livingston: Hoseney 538. Montcalm: Potter 13210. Oakland: Smith 7118, 64311; Thiers 4690. Luce: Smith 39291. Washtenaw: Smith 18639, 18846, 62638, 64625, 72610; Thiers 4519.

44. Tylopilus fumosipes (Peck), comb. nov.

Boletus fumosipes Peck, Rept. N. Y. State Mus. 50:108. 1897.
Boletus nebulosus Peck, Rept. N. Y. State Mus. 51:292. 1898.

Illus. Figs. 40-43, 45-46.

Pileus 2-5 cm broad, convex to nearly plane, minutely tomentose, sometimes minutely rivulose, dark olive-brown; flesh whitish; tubes at first nearly plane, becoming convex with age, their pores whitish when young, becoming yellowish brown, changing to bluish black where bruised; stipe 2-5 cm long, 6-8 mm thick, equal, solid, smoky brown, minutely scurfy under a lens. Spore deposit purplish brown.

Type study: Two basidiocarps studied and both represented the same species.

Spores 12-14 (18) $\times$ 4-5 (6) μ, smooth, apex lacking a pore but a very minute thin spot evident in some spores; shape in face view subovate to obtusely fusiform, in profile more or less inequilateral, the suprahilar depression broad and varying from shallow to deep; color in KOH dingy ochraceous becoming paler, in Melzer's pale dingy tan with a darker border (possibly indicating a slight amyloid reaction), no amyloid content seen in any spores; wall 0.3-0.4 μ thick, thicker in KOH than in Melzer's.

Basidia 4-spored, 24-35 $\times$ 8-12 μ, hyaline in KOH, yellowish in Melzer's. Pleurocystidia scattered, 33-42 $\times$ 9-13 μ, subfusoid, content dull brown and homogeneous in Melzer's, more refractive and granular in KOH (not typical chrysocystidia). Cheilocystidia smaller than pleurocystidia; tube edge gelatinizing readily in KOH. Caulocystidia 30-60 $\times$ 10-18 μ, mostly clavate but some globose and some elongate-subfusoid, thin-walled, hyaline in KOH, not incrusted.

Tube trama of somewhat divergent hyaline tubular hyphae 6-12 μ in diameter and showing no amyloid reactions. Pileus cuticle a trichodermium of elements 10-15 μ in diameter, the cells tubular to somewhat inflated, walls thin, hyaline, not incrusted but a few with a slight gelatinous sheath, cell content merely yellowish in Melzer's, no amyloid particles seen in or outside of the cells, end-cells subclavate to subfusoid (apex obtuse), other cells in the hypha mostly short (less than 5 times as long as broad), and often more or less inflated. Context hyphae near cutis having deeply colored (ochraceous to orange-ochraceous) content in Melzer's. Clamp connections absent.

Observations.—The exceptionally long spores (15-18 μ) are mostly irregularly shaped and represent an abnormal type of spore found in many groups of colored-spored *Agaricales.* They are not thin-walled and not exceptionally wide—in fact they may be narrower than those in the range considered typical of the species. No wide spores of the type found in *T. sordidus* were found. There was no sign of ornamentation on or in the wall of any of the spores. The basidiocarps studied were unquestionably the same species. In the type collection the basidiocarps are glued to pieces of paper. Smith sampled from the groups, one of which, according to comments written on the paper, represented the species with striate spores (Snell, 1936), and the other the group with nonstriate spores. Both had smooth spores. The discrepancy cannot now be explained. Certainly, Snell (1c) found the 2 types, as he puzzled long over this situation since it concerned spore ornamentation. Fortunately, he cited one of his own collections as the type for *B. chrysenteroides,* and the typification of the latter species is not involved. However, we think Peck's comment about the cracks in the pileus becoming reddish very likely applies to the specimens with ornamented spores, and this feature is purposely left out of the above diagnosis. All the basidiocarps now forming the holotype appear to be the same species and clearly belong to the stirps *Pseudoscaber.*

We append here a detailed account of the microscopic features of *B. nebulosus* for comparison (Figs. 29-30). The basidiocarps in the type were larger than those in the type of *B. fumosipes,* but we believe they are the same species.

Pileus 5-10 cm broad, convex, dry, snuff-brown or smoky brown; context white, unchangeable; tubes convex, depressed around the stipe, pallid or brownish becoming purplish brown where wounded, the pores small rotund; stipe 7-10 cm long, 7-12 mm in diameter, enlarged toward base, solid, scurfy, colored like the pileus.

Spores 12-15 (18) $\times$ 4.5-6 μ, smooth, lacking an apical pore, subelliptic to somewhat inequilateral, color in KOH ochraceous, paler on standing, in Melzer's weakly dextrinoid, no amyloid inclusions present; wall up to 0.4 μ thick; spore outline in exceptionally large spores tending to angularity.

Basidia 4-spored, 8-12 μ in diameter, hyaline in KOH. Pleurocystidia rare, more or less fusoid-ventricose, 33-47 $\times$ 9-15 μ, apex acute to subacute, content dull brown in Melzer's, hyaline or nearly so in KOH (after standing 20 minutes or more).

Tube trama lacking hyphae with amyloid inclusions or amyloid walls. Pileus cuticle of interwoven hyaline hyphae 6-12 μ in diameter with long to short tubular to slightly inflated cells but mostly the cells elongate, content hyaline in KOH, and in Melzer's ochraceous, walls thin and lacking incrustations, end-cells merely blunt but varying in shape from subcylindric to fusoid-ventricose. Hyphae next to cuticle merely yellow in Melzer's. Clamp connections absent.

Observations. – The changing pileus context to blue and then blackening should distinguish *T. fumosipes* from *T. umbrosus* if the rooting base is poorly developed. In Smith 62850 the injured context stained blue slowly.

45. Tylopilus umbrosus (Atkinson) Smith & Thiers

Mycologia 60:950. 1968

Boletus umbrosus Atkinson, Journ. Mycol. 8:112. 1902.

Illus. Figs. 27-28.

Pileus 5-9 cm broad, convex then expanded, subtomentose, in age becoming finely areolate, mummy-brown to walnut-brown, fleshy; flesh whitish, very slowly changing to flesh color and then brown. Tubes convex, white at first, then becoming pale brown, in age deeper brown, when bruised becoming dark brown. Stipe colored like the pileus, 8-10 cm long, 1.5-2 cm thick, broadly and irregularly furrowed to rugose longitudinally, with very minute dark points under a lens; base tapered into a short root.

Spores 11-14 $\times$ 4.5-5.5 μ, smooth, pale in KOH but not hyaline (the darkest a pale bister) smooth, thin-walled, lacking apical differentiation, wall cyanophyllic in young stages but not at maturity or staining blue only slowly on standing a day or so; shape in face view elliptic to oval and in profile subelliptic to very obscurely inequilateral (note that in lactic acid mounts some young spores inflated to 6-7 μ but in KOH or Melzer's mounts from the same hymenophore no such wide spores were evident).

Basidia 4-spored, 9-12 μ broad, clavate, lacking a clamp at basal septum, content not distinctive in KOH or Melzer's. Pleurocystidia 36-48 X 9-15 μ, fusoid-ventricose, thin-walled, hyaline in KOH and Melzer's (or some with brown granules when mounted in the latter medium), readily collapsing and difficult to demonstrate in revived mounts. Caulohymenium of cells 8-12 μ broad, clavate, variable in length, content hyaline to dingy ochraceous in KOH, no cystidia located other than the clavate sterile cells.

Pileus cuticle a collapsing trichodermium (the basidiocarps had been pressed), the end-cells 10-15 μ wide with a rounded apex, smooth, walls lacking incrusting material, content colorless or in some as revived in Melzer's with bister granules; trichodermial elements 3-5 cells long usually, cells not appreciably inflated. Context hyphae next to the cuticle hyaline or nearly so in Melzer's or KOH, walls thin and width 9-15 μ. Clamp connections none.

Observations.—It is possibly noteworthy that in the type study only one possible "amyloid" cell-inclusion was found, and it was of a doubtful nature. It may be that this species is not distinct from *B. porphyrosporus* Fr., but before concluding that it is not it would be well to have a more critical study of both, in the light of the present classification, as a basis for conclusions.

Subgenus TYLOPILUS

Stirps ALBOATER

KEY

1. Spores 11-17 μ long, 3.5-4.5 (5) μ wide. . . . . *T. eximius*
1. Spores shorter (9-12 μ long) . . . . *T. alboater*

46. Tylopilus eximius (Peck) Singer

Amer. Midl. Nat. 37:109. 1947

Boletus eximius Peck, Journ. Mycol. 3:54. 1887.
Leccinum eximium (Peck) Pomerleau, Bol. Bull. du Cercle des Myc. Amat. de Québec 6:117. 1959.
Boletus robustus Frost, Bull. Buff. Soc. Nat. Sci. 2:104. 1874. (not *B. robustus* Fr.)

Pileus 5-12 cm broad, hemispheric, becoming broadly convex to nearly plane, surface uneven to pitted, dry, unpolished to subtomentose (viscid to slimy to the touch when wet and staining the fingers yellow—Coker), color evenly chocolate-brown at all stages or a purplish tone evident, with a distinct bloom at first, margin sterile. Context 1-2 cm thick, firm, pallid to avellaneous or with more purplish tints, becoming dark around the wormholes ("odor of rotting wood or rancid oil and taste similar" or "bitterish"—Coker).

Tubes 10-17 mm deep, dark brownish purple (darker than the context of the pileus), depressed or at times nearly free; pores small (about 3 per mm), dissepiments thick, stuffed when young, concolorous with pileus or more purplish, not changing readily when bruised but nearly black in age.

Stipe 4.5-9 cm long, 1-3 cm thick, straight or curved, equal or enlarged upward, often somewhat compressed, solid, colored like the surface, surface about concolorous with the tubes and furfuraceous-punctate, the particles of roughness purplish, even or at times longitudinally striate to grooved.

Spore deposit reddish brown to more vinaceous. Spores 11-17 X 3.5-5 μ, narrowly subfusoid in face view, in profile narrowly inequilateral with a broad suprahilar depression, smooth, wall thickened somewhat, dingy yellowish in KOH, dingy tan to dull rusty brown in Melzer's.

Basidia 23-30 X 7-9 μ, 4-spored, clavate. Pleurocystidia scattered, 27-42 X 8-12 μ, fusoid-ventricose, thin-walled, in KOH and Melzer's with a dingy yellow wrinkled content similar to that in many basidia. Cheilocystidia 20-30 X 7-8 μ, narrowly fusoid, yellowish to hyaline in KOH.

Cuticle of pileus a trichodermium of hyphae 3-6 μ wide, the cells greatly elongated (septa sparse), in KOH and Melzer's the content mostly dingy ochraceous to pale bister when revived, end-cells mostly cylindric to clavate and obtuse, thin-walled, with much pale tawny intercellular debris in Melzer's. Clamp connections absent.

Habit, habitat, and distribution.—This species has been reported from Michigan, but all the collections we have located were misidentified. As for *T. alboater,* we expect that it does occur here and for that reason include it.

Observations.—A specimen of *B. robustus* Frost on deposit in the Peck collections at Albany has been studied. It has the typical elongate narrow spores nearly hyaline to dingy yellowish in KOH and the somewhat dextrinoid content in Melzer's which feature Peck's collections.

The microscopic data in our description are from a collection by Snell August 24, 1937, near Reading, Vermont. Concerning whether or not the species should be placed in *Leccinum,* we can only say that the color of the stipe ornamentation is merely a reflection of the color of the stipe generally and that it does not change color in a characteristic pattern as it ages. For this reason we exclude it from *Leccinum* and agree with Singer that it is a *Tylopilus.*

47. Tylopilus alboater (Schweinitz) Murrill

Mycologia 1:16. 1822

Boletus alboater Schweinitz, Schr. Naturf. Ges. Leipzig 1:195. 1822.
Suillus alboater (Schweinitz) Kuntze, Rev. Gen. Pl. 3(2):535. 1898.
Porphyrellus alboater (Schweinitz) Gilbert, Bolets, p. 99. 1931.

Pileus (3) 4-8 (15) cm broad, convex becoming plane or remaining broadly convex; surface dry, appearing minutely granular under a lens, glabrous but with a hoary bloom at first, color evenly "dusky drab" (dark gray) overall, margin extended slightly beyond the tubes but not forming a conspicuous sterile band. Context thick and firm, pale grayish white, soon changing to near "cinnamon-drab" (vinaceous-gray) or more reddish and finally blackish, odor and taste not distinctive.

Tubes 5-8 (10) mm deep, pallid ("tilleul-buff"), soon flesh color, slightly adnate at first, finally deeply depressed; pores white at first, soon "pale vinaceous-fawn" and staining black where bruised, about 2 per mm.

Stipe 4-10 cm long, 2-4 cm thick, clavate becoming equal, solid, flesh pale grayish and changing to reddish and then black when bruised, black in the base when collected, surface concolorous with pileus or apex paler, reticulate at apex or unpolished overall (surface similar to that of pileus).

Spore deposit vinaceous; spores 7-11 X 3.5-5 μ, smooth, wall scarcely thickened, narrowly ovate in face view, somewhat inequilateral in profile, nearly hyaline in KOH, pale dingy yellow-brown to dull rusty brown in Melzer's.

Basidia 16-20 X 7-8 μ, 4-spored, short and fat, yellowish to pale brownish in KOH, scarcely darker in Melzer's. Pleurocystidia 36-50 X 10-16 μ, fusoid-ventricose, thin-walled, smooth, content pale bister in KOH, darker bister in Melzer's. Caulocystidia 32-48 X 9-14 μ, clavate, mucronate or somewhat fusoid-ventricose, walls smooth and thin, content pale to dark bister in KOH and homogeneous.

Tube trama of the *Boletus* subtype, many of the hyphae with dark granules in KOH or in Melzer's but nonamyloid. Pileus cutis an intricately interwoven layer of hyphae 4-7 μ wide having short and long, tubular cells, the end-cells tubular to narrowed slightly to the apex, walls thin and smooth, content in KOH dark smoky ochraceous, in Melzer's the same layer having a homogeneous to granular content slightly paler than in KOH mounts, the colored layer forming an epicutis 4-6 hyphae deep, but beneath this is a thick layer of hyphae similar anatomically but not as highly colored in either medium. Context hyphae adjacent to the cutis sometimes with orange-brown homogeneous content in Melzer's. Clamp connections absent.

Habit, habitat, and distribution.—Solitary to scattered in hardwood forests especially along roads, late summer, not recorded for certain in Michigan.

Observations.—The only specimen we have located among Kauffman's collections is misidentified. Hence we cannot claim for certain that the species occurs here. It is to be expected in our southern tier of counties, however, so a description is included based on fresh material (Smith 73072) collected in Maryland. The darkening of the flesh is much like that observed in many species of *Leccinum* and it is curious that in that genus certain species have a spore-deposit color not too different from that observed in some species of *Tylopilus.*

No thick-walled elements were found on the pileus or stipe (see Singer, 1947). It is likely that we have here a cluster of variants needing further study. We have not attempted this since the species and its variants are, to say the least, rare in our area, and such a study should be based on fresh material.

Stirps FELLEUS

KEY

1. Pileus whitish and surface wrinkled like parchment . . . . *T. intermedius*
1. Pileus not as above 2

   2. Pileus and stipe violaceous when young; stipe mostly only slightly reticulate at the apex. . . . . *T. plumbeoviolaceus*
   2. Pileus pale to dark tan to dull cinnamon 3

3. Pleurocystidia with a large hyaline globule as seen in KOH mounts; habitat under pine or spruce on swampy or barren sandy soil. . . *T. felleus* var. *uliginosus*
3. Pleurocystidia lacking a central globule; often growing around old conifer stumps but terrestrial as well . . . . *T. felleus* var. *felleus*

48. Tylopilus intermedius Smith & Thiers, sp. nov.

Illus. Pl. 51.

Pileus 6-12 (15) cm latus, late convexus demum subplanus, siccus, subrugulosus, albidus, tactu tarde brunneus, amarus; tubuli albidi, demum vinacei; stipes 8-14 cm longus, 1-4 cm crassus, clavatus, albidus, tactu sordide brunneus, sursum indistincte reticulatus; sporae 10-15 X 3-4.5 (5) μ. Specimen typicum in Herb. Univ. Mich. conservatum est; prope Highlands, legit Smith n. 64301.

Pileus 6-12 (15) cm broad, broadly convex with an incurved sterile margin, expanding to broadly convex or finally nearly plane; surface dry and uneven, often wrinkled as in parchment paper, unpolished to pruinose, whitish ("tilleul-buff") young, remaining so or in age slowly becoming pale tan ("pinkish buff"), slowly staining brown where injured. Context thick, hard, white, discoloring brown around larval tunnels, odor fungoid, taste very bitter, $FeSO_4$ pinkish on young pilei; KOH causing no color change at all.

Tubes deeply depressed around the stipe or free, white, becoming vinaceous, 10-15 mm deep; pores white then pinkish vinaceous, round or nearly so, small (1-2 per mm).

Stipe 8-14 cm long, 1-4 cm thick, enlarged downward (clavate), solid, white within, slowly brownish where cut, yellow-brown around larval tunnels; surface white, staining dingy yellow-brown where handled, apex faintly reticulate, base whitish and practically glabrous.

Spore deposit "fawn color" (dingy vinaceous); spores 10-15 X 3-4.4 (5) μ, in profile elongate-inequilateral with a distinct suprahilar depression, in face view narrowly boat-shaped to nearly oblong, smooth, wall only slightly thickened, mostly yellowish in Melzer's but a few dark rusty brown, nearly hyaline in KOH.

Basidia 26-32 X 9-12 μ, clavate, hyaline in KOH, yellowish in Melzer's. Pleurocystidia 36-50 X 8-13 μ, narrowly fusoid-ventricose with narrow neck and subacute apex, hyaline, thin-walled, readily collapsing, content not distinctive in KOH, in Melzer's those near the pores with deep cinnabar-red oily-globular content and many large red globules free in the mount, farther back from the edge more of the cystidia having merely a pinkish more homogeneous content.

Tube trama of slightly divergent hyphae which in the region near the pores when revived in Melzer's have granular to globular content red to orange and this also present in the basidia and basidioles to some extent. Pileus cutis of narrow (2-4.5 μ) interwoven hyaline smooth, hyphae not gelatinous in KOH, the layer very thin in older pilei. Context hyphae 7-12 μ wide, tubular to somewhat inflated and with dark red homogeneous content as revived in Melzer's. Clamp connections absent.

Habit, habitat, and distribution.—Gregarious under oak during late summer and fall in the southern part of the state, rare but abundant during seasons with heavy rainfall in late August.

Observations.—Bruised spots on the pileus slowly stain brown, a feature of *T. peralbidus,* but our species never has the colors of the pileus given by Singer for *T. peralbidus,* the spores are larger, and the pileus is wrinkled like parchment. *T. intermedius* is also close to *T. rhoadsiae,* but is readily distinct by having only faint reticulation over the apex of the stipe in contrast to being strongly reticulate, and again the wrinkled pileus is an added difference. Singer lists *T. rhoadsiae* as extending as far north as Michigan, but we have seen no material which fits the description. Murrill (1c) described the stipe as beautifully reticulate and Singer's account bears this out, although he apparently accepts Snell's *B. felleus* f. *albiceps* as a Michigan record. However, the basidiocarps of the latter in the part of Kauffman's collection preserved at Michigan do not have any sign of a reticulum on the stipe and in this respect differ sharply from *T. felleus* f. *felleus.* We regard *T. rhoadsiae* as southern in distribution and not yet known in the central Great Lakes area, and *T. intermedius* as an autonomous species endemic to this region.

Material examined.—Oakland: Smith 64301. Washtenaw: Hoseney 585, 630; Kauffman, type of *B. felleus* f. *albiceps;* Smith 64186, 6450.

49. Tylopilus plumbeoviolaceus (Snell & Dick) Singer

Amer. Midl. Nat. 37:93. 1947

Boletus plumbeoviolaceus Snell & Dick, Mycologia 33:32. 1941.

Illus. Pl. 52.

Pileus 4-15 cm broad, convex at first, becoming broadly convex to nearly plane at maturity, margin slightly inrolled at first; surface dry and unpolished, occasionally becoming rimulose in age; color at first distinctly violaceous, becoming darker slate in the young pilei and then changing to browner at maturity (from "heliotrope-gray" through "dark vinaceous-drab" to near "drab" at maturity), or finally dingy cinnamon. Context firm, white, only slightly if at all discoloring, taste very bitter, odor slight, not distinctive.

Tubes 1-2 cm deep, depressed around the stipe, cream colored at first, near pale vinaceous-tan at maturity, pores small and nearly round at first, 1-2 per mm, broader and slightly irregular at maturity.

Stipe 8-12 cm long, 10-17 mm thick at the apex, somewhat swollen centrally at first, becoming equal, at times slightly and evenly enlarged downward at maturity; surface glabrous and only with a faint reticulation at the apex or not reticulate; color at first a mottled dark violaceous, the base white, occasionally developing olivaceous stains in or on the whitish base; context firm, white, unchanging when cut.

Spore deposit between "russet-vinaceous" and "vinaceous-fawn" (flesh color); spores 10-13 X 3-4 μ, smooth, thin-walled, obscurely inequilateral in profile view, with a slight suprahilar depression, nearly hyaline in KOH, only slightly tan in Melzer's.

Basidia 21-24 X 7-9 μ, 4-spored, (rarely 2-spored), hyaline in KOH, short-clavate. Pleurocystidia 35-55 X 8-12 μ, abundant, ventricose with a long attenuate tip, frequently covered by adhering spores, hyaline but filled in the ventricose part by pale yellow-brown amorphous material which becomes reddish brown in material revived in Melzer's.

Tube trama bilateral. Cuticle of pileus made up of a tangled trichodermium of smooth-walled narrow brownish hyphae containing a material which beads up in KOH or in Melzer's to form pigment globules. Clamp connections absent.

Habit, habitat, and distribution.—Gregarious to cespitose on sandy soil in rather open oak-hickory-aspen woods, late summer and early fall in the southern part of the state.

Observations.—This is a truly violet-colored bolete when it is young and fresh, but the colors fade and old specimens may at times be difficult to identify.

Material examined.—Barry: Smith 73211, 73117, 73267, 73296. Washtenaw: Hoseney 458.

50. Tylopilus felleus (Fries) Karsten

Rev. Myc. 3:16. 1881

Boletus felleus Fries, Syst. Myc. 1:394. 1821.
Dictyopus felleus (Fries) Quélet, Enchir. Fung. p. 159. 1886.
Rhodoporus felleus (Fries) Quélet, Fl. Myc. Fr. p. 421. 1888.
Boletus felleus var. *obesus* Peck, Bull. N. Y. State Mus. 2(8):154. 1889.

var. felleus

Illus. Pls. 53-54.

Pileus 5-15 (30) cm broad, hemispheric to convex when young, becoming broadly convex to plano-convex to plane when mature, occasionally depressed when very old, margin entire, incurved becoming decurved to straight or flared with age; surface glabrous or pruinose to distinctly tomentose when young, glabrous with age, dry, sometimes sticky to the touch when wet (subviscid) especially in old basidiocarps, smooth, occasionally split or rimose with age especially toward the margin, sometimes surface near margin becoming pitted with age, color pale to dark alutaceous to vinaceous-brown ("wood-brown" to "Verona-brown" when young), unchanging or becoming cinnamon ("cinnamon-brown" to "cinnamon-buff") with age, often with paler (pallid) margin. Context very thick (up to 3 cm), white, unchanging or becoming pinkish brown when exposed, often becoming discolored around the larval tunnels; odor none or occasionally slightly farinaceous, taste very bitter.

Tubes 1-2 cm deep, adnate to subdecurrent when young becoming deeply depressed around the stipe with age, whitish to as dark as "wood-brown" when young, soon becoming vinaceous to deep flesh color ("light russet-vinaceous") in age, staining brownish ("cinnamon" to "ochraceous-tawny") when bruised; pores at maturity about 1 mm in diameter, whitish to dingy pallid when young, staining dingy brown when bruised.

Stipe 4-10 cm long, rarely up to 20 cm or more long, 1-3 cm thick at apex, rarely up to 6 cm thick below, clavate to bulbous, rarely equal when young, becoming nearly equal to ventricose or remaining bulbous to clavate, typically reticulate especially over the upper portion, often smooth lower down, at times the reticulation prominent over most of its length, solid, firm, dry, fleshy, usually pallid upward and becoming pale brown or alutaceous to darker below, often with olivaceous stains from handling.

Spore deposit deep vinaceous; spores 11-15 (17) $\times$ 3-5 μ, narrowly subfusoid in face view, in profile somewhat inequilateral; smooth, hyaline or nearly so in KOH, yellowish in Melzer's, strongly dextrinoid when fresh.

Basidia 4-spored, 18-24 $\times$ 8-10 μ, hyaline in KOH, yellowish in Melzer's. Pleurocystidia numerous to abundant, prominently projecting, 36-50 $\times$ 8-14 μ, fusoid-ventricose with subacute to submucronate or tapered to an acute apex, thin-walled, content hyaline to frequently bister or yellowish in KOH, in Melzer's often dark red to reddish brown (dextrinoid), the content granular. Cheilocystidia often crowded, similar to the pleurocystidia or neck less developed. Tube trama divergent and gelatinous from a central strand, hyphae nonamyloid in all parts.

Pileus cuticle a trichodermium of matted hyphae 4-7 μ wide, yellow in KOH, tubular, the cells elongate, nonamyloid both as to wall and content. Context of floccose, interwoven, hyaline hyphae broader than in the cuticle, no incrustations seen, all hyphae nonamyloid. Clamp connections absent.

Habit, habitat, and distribution.—Solitary to gregarious or rarely in large clusters, on rotten debris of hemlock and on humus in conifer woods or in woods of hardwoods. It is one of the very common late summer boletes in the state.

Observations.—In the past this species has been confused with *T. rubrobrunneus* as a "form" with reticulation only at the stipe apex. Both develop olive to olive-brown stains on the stipe from handling, and both are very bitter. Mild tasting boletes of similar appearance are most likely to be *T. indecisus* or *T. ferrugineus* (if the two are distinct). In *T. felleus* the reticulation is raised and conspicuous. In Shaffer 4888 from France the content of the pleurocystidia is granular and ochraceous in KOH but in Melzer's is deep red to reddish brown. In var. *obesus* Peck, as revived in KOH the content of the cystidia is homogeneous.

Material examined.—Alger: Mains 32-103; Smith 9-19-29. Barry: Smith 73227, 73297. Cheboygan: Chmieleweski 70, 100; Shaffer 1447; Smith 25983; Thiers 3097, 3610, 3830, 3968, 4203. Chippewa: Smith 71892. Crawford: Thiers 3398. Emmet: Smith 57399, 62965, 63043, 71859, 74703; Thiers 903, 3230, 3528, 3710, 4414. Gratiot: Potter 3507, 9909b, 12299, 13939. Luce: Smith 37294, 37821, 39310, 67156, 71877, 72867; Thiers 3332. Marquette: Bartelli 2520, 2524. Montcalm: Potter 13186. Oakland: Smith 6401, 6574, 6777. Ontonagon: Shaffer 3744. Washtenaw: Kauffman 1905; Smith 1509, 62673, 64570; Thiers 4605.

50a. Tylopilus felleus var. uliginosus, var. nov.

A typo differt: Cystidia in "KOH" cum globulis rotundis. Typus Smith 72709 (MICH).

Pileus 5-10 cm broad, broadly convex, margin even, surface minutely areolate overall—rather conspicuous in dried pileus, dull leather-brown and when dried near sepia. Context thick, pallid, dingy brownish where cut and as dried grayish brown on fractured surfaces.

Tubes about 1 cm deep, adnate; pallid to whitish, becoming vinaceous when mature, pores 1-2 per mm, staining brownish slightly when injured.

Stipe 5-7 cm long, 12-18 mm thick at apex, enlarged downward, solid; surface conspicuously reticulate to base and as dried the lines of the reticulum blackish, near base dingy yellow-brown and becoming darker from handling, paler brownish above and discoloring to olivaceous on standing.

Spores 13-17 (18) × 4-5 μ, smooth, nearly hyaline in KOH, tawny in Melzer's, no apical differentiation; in profile narrowly inequilateral and "sway-backed," in face view subfusiform, wall relatively thin.

Basidia 4-spored, 8-11 X 17-25 μ, clavate, hyaline. Pleurocystidia abundant, 40-67 X 9-15 μ, subventricose-rostrate to subfusoid, apex acute, hyaline in KOH but content showing in the broad part as a large (8-12 μ) globule reminding one of the content of a chrysocystidium as revived in KOH, in Melzer's the globule disappearing but reddish pigment globules 2-8 μ in diameter are present, the latter are also scattered through some hyphae of the hymenophore or hymenial layer. Context hyphae yellowish to hyaline in Melzer's but scattered reddish pigment globules 2-8 μ in diameter present in some.

Pileus cutis of appressed hyphae 4-9 μ wide. The content reddish brown to yellowish in KOH and darker in Melzer's with some small globules present in the latter as revived in Melzer's; end-cells tubular with blunt apex; in mounts in Melzer's there is much extraneous granular material in the layer. Clamp connections absent.

Habit, habitat, and distribution. – Solitary on sand under pine, Marquette, September 21, 1965, Mrs. Ingrid Bartelli (Smith 72709).

Observations. – This variety is distinguished by the reddish pigment globules in mounts of the hymenophore in Melzer's, in the large globule in the pleurocystidia as revived in KOH, and the terrestrial habitat among lichens and moss on sandy low soil under pines. Smith 72805 was found on sphagnum in a spruce bog in Washtenaw County. It also belongs here and indicates a generalized habitat for the variety.

Stirps RUBROBRUNNEUS

51. Tylopilus rubrobrunneus Mazzer & Smith

Mich. Bot. 6:58. 1967

Illus. Pls. 55-59.

Pileus 8-20 (30) cm broad, broadly convex expanding to nearly plane or at times the disc slightly depressed, margin inrolled at first; surface dry and unpolished, at times areolate in age (in dry weather); color when young dark violaceous-brown to chocolate, slowly changing to deep vinaceous-brown and finally a dingy cinnamon in age ("dark livid-brown" to "dark vinaceous-brown" to "natal-brown" to "army-brown" or "Sayal-brown"). Context white, firm, taste very bitter, odor not distinctive, with $FeSO_4$ vinaceous-buff, in KOH no reaction to merely pallid yellow or vinaceous-buff near cuticle, borders of larval tunnels stained olivaceous.

Tubes 1-2 cm deep, adnate, becoming depressed around the stipe, when young avellaneous and slowly becoming whitish before being colored vinaceous by the spores; pores small and round at first, 1-2 per mm when mature, dingy brownish becoming pallid, staining dingy brown when bruised.

Stipe (6) 8-20 cm long, 1-5 cm thick at apex, up to 8 cm at the base in large specimens, clavate becoming equal, solid, white within but staining olive around larval tunnels at times; surface not reticulate or only obscurely so at very apex, whitish to pallid above, vinaceous-brown below but base whitish; surface often staining olivaceous from handling and base near groundline often dingy olive-brown in age.

Spore deposit "russet-vinaceous" (dark vinaceous-red); spores 10-14 X 3-4.5 μ, smooth, walls thin to scarcely thickened, in profile narrowly inequilateral with a distinct suprahilar depression, in face view suboblong to narrowly and obscurely fusoid, nearly hyaline in KOH, in Melzer's pale tan with occasional individuals dark red-brown (dextrinoid).

Basidia 20-26 X 8-11 μ, 4-spored, hyaline in KOH, short-clavate. Pleurocystidia abundant to scattered, 36-52 X 9-14 μ, ventricose in basal half, apical half a narrow (3-4 μ in diameter) neck ending in a subacute apex, as revived in KOH with an amorphous granular-refractive content (resembling that of chrysocystidia only occupying more of the cell content), this content dark reddish brown in Melzer's (dextrinoid).

Tube trama of hyphae divergent from a central strand, but at maturity nearly parallel, hyaline, thin-walled, smooth, not gelatinous as revived in KOH.

Cuticle of pileus a tangled trichodermium of hyphae pale cinnamon as revived in KOH, with smooth walls, narrow (4-6 μ), with end-cells slightly enlarged to 6-9 μ near the apex, all hyphae non-amyloid. Clamp connections absent.

Habit, habitat, and distribution. — Scattered, gregarious, or cespitose in open woods on thin sandy soil, with us mostly under second-growth oak, very abundant after heavy rains in the summer and early fall in the southern part of the state.

Observations. — A feature of this species along with several others in this genus is that apparently as revived in Melzer's the content of the cystidium becomes viscous to some degree because one often finds globules of orange-red pigment present in the mounts just as one observes them for most species of *Rhizopogon*.

We have one collection (Smith 72734) from the Highlands Recreation Area, Livingston County, September 29, 1965, in which the pores were cinnamon-brown at all stages, even when the sides were perfectly white. The pileus was about "Sayal-brown" (a dull cinnamon), the taste was very bitter, the apex of the stipe slightly reticulate, the spores 10-13 X 3-3.5 μ but with a fair number 12-15 X 5-6 μ, and the pleurocystidia had a dark red content as revived in Melzer's. The stipe stained olive-brown from handling. More collections are desirable before forming an opinion on the status of this variant.

This species is frequently parasitized. Pls. 57-59 show diseased basidiocarps. The last figure is of interest because the parasitized basidiocarps answer almost perfectly the description of the aborted bodies usually accompanying the lamellate basidiocarps of *Rhodophyllus abortivus*. They were about the size of a man's brain. It can be argued, and we think correctly, that the genus *Acurtis* can now be considered as based on a monstrosity, at least it will be difficult to be sure that the original material was not some fleshy fungus parasitized in the same manner as our bolete and with the same result.

Material examined.—Cheboygan: Smith 63778, 63887; Mycol. class 1951. Emmet: Smith 63851, 63913. Jackson: Mazzer 4060, 4065, 4081; Smith 62543. Livingston: Homola 862; Smith 64212, 64213, 73291; Whitlock 1933. Oakland: Smith 72734, 72740; Thiers 4594. Washtenaw: Hoseney 29, 31, 107, 113, 213, 436, 456; Mains 32-94; Shaffer 2497a & b; Smith 6653, 62570 (type), 64131; Thiers 4506, 4514, 4533, 4551, 4553, 4573, 4590.

Stirps INDECISUS

KEY TO SPECIES

1. Stipe 2-4 cm long, 5-8 mm thick; end-cells of cuticular hyphae 50-90 × 10-25 μ; growing on sphagnum . . . . *T. sphagnorum*
1. Not as above 2
    2. Pileus cuticle a compact palisade of pileocystidia the cystidia 30-40 × 7-10 μ . . . . *T. badiceps*
    2. Not as above 3
3. Cells of cuticular hyphae when mounted in Melzer's showing pigment globules 1-4 μ wide . . . . *T. subpunctipes*
3. Cells of cuticular hyphae lacking pigment globules when mounted in Melzer's . . . . *T. indecisus*

52. Tylopilus indecisus (Peck) Murrill

Mycologia 1:15. 1909

Boletus indecisus Peck, Ann. Rept. N. Y. State Mus. 41:76. 1888.

Illus. Figs. 44, 47; Pl. 60.

Pileus 5-15 (25) cm broad, convex becoming broadly convex to nearly plane, at times somewhat irregular; surface slightly viscid when moist but soon dry and appearing merely unpolished to finely tomentose, dingy pale cinnamon in color ("Sayal-brown," "snuff-brown," "Verona-brown," or rarely as dark as "warm sepia" when young). Context pure white fresh, slowly assuming a vinaceous-buff tint when injured; taste mild, odor none.

Tubes 9-12 mm deep, adnate at first, then somewhat depressed, white when young but soon "vinaceous-buff" or darker vinaceous and when mature staining brown where bruised or touched; pores concolorous with sides, stained brown in age, 2-2.5 per mm, angular, dissepiments entire.

Stipe 4-10 cm long, 1-2.5 (3.5) cm thick at apex, clavate becoming equal or equal from the first, solid, whitish within, brownish where bruised; surface minutely furfuraceous, finely reticulate above in addition, pallid above, soon brown below and especially so after being handled.

Spore deposit vinaceous to vinaceous-fawn; spores 10-13 X 3-4 μ, nearly hyaline in KOH, in Melzer's pale tan or a few reddish tawny, wall only very slightly thickened, smooth, narrowly subfusiform in face view, narrowly inequilateral with a broad suprahilar depression in profile, wall about 0.2 μ thick.

Basidia 18-25 X 6.5-8 μ, 4-spored, hyaline in KOH. Pleurocystidia scattered, ventricose with a projecting slender neck, content lemon to golden yellow in KOH, somewhat dextrinoid in Melzer's, thin-walled, lacking incrustations. Pileus cuticle a tangled trichodermium of hyphae 3-9 μ wide, thin-walled, collapsing, smooth, hyaline or with ochraceous content in KOH (as in some oleiferous hyphae), the end-cells mostly cylindric and obtuse, their content often dextrinoid; both the hyphae of the cuticle and the context reddish to some extent in Melzer's. Clamp connections absent.

Habit, habitat, and distribution.—Gregarious to scattered in low hardwoods in southern Michigan after heavy rains in July and August, fairly common if the weather is favorable.

Observations.—*Boletus indecisus* Peck differs from Singer's concept of *T. ferrugineus* in having vinaceous spores in a deposit, not "Isabella color" to "wood-brown." We have as yet not found a species in Michigan answering his description. The pleurocystidia of *T. indecisus* are not as strongly dextrinoid as in *T. rubrobrunneus*. The latter differs further in a very bitter taste and in developing olive stains on the stipe from handling. The stipe of *T. indecisus* is never as reticulate as in *T. felleus:* in Michigan, in fact, one usually must observe the apex rather closely to see it at all.

The following data were obtained from a study of the type of *B. indecisus:* Spores 10-13.5 × 4-4.5 μ, smooth, lacking a pore or thin spot at apex, shape in face view narrowly elliptic to subfusiform, in profile somewhat inequilateral, varying to very obscurely inequilateral, in KOH yellowish to nearly hyaline and many with dark violaceous particles in the interior (often in the aggregate resembling chromosomes in anaphase), in Melzer's weakly dextrinoid and lacking violet or dark inclusions; wall thin (–0.2 μ).

Basidia 4-spored, 7-9 μ broad, reviving very poorly. Pleurocystidia scattered, 30-52 × 10-18 μ, clavate-mucronate or the neck 4-6 μ wide and elongated to 15-30 μ, apex obtuse, content reddish tawny and amorphous in KOH (reminding one of that of chrysocystidia). Cheilocystidia gelatinizing like the hymenium.

Tube trama of tubular nonamyloid hyphae and laticiferous elements 5-11 μ wide, which are bright yellow in Melzer's; no amyloid walls or content observed in any hyphae. Pileus cuticle a tangled mass of narrow (3-6 μ) flexuous tubular long-celled hyphae yellowish in KOH, walls thin and lacking incrustations, the subcuticular hyphae reddish tawny in Melzer's from color in the cell wall. Clamp connections absent.

Observations.—We have no explanation for the dark granules observed in the spores when mounted in KOH.

Material examined.—Cheboygan: Smith 39216. Gratiot: Potter 10896, 11789. Livingston: Hoseney 390; Warren 8-1956. Marquette: Bartelli 2530. Washtenaw: Kauffman 1907, 1914, 1923; Smith 62598, 64857.

53. Tylopilus subpunctipes (Peck) Smith & Thiers, comb. nov.

Boletus subpunctipes Peck, Bull. N. Y. State Mus. 119:19. 1907.

Illus. Pl. 61.

"Pileus fleshy, broadly convex, often uneven on the surface, becoming soft with age, brown, reddish brown when dry, flesh white slowly becoming dingy where cut or broken, taste mild; tubes nearly plane in the mass, adnate or but slightly depressed around the stem, the mouths small, round, whitish or grayish white, changing to reddish brown where wounded; stem equal or nearly so, solid, slightly reticulate at the top, very minutely dotted, sometimes obscurely squamulose at the top, grayish or pallid; spores rusty brown or cinnamon-brown, oblong or subfusiform, .0004-.0005 of an inch long, .0002-.00024 broad.

"Pileus 2-4 inches broad; stem 2-3 inches long, 4-6 lines thick. Shaded sandy soil. Menands, Albany Co. August.

"The surface of the pileus is rendered unevenly coarse by shallow depressions. The species belongs to section Versipellis. The dots on the stem are nearly like those on the stem of Boletus chromapes Frost."

The original description is quoted here. Microscopic data from the type are as follows: Spores 9-12 X 3-4.5 μ, smooth, lacking an apical pore; shape in face view narrowly subelliptic to subfusoid, in profile suboblong to obscurely inequilateral, color in KOH practically hyaline, in Melzer's yellowish but a few reddish tawny, wall thin (-0.2 μ).

. Basidia 4-spored, 8-10 μ broad, clavate, hyaline in KOH. Pleurocystidia 36-48 X 10-14 μ, fusoid-ventricose, in Melzer's the content usually vinaceous-red and amorphous, in KOH "empty," walls thin, smooth, and hyaline. Cheilocystidia rather indistinct but apparently mostly basidiole-like with yellow content in KOH and some amorphous debris in the layer. Caulocystidia not studied.

Tube trama of the *Boletus* subtype, nonamyloid in all parts, numerous laticiferous elements present, these 4-10 μ wide and with orange-ochraceous content evenly distributed or granular. Pileus cutis of interwoven hyphae 4-10 μ wide, tubular, end-cells subcylindric to subcystidioid, thin-walled, much debris in the layer, cell content of the hyphae dull orange in Melzer's and content aggregating to form small globules or irregular masses almost exactly as in *Rhizopogon* species (the slides should be allowed to stand for about ten minutes), no amyloid debris observed in or on the cells, laticiferous elements numerous in the cuticle. Hyphae of subcutis merely yellowish in Melzer's and mostly "empty." Clamp connections absent.

Observations.—We did not observe pigment globules in the cuticular hyphae of *T. indecisus*. The presence of globules in the cuticular hyphae of *T. subpunctipes,* the alveolate pileus, and the stipe ornamentation as described by Peck distinguish it from *T. indecisus.*

The following is a description of Michigan collections:

Pileus 5-15 cm broad, broadly convex expanding to plane or nearly so, rarely depressed; surface resinous to the touch but not truly viscid, at maturity dry and subgranulose to unpolished, color evenly "russet," "warm sepia," or "Mars-brown" (dark rusty brown), slowly becoming paler yellow-brown ("buckthorn-brown"). Context white, firm, taste sweetish to mild, when cut slowly staining "vinaceous-fawn" to "avellaneous" or finally near "Verona-brown" (vinaceous to grayish to dull cinnamon), with $FeSO_4$ instantly blue, with KOH slowly yellow, odor not distinctive.

Tubes white, stuffed when young, becoming nearly avellaneous and where injured staining "Verona-brown," depressed around the stipe, about 1 cm. deep; pores about 1-2 per mm.

Stipe 2-6 cm long, 1-3 cm thick, equal or nearly so, solid, white within staining avellaneous where cut; surface pallid above, about concolorous with the pileus below or a grayer brown, staining dull cinnamon overall in age, finely reticulate over the upper half.

Spore deposit "Verona-brown" (dingy vinaceous-cinnamon), slightly yellower as moisture escapes; spores 9-12 X 4.5-5 μ, subelliptic to subventricose in face view, slightly inequilateral in profile and with a distinct suprahilar depression, wall thickened (about 1 μ) as measured in KOH, in Melzer's pale tan to darker, in KOH ochraceous-hyaline singly or cinnamon-buff in groups.

Basidia 18-23 X 7-9 μ, 4-spored, short-clavate, hyaline in KOH. Pleurocystidia scattered, 30-48 X 9-16 μ, ventricose in lower half, neck long and tapered to a subacute apex or scarcely any neck present, hyaline in KOH or some with an amorphous bister pigment inside, thin-walled, smooth, in Melzer's many with a reddish brown (dextrinoid) amorphous internal wrinkled mass.

Hymenium in Melzer's with numerous dark violet granules variously dispersed in some collections. Tube trama of divergent hyphae from a central strand, very gelatinous. Pileus cuticle a trichodermium, the hyphae 4-5 μ wide, gelatinous, having few cross walls and a brown pigment diffusing out in KOH mounts, the layer very soon collapsing, dark reddish in Melzer's from the colored cell content, orange-brown pigment globules numerous in the mount (as in many species of *Rhizopogon*). Clamp connections none.

Habit, habitat, and distribution.—Scattered in grassy oak woods, early fall, abundant after late summer rains in southeastern Michigan on heavy clay soil which is packed rather hard.

Observations.—Singer's (1947) disposition of *B. subpunctipes, B. ferrugineus,* and *B. indecisus* is not to be relied on. He stated of the type of *B. subpunctipes:* "... obviously a specimen of *T. ferrugineus* with entirely non-reticulated stipe." Peck described his species as having a slight reticulation at the top, and he was in a more favorable position than Singer to make the observation. Of *T. ferrugineus* Singer states the flesh changes to pink and then brown, whereas Frost who described the species described the flesh as unchangeable. Because of the confusion in this group we do not care to sponsor a "new" species in it at present. In the Michigan material the stipe, in addition to the reticulation, may be more or less pruinose to pruinose-furfuraceous at first, which is not too different from the condition Peck described. Also Peck indicated some color change in the context. Consequently, we place our collections here pending more accurate studies of the New England flora.

54. **Tylopilus badiceps** (Peck) Smith & Thiers, comb. nov.

Boletus badiceps Peck, Torrey Club Bull. 27:18. 1900.

Illus. Figs. 52, 54-55.

Pileus 4-8 cm broad, firm, convex to somewhat centrally depressed when mature, dry, velvety, obliquely truncate on the margin, bay-red or dark maroon. Context white, unchangeable but when bruised brown spots slowly developing on standing, taste and color mild, sweet, suggestive of molasses. Tubes plane, adnate, white or whitish, becoming dingy with age, the mouths minute. Stipe 4-5 cm long, 1.5-3 cm thick, equal or slightly swollen in the middle, radiating, glabrous, solid, brownish.

The above description is adapted from McIlvaine's (1900) account. He sent the specimens to Peck who described the species. Such a collection made by McIlvaine was found in the Peck collections by Stanley J. Smith. It is without question the holotype. The following microscopic data were taken from it.

Spores 8-10 X 3.5-4.5 μ, smooth, lacking an apical pore; shape in face view narrowly elliptic, in profile subelliptic or a weak suprahilar depression causing the ventral line of the spore to be straight to slightly concave; color in KOH hyaline to yellowish, in Melzer's yellowish, walls 0.2 μ thick.

Basidia 4-spored, 20-26 X 6-8 μ, hyaline in KOH, clavate. Pleurocystidia abundant, 27-38 X 8-12 μ, narrowly ventricose-mucronate to clavate-mucronate or more rarely fusoid-ventricose, content hyaline in KOH, when well revived in Melzer's only weakly yellowish and homogeneous. Cheilocystidia abundant and prominent, similar to pleurocystidia but usually larger and walls distinctly yellow in KOH—possibly from content adhering along the interior surface (the bright color present in poorly revived masses of cystidia). Caulocystidia not studied.

Tube trama of hyphae 4-8 μ wide, tubular, hyaline, thin-walled, transverse septa highly refractive and no amyloid reactions evident (the tissue weakly yellow), hyphal arrangement of the *Boletus* subtype. Pileus cuticle a compact palisade of pileocystidia 30-40 X 7-10 μ, obtusely fusoid-ventricose varying toward utriform or subclavate-mucronate to clavate, the content yellow in KOH but this soon dissolving into the mount, in Melzer's dull reddish orange fading slowly to dingy ochraceous, walls thin and with some debris adhering but no characteristic pigmentation or wall roughening showing. Hyphae beneath the palisade yellowish in Melzer's and laticiferous elements 4-7 μ in diameter with yellow content (in KOH or Melzer's) fairly numerous. Clamp connections absent.

Observations.—This, in our estimation, is clearly a *Tylopilus* even though the color of the spore deposit remains to be recorded. It has the aspect of *T. rubrobrunneus*, but is readily distinct because of its small spores, mild taste, conspicuous cheilocystidia, and turf of pileocystidia as a cuticle over the pileus. *T. badiceps* is one of the most easily recognized species in the genus.

55. Tylopilus sphagnorum (Peck) Smith & Thiers, stat. & comb. nov.

Boletus chrysenteron var. *sphagnorum* Peck, Bull. N. Y. State Mus. 150:64. 1910.

Illus. Figs. 50-51, 53.

"Pileus hemispheric or very convex, reddish-brown, the extreme margin thin, slightly surpassing the hymenium, incurved, flesh white or whitish; tubes longer than thickness of flesh.

"Pileus 2-3 cm broad; stem 2-4 cm long, 5-8 mm thick.

"Among sphagnum. Stow, Massachusetts, September. S. Davis."

The following data were obtained from the type: Spores 11-14 × 4-5 μ, smooth, lacking an apical pore, shape in face view narrowly elliptic to suboblong, in profile suboblong to somewhat inequilateral, suprahilar depression usually broad and shallow, color in KOH yellowish hyaline, in Melzer's yellowish to pale tawny, wall about 0.2 μ thick.

Basidia 4-spored, 8-10 μ broad, hyaline in KOH and yellowish in Melzer's. Pleurocystidia scattered, 28-42 × 8-14 μ, clavate to clavate-mucronate or with a narrow short neck ending in a subacute apex, walls thin, smooth, and hyaline, content homogeneous and ochraceous in KOH, brown in Melzer's but soon fading to ochraceous, amorphous. Cheilocystidia smaller than pleurocystidia but with the same yellow content.

Pileus cutis a tangled layer of broad (8-15 μ) hyphae, the cells mostly somewhat inflated, the end-cells voluminous (50-90 × 10-25 μ), clavate to fusoid-ventricose, thin-walled and smooth, content hyaline to yellowish in either KOH or Melzer's, the cells not disarticulating and typically are elongate. Hyphae of subcutis 4-10 μ wide and with ochraceous content in Melzer's. Throughout the mounts of the cuticle were found numerous patches of "amyloid" debris. Clamp connections absent.

Observations.—There can be little doubt that this fungus is a *Tylopilus*. The gigantic end-cells of the cuticular hyphae indicate it is an autonomous species. No pigment globules formed in Melzer's mounts of the cuticular hyphae.

56. **Tylopilus tabacinus** (Peck) Singer

Mycologia 36:362. 1944

Boletus tabacinus Peck, Bull. Torrey Club 23:418. 1898.

Pileus 6-12.5 cm broad, convex, fleshy, finally nearly plane, sub-glabrous, often rimose-areolate, tawny-brown. Context when mature soft and tawny-brown.

Tubes concave or nearly plane, depressed around the stipe, their mouths small, angular, colored like the pileus.

Stipe 3.5-7.5 cm long, 5-8 mm thick, subequal, solid, reticulated, concolorous.

Data from type: Spores 11-13 × 4-5 μ, smooth, nearly hyaline in KOH, in Melzer's pale tan, shape in profile somewhat obscurely inequilateral, in face view narrowly oval to suboblong, wall thin (-0.2 μ thick), lacking a pore at apex.

Basidia 4-spored, clavate, 6-9 μ at apex, content not distinctive either in KOH or in Melzer's. Pleurocystidia (36) 44-75 × 9-15 μ, ventricose with a long flexuous neck and obtuse to subacute apex, thin-walled, smooth, content coagulated and reddish in Melzer's (dextrinoid), in KOH the content refractive and nearly hyaline. Cheilocystidia and caulocystidia not studied.

Cuticle of pileus of a tangle of more or less appressed hyphae 4-8 μ wide, tubular, end-cells tubular to narrowly clavate or weakly cystidoid, walls smooth, thin, and not gelatinous, content dull orange-buff in KOH and homogeneous, in Melzer's slightly redder and homogeneous, no short cells noted in the filaments (cells mostly 4 times as long as broad or longer). Context hyphae reddish in Melzer's near cuticle. No clamps observed anywhere. A few hyaline globules 2-6 μ in diameter were seen free in mounts made in Melzer's.

Observations.—We have not as yet found this species in Michigan, but include an account of it based on the type for comparison with closely related species such as *T. subpunctipes.*

On the basis of the type the following features are clearly associated with this species: (1) Dark brown pores—reminding one of those of *T. pseudoscaber,* (2) the dextrinoid content of the pleurocystidia, (3) the characteristically elongated necks of the pleurocystidia, and (4) smooth hyphae making up the colored (epicuticular) region of the cutis.

It is clear to us from a study of the type of *Gyroporus pisciodorus* that the latter is not the same as Peck's species. Thus, Murrill's opinion (Mycologia 37:794) is supported. It is not clear to us where Singer obtained his information on the change to slate-violet from white which he described for the context of *T. tabacinus*. This, and his variable spore size (8.8-17 μ long), could easily indicate a different species. His measurements of the cystidia are clearly too small for *T. tabacinus* (18-45 $\times$ 4.5-8.5 μ). We found no significant hyphal incrustations in the type of *T. tabacinus*. The only thick-walled hyphal elements we found on the type belonged to a member of the Fungi Imperfecti which had invaded the dermal zones. The walls of its hyphae varied from hyaline near the tip to dark brown and thick farther back, and conidiophores projected in a manner sometimes resembling setae.

LECCINUM S. F. Gray

Nat. Arr. Brit. Pl. 1:640. 1821. Emended Snell, 1942

Stipe ornamented with lines, points, dots, or squamules which may be pallid or variously colored at first but by maturity becoming darker, usually very dark brown to black, or in some species these colors are present from the first; hymenophore typically white or pallid at first, rarely yellow, and tube trama typically of the *Boletus* subtype.

Type species: *Leccinum aurantiacum.*

Certain species in *Tylopilus, Boletellus,* and *Boletus* sect. *Subtomentosi* have rather pronounced stipe ornamentation as described above with the exception that it does not blacken or become dark brown as the stipe ages. The species of *Leccinum* in Michigan are numerous and varied. Much still remains to be done relative to their taxonomy and classification. Only the more readily recognized species are included here.

KEY TO SECTIONS

1. Pileus margin extending beyond the tubes as a sterile membrane at first and breaking into segments as the pileus expands . . . . Sect. *Leccinum*
1. Young pileus not with a distinct sterile margin extending beyond the tubes. 2

   2. Pileus trichodermium containing elements with rather numerous short ellipsoid to subglobose cells, or the cuticular layer appearing cellular under the microscope; or the layer composed of filamentous hyphae with some of the cells markedly inflated. . . . . .Sect. *Luteoscabra*
   2. Pileus cuticle of filamentous elements, only very rarely with an inflated cell, if many short cells are present these are not greatly inflated. . . . . .Sect. *Scabra*

Section LECCINUM

KEY TO STIRPES

1. An outer veil present on young pilei which breaks into flat patches as the pileus expands, in age pileus often glabrous . . . . Stirps *Potteri*
1. Lacking an outer veil. 2

   2. Pileus dark gray to black or dark yellow-brown. . . . . Stirps *Obscurum*
   2. Pileus white, yellow, orange, red, rusty red, liver color, dull cinnamon, or dark vinaceous-brown to vinaceous-tan. 3

3. Cut context not staining appreciably on exposure to air . . . . . .Stirps *Vulpinum*
3. Cut context in immature basidiocarps staining red to reddish cinnamon before coming gray to blackish. . . . . .Stirps *Aurantiacum*
3. Cut context in immature basidiocarps at first staining lilaceous to violaceous or "avellaneous" (grayish) before darkening further (lacking a preliminary red stage in the color change). . . . . Stirps *Insigne*

Stirps POTTERI

57. Leccinum potteri Smith, Thiers, & Walting

Mich. Bot. 5:138. 1966

Illus. Pls. 62-63.

Pileus 4-12 cm broad, convex, becoming broadly convex; surface at first coated with a thin submembranous pallid veil which soon breaks up into small areolate patches that gradually disappear, margin with sterile segments of the cuticle as in *L. aurantiacum* and some veil material adhering to outer surface of these segments at times, the surface beneath the veil more or less matted-fibrillose and dull brick-orange to dingy pale orange-tawny, evenly colored. Context pallid, when cut turning vinaceous then lavender and then gray, staining blue in the stipe base; odor slight, taste not distinctive, $FeSO_4$ bluish on context, KOH yellow on the tubes.

Tubes white when young, becoming wood-brown at maturity, yellowish then brownish when pores are bruised; pores pallid where undamaged (when young), small, round.

Stipe 5-10 cm long, 1-2.5 cm thick at apex, equal or enlarged downward, solid, white and scabrous when fresh, base slowly bluish and scabrosity becoming finally near blackish brown ("mummy-brown") but usually more nearly dark yellowish brown, when dried the surface over basal part lime-green to sulphur-yellow, not so yellow in the interior, upper parts drying pallid to grayish except for the dark ornamentation.

Spore deposit tawny-brown (Potter), olive-brown (Smith 72506); spores 13-16 × 4-5 μ, smooth, narrowly fusoid in face view, in profile narrowly subinequilateral, suprahilar depression broad but shallow, when revived in KOH dingy ochraceous and in Melzer's pale tawny to tawny, no "fleeting-amyloid" reaction present.

Basidia 4-spored, 18-22 × 8-11 μ, hyaline in KOH, clavate. Pleurocystidia—none observed. Cheilocystidia 18-36 × 4-9 μ, clavate to elongate-fusoid or fusoid-ventricose, mostly ochraceous in KOH, soon gelatinizing. Caulocystidia in fascicles, clavate to ellipsoid or balloon-shaped,

40-90 × 12-20 μ, with smoky brown content in KOH, walls smooth and thin.

Tube trama very soon gelatinizing in KOH, hyphal arrangement obscurely bilateral. Pileus cuticle of appressed hyphae, the elements with clavate end-cells 7-12 μ wide when young, in age many of them show a tapered neck, some hyphae with roughened and some with smooth walls; with a tendency for cells to disarticulate at the septa, cell content in KOH lemon-yellow to dingy brownish ochraceous, in Melzer's the content merely yellowish and homogeneous or in a few hyphae red pigment globules present in small numbers. Context hyphae hyaline in KOH, thin-walled, interwoven, 9-12 (15) μ wide, as revived in Melzer's with masses of pinkish red granular material in some of the cells of some hyphae, remaining cells hyaline to yellowish. Veil material lemon-yellow in KOH from dissolved pigment, the hyphae interwoven, thin-walled and 4-10 μ wide. Clamp connections absent. All hyphae with inamyloid content.

Habit, habitat, and distribution.—Scattered on the ground in mixed woods especially with large-toothed aspen, but also with scrub oak, common at times.

Observations.—This fungus has passed as a variant of *L. aurantiacum* for years in Michigan, but is constant in the KOH reactions, in the veil features, and in the greenish yellow stipe base which is often quite intensively colored when dried. Specimens of *L. aurantiacum* attacked by the common white bolete mold may at times have some of the white mold on the pileus as patches or along the pileus margin, but basidiocarps so infected seldom mature properly and do not have the above-mentioned features. The white fragments of mycelium from the parasite of *L. aurantiacum* can be cultured readily whereas the veil remnants on *L. potteri* cannot, a further indication that the two are in no way homologous. The KOH reaction of the cuticular hyphae will separate *L. potteri* from *L. aurantiacum* if veil material is lacking.

Material examined.—Cheboygan: Smith 63856. Chippewa: Smith 72435. Gratiot: Potter 12480. Oakland: Smith 72575. Ogemaw: Smith 64492, 67445. Washtenaw: Smith 72504, 72506, 72529, 72657; Thiers 4584. Wayne: Smith 72656.

Stirps OBSCURUM

KEY

1. Cut context in stipe apex staining pinkish but not darkening to gray or fuscous; spores 16-21 × 5-7 μ . . . . *L. subatratum*
1. Not as above 2

2. Some of the cuticular hyphae with brownish bands of incrusting pigment . . . . *L. obscurum*
2. Cuticular hyphae not incrusted with bands of colored pigment. 3

3. Cuticular hyphae with pigment globules as seen revived in Melzer's . *L. uliginosum*
3. Cuticular hyphae lacking pigment globules as seen mounted in Melzer's
. . . . *L. subspadiceum*

58. **Leccinum subatratum** Smith, Thiers, & Watling

Lloydia 31:260. 1968

Pileus 7-9 cm broad, obtuse, expanding to broadly umbonate, sterile margin about 2 mm wide and with a tendency to become crenate-appendiculate, viscid but granulose-appearing and subrimose at the same time, blackish overall fresh and also as dried. Context pallid, unchanging except for apex of stipe where it slowly stains pink.

Tubes pallid, depressed around the stipe; pores minute and pallid, staining dingy ochraceous to "Isabella color" when bruised.

Stipe 8-11 cm long, 8-12 mm thick at apex and up to 2 cm at the base, solid, white within and drying white, but when cut in the apex staining slightly pinkish (no other color change taking place); surface with coarse black ornamentation that is finer above and coarser below, pallid ground color showing slightly in places, base pallid and with faint greenish stains showing in places but these not becoming yellow in drying.

Spores (14) 16-21 (23) X 5-7 μ, smooth, in face view fusoid to subfusoid, in profile inequilateral with a distinct suprahilar depression, color near buckthorn-brown in KOH, more tawny in Melzer's at least in the distal half, with a pallid spot at apex (using oil-immersion lens).

Basidia 4-spored, clavate, 24-30 X 9-12 μ, hyaline in KOH but with oil droplets present. Pleurocystidia scattered, 52-68 X 10-16 μ, ventricose-rostrate to fusoid-ventricose, hyaline in KOH, thin-walled, smooth. Cheilocystidia clavate to fusoid, 25-35 X 8-12 μ, in many areas with ochraceous to fulvous internal pigment. Caulocystidia clavate to fusoid-ventricose, some up to 20 μ wide, content bister in KOH, apex not filamentous; patches of bright fulvous pigment found in places among the cystidia but their origin not evident.

Pileus cutis basically a trichodermium but collapsed by maturity, the hyphae of the epicutis with smoky-brown content in KOH, 9-15 μ wide (rarely up to 30 μ), the hyphal cells often less than 5 times as long as wide and inflated considerably to being ellipsoid or rarely nearly globose, cells readily disarticulating, the end-cells bullet-shaped to ellipsoid to subglobose, rarely fusoid, walls smooth to irregular, some lens-shaped interior thickenings evident and these highly refractive in KOH, content in Melzer's bister or darker and in some cells aggregating into large to small masses or globules. Hyphae of subcutis hyaline or nearly so in Melzer's and not gelatinous. Clamp connections none.

Habit, habitat, and distribution.—On a seepage area near birch and aspen, Lower Falls, Tahquamenon Falls State Park, Chippewa County, September 4, 1965, Smith 72835.

Observations.—This species is intermediate between sections *Luteoscabra* and *Leccinum* in that the sterile margin became somewhat divided into segments but the spores and pileus color are more like those of subsection *Pseudoscabra* of section *Luteoscabra.* Its diagnostic features are the blackish pileus, its sterile margin, white context as dried (in both pileus and stipe), lack of gray stains when fresh context is cut, and in the caulocystidia lacking an apical proliferation of the filamentous type. In *L. obscurum* the context stains gray eventually, and its spores are narrower.

59. Leccinum obscurum Smith, Thiers, & Watling

Mich. Bot. 5:166. 1966

Illus. Pl. 65.

Pileus about 7 cm broad, convex, incurved margin breaking into sterile segments as in *L. aurantiacum*; surface dry, glabrous, uneven (no fibrils on the immature pileus), color near "hair-brown" (grayish fuscous) and drying dark gray-brown. Context pallid, when cut staining vinaceous then violaceous-fuscous, similar changes occurring in the stipe, in addition near the stipe-base staining blue in some areas and red in others, with $FeSO_4$ bluish fuscous, KOH yellow in the base of the stipe.

Tubes white, slowly staining pinkish when cut, deeply depressed around the stipe; dull cinnamon-brown when dried; pores pallid, staining vinaceous to fuscous when bruised in young specimens.

Stipe about 14 cm long, 2.5 cm thick, equal, solid (see above under context for color changes), surface pallid, stained red over the base, ornamentation coarse and blackish to the pallid apex.

Spores 12-16 (19) X 4.5-5.5 (6) μ smooth, in face view subfusoid, obscurely elongate-inequilateral in profile, wall slightly thickened, dingy ochraceous in KOH, in Melzer's pale dingy tan.

Basidia 4-spored, hyaline to yellowish in KOH and Melzer's, 16-20 X 7-9 μ, clavate. Pleurocystidia 34-48 X 8-13 μ, scattered, hyaline in

KOH and Melzer's, fusoid-ventricose, subacute at apex. Cheilocystidia similar to pleurocystidia but many with brown content in KOH. Caulocystidia in large patches of caulobasidia, 36-56 X 9-18 μ, a fair number fusoid-ventricose but many clavate to mucronate, with dull brown walls; caulobasidia numerous and sporulating, 2-, 3-, or 4-spored, dull brown in KOH; spores from these basidia 8-10 X 4-5.5 μ, ellipsoid to ovoid, hyaline to slightly colored and very different from those borne on the basidia of the hymenophore.

Pileus cutis a layer of tangled hyphae with hyaline to brownish bands of incrusing material on some of the hyphae but mostly roughened as in *L. insigne* (fig. 15), with some cells merely asperulate, content smoky brown in KOH from a homogeneous content, in Melzer's dingy orange-brown and homogeneous to reticulate-coagulated, no pigment globules seen, the hyphae tubular to the inflated fusoid end-cell or the latter bullet-shaped, the enlarged cells up to 20 μ broad, some cells disarticulating, walls variously ornamented but a minority smooth; amyloid particles scattered in the mount. Context hyphae hyaline or with yellow walls in Melzer's, the content not distinctively colored. Clamp connections absent.

Habit, habitat, and distribution.–Solitary under aspen, Haven Hill, Oakland County, October, apparently rare.

Observations.–For a comparison with *L. subatratum* see that species.

60. Leccinum uliginosum Smith & Thiers, sp. nov.

Illus. Pl. 64.

Pileus 5-15 cm latus, convexus, siccus, demum rimulosus, fuscus. Contextus albidus tactu roseolus, tarde fuscus. Tubuli 1-1.5 cm longi, pallidi; pori pallidi tactu ochracei demum sordide brunnei. Stipes 9-16 cm longus, 1-2.5 cm crassus, punctato-squamulosus, squamulae fuscae. Sporae 14-17 (18) X 3.5-5 μ. Specimen typicum in Herb. Univ. Mich. conservatum est; prope Wycamp Lake, Emmet County, August 3, 1968, legerunt Thiers et Smith 75837.

Pileus 5-15 cm broad, convex becoming broadly convex; surface dry and unpolished and in age often rimulose, color pale to dark fuscous ("hair-brown" to "benzo-brown"–a violaceous-fuscous), in age finally merely drab-gray near the margin; margin exceeding the tubes by 1-2 mm (but not becoming lobed in the typical manner as in species referred to in section *Leccinum*). Context white, when cut slightly reddish (young specimens), then changing to bluish fuscous, $FeSO_4$ causing a change to olivaceous; odor and taste not distinctive.

Tubes 1-1.5 cm deep, depressed around the stipe, pallid to olivaceous-pallid at first, slowly becoming grayish as they mature, staining yellowish where lightly bruised and the yellow darkening to dingy brown.

Stipe 9-16 cm long, 1-2.5 cm thick, equal or tapered downward, solid, when cut staining as in pileus but in addition at the base staining blue and/or yellow but the yellow soon fading; surface with charcoal-gray to (finally) black ornamentation as punctae but these soon stretched into lines forming ridges and in some the latter fused to form an obscure reticulum, ground color pallid or lower down blackish to yellowish.

Spore deposit a dark dingy cinnamon-brown; spores 14-17 (18) X 3.5-5 μ, smooth, apex lacking a distinct pore, in face view narrowly fusoid, in profile narrowly and somewhat inequilateral, in KOH dull tawny-brown, slightly yellower in Melzer's, walls less than 0.3 μ thick.

Basidia 4-spored, 9-11 μ broad, variable in length, hyaline in KOH and yellowish in Melzer's, as revived in KOH showing a large hyaline globule in the central part. Pleurocystidia rare and inconspicuous, content hyaline to dingy ochraceous revived in KOH. Cheilocystidia numerous, similar to pleurocystidia or smaller. Caulocystidia (28) 36-55 (80) X (8) 10-15 (22) μ, mostly fusoid-ventricose and with narrow often proliferated necks, the apex subacute, content dingy yellow-brown in KOH, walls thin and smooth.

Tube trama of hyaline, tubular, smooth hyphae 4-7 μ wide, weakly divergent in mounts of fresh material in water, gelatinization of walls not pronounced. Pileus cutis of appressed hyphae 5-11 μ wide, mostly nearly tubular but end-cells often clavate and up to 15 μ wide in broadest part, walls thin and smooth to minutely roughened (as in *L. insigne*), the cells mostly 4 times longer than broad and often disarticulating, content yellow-brown in KOH and in this medium often rounding into more or less globose concentrations (but boundaries not sharp), in Melzer's forming beads 2-6 μ in diameter in many hyphae as well as in some less distinct larger bodies. Hyphae of subcutis and adjacent context not highly colored as revived in Melzer's. Clamp connections absent.

Habit, habitat, and distribution.—Gregarious under aspen and willow in a marshy area at Wycamp Lake, Emmet County August 3, 1968, H. D. Thiers collector, Smith 75837.

Observations.—The outstanding features of this species are the change to reddish and then fuscous when the flesh is cut, the narrow spores, proliferated caulocystidia, pigment globules in the cuticular hyphae as mounted in Melzer's, the lack of color in the subcuticular hyphae as well as in those in the adjacent context as revived in Melzer's, and the dark color of the fresh pileus. The olive $FeSO_4$ reaction may also be significant. It appears to be close to *L. murinaceo-stipitatum*, but the subiculum hyphae of the stipe ornamentation are not colored as they are in that species, and the color of the pileus is bluish fuscous, not dark cinnamon-brown. No incrusting pigment was noted on the cuticular hyphae of *L. uliginosum*.

In the dried basidiocarps of *L. uliginosum* the lower part of the stipe dried pallid to yellowish. The sterile band forming the margin of the pileus almost disappeared in the process of drying. *L. obscurum* has brownish bands of incrusting pigment on at least some of the cuticular hyphae.

61. Leccinum subspadiceum Smith, Thiers, & Watling

Lloydia 31:263. 1968

Illus. Pl. 66.

Pileus 8-15 cm broad, convex, expanding to broadly convex, margin exceeding the tubes and the sterile portion becoming crenate, surface dry and unpolished, color "bister" (a medium date-brown) to "snuff-brown" (a pale date-brown) when young, when old evenly snuff-brown, a few with darker streaks. Context pallid, in buttons staining avellaneous to fuscous directly in older ones, slowly dull lilac above the tubes and then darker, with $FeSO_4$ pale gray; odor and taste not distinctive.

Tubes depressed around the stipe, 1-2 cm deep, pallid becoming dark dingy brown; pores small, "cinnamon-buff" when immature and color not changing much in age, when lightly bruised staining olive, if severely bruised soon fuscous.

Stipe 10-18 cm long, 1-2 cm thick at apex, solid, streaked yellow in old specimens and also staining blue, yellowish as dried, in young stipes merely watery streaked and soon staining blackish; surface with dull yellow-brown ornamentation to near the pallid apex, finally darkening to blackish.

Spore deposit dingy olive-brown on the debris around the basidiocarps (none obtained on white paper); spores 13-16 × 4.5-5.5 (6) μ, smooth, some with a thin spot at apex, wall slightly thickened elsewhere (0.5 μ); shape in face view subfusiform to fusiform, in profile ventricose and inequilateral to somewhat inequilateral, dingy ochraceous in KOH or Melzer's.

Basidia 23-36 X 8-11 μ, 4-spored, clavate, hyaline in KOH and merely weakly yellowish in Melzer's. Pleurocystidia scattered, 38-65 X 10-18 μ, ventricose toward the base and with a long tubular neck and the apex slightly enlarged, content of the ventricose part often smoky brown in KOH or Melzer's. Cheilocystidia 25-40 X 8-12 μ, more or less fusoid-ventricose, neck narrow and apex subacute to obtuse. Caulocystidia elongate-subfusiform to clavate, 60-120 X 10-25 μ, some with a drawn-out neck and obtuse apex, some merely mucronate, content bister in KOH, walls thin and smooth; caulobasidia also with bister content.

Tube trama of more or less tubular hyphae slightly divergent to the subhymenium; yellow laticiferous hyphae occasional throughout the trama. Pileus cutis of appressed hyphae 6-12 μ broad, content dark yellow-brown in KOH and Melzer's, end-cells narrowly clavate to evenly tapered to an obtuse apex, remainder of cells in the hyphae mostly elongate but a few short to subglobose (usually only one in a hypha), content in Melzer's bister and mostly granular but an occasional hypha with beadlike globular aggregations of pigment (but no characteristic pigment globules noted), walls smooth to roughened slightly as in *L. insigne* or with an irregular gelatinous coating. Hyphae of the subcutis hyaline or nearly so in Melzer's. Context hyphae also hyaline in Melzer's. Clamp connections none.

Habit, habitat, and distribution.—Gregarious in a mixed hardwood-conifer forest west of Vanderbilt, June 28, 1967, Samuel Mazzer (Smith 74341).

Observations.—This species differs from *L. cinnamomeum* in not staining reddish at first when the context is cut, by the "cinnamon-buff" pores of young specimens, and the stipe not staining brown from handling. There is also a possible difference in the shape of the pleurocystidia. From *L. insolens* var. *brunneo-maculatum* it differs in the colored pores when young, in the yellow-brown stipe ornamentation, and in the elongate subfusoid caulocystidia.

Stirps VULPINUM

62. Leccinum vulpinum Watling

Trans. Proc. Bot. Soc. Edinb. 39:197. 1961

Pileus 7-10 cm broad, flat when expanded; surface dry and obscurely roughened but not squamulose; dark brick-red ("ferruginous" or darker), when dried dull vinaceous-brown; margin appendiculate when young. Context pallid when cut and scarcely changing or merely slightly brownish in the pileus, with a dark olive watery line above the tubes, when dried mostly pallid but with some gray streaks or patches, with $FeSO_4$ bluish olive.

Tubes pallid becoming avellaneous, slightly depressed around the stipe; pores pallid staining yellow and finally yellowish brown, when severely bruised not becoming fuscous or vinaceous-gray, when dried dingy cinnamon.

Stipe 10-13 cm long, 1-1.5 cm thick at apex, nearly equal, solid, fibrous, pallid within and when cut slowly very pale vinaceous in places, soon staining rich indigo-blue in the base; surface pallid above under the brown ornamentation, extreme apex whitish, lower down the ornamentation moderatly coarse and brown (but drying blackish), surface flushed yellowish to greenish in the midportion and drying with these tones evident, as in *L. potteri*.

Spore deposit dingy clay color as air-dried, but on standing darkening to dingy cinnamon; spores 14-17 (19) X 3.5-4.5 μ, smooth, wall scarcely thickened, narrowly subfusoid in face view, narrowly inequilateral in profile, suprahilar depression broad and shallow, in KOH clay color, in Melzer's yellowish brown.

Basidia 4-spored, 16-20 X 8-10 μ, yellowish hyaline in KOH and yellowish in Melzer's. Pleurocystidia—none seen. Cheilocystidia mostly clavate, 20-30 X 7-9 μ, ochraceous in KOH to hyaline, walls thin and soon gelatinizing. Caulocystidia 34-50 (60) X 8-15 μ, yellowish hyaline in KOH and Melzer's or with smoky brown content in KOH.

Pileus cutis of tangled hyphae 5-8 (12) μ wide, cells 30-70 μ long or longer, end-cells not distinctively differentiated, walls smooth, thin, no disarticulation of cells noted, content dull orange in KOH, in Melzer's the content soon forming minute granules, very few refractive thickenings seen against interior wall surfaces. Subcutis of interwoven gelatinous hyaline hyphae 2.5-5 μ wide, not colored in KOH or Melzer's. Context hyphae hyaline, interwoven, 6-15 μ wide, lacking colored content in Melzer's or refractive content in KOH. Clamp connections absent. No colored content in Melzer's or refractive content in KOH. Clamp connections absent. No colored pigment globules present in any hyphae when mounted in Melzer's.

Habit, habitat, and distribution.—Solitary in mixed woods with pine, Cheboygan and Barry counties in the fall; rare.

Observations.—This is a distinctive species related to *L. potteri* in the narrow spores and the color of the stipe especially as dried, but the cuticular hyphae are not lemon-yellow in KOH and no veil material was found. There was no change to gray on fresh material though as dried some does show on cut surfaces.

Stirps AURANTIACUM

KEY

63. Leccinum vinaceo-pallidum Smith, Thiers, & Watling

Lloydia 31:265. 1968

Pileus 4-14 cm broad, convex becoming nearly plane, surface appressed-squamulose to glabrous, dry at first but subviscid in age, colors pallid with a flush of pink or near "vinaceous-buff," with a pallid margin; margin appendiculate. Context white staining reddish lilac then fuscous, $FeSO_4$ olive, in the stipe when cut or bruised both blue and red stains prominent near the base.

Tubes whitish then olive-pallid, depressed around the stipe, near wood-brown in age; pores pallid, minute, staining olive to olive-brown or if severely bruised avellaneous to fuscous.

Stipe 6-15 cm long, 1-2 cm thick, equal, solid, pallid, staining fuscous (and both blue and red in places); surface white or blue stained near the base; ornamentation pallid at first then dull brown and finally dark brown, apex remaining pallid.

Spore deposit dingy yellow-brown (about bister as dried); spores 13-16 X 4-5 μ, smooth, with a thin spot at the apex, wall less than 0.5 μ thick, shape in face view fusiform to subfusiform, in profile inequilateral, color dingy pale ochraceous to pale ochraceous-brown in KOH, in Melzer's many becoming dull tawny.

Basidia 22-34 X 9-14 μ, fusoid-ventricose, walls thin and smooth, content usually smoky brownish as revived in KOH and in Melzer's. Cheilocystidia smaller than pleurocystidia but more or less fusoid-ventricose. Caulocystidia (30) 40-60 X (10) 15-30 μ, clavate to saccate, clavate-mucronate or some fusoid-ventricose, thin-walled, content about "snuff-brown" as revived in KOH (pale date-brown); caulobasidia often with pale yellow-brown content; subiculum hyphae hyaline in KOH, tubular, thin-walled but many with one or more refractive hyaline lens-shaped deposits against the wall of the interior.

Tube trama typical for the section (weakly bilateral); some laticiferous elements present. Pileus cutis of appressed hyphae 7-12 μ wide, tubular, smooth or rarely minutely roughened, end-cells 60-100 X 10-15 μ, subfusoid-ventricose or merely tapered to apex, content of many cells orange reddish in Melzer's and some hyphae filled with small (about 1-2 μ in diameter) pigment beads as revived or when fresh, in KOH the content ochraceous brownish and stringy to homogeneous, no short cells seen but the cells do disarticulate. Hyphae of subcutis not appreciably colored in Melzer's. Clamp connections absent.

Habit, habitat, and distribution. — Scattered under birch and aspen, Montmorency County, July 24, 1967, Smith 74633.

Observations. — This is a pallid species with reflections or slight tinges of pink in the pileus or the disc finally more or less vinaceous-buff, the context when cut at first showing reddish tints, and a characteristically dull brown ornamentation on the stipe in age. The large refractive hyaline lens-shaped bodies adhering to the interior of the walls of the subiculum hyphae of the stipe are also unusual and possibly characteristic. The colors are too pale for *L. aurantiacum* and the color changes are not typical for that species.

64. Leccinum ambiguum Smith & Thiers, sp. nov.

Illus. Pl. 67.

Pileus 6-15 cm latus, convexus, siccus, fibrillosus, sordide olivaceo-brunneus. Contextus pallidus, tactu sordide cinnamomeus demum violaceofuscus. Tubuli olivaceo-pallidi tactu sordide brunnei. Stipes 6-15 cm longus, 1-2.5 cm crassus, clavatus, atrosquamulosus. Sporae 14-17 X 4-5.5 μ. Specimen typicum in Herb. Univ. Mich. conservatum est; prope Spectacle Lake, Chippewa County, July 13, 1968, legit Smith 75571.

Pileus 6-15 cm broad, convex becoming nearly plane, with a distinct sterile margin which becomes broken into segments, surface dry and appressed-fibrillose, not showing any separation of fibrils into patches in oldest pilei; color olive-sepia (dull brown to grayish brown) when young, nearly olive-buff in age or with browner areas where bruised. Context pallid when cut, in buttons staining dingy cinnamon and then lilac but slowly progressing to violaceous-fuscous, in mature specimens staining lilac-fuscous directly; readily staining bluish on application of $FeSO_4$.

Tubes pallid, soon olive-buff, depressed around the stipe; pores very small (2-3 per mm), pallid young, becoming olive-buff and when injured staining dull brown.

Stipe 6-15 cm long, 1-2.5 cm thick near apex, solid, clavate when young; context changing color like that of pileus; surface of young specimens with coal-black rather coarse ornamentation overall except the whitish base, becoming more or less fibrillose-squamulose by maturity with the paler ground color showing, base remaining whitish.

Spores 14-17 X 4-5.5 μ, smooth, apex lacking a distinct pore, shape in face view fusoid, in profile ventricose and narrowly inequilateral, color in KOH dull yellow-brown and in Melzer's not significantly different, wall about 0.2-0.3 μ thick, color in deposit not obtained.

Basidia 4-spored, about 30 X 10 μ, hyaline in KOH and merely weakly yellowish in Melzer's. Pleurocystidia scattered, 35-47 X 9-16 μ, fusoid-ventricose, apex pointed to subacute, content revived in KOH hyaline to dull yellow-brown and in Melzer's hyaline to dark brown and amorphous—but evenly distributed. Cheilocystidia subfusoid to clavate, content often yellowish brown in KOH, 23-32 X 5-11 μ. Caulocystidia

(35) 45-80 X (90) 10-20 μ, clavate to fusoid-ventricose or utriform, content yellow-brown in KOH, arising from a dense subiculum of hyphae 4-8 μ wide and many having a colloidal content in KOH.

Pileus cutis of appressed hyphae 7-16 μ wide, many with yellow-brown content in KOH, in Melzer's not forming pigment globules (instead remaining diffused and amorphous) walls smooth to uneven as in *L. insigne* and the cells often disarticulating, the cells mostly 4 times as long as broad or longer. Hyphae of subcutis and adjacent context not red or orange in Melzer's. Clamp connections absent.

Habit, habitat, and distribution.—Gregarious under birch above Spectacle Lake, Chippewa County, July 13, 1968, H. D. Thiers and W. Patrick (Smith 75571).

Observations.—This species is similar to *L. atrostipitatum* in having black stipe ornamentation when young and in its robust stature. It differs in the color of the pileus in both the fresh and dried condition. The color change on the cut flesh is ambiguous since the change to cinnamon does not show on older basidiocarps.

65. Leccinum pellstonianum Smith & Thiers, sp. nov.

Pileus 4-9 cm latus, late convexus, siccus, demum subrimosus, subalutaceus. Contextus albidus tactu fulvo-rubellus demum atratus. Tubuli 1-1.5 cm longi, pallidi, tactu ligno-brunnei; pori pallidi tarde lutescens, tactu olivaceo-brunnei. Stipes 8-12 (16) cm longus, 1-3 cm crassus, clavatus, tactu rubescens demum atratus, atrosquamulosus. Sporae 12-15 X 3.5-4.5 μ. Specimen typicum in Herb. Univ. Mich. conservatum est; prope Pellston, Emmet County, August 15, 1968, legit W. Patrick (Smith 76025).

Pileus 4-9 cm broad, convex expanding to broadly convex or nearly plane in age; margin appendiculate; surface dry and matted-fibrillose, when old finely rimulose (but not in patches as in *L. atrostipitatum*); colors at first pinkish buff to cinnamon-buff becoming subalutaceous but as dried grayish buff. Context white when cut but soon changing to reddish cinnamon and finally going to black; taste mild, odor slight, $FeSO_4$ quickly green.

Tubes 1-1.5 cm deep, depressed around the stipe, pallid at first, becoming wood-brown or darker, when in the pallid stage staining wood-brown when cut; pores minute, white and staining dingy yellow-brown when injured, gradually becoming yellowish overall in the process of maturing and then staining olive-brown when injured, pore-surface uneven.

Stipe massive, 8-12 (16) cm long, 1-3 cm at apex, clavate, solid, pallid within and when cut staining reddish and then black; surface covered by coarse black fibrils often in zones giving a scaly rather than punctate effect, pallid beneath the scales.

Spore deposit near "bister" as dried (a dingy yellow-brown); spores 12-15 X 3.5-4.5 μ, smooth, apex lacking a pore or thin spot, shape in face view fusiform, in profile somewhat inequilateral, yellowish hyaline to dingy ochraceous in KOH, in Melzer's slightly browner, wall about 0.2 μ thick.

Basidia 4-spored, clavate, hyaline in KOH. Pleurocystidia scattered, clavate-rostrate, neck narrow and abruptly set off from the ventricose part, apex subacute, thin-walled, content ochraceous in KOH or hyaline. Cheilocystidia 23-37 X 7-12 μ, fusoid-ventricose, apex subacute to obtuse, content hyaline to yellowish as revived in KOH. Caulocystidia clavate, elliptic-pedicellate, mucronate or fusoid-ventricose, 35-70 X 10-25 μ, content dingy ochraceous in KOH, wall thin and apex often remaining collapsed, subiculum hyphae often colored like the cystidia, caulobasidioles and basidia also dingy ochraceous in KOH.

Tube trama typical for the genus; laticiferous hyphae rare, their content ochraceous in KOH and Melzer's and not standing out prominently in the mount. Pileus cutis of appressed hyphae 4-8 (10) μ wide, mostly tubular or nearly so, content weakly ochraceous in KOH but in Melzer's many with a rusty red homogeneous content and the others with yellow to orange homogeneous content, walls mostly smooth, apical cell tubular to somewhat cystidioid. Hyphae of subcutis and adjacent context yellow to ochraceous-orange in Melzer's. Clamp connections none.

Habit, habitat, and distribution.—Gregarious under aspens, Pellston, Emmet County, August 15, 1968, collected by W. Patrick (Smith 76025).

Observations.—Although obviously related to *L. atrostipitatum*, this species is at once distinguished by the duller color, more fibrillose appearance of the stipe ornamentation, and by the green FeSO$_4$ reaction. The colors are much duller than in *L. testaceoscabrum*. Oak is excluded as a tree-associate, and the cuticular hyphae of the pileus are narrower. At maturity it most closely resembles *L. insolens* in color.

66. Leccinum testaceoscabrum (Secretan) Singer

Amer. Midl. Nat. 37:123. 1947

Boletus testaceus scaber Secretan, Mycogr. Suisse 3:8. 1833.
Boletus rufescens scaber Secretan, Mycogr. Suisse 3:11. 1833.
(*B. rufescens* Konrad, Bull. Mens. Soc. Linn. Lyon p. 10. 1932.)

Krombholzia rufescens Singer, Rev. de Mycol. 3:189. 1938.

Pileus 4-10 cm broad, convex becoming broadly convex, the margin appendiculate at first; surface dry and obscurely fibrillose when young, practically glabrous as well as being subviscid at maturity, color dull orange with a tint of rose and when dried a dull "orange-cinnamon." Context white, firm when young, when cut staining reddish and then changing to avellaneous and finally fuscous, odor and taste not distinctive.

Tubes 1.5 cm deep, depressed around the stipe, whitish then olivaceous-pallid; pores olive-pallid before becoming dingy brown (near wood-brown), when lightly bruised staining olive, if severely bruised changing to fuscous.

Stipe 5-10 cm long, 1-1.5 cm thick, equal or nearly so, solid, cut surfaces finally staining fuscous; surface with a pallid to white ground color, ornamentation typically blackish in buttons as in *L. atrostipitatum,* with a slight flush of reddish around the dark squamules.

Spore deposit not obtained; spores (12) 13-16 (18) $\times$ 3.5-4.5 μ, smooth, wall -0.5 μ thick, apex lacking a pore, shape in face view subfusoid to elongate-subelliptic, in profile narrowly inequilateral or the suprahilar area broad and not deeply depressed, color in KOH or Melzer's merely ochraceous, young spores rather pale.

Basidia 4-spored, 22-30 $\times$ 9-11 μ, hyaline, clavate. Pleurocystidia 33-45 $\times$ 9-14 μ, fusoid-ventricose with subacute apex, walls thin and hyaline, content in KOH hyaline to brownish, in Melzer's dark brown (not red). Cheilocystidia more or less like the pleurocystidia but smaller and many basidiole-like cells also present. Caulocystidia 40-80 $\times$ 10-18 μ, clavate, clavate-mucronate or fusoid-ventricose, their content pale fuscous in KOH or some pigment also in the wall.

Tube trama typical of the section; weakly bilateral and with some laticiferous hyphae present which are yellowish to brownish (in places) in KOH or Melzer's. Pileus cutis of appressed hyphae 6-15 (18) μ wide, the walls thin and often minutely roughened with hyaline plates (in KOH and Melzer's); cells readily disarticulating, mostly more than 5 times as long as broad, some short cells more or less oval in outline also present, end-cells cystidioid (ventricose in the midportion) to bullet-shaped. Content of cells slowly forming irregular pigment masses and finally globules 3-10 μ in diameter in Melzer's. Hyphae of subcutis orange to hyaline in Melzer's and many with an irregular gelatinous coating. Clamp connections absent.

Habit, habitat, and distribution.—Gregarious under aspen, birch, oak, and mixed conifers, east of Cheboygan, August 22, 1967, Samuel Mazzer (Smith 75139).

Observations.—This is an interesting species in the light of the data given in the original description of *Boletus testaceus-scaber* Secretan. This American collection could have been associated with oak, it does have charcoal-black stipe ornamentation, and it does have a reddish orange pileus which is certainly not inconsistent with tile-red—to translate Secretan's color designation. It meets Singer's interpretation of Secretan's species in having occasional oval cells in the cuticular hyphae of the pileus. Moser (1967) indicates that *L. tastaceoscabrum* stains reddish then fuscous when the basidiocarp is cut open. The change we observed in no. 75139 is redder than observed for *L. atrostipitatum*. In addition the latter is definitely associated with birch, the caulocystidia more frequently are fusoid-ventricose with elongated necks, and the pileus has a much more patchy appearance at maturity.

67. Leccinum atrostipitatum Smith, Thiers, & Watling

Mich. Bot. 5:155. 1966

Illus. Figs. 56-58, 83; Pls. 68-69.

Pileus 6-13 (18) cm broad, hemispheric becoming broadly convex with the margin exceeding the tube line by 5 mm or often more and this band soon breaking up into segments; surface dry and appressed-fibrillose, becoming obscurely to distinctly appressed-squamulose by maturity, the outer fibrils often grayish; color pale dull orange-buff ("pale apricot-buff"), slowly becoming pale orange brownish and finally a dingy pinkish tan, the squamules finally merely brownish. Context white, when cut soon vinaceous-buff and this progressing to violaceous-fuscous; with $FeSO_4$ bluish gray, with NH_4OH bluish, color changes in the stipe the same as in the pileus but sometimes staining blue-green in addition.

Tubes 1-1.5 cm long, adnate-seceding, olivaceous-pallid when young, then dingy honey-brown and finally wood-brown; pores minute, staining olive to olive-brown and if severely bruised avellaneous or darker.

Stipe massive, 8-15 cm long, 1-2.5 cm thick at apex and up to 4 cm at the base, clavate becoming equal in age, solid, consistency hard and fibrous, white within, when cut staining pinkish avellaneous to violaceous-fuscous, finally toward the base staining bluish green in some either within or on the surface, the surface with coarse black ornamentation in the form of points and squamules from button stages to maturity and from base to apex, with a cottony-fibrillose layer beneath this, dry at base in young specimens.

Spore deposit dull yellow-brown ("snuff-brown"); spores 13-17 X 4-5 μ, smooth, wall slightly thickened, in face view subfusiform, in profile very obscurely inequilateral, suprahilar depression broad and shallow, pale yellow-brown in KOH and not changing much in Melzer's.

Basidia 4-spored, 8-12 μ broad, hyaline in KOH. Pleurocystidia scattered, 30-40 X 8-12 μ, fusoid-ventricose, with dull yellow-brown content when revived in KOH or in Melzer's, smooth, thin-walled. Cheilocystidia 26-35 X 6-10 μ, fusoid-ventricose to clavate, ochraceous in KOH, soon gelatinizing in KOH. Caulocystidia 40-120 X 10-25 μ, clavate to fusoid-ventricose, content bister in KOH.

Cutis of pileus of tangled filaments, the hyphae with lemon-yellow content in KOH when fresh, yellow to nearly hyaline when revived in KOH and in Melzer's the content granular to homogeneous and yellowish (no pigment globules seen); the hyphae with cells tubular to somewhat inflated, mostly 6-11 μ wide, some rather short, finally disarticulating, the end-cells tubular to either narrowly clavate or tapered to apex, walls smooth to minutely asperulate. Context hyphae when revived in Melzer's with orange to red content in some. Clamp connections absent.

Habit, habitat, and distribution.—Solitary to scattered in the hardwood forests of the state if birch is present. Not uncommon during some seasons.

Observations.—This is one of the most distinctive species in the genus by virtue of its pale rather dull colored pileus which typically appears patchy in age from appressed aggregations of fibrils, the black ornamentation covering the entire stipe except possibly the extreme base, and the medium-sized spores. It is very close to *L. testaceoscabrum.* For further comment see that species.

Material examined.—Cheboygan: Smith 36390, 67225, 71535, 72744, 74305, 74454, 75401; Thiers 2688, 3094, 3374, 4258. Chippewa: Smith 71832, 71912, 71913, 72421, 72484; Thiers 3023, 2406, 3865, 3866, 4355. Emmet: Smith 71559, 71670; Thiers 622, 3073, 3529, 3729, 4279, 4386. Luce: Smith 36969, 37442, 50240, 72012, 74439; Thiers 3402. Mackinac: Smith 71902; Thiers 3354, 3618. Marquette: Bartelli 2173; Smith 6654, 72767. Montmorency: Smith 66998.

68. Leccinum aurantiacum (Bull. ex St. Amans) S. F. Gray

Nat. Arr. Brit. Pl. I:646. 1821

Illus. Pl. 70.

Pileus 5-15 (20) cm broad, convex to broadly convex, finally plane or broadly convex with an upturned margin, rarely with a low umbo, margin at first with a sterile band of tissue breaking up into segments; surface dry and uneven, roughened or appressed-fibrillose, in age more or less glabrous and subviscid; color bright to dull ferruginous-red but often with a pallid area where covered by leaves or debris. Context thick, white, when cut slowly staining vinaceous then grayish and then slowly fuscous, with $FeSO_4$ very pale bluish; in the stipe when cut pallid at first but staining to fuscous in the upper part, sometimes staining caerulean blue in the base and in places strongly reddish to red-brown.

Tubes 1-2 cm long, depressed around the stipe to nearly free, pale olive-buff slowly becoming darker "wood-brown"; pores minute, oliva-ceous-pallid, when lightly bruised staining olive to olive-brown but where severely bruised stained avellaneous or darker.

Stipe 10-16 cm long, 2-3 cm thick, dry, narrowly clavate to fusi-form becoming equal, solid, fibrous and hard lower down, pallid within, when cut staining as indicated above, surface scabrous-roughened, pallid to whitish at first but ornamentation soon staining brown and finally blackish at least over the basal half, apex often remaining pallid, in places yellowish but ground color mostly pallid to white, when dried often yellow over the base.

Spore deposit dark yellow-brown to dull cinnamon; spores 13-16 (18) $\times$ 3.5-4.5 (5) μ, smooth, in face view subfusiform, in profile nar-rowly somewhat inequilateral, in KOH pale ochraceous-brown, in Melzer's merely yellowish brown.

Basidia 4-spored, 8-9 μ broad, clavate. Pleurocystidia rare to scat-tered, hyaline in KOH to merely yellowish, 35-60 $\times$ 8-12 μ, apices obtuse. Cheilocystidia clavate to subfusoid, 18-26 $\times$ 5-8 (10) μ, ochra-ceous to brownish in KOH. Caulocystidia mostly clavate, 9-30 μ in diameter, a few cylindric, smooth, content dingy yellowish in KOH.

Cutis of pileus of appressed interwoven hyphae 5-12 (15) μ wide, cells elongate (mostly more than 5 times as long as broad), walls thin, smooth, some minutely roughened, the cells disarticulating to some extent; the end-cells narrowly clavate to tubular or near apex slightly tapered, the content yellow to orange-buff in KOH (in young pilei distinctly reddish), in Melzer's the content orange-brown and forming globules of various sizes often filling the cells. Context hyphae hyaline in Melzer's. Clamp connections absent.

Habit, habitat, and distribution.—Scattered under pine and aspen, not uncommon during summer and fall throughout the state where aspen and pine grow.

Observations.—As we have observed this species in Michigan the cuticular hyphae are relatively smooth in contrast to those of *L. insigne,* which is our common species under aspen in June and July. In *L. insigne* the stipe is ornamented with black points and squamules by maturity, whereas in *L. aurantiacum* it is not uncommon to find basidiocarps with the stipe nearly white when half expanded and with the ornamentation becoming orange-tan or more reddish before finally going to black. In Michigan it is not limited to a single tree-associate.

69. Leccinum subrobustum Smith, Thiers, & Watling

Lloydia 31:261. 1968

Pileus 6-12 cm broad, broadly convex to convex, in age nearly plane; surface dry at first and more or less distinctly appressed-fibrillose, in age nearly glabrous and subviscid, at times becoming distinctly fibrillose-squamulose; color dark rusty orange ("burnt-sienna" to "Sanford's brown"), paler to the ochraceous-orange margin; margin appendiculate. Context pallid when cut but soon tinged reddish cinnamon and then slowly changing to fuscous, odor and taste not distinctive, the $FeSO_4$ reaction olive blackish finally.

Tubes white to pallid, depressed around the stipe, slowly becoming wood-brown to sepia; pores minute, staining olive if lightly bruised and olive-brown when severely bruised but stains eventually fuscous.

Stipe 9-12 (15) cm long, 1-3 cm thick at apex, enlarged downward, solid, pallid within when cut but soon staining reddish cinnamon and finally fuscous, in the lower part often stained blue and/or deep red also; surface ornamented at first with reddish to reddish brown squamules which by maturity are usually blackish, ornamentation rather coarse.

Spore deposit olive to olive-brown, olive-brown as air-dried; spores 13-16 × 4-5 μ, smooth, wall 0.5 μ thick, apex with a thin spot but no true pore; shape in face view subfusiform, in profile elongate-inequilateral; color dingy ochraceous to brownish ochraceous in KOH; in Melzer's ochraceous to weakly dextrinoid.

Basidia 4-spored, 20-30 X 9-11 μ, clavate, hyaline in KOH. Pleurocystidia 28-46 X 8-13 μ, fusoid-ventricose, thin-walled, mostly with hyaline content, not projecting prominently from the hymenium, apex subacute. Cheilocystidia about like the pleurocystidia but many merely mucronate. Caulocystidia elongate-subelliptic to clavate to mucronate to fusoid-ventricose but the latter not numerous, 34-76 X (7) 10-20 (25) μ, walls smooth and thin, content sepia in KOH; subiculum hyphae hyaline, tubular, when revived in KOH having a granular-colloidal content and lacking refractive inclusions.

Tube trama typical of the section, some laticiferous elements present. Pileus cutis of appressed hyphae 7-15 μ wide, the cells readily disarticulating, the walls minutely roughened as in *L. insigne,* many cells merely short-elliptic, end-cells short and bullet-shaped to elongate and cystidioid, content merely dingy brownish in KOH, in Melzer's granular to occasionally forming beads or globules in some cells but often remaining perfectly homogeneous. Hyphae of subcutis merely weakly yellow in Melzer's and lacking any "colloidal" content. Clamp connections none.

Habit, habitat, and distribution. — Abundant under aspen and birch, northwest corner of Montmorency County, July 24, 1967, Smith 74632.

Observations. — This is a robust reddish orange species usually with the pileus distinctly fibrillose in which the context stains reddish cinnamon before it changes to fuscous and in which the stipe ornamentation is reddish to reddish brown before it darkens. It is very similar in appearance to *L. insigne* var. *insigne,* but the species lacks a preliminary change to reddish cinnamon before the cut surface darkens to fuscous. It is closely related to *L. aurantiacum*, but the pileus is not as red and it lacks the characteristic formation of globules in the cuticular hyphae as seen in mounts made in Melzer's which characterize that species.

70. Leccinum cinnamomeum Smith, Thiers, & Watling

Mich. Bot. 5:157. 1966

var. cinnamomeum

Illus. Pl. 71.

Pileus 6-12 (18) cm broad, convex, broadly convex and finally nearly plane, margin curved in at first and soon appendiculate as the sterile band becomes segmented, surface dry and unpolished, glabrous to appressed-fibrillose, or squamulose from spotlike squamules, subviscid in older material and then glabrous or nearly so; colors dingy cinnamon over the center (overall in buttons) and grayish cinnamon-buff toward the margin, in age somewhat alutaceous with olive tones along the margin if wet. Context firm, hard, and white, soft in age, when cut staining quickly to vinaceous then to violaceous-fuscous, with $FeSO_4$ blue-green, with KOH no reaction in or on the stipe-base but on tubes pale cinnamon.

Tubes pallid to olivaceous yellowish and on maturing becoming avellaneous to wood-brown, about 1.5 cm long, depressed around the stipe; pores minute, olive whitish when young and soon staining yellowish if lightly bruised but sepia if severely bruised.

Stipe 8-10 cm long, 2-2.5 cm thick, solid, firm, hard, fibrous, pithy in the center, when cut slowly changing to fuscous, surface white, ornamented with fuscous-brown squamules and points, slowly staining brownish around base from handling.

Spore deposit between cinnamon-brown and bister; spores 12.5-15 $\times$ 4-5 μ, smooth, walls thickened slightly, subfusoid (with blunt ends in face view), somewhat inequilateral in profile, suprahilar depression often broad and distinct, dingy ochraceous in KOH, yellowish in Melzer's.

Basidia 4-spored, clavate, 8-9 μ broad, hyaline in KOH and yellowish in Melzer's. Pleurocystidia 37-66 $\times$ 9-14 μ, fusoid-ventricose, apices obtuse, thin-walled, content pale cinnamon in KOH when fresh, darker cinnamon when revived in KOH, near sepia in Melzer's. Cheilocystidia 24-33 $\times$ 6-10 μ, clavate to ventricose-rostrate or fusoid-ventricose, ochraceous in KOH, soon gelatinizing. Caulocystidia varying from clavate to fusoid-ventricose, the content sepia in fresh material, all smooth and thin-walled.

Pileus epicutis of tangled hyphae (5) 7-12 (22) μ in diameter, many slender hyphae 5-10 μ wide, tubular, with end-cells tapered slightly, walls thin and smooth to minutely roughened, content dingy ochraceous to orange-brown in KOH, content reddish brown and granular in Melzer's; the broader hyphae (10-22 μ wide) with short cells 15-60 μ long, their walls smooth or roughened and the walls parallel as seen in optical sections, or only slightly bulging (not sphaerocyst-like), end-cells often short and bullet-shaped; subcutis of hyaline hyphae in either KOH of Melzer's, smooth, finally subgelatinous. Context of hyaline smooth interwoven hyphae both in KOH and Melzer's. Clamp connections none.

Habit, habitat, and distribution.—In low woods of white birch and aspen, Upper Peninsula, but most likely throughout the aspen-white birch belt of the state, in early summer.

Observations.—The dull cinnamon-colored pileus distinguishes this species from its close relatives, which show a reddish stain before the color change to fuscous sets in.

70a. Leccinum cinnamomeum var. fibrillosum Smith & Thiers, var. nov.

Pileus 6-12 cm latus, late convexus, siccus, rufo-fulvus vel fulvus, fibrillose squamulosus. Stipes atropunctatus vel squamulosus. Sporae 13-16 × 3.5-5 μ. Specimen typicum in Herb. Univ. Mich. conservatum est; prope Brutus, Emmet County, July 12, 1968, legit Smith 75560.

Pileus 6-12 cm broad, broadly convex expanding to nearly plane, margin appendiculate, surface dry and distinctly fibrillose-squamulose from "tawny" to "ochraceous-tawny" squamules (having much the appearance of *Agaricus subrufescens*), when dried dull cinnamon in color and spotted from the squamules. Context pallid when cut but soon staining reddish cinnamon and finally blackish and the latter persistent in the dried basidiocarps, odor and taste not distinctive.

Tubes pallid when young, slowly dull brown by maturity but passing through an olive-gray stage; pores minute, pallid to olive-pallid and when bruised staining olive-brown going to fuscous.

Stipe 9-11 cm long, 12-15 cm thick, equal, solid, context staining as in the pileus; surface pallid with an overlay of black rather coarse ornamentation, apex pallid.

Spores 13-16 × 3.5-4.5 (5) μ smooth, lacking an apical pore or thin spot, shape in face view narrowly fusoid, in profile narrowly inequilateral, wall about 0.2 μ thick, color in KOH ochraceous to weakly yellow-brown, not appreciably changing in Melzer's.

Basidia 4-spored, 8-11 μ broad at apex, clavate, hyaline in KOH and merely yellowish in Melzer's. Pleurocystidia scattered, more or less fusoid-ventricose, 28-46 × 8-13 μ, apex subacute, hyaline in KOH, content often brownish as revived in Melzer's. Cheilocystidia 18-28 × 6-11 μ, subfusoid, content yellowish revived in KOH. Caulocystidia 30-80 × 9-22 μ, clavate or mucronate, content dingy yellow-brown in KOH, wall thin and smooth, subiculum hyphae not colored.

Tube trama of weakly divergent hyphae with refractive septa, hyphae 4-6 μ wide, not obviously gelatinous in KOH. Pileus cutis of appressed hyphae much as in *L. insigne*—some very wide (up to 25 μ) and their end-cells usually bullet-shaped, most hyphae 4-10 μ wide and tubular, the cells often disarticulating, content dingy yellow-brown in KOH, walls smooth to minutely ornamented (roughened), in Melzer's the content remaining homogeneous or more rarely becoming granular, the cells mostly more than 4 times as long as broad. Hyphae of subcutis and adjacent context not highly colored in Melzer's. Clamp connections none.

Habit, habitat, and distribution.—Gregarious under aspen and birch, lower Maple River, Emmet County, July 12, 1968, Smith 75560.

Observations.—This appears to be an extreme variant as regards the development of fibrillose squamules on the pileus which impart a tawny color to the pileus—a color much richer than the dingy cinnamon of the type variety. In addition the base of the stipe did not stain brown when handled.

71. Leccinum laetum Smith, Thiers, & Watling

Mich. Bot. 5:154. 1966

Pileus 5-15 cm broad, oval to convex, expanding to broadly obtuse to broadly convex, margin appendiculate; surface dry and appressed fibrillose-squamulose, becoming glabrous; color when young and fresh orange to orange-ochraceous but becoming dull ochraceous, the appressed fibrils or squamules yellow-brown ("buckthorn-brown"), glabrescent and in age subviscid and at this time with olive tones on the margin and the disc dingy ochraceous to dull orange-brown. Context pallid, when cut soon staining vinaceous then gray and then fuscous; in the stipe staining in addition blue and yellow and reddish (in different areas), $FeSO_4$ blue; KOH on cutis no reaction to slightly rusty brown; odor none, taste mild to slightly acid.

Tubes 1-2 cm deep, depressed around the stipe to nearly free, white staining avellaneous when cut; pores minute, round, white to olive-white when young, staining olive when slightly bruised, but when severely bruised staining near grayish vinaceous-brown (near "wood-brown").

Stipe 8-16 cm long, 1-2.5 cm thick, equal, solid, when cut changing color as noted under context above, soon fuscous-black around the larval tunnels; surface faintly ornamented by pallid then brownish and finally darker points, lines, or squamules, apex pallid, viscid over the base when young as in *Boletellus russellii.*

Spore deposit olive when wet and olive-brown when air-dried ("dark olive-buff" to "olive-brown"); spores 13-16 (17) $\times$ 4-5 μ, smooth, wall slightly thickened, shape in profile somewhat elongated-inequilateral with a broad shallow suprahilar depression, in face view subfusoid, pale ochraceous in KOH, in Melzer's a few dextrinoid, the rest pale yellow-brown (in mounts made from hymenial tissue when revived in Melzer's showing a faint suggestion of the fleeting-amyloid reaction).

Basidia 4-spored, 17-23 X 8-11 μ, hyaline in KOH. Pleurocystidia abundant, 40-60 X 10-15 μ, fusoid-ventricose with blunt apex but neck often elongated and flexuous, thin-walled, hyaline in KOH at first but when revived in KOH having a dingy yellow-brown content, and in Melzer's dark yellow-brown. Cheilocystidia 23-34 X 6-11 μ, mostly fusoid-ventricose with narrow necks (about 2 μ wide) and apex subacute, ochraceous in KOH. Caulocystidia a mixture of ventricose, mucronate, and clavate cells 27-40 X 8-12 μ and larger ones up to 15-20 μ wide scattered in each fascicle, their content pale brownish ochraceous in KOH.

Cutis of pileus a tangled layer of epicuticular hyphae with subcutis of merely interwoven hyphae; the coarse fibrils of the epicutis 8-15 μ wide with some cells 40-120 μ or more long and some of these with minutely roughened walls, the end-cells often ovate-pointed (bullet-shaped), no pigment globules present in Melzer's, the colored cells with content amorphous-reticulate and orange-brown, the cells tending to disarticulate in age and at this time more of them showing slight wall irregularities. Context of hyaline hyphae in KOH and Melzer's. Clamp connections absent.

Habit, habitat, and distribution.—Scattered in an aspen thicket which had a ground cover of willows, Highland Recreation Area, Oakland County.

Observations.—This species is probably much more frequent in occurrence than the material to date indicates. *L. laetum* is near *L. insigne* by virtue of the features of the broad epicuticular hyphae, but is distinct by the olive-colored spore deposit, by the viscid base of the stipe, the relatively weak stipe ornamentation, and the basically orange-ochraceous of the pileus. About 50 basidiocarps represent the type collection.

72. Leccinum ochraceum Smith, Thiers, & Watling

Mich. Bot. 5:162. 1966

Illus. Pl. 72.

Pileus 5-12 cm broad, convex becoming broadly convex, the margin appendiculate at first; surface dry and unpolished to appressed-fibrillose, in some with appressed-fibrillose squamules near the margin; color varying from ochraceous-yellow with grayish overtones from fibrils or dingy tan ("pinkish buff") over the disc, or becoming pinkish cinnamon to pale dingy cinnamon. Context thick, white, when cut changing to vinaceous then avellaneous and finally violaceous-fuscous, with $FeSO_4$ bluish; odor and taste not distinctive.

Tubes 1-1.5 cm long, adnate at first, becoming depressed, pallid-olivaceous to dull brown where cut; pores minute, olivaceous-pallid, staining dark brown where severely bruised but olive yellowish where lightly bruised.

Stipe 11-15 cm long, 12-23 mm thick, solid, fibrous, hard, pallid within, when cut slowly staining fuscous in streaks, no blue or red stains developed on specimens observed; surface whitish with coarse blackish reticulation and squamules or points, when dried with an overall grayish ground color.

Spores dark yellow-brown ("snuff-brown") in deposits, 14-16 X 4.5-5.5 μ, subfusoid in face view, in profile inequilateral (often rather ventricose), suprahilar depression broad; walls slightly thickened, color in KOH dingy brownish ochraceous, not much different in Melzer's.

Basidia 4-spored, 18-22 X 9-10 μ, clavate, hyaline in KOH and Melzer's. Pleurocystidia 38-52 X 9-14 μ, fusoid-ventricose, content usually brownish to bister in KOH and bister to violaceous-bister in Melzer's, large bister or violaceous-bister laticiferous elements present in mounts revived in Melzer's. Cheilocystidia 30-45 X 5-9 μ, clavate to fusoid-ventricose, ochraceous in KOH. Caulocystidia voluminous, up to 50-60 X 20 μ, clavate to obclavate or mucronate more rarely fusoid-ventricose, content bister in KOH.

Hyphae of pileus brownish at times with some cells with dingy brown walls as seen in KOH, hyphae commonly disarticulating; hyphal width, cell size, and ornamentation as in *L. insigne*. Context hyphae when revived in Melzer's typically orange to red but also with a few fuscous (amyloid ?) masses or particles. Clamp connections absent.

Habit, habitat, and distribution.—Solitary to scattered, Sugar Island, and Tahquamenon Falls State Park, Chippewa County, July to September.

Observations.—The pilei dry to an olive-gray. The fresh pilei are ocher-yellow to paler at first with a gray shadow from the surface fibrils, but gradually get duller as they mature. When old they resemble those of *L. cinnamomeum.* It differs slightly from *L. insigne,* to which it is most closely related, in the color of the fresh young pileus, in the KOH reaction of the content of the cuticular cells, and in the dark-colored laticiferous elements as mounted in Melzer's.

73. Leccinum pseudoinsigne Smith & Thiers, sp. nov.

Pileus 4-13 (16) cm latus, convexus, siccus demum subviscidus, luteus vel aurantioluteus. Contextus albus, tactu subroseus demum fuscus. Tubuli 1-1.5 cm longi, pallidi; pori pallidi tactu olivacei. Stipes 6-15 cm longus, 10-20 mm crassus, minute squamulosus vel punctatus; squamulae pallidae demum tarde aurantio-brunneae, denique atratae. Sporae (13) 14-17 × 3.5-5 μ. Specimen typicum in Herb. Univ. Mich. conservatum est; prope Brutus, Emmet County, July 12, 1968, legit Smith 75559.

Pileus 4-13 (16) cm broad, obtuse to convex, the margin appendiculate; surface dry at first, becoming subviscid to viscid later depending on the weather; color "deep chrome" to "ochraceous-orange" (bright yellow, orange-yellow, or bright orange to orange-red), fading to orange-buff, on standing slowly orange brownish but never dull rusty red. Context white, when cut in the young basidiocarps becoming pale reddish cinnamon and finally bluish fuscous, in mature basidiocarps the color change often going directly to violet-gray; odor and taste not distinctive; with $FeSO_4$ olive-gray.

Tubes 1-1.5 cm deep, depressed around the stipe, white then olive-pallid and becoming near wood-brown; pores pallid to pale olive-buff and finally wood-brown, when lightly bruised staining olive, when severely bruised staining olive-fuscous.

Stipe 6-15 cm long, 10-20 mm at apex, equal or nearly so, solid, when cut staining weakly vinaceous in apex and finally changing to violaceous-fuscous—or in older stipes going to violaceous-fuscous immediately, slowly staining reddish cinnamon in the base in some on standing; surface covered with fine to moderately coarse ornamentation pallid at first but slowly changing to orange-brown and long remaining this color, finally blackish, ground color pallid.

Spore deposit near "bister" (deep yellow-brown as air-dried); spores (13) 14-17 × 3.5-5 μ, smooth, apex with an obscure thin spot, narrowly fusoid in face view, narrowly somewhat inequilateral in profile, wall about 0.2 μ thick, color ochraceous to pale tawny in KOH and about the same in Melzer's.

Basidia 4-spored, clavate, 7-11 μ broad, hyaline in KOH and weakly yellowish in Melzer's. Pleurocystidia scattered, clavate, mucronate, or broadly fusoid-ventricose with short neck and subacute apex, content ochraceous as revived in KOH and sometimes aggregated into an amorphous mass; walls thin and smooth. Cheilocystidia clavate to subfusoid, 20-35 × 6-12 μ, content ochraceous and pigment evenly distributed. Caulocystidia 33-70 × 9-20 μ, clavate, ovoid-pedicellate, mucronate, or broadly fusoid with a short neck and subacute to obtuse apex, content ochraceous in KOH and evenly distributed.

Tube trama typical for the genus, some laticiferous elements with ochraceous content (in KOH) present. Pileus cutis of hyphae 4-10 (12) μ broad, tubular to somewhat inflated, cells frequently disarticulating but usually 4 times or more as long as broad (but secondary septa may form in some terminal cells thus producing shorter segments), wall smooth or minutely rough as in *L. insigne,* content ochraceous in KOH and more yellow-brown in Melzer's but not forming globules in the latter medium or a globule forming slowly in an occasional cell. Hyphae of subcutis and adjacent context yellow to orange in Melzer's. Clamp connections none.

Habit, habitat, and distribution.—Scattered under aspen and birch in a swampy area, lower Maple River, Emmet County, July 12, 1968, Smith 75559.

Observations.—This species is paler than typical *L. aurantiacum* and for all practical purposes lacks the formation of globules in the cuticular hyphae as the latter are mounted in Melzer's. The stipe ornamentation remains pallid longer than in *L. insigne* var. *insigne* and is often rather bright orange-brown for a long time. The color change to reddish cinnamon separates it from the *L. insigne* group but old basidiocarps do not always show it.

Stirps INSIGNE

KEY

1. Caulocystidia (or many of them) with a caplike wall thickening at the apex . . . . *L. areolatum*
1. Not as above 2

   2. Pileus dull white to weakly grayish but not spotting brown. . . . *L. broughii*
   2. Pileus more highly colored or becoming so when mature, sometimes spotted brown where bruised 3

3. Pileus whitish spotting brown when bruised and in age becoming more or less cinnamon-buff overall . . . . *L. insolens* var. *brunneo-maculatum*
3. Not spotting as in above choice 4

   4. Pileus rusty red to liver color; many cuticular hyphae in Melzer's having the content rounding into pigment globules (see *L. sublutescens* also) 5
   4. Not as above 6

5. Pores cinnamon-brown young; spores 4-5 μ broad . . . .
. . . . *L. subtestaceum* var. *subtestaceum*
5. Pores pallid to grayish; spores 3-4 μ wide (see *L. sublutescens* also) . . . . . .
. . . . *L. subtestaceum* var. *angustisporum*

6. Spores 5-7 μ wide . . . . *L. imitatum*
6. Spores narrower 7
7. Pileus yellow with an overlay of cinnamon-brown fibrils . . . . *L. fuscescens*
7. Pileus not as above 8
   8. Pileus pale grayish cinnamon-buff then snuff-brown . . *L. insolens* var. *insolens*
   8. Not as above 9
9. Pileus context yellowish when cut; cuticular hyphae mounted in Melzer's
   often showing pigment globules . . . . *L. sublutescens*
9. Pileus yellow to orange or bright red; context not staining yellow when cut;
   cuticular hyphae lacking pigment globules when revived in Melzer's . . . . .
   . . . . *L. insigne* and *variants*

74. **Leccinum areolatum** Smith & Thiers, sp. nov.

Illus. Fig. 121.

Pileus 6-15 cm latus, convexus, siccus, mollis, demum areolatus, pallide incarnato-alutaceus. Contextus albus, tactu lilaceus demum fuscus. Tubuli 1-1.5 cm longi, pallidi; pori pallide lutei tactu brunnei. Stipes 6-12 cm longus, 1.5-2 cm crassus, clavatus; brunneo-punctatus. Sporae 14-16 × 3.5-4.5 (5) μ. Specimen typicum in Herb. Univ. Mich. conservatum est; prope Spectacle Lake, Chippewa County, July 13, 1968, legerunt W. Patrick et H. D. Thiers (Smith 75570).

Pileus 6-15 cm broad, convex becoming broadly convex to nearly plane, margin appendiculate; surface dry but soft, conspicuously areolate by maturity, color dull pinkish cinnamon on the areolae, paler between. Context white when cut but soon staining lilaceous then lilac-fuscous, lacking a red stage; odor and taste not distinctive; with $FeSO_4$ quickly bluish olive.

Tubes olive-pallid, long (up to 1.5 cm), depressed around the stipe, staining fuscous when cut; pores very minute (2-3 per mm), yellowish overall at maturity or brownish over some areas, staining fuscous if severely bruised.

Stipe 6-12 cm long, 1.5-2 cm thick at apex, clavate, solid, interior staining like the pileus context; surface pallid at the base; midportion covered by dark brown coarse ornamentation as streaks or points, apex pallid and ornamentation finer.

Spores 14-16 × 3.5-4.5 (5) μ, smooth, apex lacking an apical pore or thin spot, shape in face view fusiform, in profile somewhat inequilateral, ochraceous-brown in KOH and not much different in Melzer's, wall about 0.2 μ thick.

Basidia 4-spored, 8-10 μ broad, clavate, hyaline in KOH and yellowish in Melzer's. Pleurocystidia scattered, 40-60 X 10-16 μ, fusoid-ventricose, thin-walled content dull yellow-brown in KOH, in Melzer's often dark reddish brown. Cheilocystidia more or less like the pleurocystidia but smaller. Caulocystidia clavate-mucronate to fusoid-ventricose, less commonly the wall over the apex thickened to form a cap, smooth, content snuff-brown to bister in KOH and the subiculum hyphae also often colored similarly.

Tube trama typical for the genus, the divergent hyphae distinctly gelatinous as revived in KOH; laticiferous elements rare, with ochraceous content as revived in either KOH or Melzer's. Pileus cutis of appressed interwoven hyphae 4-10 μ wide, tubular or nearly so, content weakly ochraceous revived in KOH, dull ochraceous to orange-ochraceous and homogeneous as revived in Melzer's, walls thin and typically smooth, end-cells tubular to weakly cystidioid, cells disarticulating to some extent. Clamp connections absent.

Habit, habitat, and distribution.—Mixed aspen and birch, Spectacle Lake, Chippewa County, July 13, 1968, collected by H. D. Thiers and W. Patrick (Smith 75570).

Observations.—The conspicuously areolate pileus at maturity, pinkish cinnamon color, lack of a reddish stage in the color changes when the context is cut, the presence of coarse dark brown (not black) stipe ornamentation, and the thick-walled apex of many of the caulocystidia are distinctive. In pileus features, in color changes, and in the capped caulocystidia it is distinguished from *L. pellstonianum* to which it is otherwise most similar.

75. Leccinum broughii Smith & Thiers, sp. nov.

Pileus circa 7.5 cm latus, convexus, subudus, sordide albidus. Contextus albidus, tactu vinaceo-fuscus. Tubuli pallidi; pori pallidi tactu violaceo-brunnei. Stipes circa 13 cm longus, 2 cm crassus, albidus, squamulose punctatus; punctae sordide albidae, tarde brunneae. Sporae (12) 13-15 X 4-5 μ. Specimen typicum in Herb. Univ. Mich. conservatum est; prope Wildwood, Emmet County, July 14, 1963, legit Sherman Brough (Smith 66914).

Pileus 7.5 cm broad, broadly convex; surface moist but not viscid, with inconspicuous innate appressed fibrils, dull white overall (and exposed to light during its development), on standing over night becoming pale avellaneous over the disc but pallid as dried; margin sterile as in *L. aurantiacum.* Context white soon staining vinaceous-fuscous and finally fuscous (no intermediate truly reddish state); $FeSO_4$ staining flesh olivaceous.

Stipe 13 cm long, 2 cm thick at apex, 3.5 cm at base, whitish overall, when cut soon "dusky brown" (violaceous-brown), lacking a red stage in the color change, surface whitish and conspicuously roughened by whitish points or squamules to almost reticulate in places, the ornamentation slowly brown to blackish on aging (no green, yellow, or blue stains developing).

Spores (12) 13-15 $\times$ 4-5 μ, subfusiform in face view, rather obscurely inequilateral in profile, suprahilar depression shallow; smooth; wall thickened somewhat, brownish ochraceous in KOH, in Melzer's yellowish or a few reddish tawny, no fleeting-amyloid reaction present.

Basidia 4-spored, 20-27 $\times$ 9-11.5 μ, some with dark bister content in KOH. Pleurocystidia rare, similar to cheilocystidia. Cheilocystidia fusoid-ventricose, 26-35 $\times$ 8-12 μ, apex obtuse, content either colored or hyaline in KOH. Caulocystidia mostly clavate, 36-50 (70) $\times$ 10-20 μ, content smoky yellow in KOH, thin-walled, smooth, in fascicles. Caulobasidia rare.

Cutis of pileus a tangled layer of thin-walled, smooth, hyaline hyphae 4-7 μ wide, the terminal cells equal in diameter throughout their length and obtuse at apex or varying to only slightly cystidioid, content hyaline to pale dingy ochraceous in KOH, in Melzer's the content homogeneous (no sign of pigment globules), lacking truly inflated cells in the cutis hyphae. Context of hyphae having red to orange-red content in Melzer's in those near the cutis. Clamp connections none.

Habit, habitat, and distribution.—Under birch and aspen, Wildwood, Emmet County, July 14, 1963, Sherman Brough (Smith 66914).

Observations.—As dried the cutis of the pileus is somewhat cottony and so also is the ornamentation of the stipe. The pale color was not an accident of development (such as the pileus being covered by leaves). The species is closely related to *L. clavatum,* but differs in the tree-associate, slightly smaller spores, different stature and in the hyphae of the context next to the cutis being orange-red in Melzer's.

76. Leccinum subtestaceum Smith, Thiers, & Watling

Mich. Bot. 5:145. 1966

var. **subtestaceum**

Illus. Pls. 73-74.

Pileus 4-12 cm broad, convex to broadly convex, surface dark ferruginous to liver color ("Hays-russet") and glabrous to finely fibrillose, in age becoming obscurely areolate, sterile margin soon broken into segments. Context white, changing to vinaceous-gray to slate color, soon blue in base of stipe; odor and taste mild; with $FeSO_4$ greenish blue.

Tubes whitish becoming gray to wood-brown, 1-2 cm long, depressed around the stipe in age, adnate at first, slowly vinaceous-gray when cut; pores small round and "cinnamon-brown" when young, paler in age.

Stipe 7-12 cm long, 2-3 cm thick at apex, equal or evenly enlarged downward, solid, white within, staining vinaceous-slate when cut, bluish in and around the base in some; surface white but overlaid with a black reticulate scabrous coating of lines and points.

Spores 11-15 X 5-5.5 μ, smooth, apical pore lacking, walls only slightly thickened, shape in face view subfusoid, in profile inequilateral-elongate, brownish ochraceous when revived in KOH, in Melzer's ochraceous-tan (rather pale and dingy).

Basidia 4-spored, clavate, 22-27 X 8-11 μ, hyaline to yellowish in KOH. Pleurocystidia—none seen. Cheilocystidia 23-24 X 5-9 μ, subfusoid with obtuse apex to clavate, dingy ochraceous in KOH. Caulocystidia mostly fusoid-ventricose or clavate-mucronate, some remaining clavate, thin-walled, smooth, content dark smoky brown as revived in KOH. Pileus cutis of appressed interwoven hyphae 4-10 (12) μ wide, the end-cells mostly narrowly clavate to cylindric, tips obtuse, content in KOH orange-brown to orange-ochraceous, in Melzer's the content breaking up into pigment globules or masses so that end-cells often simulate a tube filled with beads. Context hyphae lacking a highly colored content when revived in Melzer's. Clamp connections absent.

Habit, habitat, and distribution.—On sandy soil under aspen, late summer, rarely collected to date.

Observations.—This species differs from *L. insigne* in having the dark brown pores when young and in the pigment globules observed when cuticular hyphae are mounted in Melzer's either on fresh or dried material. It lacks a red color-change before the changes progress to fuscous, and has dark brown pores; both features distinguish it from *L. aurantiacum.* The dark liver-red pileus distinguishes it from both *L. aurantiacum* and *L. insigne.*

76a. Leccinum subtestaceum var. angustisporum Smith & Thiers, var. nov.

A typo differt: Sporae 14-18 X 3-4 (4.2) μ; pori grisei. Specimen typicum in Herb. Univ. Mich. conservatum est; prope Hardwood Lake,

Otsego County (sed in Montmorency County), August 10, 1967, legit Smith 74904.

Pileus 6-12 cm broad, convex expanding to broadly convex, surface dry at first and only inconspicuously fibrillose, color dull rusty red to dark dull orange-red (subtestaceous), glabrous and subviscid in age. Context white, the cut surface staining violaceous-gray directly and soon the change progressing to a darker tone but fading out in drying, odor and taste not distinctive.

Tubes pallid young, depressed around the stipe, dark brown in age; pores small, grayish when young and fresh, staining brownish avellaneous when bruised.

Stipe 9-14 cm long, 1.5-3 cm thick, enlarged downward slightly, solid, cut surface staining pale fuscous slowly with no intermediate red stage; surface ornamented with charcoal-black ornamentation at an early stage, ground color pallid, in drying with a yellowish tone over basal area by fluorescent light.

Spores 14-18 X 3-4 (4.2) μ, the shorter ones often widest, smooth, wall 0.5 μ thick, apex lacking a pore; shape elongate-inequilateral in profile, in face view elongate-subfusoid, color pale ochraceous in KOH, merely weakly brownish in Melzer's.

Basidia 4-spored, 23-28 X 7-9 μ, clavate, hyaline in KOH. Pleurocystidia 28-40 X 8-12 μ, fusoid-ventricose with only the neck projecting, apex acute, content hyaline or brownish in KOH (the same as in Melzer's). Caulocystidia 36-50 X 10-30 μ, clavate to clavate-mucronate, the content gray-brown in KOH.

Pileus cutis of appressed smooth tubular hyphae 4-11 μ wide, the cells long and the content dingy ochraceous in KOH, very little disarticulation of cells noted; content in Melzer's soon rounding up into pigment globules often variously shaped, on standing, however, much of the pigment becoming a mass of dispersed granular material. Hyphae of the subcutis merely weakly yellowish in Melzer's. Clamp connections absent.

Habit, habitat, and distribution.—Solitary under birch-aspen and balsam-spruce mixed, Montmorency County, August 10, 1967, Smith 74904.

Observations.—If the combination of characters featured in the type collection prove to be constant, this varient should be ranked as an autonomous species.

77. Leccinum imitatum Smith, Thiers, & Watling

Mich. Bot. 5:136. 1966

Pileus 6-10 cm broad, convex to broadly convex or finally nearly plane; surface dry and matted-fibrillose, becoming more conspicuously fibrillose in age, color when young evenly dingy rufous, becoming duller and in age or when dried dull cinnamon-brown, margin appendiculate at first. Context white, firm becoming soft, slowly bluish gray to fuscous when cut, odor and taste not distinctive, greenish with $FeSO_4$.

Tubes pale yellow in young specimens with the pores dull yellow-brown (bister), depressed around the stipe in age, 1.5 cm long in ventricose part, becoming grayish to near wood-brown at maturity; pores small, round, concolorous with the sides in age.

Stipe 6-12 cm long, 1-2 cm thick, equal to evenly enlarged downward, solid, white, within finally staining blackish when cut or in drying, weakly yellowish in basal area in some as dried, staining blue in base when fresh; surface whitish with blackish ornamentation to apex.

Spore deposit dull rusty brown ("cinnamon-brown"); spores 14-18 $\times$ 5-7 μ, smooth, wall thickened, in KOH dingy ochraceous to brownish ochraceous, in Melzer's dingy yellowish to pale tan, rarely a few dextrinoid; shape in face view subfusiform, in profile somewhat inequilateral, suprahilar depression broad.

Basidia 23-28 $\times$ 10-12 μ, 4-spored, short-clavate, hyaline to yellowish in both KOH and Melzer's. Pleurocystidia abundant, 38-52 $\times$ 10-15 μ, fusoid-ventricose, the apex subacute to obtuse, wall thin and smooth, content hyaline in some both in KOH and Melzer's. Caulocystidia clavate to a few mucronate or fusoid-ventricose, 9-22 μ wide, content bister in KOH, walls thin and smooth. Tube trama of hyphae somewhat divergent from a central strand.

Pileus cutis a tangled layer (possibly a collapsed trichodermium) of hyphae 4-9 μ wide, the end-cells usually tapered somewhat and rarely inflated in the midportion; in KOH the content orange-brown to dingy pale ochraceous-brown and homogeneous to granular when revived in Melzer's the content reddish and homogeneous at first but slowly becoming paler and somewhat granular; walls thin and smooth. Context hyphae next to the cutis with orange to orange-red content as revived in Melzer's. Clamp connections none.

Habit, habitat, and distribution.—Under aspen, Pellston, the hills west of town, June 21, 1963, Smith 66706.

Observations.—The distinguishing features of this species are the wide spores, the tubes yellow at first and with their mouths dark yellow-brown, and the orange to red hyphae of the context adjacent to the pileus cutis as these are revived in Melzer's. The pileus colors are intermediate between the ferruginous and the dull cinnamon species. The spores are wider than in *L. aurantiacum* and the Melzer's reactions are different.

78. Leccinum fuscescens Smith, Thiers, & Watling

Lloydia 31:255. 1968

Pileus 4-9 cm broad, convex becoming broadly convex; surface dry at first, color yellowish with an overlay of cinnamon-brown to "Mars-brown" fibrils forming streaks, in age more or less dull brownish, subviscid and nearly glabrous, margin appendiculate. Context firm, white, when cut staining lilac-gray to fuscous directly, no reddish preliminary stage present, $FeSO_4$ spotting bluish on white flesh, odor and taste mild.

Tubes pallid, depressed around the stipe, staining fuscous when cut, finally wood-brown to bister where slightly bruised; pores small, olivaceous-pallid, where slightly bruised staining olive to olive-brown, where severely bruised finally staining fuscous.

Stipe 8-11 cm long, 12 mm thick, equal or nearly so, solid, pallid, where cut staining like the pileus or in some with blue stains at the base; ground color on surface whitish, ornamentation in the form of coarse blackish reticulations and squamules to the apex much as in *L. atrostipitatum.*

Spore deposit dark snuff-brown to bister as air-dried; spores 12-15 X 3.8-4.5 μ, smooth, wall less than 0.5 μ thick, apex lacking a pore; shape in face view subfusiform, in profile somewhat inequilateral; color in KOH ochraceous to pale ochraceous-brown, in Melzer's duller brown but pale.

Basidia 4-spored, 22-30 X 8-11 μ, clavate, hyaline in KOH. Pleurocystidia scattered, 32-47 X 9-14 μ, fusoid-ventricose, the neck narrow and apex subacute, walls thin and smooth, content hyaline to bister in KOH, in Melzer's often dark sepia. Cheilocystidia similar to pleurocystidia but smaller. Caulocystidia 30-70 X 8-18 μ, mostly elongate-mucronate, varying to clavate-mucronate, fusoid-ventricose or simply clavate, numerous, content·pale bister in KOH; subiculum well developed beneath patches of the caulohymenium, hyphae lacking refractive bodies adhering to the inner walls.

Tube trama typical for the section; laticiferous hyphae present but rather rare, content dingy ochraceous in Melzer's. Pileus cutis of narrow to broad (up to $20\,\mu$) tubular hyphae with the cells readily disarticulating and end-cells often bullet-shaped, walls smooth to minutely ornamented (as in *L. insigne*), content dingy ochraceous to brownish ochraceous in KOH, homogeneous; in Melzer's the content homogeneous to granular but in some hyphae after 15 minutes forming some globules 4-$10\,\mu$ in diameter, but these not highly colored. Hyphae of subcutis not distinctively colored in Melzer's. Clamp connections absent.

Habit, habitat, and distribution.—Scattered over a small area, under birch and aspen, northwest corner of Montmorency County, August 9, 1967, Smith 74570, 74887, 74889.

Observations.—Dried pilei are dull brown (near "wood-brown") and slightly spotted, the cuticular hyphae are those of *L. insigne,* the stipe ornamentation is coarse and dark, the fresh pileus is a pale dingy yellow with an overlay of dark rusty brown fibrils, and there is no preliminary change to red when the context of the pileus or apex of stipe is cut.

79. Leccinum insolens Smith, Thiers, & Watling

Lloydia 31:257. 1968

var. insolens

Illus. Pl. 75.

Pileus 5-15 cm broad, broadly convex, becoming broadly convex to nearly plane, margin sterile and appendiculate and incurved; surface dry at first, subviscid in age, more or less fibrillose to spotted-squamulose, color a pale grayish cinnamon-buff with a dingy pinkish buff margin, in age pale "snuff-brown" and nearly glabrous. Context pallid, soon stained lilaceous-gray then fuscous, with $FeSO_4$ bluish, odor and taste not distinctive.

Tubes depressed, whitish then olive-buff becoming avellaneous to wood-brown; pores minute, white, staining olive when lightly bruised but avellaneous to fuscous if severely bruised.

Tubes depressed, whitish then olive-buff becoming avellaneous to wood-brown.

Stipe 8-15 cm long, 15-30 mm thick at apex, solid, staining avellaneous to fuscous when cut (reddish only in the base); surface with coarse ornamentation on a cottony subiculum, the points and squamules grayish avellaneous, pallid in buttons but blackish before maturity.

Spore deposit "argus-brown" when freshly air-dried but on standing slowly becoming bister or more olivaceous-brown; spores 12-15 X 3.8-8.3 μ, lacking an apical pore, wall only slightly thickened (less than 0.5 μ thick); in face view elongate-subelliptic to subfusiform, in profile narrowly somewhat inequilateral, pale dingy ochraceous in both KOH and Melzer's.

Basidia 22-34 X 9-11 μ, clavate, hyaline in KOH and merely weakly yellowish in Melzer's. Pleurocystidia abundant, 30-48 X 9-15 μ, fusoid-ventricose with the neck narrowed to a subacute apex, content hyaline to smoky brown (in either KOH or Melzer's), brown ones most numerous in areas where discoloration had occurred on the fresh specimen. Cheilocystidia 18-29 X 5-11 μ, clavate, clavate-mucronate, or obscurely fusoid-ventricose with obtuse apex, walls yellowish brown in KOH to hyaline, thin, collapsing readily. Caulocystidia numerous in the caulohymenium 28-50 X (8) 10-20 μ, clavate-pedicellate to clavate-mucronate, thin-walled, smooth, content bister in KOH.

Tube trama subgelatinous, somewhat divergent to the subhymenium, hyphae hyaline, more or less tubular and thin-walled, oleiferous hyphae very rare. Pileus trama near the cuticle with dextrinoid patches of pigment along the hyphae, the hyphal content in some dull orange. Cutis of appressed hyphae 5-10 μ wide, with thin-walled cells having a smoky ochraceous content as revived in KOH, walls smooth or in some irregularly roughened from a thin gelatinous sheath, end-cells merely tapered somewhat to apex; the cell contents in Melzer's merely yellowish orange or paler and granular, in a few hyphae rounding up into beadlike globules but these not highly colored, no characteristic formation of pigment globules noted. Subcuticular hyphae orange as revived in Melzer's. Clamp connections absent.

Habit, habitat, and distribution.—Scattered under birch and aspen, northwest corner of Montmorency County, July 17, 1967, Smith 74546.

Observations.—This species is characterized by the pale dingy colors, the almost glandular-dotted stipe of young basidiocarps, the narrow cuticular hyphae with only a suggestion of pigment-globule formation in Melzer's, and the orange subcuticular hyphae as revived in Melzer's. It is closest to *L. cinnamomeum,* but the context of the latter stains vinaceous before going to fuscous, and the cuticular hyphae, at least some of them, are very broad and with short cells—much as in *L. insigne.* In the dried condition the pilei of *L. insolens* var. *insolens* are a dingy grayish brown.

79a. Leccinum insolens var. brunneo-maculatum Smith, Thiers, & Watling

Lloydia 31:259. 1968

Pileus 7-9 cm broad, convex becoming broadly convex, the margin appendiculate with segments of the sterile margin which extends beyond the tubes, surface dry and minutely fibrillose-squamulose at first, subviscid and glabrescent in age; color whitish to "pale pinkish buff," the squamules dingy cinnamon-buff, staining dull brown where injured. Context white, slowly staining bluish fuscous (no preliminary reddish stage evident), odor and taste mild; with $FeSO_4$ bluish on the context.

Tubes white, depressed around the stipe at maturity becoming avellaneous to wood-brown; pores minute, white, staining dingy yellow-brown when bruised and the injured portions finally avellaneous or darker.

Stipe 8-13 cm long, 1-2.5 cm thick at apex, narrowly clavate, solid, white within, staining violaceous-fuscous, in some staining blue near base also; surface dull white with darker brown points and dots from near base to apex.

Spore deposit olive-brown moist, near bister as air-dried; spores 12-15 X 4-5 μ, smooth, wall about 0.2 μ thick, lacking an apical pore; shape in face view subovate, in profile elongate-inequilateral, ochraceous to pale ochraceous-brown in KOH, in Melzer's weakly dextrinoid to yellowish.

Basidia 22-30 X 8-10 μ, 4-spored, clavate, hyaline in KOH. Pleurocystidia scattered, 32-46 X 8-14 μ, narrowly to broadly fusoid-ventricose, neck elongated and apex acute to subacute, content smoky brown to hyaline in either KOH or Melzer's, wall thin and smooth. Cheilocystidia more or less similar to pleurocystidia but smaller and a larger number with colored content. Caulocystidia mostly clavate-mucronate to fusoid-ventricose, 36-50 X 9-15 μ, few large clavate to saccate-pedicellate cells also present, content brownish ochraceous in KOH; subiculum hyphae lacking refractive lens-shaped bodies on the interior walls.

Tube trama typical of the section (weakly bilateral). Some laticiferous elements present. Pileus cutis of appressed hyphae many 10-18 μ wide and with some short cells in many of the hyphae, these cells rarely almost sphaerocyst-like, their walls ornamented by small plates of material as in *L. insigne*, the end-cells often short and bullet-shaped, content pale dingy brownish in KOH, in Melzer's becoming granular or rarely a few globules present but formation of globules not constant or characteristic, after standing for 15 minutes some hyphae filled with minute beadlike droplets. Hyphae of subcutis merely yellowish in Melzer's but some with irregular gelatinous material coating the surface. Hyphae of the context next to the subcutis in part with orange to red content as revived in Melzer's. Clamp connections absent.

Habit, habitat, and distribution.—Scattered under birch and aspen, northwest Montmorency County, July 24, 1967, Smith 74630.

Observations.—The distinctive features of this variety are the whitish pileus which when bruised become brown-spotted, the gradual changes of the color finally to sordid alutaceous, the lack of a preliminary red color-change before the context stains fuscous when it is cut, the wide, ornamented, cuticular hyphae very similar to those of *L. insigne,* and a fair number of fusoid-ventricose caulocystidia among the clavate to mucronate individuals. This variety fruits when *L. insigne* is abundant, and the two maintain their identity even when growing in the same local area. As successive fruitings were observed apparently from the same mycelia the lack of red pigment cannot be attributed to a chance failure of the pigment to develop.

Material examined.—Montmorency: Smith 74363, 74630, 74661, 74759, 75247.

80. Leccinum sublutescens Smith, Thiers, & Watling

Mich. Bot. 5:139. 1966

Pileus about 7 cm broad, convex, margin appendiculate, surface dry and conspicuously areolate from the separation of the cutis, orange-cinnamon and drying orange-brown, the pallid context showing in the cracks. Context when cut slowly yellowish in both pileus and in the stipe-apex, in the apex of the stipe soon pale vinaceous-buff, dull pinkish tan downward, the lower two-thirds white and remaining so even when dried.

Tubes about 1 cm long, depressed around the stipe, white becoming dull cinnamon in drying; pores small, round, pallid at first, staining ochraceous when lightly bruised and this progressing to olivaceous.

Stipe about 8 cm long, about 1 cm thick, nearly equal, solid, for color changes see under context above, surface whitish beneath a fine sparse blackish ornamentation which fades upward to the pallid only faintly ornamented apical region.

Spores 12-15 X 3-4.5 μ, narrowly subfusoid in face view, narrowly inequilateral in profile, smooth, pale ochraceous in KOH, pale tan in Melzer's.

Basidia 4-spored, clavate, hyaline in KOH, 18-24 × 1-10 μ, a large hyaline globule frequent in the interior as revived in KOH. Pleurocystidia 39-45 × 8-12 μ, ventricose below, apex flexuous and subacute, 2-4 μ wide, hyaline. Cheilocystidia clavate to subfusoid, 22-32 × 7-10 μ, content hyaline to smoky brown in KOH. Caulocystidia 36-60 × 10-17 μ, clavate mucronate or fusoid-ventricose, content finally smoky brown but caulobasidioles and supporting hyphae typically lemon-yellow in KOH.

Pileus cutis of appressed hyphae 5-12 μ wide, the cells disarticulating to some extent, walls smooth to rarely obscurely asperulate, content of cells orange-brown to ochraceous-brown in KOH and dark orange-brown in Melzer's, in the latter medium large orange-brown to dull brown pigment globules forming in many cells, the cells mostly more than 5 times as long as wide; hyphae of subcutis interwoven, subgelatinous, lemon-yellow in KOH, nearly the same color in Melzer's. Context hyphae with hyaline to pale yellow content when revived in Melzer's. Clamp connections absent.

Habit, habitat, and distribution. – Solitary in mixed hardwood-conifer area, Mackinaw City hardwoods west of town, October 2, 1965, Smith 72765.

Observations. – The distinctive features appear to be the areolate pileus, general lack of gray stains when fresh basidiocarps are sectioned (though the context is very pale gray as dried), the yellow staining of the cut context of the pileus and stipe apex, and the bright yellow of the hyphae of the subcutis. The yellow stains in KOH (yellow pigment in the hyphae of the subcutis as revived) clearly distinguish it from *L. vulpinum*. The hyphae supporting the stipe ornamentation and the elements themselves when young are lemon-yellow in KOH also. The narrow spores and KOH reactions relate the species to *L. potteri*, but no masses of pigment were found in any of the hyphae of the context, and no veil remnants were present though the material was young when collected.

81. Leccinum insigne Smith, Thiers, & Watling

Mich. Bot. 5:160. 1966

Illus. Pls. 76-77.

This is a common and variable species in our state, and is one of the first species of the genus to be collected, for it fruits from late spring into the summer. We have divided it into a number of varieties and forms identified in the following key. The term "form" is used in the sense of variant, not in its formal sense.

KEY

1. Pileus yellow or orange-yellow, at times with a sparse overlay of slightly grayish fibrils. . . . . f. *ochraceum*
1. Pileus some other color 2

  2. Pileus dull orange-brown; stipe ornamentation dingy yellow-brown; spores 4.5-5.5 (6) μ . . . . f. *obscurum*
  2. Pileus more orange shading to red; stipe ornamentation soon dark brown 3

3. Pileus typically conspicuously fibrillose-squamulose when mature or older, margin rusty orange, disc rusty red . . . . f. *squamosum*
3. Pileus more or less fibrillose to squamulose when young; usually glabrous in age 4

  4. Tubes becoming yellowish before the darkening of maturity
  . . . . var. *luteopallidum*
  4. Tubes lacking a yellow stage 5

5. Stipe ornamentation remaining reddish brown for some time before becoming blackish; pileus ferruginous brown . . . . f. *lateritium*
5. Pileus distinctly orange-red or brighter 6

  6. Pores staining dingy brown when lightly bruised . . . . f. *subferrugineum*
  6. Pores yellowish then olive when lightly bruised . . . . f. *insigne*

81a. Leccinum insigne var. luteopallidum, var. nov.

A typo differt: Tubuli luteolae. Typus: Smith 71745 (MICH).

Pileus about 8 cm broad, convex, margin appendiculate, surface dry, near the margin with appressed fibrils, elsewhere merely uneven; color dull ferruginous ("cinnamon-rufous") to orange-tan, drying to a dark wood-brown (gray-brown). Context pallid, when cut soon flushed with fuscous streaks in the stipe apex and becoming black in drying. $FeSO_4$ on fresh context bluish, with KOH brownish; odor mild; taste acid then mild.

Tubes about 1 cm long, depressed around the stipe, pale yellow young ("ivory-yellow" or slightly brighter), with KOH dull cinnamon, dull brown where cut; mouths minute, yellowish, slowly staining snuff-brown when lightly bruised, when severely bruised violaceous-fuscous.

Stipe about 13 cm long, 2 cm thick, equal or at first evenly enlarged downward, pallid within, when cut showing greenish blue streaks, slowly staining fuscous generally, with NH_4OH or KOH yellow in the base; surface pallid, with this mostly covered by an ornamentation of blackish points and lines.

Spores 12-15 X 4-5 μ, smooth, fusoid in face view, in profile elongate, somewhat inequilateral, suprahilar depression broad and shallow, ochraceous in KOH, in Melzer's pale tawny-yellow to nearly hyaline.

Basidia 23-27 X 8-9 μ, clavate, 4-spored, hyaline in KOH. Pleurocystidia abundant, 33-50 X 8-14 μ, fusoid-ventricose with acute apex, smooth, thin-walled, content hyaline to smoky brown in KOH. Cheilocystidia 24-33 X 5-9 μ, clavate to fusoid-ventricose or obventricose-mucronate, yellow-brown in KOH. Caulocystidia clavate to mucronate, both large and small types present, content smoky brown.

Pileus cutis of hyphae 4-15 (20) μ broad, the wider ones often with asperulate walls and bullet-shaped apical cells, cells disarticulating, content dingy brownish in KOH, many cells .60-150 μ long. Context hyphae when revived in Melzer's lacking yellow or orange-red content. Clamp connections none.

Habit, habitat, and distribution.—In a hardwood slashing with *Betula lutea* present, Emerson, Chippewa County, July 14, 1965, Smith 71745.

Observations.—It is important to place this variety on record as it introduces pale yellow tubes to the *L. insigne* complex.

81b. Leccinum insigne var. insigne f. insigne

Pileus 4-15 cm broad, convex, becoming broadly convex, margin exceeding the tubes by over 1 mm when young and this tissue soon breaking into segments; surface dry when young and then minutely granular to obscurely appressed-fibrillose or distinctly fibrillose-squamulose, especially near the margin, glabrescent and in age soft and subviscid; color bright ferruginous ("ferruginous") when in full color but buttons sometimes pallid where covered by leaves, at maturity or when old ferruginous-cinnamon to dingy orange-brown with olive tones near the margin if wet. Context white, thick, firm becoming soft, when cut staining vinaceous-gray going to fuscous, or staining gray slowly—but lacking a distinctly reddish stage, with $FeSO_4$ bluish; taste slight, odor mild to slightly fragrant.

Tubes 1.5-2 cm long, depressed, pallid becoming olivaceous-pallid, turning wood-brown when cut, in age more or less wood-brown with a yellowish tan tint over the pore layer; pores whitish staining slowly to dingy yellow to yellow-brown, olive-brown, or olive finally, but when severely injured staining avellaneous to fuscous readily, in age where not injured with a dingy yellowish cinnamon tone.

Stipe 8-12 cm long, 1-2 cm thick near apex, evenly enlarged to the base where it is 2.5-3 cm thick, often tapered to a point below, solid, pallid when cut but soon darkening to fuscous on cut surface, KOH usually yellow on the cut base; surface whitish and ornamented with brown punctate dots or squamules which darken to blackish by maturity or age, the surface underlying the ornamentation cottony and uneven to obscurely reticulate, base dry.

Spore deposit dark yellow-brown ("snuff-brown" to "amber-brown"); spores 13-16 X 4-4.5 μ, smooth, pale ochraceous in KOH, lemon-yellow in Melzer's but a few becoming tawny, duller in both KOH and Melzer's when revived, in profile somewhat inequilateral, in face view subfusoid, wall slightly thickened.

Basidia 4-spored, clavate, hyaline in KOH. Pleurocystidia 42-56 X 10-15 μ, fusoid-ventricose with subacute apex, thin-walled, hyaline in KOH fresh, with dark brown content when revived in KOH. Cheilocystidia similar to pleurocystidia but neck often shorter and varying to clavate not uncommonly, those in bruised areas with smoky ochraceous content. Caulocystidia clavate to ventricose-mucronate, often resembling pleurocystidia in size but giant cells 70 X 20 μ also present and with smoky brown content in KOH.

Pileus cutis of hyphae mostly 4-8 μ wide and end-cells tubular but some hyphae 12-20 μ wide, these most numerous on young pilei, with either smooth or minutely roughened, hyphal cells of the large hyphae short (often only 20-80 μ long), and the end-cells frequently bullet-shaped, hyphae of the cutis showing a tendency for the cells to disarticulate, content of cells ochraceous to brownish in KOH and in Melzer's more ochraceous-orange cell content soon forming minute granules but no large pigment globules in Melzer's. Hyphae of the subcutis in Melzer's yellow to orange from colored homogeneous content. Hyphae of context hyaline in Melzer's. Clamp connections absent.

Habit, habitat, and distribution.—Common under large-toothed aspen (*Populus grandidentata*) on low ground in June and early July or throughout the summer if the weather is cool and wet. It may not be limited to an association with large-toothed aspen. This is the common early *Leccinum* with a red pileus and associated with aspen.

Observations.—This species, in the collective sense, has been mistaken for *L. aurantiacum* in North America in earlier times. We cannot say as yet whether it occurs in northern Europe, but it is to be expected there. It is a good edible species, though it is often difficult to find specimens not infested with insect larvae.

81c. Leccinum insigne f. ochraceum

Pileus 5-10 cm broad, convex becoming broadly convex, margin appendiculate at first, surface dry and nearly glabrous to thinly fibrillose with sparse grayish fibrils, yellow to orange-yellow overall, dry at first but subviscid in age. Context firm at first, lax in age, whitish, staining avellaneous and then fuscous when cut. Odor and taste mild.

Tubes about 1.5 cm deep, soon depressed around the stipe, whitish when young, when cut the exposed surface weakly yellow (in the white state), near bister-brown when mature; pores pallid at first, when lightly bruised staining yellowish.

Stipe 9-14 cm long, 1-2 cm thick, equal or nearly so, solid, whitish but soon fuscous on cut surface; stipe ornamentation fine, pallid when young but soon blackish brown and drying blackish.

Spores 11-14 X 4-4.5 μ, smooth, inequilateral in profile, subfusiform in face view. Pileus cutis of appressed hyphae 6-10 (15) μ wide, cells frequently disarticulating, walls smooth or ornamented as in var. *insigne* f. *insigne*, end-cells bullet-shaped to cystidioid, short cells not infrequent in the wider hyphae, content ochraceous in KOH, in Melzer's granular or as beads up to 1.5 μ in diameter. Hyphae of subcutis orange-red in Melzer's.

Habit, habitat, and distribution. —Scattered under aspen-birch, northwest Montmorency County, September 7, 1967, Smith 75113.

Observations. —This variant is readily confused with *L. ochraceum* but it does not stain reddish on the cut flesh before changing to fuscous. The content of the cuticular hyphae was not olivaceous as revived in KOH and no violaceous-brown laticiferous elements were observed in any part of the basidiocarp when portions were mounted in Melzer's. It appears to us to be similar to f. *insigne* in all essentials except the lack of red pigment in the cuticular hyphae.

81d. Leccinum insigne f. obscurum

Pileus 6-12 cm broad, convex expanding to plane, the sterile margin breaking into segments, surface spotted or streaked with obscure patches of fibrils, glabrous in age, subviscid at maturity, a dull orange-brown. Context soft, pallid, when cut staining lilac to grayish and then fuscous (lacking a preliminary red stage), odor and taste not distinctive, the $FeSO_4$ reaction grayish.

Tubes 1-2 cm deep, depressed around the stipe, white then olivaceous-gray, when cut slowly staining fuscous; pores minute, whitish, pale yellowish tan when mature, staining slowly to avellaneous when bruised severely, where bruised lightly staining olive.

Stipe 8-16 cm long, 1-2 cm thick at apex, evenly enlarged downward, slowly staining lilac-fuscous; surface with snuff-brown ornamentation finally becoming black, white ground color readily visible, base at times with greenish blue stains.

Spores 13-16 X 4.5-5.5 (6) μ, smooth, walls -0.5 μ thick, apex with a minute thin spot; color ochraceous in KOH to brownish ochraceous in Melzer's, many slightly more brownish; shape in face view fusoid, in profile inequilateral.

Basidia 20-30 X 8-11 μ, 4-spored, clavate, hyaline in KOH. Pleurocystidia scattered, 32-48 X 9-15 μ, fusoid-ventricose with neck elongated and apex subacute, content hyaline to dingy brown in KOH, in Melzer's darker reddish brown (but not dextrinoid). Cheilocystidia smaller than pleurocystidia and neck typically shorter, more often with colored content than not. Caulocystidia mostly saccate to clavate and with a long pedicel, 30-65 X 9-25 μ, content bister in KOH or Melzer's.

Tube trama typical of the section; some laticiferous elements present. Pileus cutis of appressed hyphae 5-9 (10-15) μ wide, usually tubular and often with minutely ornamented walls in f. *insigne,* cells short or long but mostly over 5 times as long as wide, cells disarticulating, endcells bullet-shaped to cystidioid, content dull orange-ochraceous to brownish in KOH, not distinctively colored in Melzer's but aggregating into beads 0.5-3 μ in diameter and very rarely globules almost the diameter of the narrower cells. Hyphae of the subcutis to some extent at least with orange-yellow content in Melzer's.

Habit, habitat, and distribution.—Scattered under mixed hardwoods including *Populus, Fagus, Betula, Ostrya,* and *Corylus.* Fruiting in June, Emmet County.

Observations.—This is clearly a variant of *L. insigne* in which the pileus is very dull in color, in which the stipe ornamentation is dull yellow-brown at first, which not infrequently has spores up to 6 μ wide and pigment beads present in some of the cuticular hyphae when mounted in Melzer's.

81e. Leccinum insigne f. lateritium

Pileus 4-12 cm broad, obtuse to convex, margin with a sterile band breaking into segments as pileus expand, surface dry and fibrillose, "auburn" to "Kaiser-brown" over disc, paler and more ochraceous near the margin, in age at times rather conspicuously appressed fibrillose-squamulose, when old and wet the colors a dingy brownish orange. Context thick, firm, pallid, when cut soon avellaneous changing to fuscous (lacking red tones in the change), odor and taste mild, $FeSO_4$ bluish olive.

Tubes white becoming olivaceous-pallid and finally wood-brown, depressed; pores minute, pallid, when lightly bruised staining olive, staining avellaneous if severely bruised; pore surface where undamaged often with an olivaceous sheen, at maturity duller.

Stipe 9-14 cm long, 1.5-2.5 cm thick, nearly equal, solid, slowly fuscous in apex when cut (no other color change observed); surface with coarse ornamentation which is pallid at first, then slowly becomes reddish brown and by late maturity charcoal-black.

Spore deposit olive-brown moist, dingy snuff-brown as air-dried (a dark yellow-brown); spores 12-15 X 3.8-5.2 μ, smooth, weakly yellow-brown in KOH and about the same in Melzer's, in face view obscurely to distinctly fusoid, in profile somewhat inequilateral; wall slightly thickened (-0.5 μ thick), no apical pore present.

Basidia 4-spored, 30-38 X 9-12 μ, clavate, hyaline in KOH and nearly so in Melzer's. Pleurocystidia 35-50 X 9-15 μ, scattered, fusoid-ventricose with neck tapering to an acute apex, walls thin and smooth, content of cells hyaline to weakly brownish (rarely) in KOH or Melzer's. Cheilocystidia similar to pleurocystidia or smaller and some more or less basidiole-like. Caulocystidia clavate to mucronate 10-30 μ wide, length variable, very few fusoid-ventricose cells present; walls thin and smooth, content bister to snuff-brown as revived in KOH.

Tube trama subgelatinous, hyphae more or less tubular, smooth, hyaline, thin-walled, diverging to subhymenium which is interwoven; some oleiferous hyphae with dingy ochraceous content present (as revived in KOH), in Melzer's the content nearly hyaline. Pileus trama of hyaline thin-walled hyphae with more or less inflated cells, laticiferous elements as in the tube trama, all hyphae hyaline to merely weakly yellowish in Melzer's. Pileus cutis of a sparse surface layer of hyphae with cells 10-25 (30) μ broad, and varying from nearly globose to oval to elongate, smooth or with a gelatinous layer causing irregularities in the outline; terminal cells long- or short-bullet-shaped, pigment finely granular in Melzer's. Hyphae of the subcutis hyaline in Melzer's or merely yellowish. Clamp connections absent.

Habit, habitat, and distribution.—Gregarious under aspen-birch and balsam, northeast corner of Otsego County, July 17, 1967, Smith 74568.

Observations.– The pore surface has an olive tinge over undamaged portions even in age, whereas f. *insigne* at a comparable stage has pores dingy yellowish cinnamon. In the epicutis of f. *lateritium* there is a stronger tendency to produce short oval to subglobose cells in the wide hyphae than was noted for f. *insigne*. In both after drying, the pileus is grayish brown to dingy dull cinnamon. When fresh the pileus in f. *lateritium* is distinctly fibrillose and the color more rusty orange than in f. *insigne*.

81f. Leccinum insigne f. squamosum

Pileus 5-10 cm broad, convex expanding to nearly plane, margin sterile and incurved, rusty orange on margin rusty red on disc and becoming darker red in age, buttons rusty orange overall, surface dry at first, subviscid in age, in age often conspicuously fibrillose-squamulose from distinct appressed-fibrillose scales. Context white, slowly changing directly to violaceous-fuscous when cut, $FeSO_4$ on cut surface bluish, odor and taste not distinctive.

Tubes white then olivaceous-pallid, finally dark dingy yellowish brown, depressed around the stipe; pores minute, when lightly bruised staining weakly yellow-brown, to dingy brown (staining avellaneous to fuscous when severely bruised).

Stipe 7-12 cm long, 11-18 (20) mm thick, equal, solid, pallid, staining fuscous when cut; surface with coarse ornamentation which is brown but first merely darkens, though by late maturity it is blackish.

Spore deposit near bister as air-dried; spores 13-15 × 3.5-4.5 μ. Caulocystidia mostly clavate with bister content as revived in KOH, of various sizes. Pileus cutis as in f. *insigne* only with more hyphae 12-26 μ wide, content of wide hyphae homogeneous to granular in Melzer's both in fresh and dried condition. Clamp connections absent.

*Habit, habitat, and distribution.–*Scattered to gregarious in aspen-birch conifer stands, Montmorency County, July 18, 1967, Smith 74571.

*Observations.–*Because of the dark red pileus of mature basidiocarps we at first thought of *L. aurantiacum* but the cut flesh does not change to cinnamon reddish before going to fuscous, the cuticular hyphae of the pileus are an excellent example of the *L. insigne* type, and the stipe ornamentation tends to be blackish sooner, but this is difficult to interpret especially on collections of fewer than a dozen basidiocarps in various developmental stages. The color of the pileus distinguishes it from *L. atrostipitatum* if basidiocarps are encountered which have become arrested in their development but in which the stipe ornamentation has blackened.

81g. Leccinum insigne f. subferrugineum

Pileus 4-10 cm broad, convex becoming broadly convex to plane, margin crenate; surface with an overlay of brownish fibrils which dull somewhat the color of the pileus, color dark to pale orange, surface dry at first and fibrillose, becoming glabrous and subviscid. Context white, soft, when cut slowly staining lilac-gray and finally fuscous (no preliminary reddish stage).

Tubes depressed around the stipe, white to olive-pallid, becoming weakly yellowish before becoming wood-brown, when cut staining brown and then fuscous; pores pallid then ivory-buff and finally pale wood-brown, staining dingy brown when fresh.

Stipe 4-9 cm long, 8-15 mm thick, equal, solid, white within, slowly staining lilac-gray then fuscous; surface pallid beneath cinnamon colored then dark brown and finally blackish moderately coarse ornamentation.

Spore deposit near bister as air-dried but more olivaceous fresh; spores 14-16 X 4.5-5.5 μ, smooth, wall only slightly thickened (about 0.5 μ); in face view somewhat fusiform, in profile somewhat inequilateral, color in KOH dingy ochraceous to pale yellow-brown, in Melzer's merely brownish.

Basidia 4-spored, 18-27 (32) X 8-11 μ, clavate, hyaline in KOH. Pleurocystidia abundant, ventricose with a crooked neck and subacute apex, thin-walled, content usually near snuff-brown (in KOH) more rarely hyaline. Cheilocystidia mostly 26-33 X 7-12 μ, fusoid-ventricose, neck not greatly elongated and hence not crooked, apex subacute, content often yellowish. Caulocystidia variable in size, 30-70 (80-120) X 9-12 (15-30) μ, clavate, utriform, fusoid-ventricose, and elongate-subfusoid all intermingled but mostly clavate to mucronate, thin-walled, smooth, many with exceptionally long pedicels.

Tube trama typical of the genus, some laticiferous elements present. Pileus cutis of hyphae 4-12 μ wide, tubular and the end-cell slightly tapered to an obtuse apex or merely blunt, cells often disarticulating, cell content nearly hyaline to yellowish in KOH but colloidal-granular in appearance, in Melzer's pallid to dingy pale brownish (the colloidal content not brightly colored as is usually the case), pigment globules not forming in Melzer's; the walls thin and often ornamented as in f. *insigne.* Hyphae of subcutis not highly colored in Melzer's. Clamp connections none.

Habit, habitat, and distribution.—Gregarious at edge of a swamp, lower Maple River, Emmet County, July 7, 1967, Smith 74468.

Observations.—This variant was found at the edge of a cedar swamp with ash, birch, spruce, balsam, and aspen nearby. Its distinguishing features are the almost colorless hyphae (but having a colloidal content) which form the epicutis of the pileus, the lack of any appreciable morphological differentiation of the end-cells, the dingy brown stains which develop when the tubes are bruised, the dingy orange pileus, and the lack of a reddish stain in the cut context before it turns to fuscous. The coarse, brown stipe ornamentation of young specimens is also distinctive.

Section SCABRA Smith, Thiers, & Watling

Mich. Bot. 6:124. 1967

This section features species with tubular hyphae in the cuticle of the pileus. The end-cells of the cuticular hyphae, however, may be quite enlarged and cystidioid. In occasional species some short cells may be found in cuticular hyphae, and these may be inflated somewhat, but they are too rare to be regarded as a feature of taxonomic importance. Their presence, however, does emphasize that intergradation occurs between this section and section *Luteoscabra.*

Type species: *Leccinum scabrum.*

KEY TO SUBSECTIONS

1. Pileus essentially white to pallid when young and pale colored when mature. . . . . *Subsect. Pallida*
1. Pileus more highly colored than in above choice 2

    2. Cut flesh in apex of stipe eventually staining gray to fuscous . . . . . .
    . . . . *Subsect. Fumosa*
    2. Cut flesh in stipe apex staining yellow to reddish but the change not progressing to gray or fuscous . . . . *Subsect. Scabra*

Subsection PALLIDA Smith, Thiers, & Watling

Mich. Bot. 6:124. 1967

The diagnostic characters are given in the key. The subsection merges imperceptibly with subsection *Scabra* in the presence in the latter of pallid taxa which darken to gray by maturity or in drying.

Type species: *Leccinum holopus.*

KEY TO SPECIES

1. Pseudocystidia present in the hymenium . . . . *L. olivaceo-pallidum*
1. Pseudocystidia absent from the hymenium 2

    2. Tubes staining rusty cinnamon when injured; odor strong of
       radish . . . . *L. parvulum*
    2. Not as above 3

3. Cut context in stipe apex unchanging or only slowly becoming dingy tan . . . 4
3. Cut context in stipe apex staining reddish, lilaceous (slightly) to gray 8

    4. Pileus slimy-viscid; hyphae of context next to subcutis not brightly
       colored as revived in Melzer's . . . . *L. glutinopallens*
    4. Not with both of the above features 5

5. Caulocystidia lemon-yellow in KOH; habitat under dwarf birch; northern
   in distribution . . . . *L. rotundifoliae*
5. Not as above 6

    6. Many caulocystidia proliferated; spores 18-26 $\times$ 6-7.5 μ . . . *L. proliferum*
    6. Not as above 7

7. Spores 4-5 μ wide . . . . *L. angustisporum*
7. Spores 5-6.5 μ wide . . . . *L. holopus*

    8. Cut context when stipe apex is sectioned slowly grayish or lilaceous
       (see subsect. *Fumosa* also) 9
    8. Cut context in stipe apex changing to reddish first 10

9. Spores 18-26 $\times$ 6-7.5 μ; caulocystidia often proliferated . . . . . . *L. proliferum*
9. Spores 16-18 $\times$ 5.5-6.5 μ; caulocystidia not proliferated . . . .
   . . . . see unchanging variants of *L. chalybaeum*

    10. Spores 12-19.5 (23-40) $\times$ 4.5-6 (8) μ . . . . see *L. variabile*
    10. Spores not as variable as in above choice 11

11. Cuticular hyphae 4-8 μ in diameter . . . . *L. holopus* var. *americanum*
11. Cuticular hyphae 7-12 μ in diameter . . . . see *L. variabile*

82. **Leccinum olivaceo-pallidum** Smith, Thiers, & Watling

Mich. Bot. 6:125. 1967

Illus. Pl. 78.

    Pileus 6-9 cm broad, convex, the margin at first with narrow sterile flaps of tissue which are inconspicuous and soon become obliterated, expanding to broadly convex, dry and unpolished, whitish to "pale olive-buff" when young, becoming grayer in age and slowly dark brown in spots where injured. Context pallid, soon vinaceous-gray when cut, pale blue with $FeSO_4$ and giving no reaction with KOH; odor and taste none.

Tubes about 10 mm deep, adnate becoming depressed around the stipe, pallid to olive-buff, staining dingy vinaceous-brown when injured; pores small, 2-3 per mm, olive-buff staining vinaceous-brown where injured.

Stipe 11-14 cm long, 10-13 mm thick, equal, coarsely scabrous but tips of scales mostly merely discolored dingy brownish (when dried blackish brown), base dingy olive to pallid, remainder dingy pallid beneath the ornamentation, solid, pallid within and staining less than in pileus context, olive in the base and with greenish areas in the cortex but mostly remaining pallid.

Spore deposit near sepia (as taken naturally on a pileus); spores 11-15 X 4-5 μ, smooth, narrowly fusiform in face view, somewhat inequilateral in profile, the suprahilar depression shallow, dingy ochraceous to ochraceous-tan in KOH, dingy tan to ochraceous in Melzer's only a few becoming dark reddish brown.

Basidia 4-spored, 20-25 X 7-9 μ, clavate, yellowish to hyaline in KOH and Melzer's. Pleurocystidia numerous as filamentous pseudocystidia projecting into the hymenium and with bister (dark brown) amorphous content as seen in both KOH and in Melzer's. Pileus trichodermium matted down to form an epicutis the hyphae of which are 4-7 (10) μ wide, the end-cells mostly tubular and obtuse at the tip. Clamp connections absent.

Habit, habitat, and distribution.—On sandy soil, clustered in an oak-aspen grove, Waterloo Recreation Area, Jackson County, September 10, 1961. Known only from the type locality.

Observations.—The tube mouths did not stain yellow as happens in so many of the species of *Leccinum,* and the spores are narrower than in *L. subleucophaeum.* The distinguishing combination of features is the whitish to olivaceous pileus, general tendency to stain vinaceous-brown, the numerous pseudocystidia in the hymenium best demonstrated with KOH or Melzer's, and the lack of inflated elements in the pileus cutis. The slightly appendiculate margin of the pileus indicates a possible connection to section *Leccinum,* but in view of its being so poorly developed, the species is placed here.

83. **Leccinum angustisporum** Smith, Thiers, & Watling

Mich. Bot. 6:126. 1967

Pileus 3-5 cm broad, convex, dry and plush-like to merely unpolished, vinaceous-buff over the disc, whitish over margin, when young dull whitish overall, margin irregular but not with a distinct sterile zone.

Context white, scarcely changing when sectioned, slowly bluish with FeSO$_4$, odor none, taste mild.

Tubes pallid, depressed around the stipe, staining avellaneous where bruised; pores small, whitish, staining yellowish to brownish and when severely bruised staining wood-brown.

Stipe about 7 cm long and 1 cm thick near apex, whitish and with very fine dark brown punctation, lower part faintly yellowish in places but no blue to blue-green stains present, solid, when sectioned the cortex pinkish tan, fuscous-brown around wormholes (probably changing slowly when cut), pallid in the base.

Spores 14-20 X 4-4.5 (5) μ, smooth, elongate-subfusiform in face view, in profile elongate-subfusoid, the suprahilar depression very shallow and broad (almost absent in some), pale clay color and tawny singly and in masses respectively as revived in KOH, scarcely changing color in Melzer's, wall very slightly thickened and no apical differentiation visible.

Basidia 4-spored, 18-26 (35) X 10-14 μ, clavate, hyaline to ochraceous in KOH, with some refractive granular material, yellowish in Melzer's. Pleurocystidia scattered, 38-56 X 9-14 μ, fusoid-ventricose, apex subacute, content hyaline to smoky ochraceous from dissolved pigment. Cheilocystidia mostly resembling basidioles but many more with ochraceous content. Caulocystidia fusoid-ventricose and of two subtypes, 40-70 X 15-25 μ and with relatively short neck, and 45-70 X 8-14 μ with a long tapered neck, both types with smoky ochraceous content and smooth walls.

Pileus trichodermium of tangled smooth, thin-walled filaments 5-12 μ wide with end-cells somewhat cystidioid and hyaline to ochraceous, content in KOH or Melzer's, homogeneous. Subcutis of interwoven hyaline subgelatinous hyphae 4-7 μ wide. Context hyphae 8-15 μ wide and strongly orange to orange-red in Melzer's toward the subcutis and paler away from it. Clamp connections absent.

Habit, habitat, and distribution. — Under birch, St. Ignace, Mackinac County, July, apparently rare.

Observations. — This species is distinct by reason of the long narrow spores and the slowly changing context. It is most closely related to *L. flavostipitatum,* but the spores are longer and the pileus is much paler in color.

84. Leccinum proliferum Smith, Thiers, & Watling

Mich. Bot. 6:127. 1967

Pileus 4-6 cm broad, convex becoming broadly convex; surface dry to subviscid, white, spotted with minute brownish appressed squamules, margin even and sterile for about 0.25 mm. Context white, when cut slowly grayish near stipe apex.

Tubes depressed around the stipe, pallid becoming wood-brown and when dried dark yellowish brown.

Stipe 4-6 cm long, 8-11 mm at apex, 12-18 mm at base, pallid, overlaid with dull brown ornamentation which blackens on drying, solid, white within, slowly grayish when cut but with no initial reddish state, lacking yellow stains as dried.

Spores 18-26 X 6-7.5 μ, smooth, fusoid in face view, inequilateral in profile view, pale cinnamon and dark cinnamon singly and in groups respectively in KOH, becoming reddish cinnamon in Melzer's, wall thickened slightly, apical spot distinct but not a true pore.

Basidia 4-spored, 18-28 X 10-14 μ, yellow in Melzer's, readily gelatinizing in KOH, lacking a large globule when revived in KOH. Pleurocystidia rare, mostly near tube mouths, 35-45 X 8-11 μ, hyaline in KOH. Cheilocystidia 18-50 X 4-11 μ (long and narrow), fusoid-ventricose with acute apex, yellow to hyaline in KOH. Caulocystidia dull ochraceous in KOH, clavate-mucronate to fusoid-ventricose but neck hyphal-like (some branched once or twice), ventricose part 25-40 X 8-15 μ, neck finally up to 200 μ long and 4-6 μ wide, some filamentous elements present in fascicles over lower half of stipe in addition.

Pileus trichodermium with the elements soon becoming appressed, tubular, 4-10 μ wide, the cells moderately long, walls smooth to inconspicuously roughened, with ochraceous content in KOH and hardly changing in Melzer's, end-cells tubular to slightly enlarged. Hyphae of subcutis 3-5 μ wide, gelatinous, orange to bright ochraceous in Melzer's. Hyphae of context not as bright as the subcuticular hyphae in Melzer's. Clamp connections absent.

Habit, habitat, and distribution.—In a swamp under large-toothed aspen, St. Ignace, Mackinac County, early July, not common.

Observations.—This species is similar in some respects to *L. snellii*, but the difference in shape of the cells in the elements of the pileus cutis distinguishes the two readily, to say nothing of the overall coloration.

85. Leccinum rotundifoliae (Singer) Smith, Thiers, & Watling

Mich. Bot. 6:128. 1967

Krombholzia rotundifoliae Singer, Schw. Zeitschr. Pilzk. 16:148. 1938.

Pileus 2-6 cm broad, obtuse to convex becoming broadly convex, margin not appendiculate, dry at first and unpolished, becoming somewhat viscid by maturity, white to dull white overall at first, slowly becoming pale buff to pinkish buff then cinnamon-buff to dingy clay color, rimulose in age at times or on drying. Context white, showing no appreciable color change when sectioned and pallid when properly dried, with $FeSO_4$ blue both in pileus and stipe; odor and taste not distinctive.

Tubes 10-15 mm deep, white to olive-buff, slowly becoming pale wood-brown by maturity, unchanging when cut or bruised; pores small, 2-3 per mm, whitish becoming avellaneous.

Stipe 4-10 cm long, 5-12 mm at apex, equal or nearly so, white and roughened from white scabrosity which slowly becomes brown like the pileus but was never seen to become blackish, solid, white within when cut, soon greenish blue in the base and brownish in the apex, as dried with a faint yellow tone over basal half in some.

Spores 16-20 (25) $\times$ 5.5-7.5 (8) μ, smooth, narrowly subfusoid to elongate-subovate in face view, very obscurely inequilateral in profile, the suprahilar depression often very indistinct, pale ochraceous to brownish ochraceous in KOH, many soon becoming dextrinoid when revived in Melzer's, wall thickened slightly, no distinct pore present at apex.

Basidia 4-spored, 23-28 $\times$ 9-13 μ, hyaline to yellow in KOH. Pleurocystidia 28-36 $\times$ 8-14 μ, more or less fusoid-ventricose, hyaline or nearly so. Cheilocystidia versiform but mostly fusoid-ventricose, 32-60 $\times$ 8-12 μ, the neck tapered and flexuous or irregular in outline, ending in an acute to subacute apex, with lemon-yellow content in KOH.

Pileus trichodermium consisting of tangled mostly thin-walled flexuous narrow hyphae with hyphal ends 3-5 μ wide, subgelatinous in KOH, with yellow-brown content, often flexuous, the tips of the cells obtuse to subcapitate and merely dull yellowish to orange-buff in Melzer's. Hyphae of the context 8-15 μ wide, with "colloidal" content in KOH causing them to be very boldly defined as seen under the microscope, hyaline in KOH but in Melzer's reddish to brownish red and the color located in the content which remains homogeneous. Clamp connections absent.

Habit, habitat, and distribution.—Scattered under dwarf birch in and along the edges of cold bogs. Smith has seen this species in its typical habitat along the south shore of Lake Superior, but on checking his material found that apparently no specimens were saved. Hence, although an official record for the state remains to be obtained, we include the species here.

Observations.—This is one of the truly distinctive species of *Leccinum,* as Groves (1955) has indicated. The narrow elements of the trichodermium contrast strongly both in KOH and in Melzer's with the context hyphae. The species is paler than *L. scabrum,* the caulocystidia are lemon-yellow in KOH, and the stipe ornamentation is consistently paler. Since the type has not been studied, our concept of the species is tentative.

86. Leccinum chalybaeum Singer

Mycologia 37:799. 1945

Pileus 4.2-8.4 cm broad, pulvinate becoming more or less applanate, more or less distinctly viscid in wet weather, slightly shining when dry, glabrous or subtomentose, pale buff when young ("pinkish buff"), or darker ("cinnamon"), becoming dingy yellow-brown, margin often tinged caesious or this showing in the surface cracks and frequently in part greenish blue. Context white becoming reddish violet-gray or purplish lilac ("dark vinaceous-gray") when bruised, eventually becoming blackish, at first firm, then soft and watery, fibrous and hard in lower part of stipe; with $FeSO_4$ green, with formalin red; odor none, taste mild.

Tubes whitish becoming weakly yellowish, deeply depressed around the stipe, convex beneath, up to 14 mm deep; pores concolorous, with dingy olive-gray stains when bruised, small to very small, deep reddish brown with alkali.

Stipe 4-6.5 cm long, 1.3-2.1 cm thick, solid, subequal, with a very slightly attenuate apex, or ventricose or tapered upward from the base, white becoming dingy, beset with white scabrosities at apex, squamules pale brown to darker brown below, dry.

Spore deposit olive-brown; spores 16.3-17.7 × 5.5-6.2 μ, melleous-brown, fusoid, with thin or slightly thickened walls.

Basidia 28-34 × 9-14 μ, 4-spored. Pleurocystidia 31-51 × 6-11 μ, hyaline, fusoid, thin-walled, rather numerous near pores. Hymenophoral trama bilateral. Pileus trichodermium consisting of intricately interwoven smooth, filamentous brownish to hyaline hyphae 4-10 μ wide, not arranged in erect chains. Caulocystidia 38-82 × 6.2-16.2 μ, fusoid to fusoid-ampullaceous, thin-walled or wall up to 0.7 μ thick; basidioles small, hyaline to faintly pigmented.

Habit, habitat, and distribution.—In gardens, open woods, and flat-woods with *Quercus* on sandy ground or on humus, solitary or in small groups, mostly in July, Florida.

Observations.—We have some collections which may possibly belong here, but we are withholding an account of them at this time. The data in our description are taken from Singer.

87. **Leccinum glutinopallens** Smith, Thiers, & Watling

Mich. Bot. 6:131. 1967

Pileus 6-7 cm broad, broadly convex, margin even, glabrous, slimy-viscid, pallid along the margin, dingy vinaceous-buff over the disc, margin even and fertile. Context very soft, white, when cut slowly staining pinkish tan, with a bluish green line developing above the tubes by maturity, pale bluish with $FeSO_4$ and orange-tan in Melzer's.

Tubes about 15 mm deep, ventricose and nearly free, wood-brown or darker at maturity; pores small and round, grayish pallid, when bruised staining avellaneous or finally dingy brown.

Stipe 6-8 cm long, 10-15 mm thick, equal to clavate at base, pallid becoming variously stained pinkish tan from handling, very minutely ornamented by fine points and streaks, the ornamentation pallid at first becoming pinkish tan or slightly darker, solid, pallid within and when sectioned soon stained vinaceous-tan and in places greenish blue.

Spores 15-19 X 5-6.5 μ, smooth, in face view broadly fusiform, in profile inequilateral with a distinct suprahilar depression, pale ochraceous-tawny individually in KOH, pale tan in Melzer's but some becoming reddish tawny, wall –0.5 μ thick, no apical pore observed.

Basidia 4-spored, 20-25 X 10-14 μ, clavate, hyaline in KOH. Pleurocystidia 38-56 X 11-16 μ, fusoid-ventricose, apices subacute, content hyaline to pale smoky ochraceous. Cheilocystidia basidiole-like to somewhat fusoid. Caulocystidia the usual mixture of large and small fusoid-ventricose cells but with smooth walls, content when revived in KOH only pale ochraceous, and with some debris adhering to them.

Pileus trichodermium a tangled layer of gelatinous smooth-walled hyphae 4-8 μ wide, the cells mostly 30-300 μ long, hyaline to yellowish in KOH, the cells disarticulating and the end-cells tubular to somewhat cystidioid, with hyaline to weakly ochraceous content in Melzer's or in some hyphae with darker granular material. Subcutis and context hyphae adjacent to it lacking distinctively colored contents when revived in Melzer's. Clamp connections absent.

Habit, habitat, and distribution.—Scattered along an old logging road in mixed woods of birch, aspen, balsam, and spruce, Sugar Island, Chippewa County, July.

Observations.—This is a slimy-viscid, pallid bolete with very fine inconspicuous stipe ornamentation. It is readily distinguished from *L. holopus* by the slimy pileus and pores which stain avellaneous.

88. Leccinum holopus (Rostkovius) Watling

Trans. Brit. Mycol. Soc. 43:692. 1960

var. holopus

Illus. Pl. 79.

Pileus 3-10 cm broad, pulvinate to convex, becoming broadly convex or nearly flat, margin often exceeding the tubes by 0.5 mm; surface soft and tacky to the touch (subviscid), in age or when wet slightly more viscid, typically glabrous and unpolished but some slightly areolate at times over the disc, rarely with inconspicuous streaks from appressed fibrils, dull white and remaining this color or if water-soaked olivaceous along a broad marginal area, disc often tinged buff to vinaceous-buff, rarely pea-green overall in age. Context thick, white, soft when sectioned not staining or finally brownish (especially around larval tunnels), odor and taste not distinctive.

Tubes 1-2.5 cm deep, adnate but soon deeply depressed around the stipe, pallid to pale olive-buff, in age wood-brown; pores whitish becoming brownish as spores mature, when bruised staining yellow and then brown.

Stipe 8-14 cm long, 1-2 cm thick, equal or nearly so, not staining pink in apex when flesh is cut; surface pallid, ornamented with squamules which become brownish (rarely blackish), apex merely scurfy and remaining pallid a long time, greenish to bluish in the base of the stipe when mature.

Spore deposit about cinnamon-brown; spores 14-20 × 5-6.5 μ, smooth, no apical pore evident, shape in face view subfusoid, in profile narrowly inequilateral, color pale dingy cinnamon to ochraceous-cinnamon revived in KOH, yellowish in Melzer's when immature and dark reddish brown when mature, wall slightly thickened.

Basidia 4-spored, 26-32 × 8-11 μ, clavate with hyaline to pale cinnamon content in KOH, yellowish in Melzer's. Pleurocystidia scattered to rare, fusoid-ventricose, 28-36 × 9-12 μ, almost imbedded in hymenium, content not distinctive either in KOH or Melzer's, no pseudocystidia observed. Caulocystidia mostly the fusoid-ventricose type with elongate somewhat flexuous necks and subacute apex; clavate cells also present.

Trichodermium of pileus appearing as an interwoven layer of sub-gelatinous hyphae 4-7 μ wide, thin-walled, content hyaline or in a few smoky yellowish as revived in KOH, content reddish in Melzer's in some hyphae, end-cells tubular to somewhat cystidioid with obtuse apex. Clamp connections absent.

Habit, habitat, and distribution.—Scattered to solitary in cold bogs, cedar swamps, and the like, common from the Bay City-Clare line north, late summer and fall.

Observations.—There are two variants as described in this work. From present information both are common, but the question of one being more frequent than the other during some seasons remains to be studied. No "material examined" is cited because of the danger of mixed collections. This is one of the areas in the genus still deserving critical study.

88a. Leccinum holopus var. americanum Smith & Thiers, var. nov.

Illus. Pl. 80.

A typo differt: Contextus tactu rubellus; stipes minute punctatae, punctae pallidae, tarde demum sordide brunneae. Specimen typicum in Herb. Univ. Mich. conservatum est; prope Lewiston, Montmorency County, September 7, 1967, legit Smith 75109.

Pileus 5-12 cm broad, convex to broadly convex, the margin sterile and exceeding the tubes by about 0.5-1 mm but not becoming crenate; surface glabrous or only in age minutely areolate, dry at first, but in age subviscid; color white when young, soon dull white to near vinaceous-buff (over disc first), watery olivaceous over the marginal area when water-soaked, when dried pale avellaneous (grayish) over disc and whitish to pallid over marginal area. Context firm and white at first, the cut surface staining reddish slowly and often developing grayish streaks; odor and taste not distinctive.

Tubes 1-2 cm deep, deeply depressed around the stipe, becoming wood-brown as spores mature; pores small, white at first, when lightly bruised staining yellow.

Stipe 6-12 cm long, 10-15 mm thick, equal or nearly so, solid, slowly staining reddish and finally pale grayish; surface pallid beneath a fine blackish brown ornamentation of points and squamules, not yellowish in any part as dried.

Spores 14-19 X 5-6 μ, smooth, with a very minute pore, wall -0.3 μ thick (as revived in Melzer's); shape in profile elongate-inequilateral, in face view subfusiform; color in KOH ochraceous-brown, in Melzer's reddish tawny.

Basidia 24-30 X 8-12 μ, 4-spored, clavate, hyaline in KOH. Pleurocystidia scattered, 32-46 X 7-12 μ, fusoid-ventricose with acute apex, neck often flexuous, walls smooth and hyaline, content hyaline. Caulocystidia. mostly fusoid-ventricose, (26) 35-50 (70) X 7-15 (20) μ, neck when present often flexuous, some cells clavate to clavate-mucronate; content dark yellow-brown in KOH.

Pileus cutis of appressed hyphae 4-8 μ wide, tubular, cells elongate, end-cell somewhat cystidioid to tubular and often with yellow-brown content, content of isolated cells in a hyphae yellow-brown with the remaining cells hyaline, some hyphae with a slime sheath and the wall irregular from it, in Melzer's the content of the colored cells remaining homogeneous. Hyphae beneath the cutis orange-red in Melzer's and then content homogeneous. Clamp connections none.

Habit, habitat, and distribution.—Scattered on moss or gregarious, often in and along the edges of bogs, northwest corner of Montmorency County, August and September. It is common in the state, but most collections in herbaria probably represent a mixture of both varieties.

Observations.—Because of the emphasis placed on color changes when the context of the stipe apex is cut, both traditionally and in our own work, the variant described here should be placed on record. It is intermediate between *L. variabile* and *L. holopus* var. *holopus,* but because of its variable spore size *L. variabile* is possibly a hybrid between two species, *L. holopus* and *L. chalybaeum.* The problem of speciation in this group is very complex and cannot be resolved on the material collected to date here in Michigan. In the meantime it is difficult to refrain from speculation as to how these variants arose and which are the parental types.

89. **Leccinum variabile** Smith, Thiers, & Watling

Mich. Bot. 6:130. 1967

Illus. Pl. 81.

Pileus 3-5 cm broad, convex with an incurved margin, dry and unpolished, no fibrils or squamules visible to naked eye, whitish ("tilleul-buff") varying to pale avellaneous (grayish). Context pallid, thick, firm when young, when sectioned staining reddish then avellaneous and finally violaceous, slightly bluish when touched with $FeSO_4$, taste slightly acrid to mild.

Tubes about 10 mm deep, pallid to yellowish, depressed around the stipe; pores pallid at first, staining yellow when bruised.

Stipe 8-12 cm long, about 10-15 mm thick, equal or nearly so, white and furfuraceous when young, ornamentation merely brownish at maturity, solid, white within, staining reddish then avellaneous to violaceous when sectioned.

Spores 12-20 (23-40) X 4.5-6 (8) μ, smooth, in face view typically obscurely fusoid, somewhat inequilateral-elongate in profile, pale tawny when revived in KOH, slightly more cinnamon in Melzer's or occasionally dextrinoid, with moderately thickened wall, lacking a germ pore.

Basidia typically 4-spored but some 1-spored, 26-30 X 9-12 μ, clavate, hyaline to yellow-brown in KOH, yellowish in Melzer's. Pleurocystidia 38-50 X 7-14 μ, fusoid-ventricose with elongated neck and obtuse apex, scattered, hyaline or with yellowish content when revived in KOH; oleiferous hyphae present and often with brown pigment. Caulocystidia 40-65 X 13-15 μ, fusoid-ventricose with a short or elongate neck and subacute to obtuse apex, yellow-brown in KOH.

Tube trama bilateral from a central strand. Pileus trichodermium poorly developed, consisting of appressed readily disarticulating interwoven hyphae with hyaline to pale yellow content in KOH, cells often constricted at the septa, 7- 12 μ wide, the end-cells equal in width throughout and rounded at the apex, smooth and thin-walled, content when colored merely dingy orange-buff and homogeneous to granular in Melzer's. Clamp connections absent.

Habit, habitat, and distribution.—Gregarious under *Fraxinus* in a swamp near Rose City. Apparently rare.

Observations.—This species, excluding spore size, which is variable, is close to *L. chalybaeum,* but has a paler less intensely colored pileus, yellow staining pores, and a bluish $FeSO_4$ reaction. It resembles Singer's species in the weak yellowish tint developed by the tubes as they mature. There is a rather wide gap between them ecologically.

Subsection FUMOSA Smith, Thiers, & Watling

Mich. Bot. 6:138. 1967

This group is a continuation from subsection *Pallida,* connecting with the latter in those species with a very slight change to grayish on cut flesh in the stipe apex. The group differs in having brown to gray to fuscous pilei. Because of intergradation some species have been included in both keys.

Type species: *Leccinum olivaceo-glutinosum.*

KEY TO SPECIES

1. Stipe staining pinkish in apex before changing to gray or fuscous
. . . . *L. murinaceo-stipitatum*
1. Stipe staining gray directly (or avellaneous) in apex when cut (not reddish at first) 2

   2. Spores 5-6.5 μ wide (see *L. chalybaeum* also) 3
   2. Spores 4-5 μ wide. 4

3. Pileus glabrous and viscid when young . . . . *L. olivaceo-glutinosum*
3. Pileus dry and streaked with dark fibrils . . . . *L. subleucophaeum*

   4. Pileus fuscous to dark drab at first; spores 4-5 μ wide . . . . *L. griseonigrum*
   4. Pileus pale tan to dark yellow-brown (bister, etc.) 5

5. Tubes when young yellowish. . . . . *L. huronense*
5. Tubes white to olive-pallid young 6

   6. Context hyphae next to subcutis orange-red in Melzer's. . . . . *L. proximum*
   6. Context hyphae not colored as in above choice . . . . . . *L. disarticulatum*

90. Leccinum murinaceo-stipitatum Smith, Thiers, & Watling

Mich. Bot. 6:138. 1967

Pileus 8-12 cm broad, obtuse to convex, margin even, glabrous, viscid in appearance, subgranulose when perfectly young and fresh, dark cinnamon-brown with a grayish overcast, appearing blackish brown to the naked eye when fresh and fuscous as dried, extreme margin at times dull yellow-brown. Context when young and fresh rather hard, becoming soft by maturity or on aging, white, staining slowly to pinkish then avellaneous and finally fuscous when cut, slowly olive-gray to bluish with FeSO$_4$, odor and taste mild.

Tubes 1-1.5 cm deep, depressed around the stipe, whitish becoming olive-pallid and then finally dark wood-brown; pores small, round, olivaceous-pallid staining dull olivaceous if bruised lightly and vinaceous-buff when severely bruised.

Stipe 5-10 cm long, 2.5-3 cm thick in basal part, 1-2 cm at apex, drab to cinereous overall except the buried base which is whitish, the color caused by the combined effect of colored ground hyphae as well as colored ornamentation, the apex with gray ridges, not staining when handled but in drying a few yellowish areas show on the pallid ground color at or near the base, solid, white within at the apex, staining vinaceous to fuscous when stipe is sectioned.

Spores 13-16.5 X 4-5 μ, smooth, pale clay color singly and pale tawny in groups as revived in KOH, pale tawny revived in Melzer's, narrowly subfusiform in face view, obscurely elongate-inequilateral in profile, the suprahilar depression obscure, wall only slightly thickened and with no apical pore.

Basidia 4-spored 20-26 X 10-12 μ, clavate, hyaline, and usually with a large oil drop when revived in KOH. Pleurocystidia variable, essentially fusoid-ventricose, 50-70 X 9-13 μ, near the pores 30-45 X 7-11 μ, filaments resembling pseudocystidia (but lacking granular content) also present, content of all types hyaline to ochraceous. Cheilocystidia clavate to fusoid-ventricose, 20-34 X 6-10 μ, brighter ochraceous than pleurocystidia when revived in KOH. Caulocystidia 40-90 X 9-22 μ, fusoid-ventricose to ventricose-mucronate, with pale smoky brown content in KOH, typically with a proliferated neck. Caulobasidia mostly 7-9 μ broad (narrower than hymenial basidia and with colored to hyaline content).

Hyphae of the trichodermium of the pileus consisting of tubular thin-walled often disarticulating cells 5-12 μ wide or some cells near or including the terminal cell somewhat inflated, walls mostly smooth but some with incrusting pigment and on standing in KOH a tendency for blisters to develop on the outer wall, content smoky brown in KOH and near bister in Melzer's (in the latter also showing a tendency to form globules or become stringy), terminal cells cystidioid and those on the broadest hyphae somewhat bullet-shaped. Context hyphae below the subcutis yellow or orange-red in Melzer's from the colored homogeneous content. Clamp connections absent.

Habit, habitat, and distribution.—In mixed second-growth hardwoods with birch present, Sugar Island, July and August, not common.

Observations.—In a sense *L. murinaceo-stipitatum* is the counterpart of *L. flavostipitatum,* but it stains fuscous and has a gray stipe. It differs from *L. subleucophaeum* in its narrower spores. In another closely related species, *L. griseonigrum,* the pores stained yellowish instead of olive, the pallid ground color of the stipe was readily visible between the elements of the ornamentation, and no red stage was observed in the change of the flesh to gray or fuscous when it was cut. As far as is known, the latter is associated with aspen.

91. Leccinum olivaceo-glutinosum Smith, Thiers, & Watling

Mich. Bot. 6:140. 1967

Illus. Pl. 82.

Pileus 5-10 cm broad, obtuse to convex, the margin flaring and in age sometimes uplifted but lacking a sterile extension; surface glabrous and viscid to slimy, dull olive to olive-brown young, becoming olive-green over marginal area, disc developing a cinnamon undertone in age. Context pallid, firm, slowly changing to brownish (near avellaneous) when cut, instantly blue with $FeSO_4$, odor and taste mild.

Tubes 1-2 cm deep, depressed, pallid becoming wood-brown when mature, cinnamon-stained in age; pores pallid staining dingy yellowish brown when injured.

Stipe 9-15 cm long, 9-15 mm thick at apex, evenly enlarged downward to equal, pallid tinged vinaceous and occasionally olive at apex, lower part cottony-reticulate to striate, with brown or finally blackish dots or squamules at apex, rarely with reddish stains at the base, solid, pallid but staining yellow in base at times when cut, slowly changing to olivaceous near tube line and finally at times when cut flushed olive-green in apex and below this slowly becoming brownish gray ("wood-brown").

Spore deposit rich yellow-brown (near "argus-brown"); spores 16-19 × 5-6.5 μ, smooth, rich yellow-brown singly and more so in groups as revived in KOH, duller brown in Melzer's, only a small number dextrinoid, fusoid to subfusiform in face view, inequilateral-elongate and with a distinct suprahilar depression in profile, wall slightly thickened and apex with a thin spot but no distinct pore.

Basidia 4-spored, 26-33 × 10-13 μ, clavate, with large hyaline droplets, yellowish in Melzer's. Pleurocystidia not observed. Cheilocystidia 18-26 × 6-9 μ, clavate to subfusoid and ochraceous in KOH. Caulocystidia basically fusoid-ventricose, 40-75 × 9-18 μ, in addition usually developing a hyphal-like flexuous to corkscrew-like proliferation up to 50 μ or more long and 4-5 μ wide, the cell content pale brown. Caulobasidia scattered, hyaline or a few dark bister (at times including the sterigmata) from a colored wall.

Pileus trichodermium of appressed thin-walled smooth readily disarticulating hyphae 4-9 (12) μ wide, the cells tubular and with dingy ochraceous content when revived in KOH (when fresh some were seen to have green granules), content granular and pale ochraceous to brownish revived in Melzer's, terminal cells mostly elongate-fusoid, not or only slightly inflated but tapered to apex, with no tendency to produce short cells. Clamp connections absent.

Habit, habitat, and distribution.—Scattered in mixed woods, birch present, south shore of Lake Superior in Luce County and along the St. Mary's River near Goetzville, Mackinac County, August, rare.

Observations.—The pale caulocystidia with the neck often corkscrew-like, the slimy-viscid olive pileus, the pattern of color changes—particularly the change to olive in the stipe apex—in addition to the wood-brown stains, all distinguish this species from others in the section. The color of the spore deposit is about like that of *Boletus affinis* Peck. In view of the generally dull yellow-brown spore deposit of species of section *Scabra* it is not at all surprising to find a species in which this color is accentuated. We certainly would not think of erecting a genus parallel to *Xanthoconium* for it.

The caulocystidia are peculiar in that they exhibit the same character, in a way, as the somewhat coiled pileocystidia in a number of species of *Crepidotus.*

92. Leccinum subleucophaeum Dick & Snell

Mycologia 52:453. 1960

Illus. Pl. 83.

Pileus 3-10 (15) cm broad, convex to pulvinate, expanding to broadly convex or nearly plane, rarely with a low obtuse umbo, dry and typically obscurely matted-fibrillose with dark appressed hairs or fibrils, becoming almost glabrous in age, margin even, often blackish over disc and grayish toward the margin, in age becoming dingy yellow-brown at times. Context thick, white, changing slowly to gray when cut, slowly bluish gray with $FeSO_4$, odor and taste mild.

Tubes adnate becoming deeply depressed, white when young, becoming avellaneous to wood-brown, 10-20 mm deep, when mature; pores small (about 2 per mm), round, pallid, staining yellowish and then brownish slowly after injury, orange-brown with KOH, blue with $FeSO_4$.

Stipe 5-10 (15) cm long, 10-20 mm thick at apex, clavate becoming equal, white beneath a coating of blackish scabrous points, staining blue to greenish below where handled, solid, white within and typically staining pinkish and then avellaneous to wood-brown when cut, often with a blue line in a longitudinal section.

Spore deposit near cinnamon-brown when moist, near dingy cinnamon dried; spores 13-16 (19) $\times$ 4.5-6.5 (7) μ, smooth, in face view subfusiform to fusiform, in profile view somewhat inequilateral, pale dingy ochraceous-cinnamon in KOH, some becoming dark red-brown in Melzer's, thick-walled (0.5 μ).

Basidia 4-spored, 24-30 X 9-11 μ, clavate, hyaline to yellowish in KOH. Pleurocystidia scattered, 34-46 X 9-14 μ, fusoid-ventricose with an elongated often flexuous neck and subacute apex, thin-walled, smooth, reviving poorly, content not distinctive in either KOH or Melzer's reagent; pseudocystidia not observed. Caulocystidia ventricose-subcapitate, 36-60 X 9-16 μ, or clavate and very variable in size. Context hyphae of stipe dull cinnamon in KOH.

Pileus with a trichodermium of broad hyphae 6-12 μ wide, with some cells short and inflated to 15-20 μ, with smoky brown content in KOH but paler in Melzer's, lacking amyloid granules or particles. Clamp connections absent.

Habit, habitat, and distribution.—Solitary to gregarious in the sandy aspen-birch forests of Michigan, common and widely distributed during the summer and fall. Apparently, it is associated with various species of *Betula.*

Observations.—The outstanding features of this species as we recognize it from Michigan collections are the wide spores, the staining of the flesh to gray or finally fuscous, and the tubes staining yellow on the pores when lightly bruised. We have a number of variants close to this species in the state which need further study.

93. **Leccinum griseonigrum** Smith, Thiers, & Watling

Mich. Bot. 6:139. 1967

Illus. Pl. 84.

Pileus 4-9 cm broad, pulvinate becoming convex, margin even and fertile, at first slightly incurved, obscurely fibrillose when young becoming subgranular, nearly glabrous before becoming areolate, bluish black to bluish umber, slowly becoming paler to avellaneous or in age dingy cinnamon-buff. Context white, slowly changing to blue in places when cut and in age changing to violaceous-gray, with $FeSO_4$ bluish pallid, yellow on blue or green areas with KOH; odor and taste not distinctive.

Tubes 10-20 mm deep, sharply depressed around the stipe, white when young, wood-brown in age, slightly pinkish when cut; pores minute and pallid, staining wood-brown when severely bruised but if lightly bruised staining yellowish.

Stipe 5-11 cm long, 1-1.7 cm at apex, more or less equal, crooked at base, finely ornamented with brown to dark gray squamules and points, solid, pallid within but wood-brown around the wormholes.

Spore deposit dark yellow-brown (bister); spores 13-16 X 4-5.5 μ, smooth, more or less fusoid to subfusiform in face view, elongate-inequilateral in profile, the suprahilar depression distinct, ochraceous in KOH, pale tan to ochraceous in Melzer's, wall thickened somewhat (-0.5 μ).

Basidia 4-spored, 20-24 X 9-12 μ, often with a large hyaline oil drop. Pleurocystidia not observed. Cheilocystidia fusoid-ventricose, 28-42 X 8-12 μ, neck long and often crooked, apex subacute, with hyaline to yellowish content. Caulocystidia with a filamentous elongated neck, ventricose part with bister content revived in KOH. Pileus trichodermium of hyphae 4-8 μ wide, with smoky ochraceous to brown content in KOH, the walls smooth to roughened and the cells readily disarticulating at maturity, the end-cells more or less tubular; wider hyphae 8-16 μ broad also present and with some of the cells short (2 to 4 times as long as wide), the end-cells bullet-shaped. Context hyphae hyaline to yellowish in Melzer's. Clamp connections absent.

Habit, habitat, and distribution.—Gregarious under trembling aspen, common but not recognized as a species until recently.

Observations.—The context stains gray to fuscous finally, and this shows on the dried material. The base of the stipe dries olive-yellow. In this species more than in any other the outline of the subcuticular hyphae as revived in KOH appears blistered to eroded from irregular gelatinization of the outer wall. In the trichodermial hyphae this may also show but more often one observes the roughness in bands.

94. Leccinum huronense Smith & Thiers, sp. nov.

Pileus circa 10 cm latus, convexus, subviscidus, demum areolatus, pallide spadiceus. Contextus pallidus, tactu griseus. Tubuli pallide lutei, 2 cm longi. Stipes circa 9 cm longus, 11 mm crassus, minute punctatus, punctae demum atratae. Sporae 13-16 X 4-5 μ. Specimen typicum in Herb. Univ. Mich. conservatum est; prope Yellow Dog River, Marquette County, August 21, 1967, legit Mrs. Ingrid Bartelli.

Pileus about 10 cm broad, broadly convex, margin even, surface subviscid at maturity, matted-fibrillose and when dried very obscurely areolate (scarcely spotted as dried and subcutis not showing), color pale bister fresh, drying near "hair-brown" (brownish gray). Context firm, pallid, cut surface staining grayish directly and the gray areas drying gray, odor and taste not recorded.

Tubes weakly yellow at first, then finally dark wood-brown, 2 cm deep, depressed around the stipe; cut surface staining gray then fuscous; pores small, staining olive when slightly bruised.

Stipe 9 cm long, 11 mm thick, equal, cut surface soon fuscous but also stained blue around the base; surface ornamented with fine to coarse lines and squamules on a pallid ground, the elements blackish both fresh and as dried.

Spores 13-16 X 4-5 μ, smooth, wall –0.5 μ thick, lacking an apical pore; shape in profile narrowly subfusoid, with scarcely any depression in the suprahilar area, in face view subfusoid, with scarcely any apical differentiation, color in KOH pale brownish ochraceous, in Melzer's many soon become deep tawny to bay-brown.

Basidia 4-spored, about 25-30 μ long, clavate, hyaline. Pleurocystidia 38-60 X 8-12 μ, fusoid-ventricose with greatly elongated narrow neck ending in an acute apex, smooth, thin-walled, content hyaline to brownish in KOH and in Melzer's often reddish brown from granular material (but the same pigment also present in the pedicels of some of the basidia). Cheilocystidia 28-36 X 6-11 μ, more or less fusoid and ochraceous-brown as revived in KOH. Caulocystidia variable in size and shape but mostly fusoid, fusoid-ventricose with or without a greatly elongated neck, or (rarely) clavate, 38-60 X 8-12 μ and 60-100 X 10-18 μ, walls smooth and thin, content brownish and often aggregated into denser areas (in KOH).

Tube trama typical of the genus; laticiferous elements present, their content ochraceous to darker brown in KOH or Melzer's. Pileus cutis of 2 types of appressed hyphae: (1) flexuous, smooth, tubular, and with smoky ochraceous content homogeneous in KOH, 3-6 μ wide, the end-cells tubular, blunt at apex; septa sparse (cells elongate and wall not or scarcely constricted at septa), cells not readily disarticulating, in Melzer's some cells showing pigment globules 3-6 μ in diameter but usually only 1-4 per cell if present at all; (2) hyaline smooth hyphae 5-9 μ wide also with tubular, blunt end-cells and in Melzer's "empty" or with finely granular material dispersed in the central part; the first type often occurring as single cells or series of cells somewhere in an otherwise hyaline hypha. No inflated cells evident in the layer. Hyphae of the subcutis yellowish hyaline in Melzer's and gelatinized somewhat but lacking pronounced warts and blisters (but with a slight amount of adhering slime). Clamp connections none.

Habit, habitat, and distribution.—Solitary along the Yellow Dog River, near Marquette, August 21, 1967, Mrs. Ingrid Bartelli.

Observations.—This is a distinctive species in this subsection because of the pale yellow tubes when young and the narrow spores. The stipe does not become yellow as in *L. flavostipitatum.* The spores are too narrow for *L. subleucophaeum.*

95. Leccinum proximum Smith & Thiers, sp. nov.

Pileus 3-7 cm latus, convexus, obscure fibrillosus demum glaber et subviscudus, pallide spadiceus demum tarde ad centrum spadiceo-niger. Contextus albus tactu tarde subgriseus. Stipes 4-9 cm longus 8-13 mm crassus, atropunctatus vel squamulosus. Sporae 14-18 (20) X 4.5-6 μ. Specimen typicum in Herb. Univ. Mich. conservatum est; prope Onaway, Michigan, legit N. J. Smith (A. H. Smith 63545).

Pileus 3-7 cm broad, convex becoming broadly convex or nearly plane, surface glabrous to obscurely fibrillose at first and dry, becoming glabrous and subviscid, color a dingy yellow-brown ("bister") when young, "sepia" to "mummy-brown" on the disc (blackish brown) in age or when dried, some pilei remaining paler in age. Context white, thick, firm, becoming soft in age; unchanging when cut or very slowly grayish, bluish on stipe and tubes as well as the context with $FeSO_4$.

Tubes 8-12 mm deep, depressed around the stipe in age, adnate at first, white, slowly grayish brown and as dried dull cinnamon-brown; pores small round, whitish, when bruised staining yellow and then slowly changing to snuff-brown or bister.

Stipe 4-9 cm long, 8-13 mm apex, nearly equal, solid, white within, slowly staining vinaceous-gray when cut, except at base which stained orange-buff slightly in one, surface ornamented with coarse patches and points, the ornamentation brownish black on a pallid ground color and retaining these colors or becoming paler in drying, no bluish or greenish stains evident at any time.

Spore deposit chestnut-brown to "Prouts-brown"; spores 14-18 (20) X 4.5-6 μ, smooth, lacking an apical pore, shape in face view elongate-fusiform, in profile elongate-inequilateral; dull ochraceous-brown (near pale snuff-brown) in KOH, not changing much in Melzer's.

Basidia 4-spored, yellowish to hyaline in KOH and yellowish in Melzer's, 9-12 μ in diameter. Pleurocystidia scattered 46-52 X 8-12 μ, fusoid-ventricose with curved neck and acute apex, the neck about 3 μ in diameter, mostly hyaline in KOH. Caulocystidia 36-60 X 9-14 μ, mostly fusoid-ventricose to mucronate or clavate, content smoky brownish in KOH, walls thin and smooth.

Pileus cuticle with a subcutis of interwoven, narrow (3-5 μ), subgelatinous, thin-walled, smooth hyphae giving rise to a poorly formed epicutis of tangled, broad (6-15 μ), nongelatinous hyphae with cells which disarticulate to a considerable degree, have smoky brownish content in KOH, are often somewhat inflated (but not sphaerocyst-like), and often show a thin hyaline outer coating of irregular thickness (a gelatinous substance ?); end-cells typically over 25 μ long and often up to 100 μ, typically somewhat cystidioid. Context hyphae 8-15 μ wide, with orange-red content as seen revived in Melzer's. Clamp connections absent.

Habit, habitat, and distribution.—Under birch, Onaway, July 7, 1961, Nancy Jane Smith collector (A. H. Smith 63545).

Observations.—This essentially is a dark colored *L. scabrum* in which the cut flesh stains a vinaceous-gray finally and in which the context hyphae adjacent to the subcutis have bright orange-red content when revived in Melzer's.

96. **Leccinum disarticulatum** Smith & Thiers, sp. nov.

Illus. Pl. 85.

Pileus 4-10 cm latus, convexus, obscure virgatus, subviscidus, pallide argillaceus vel sordide cinnamomeus. Contextus albus tactu vinaceogriseus. Stipes 7-12 cm longus, 10-20 mm crassus, sparse brunneopunctatus vel atropunctatus. Sporae 12-16.5 $\times$ 4.5-6 μ. Specimen typicum in Herb. Univ. Mich. conservatum est; prope Carp Lake, July 13, 1961, legit Smith 63555.

Pileus 4-10 cm broad, obtuse to convex, becoming obscurely streaked from short appressed hyphae, in age soft and subviscid to the touch, surface dry and unpolished at first, color dingy pinkish buff along the marginal area, grayish alutaceous to dull cinnamon ("Sayalbrown") over the central part. Context white, soon pinkish avellaneous on cut surface, with $FeSO_4$ blue-gray, around the larval tunnels grayish brown in age; no green or blue stains developed anywhere from bruising in the material studied.

Tubes 10-12 mm long, white, depressed, soon avellaneous and finally wood-brown; pores white but sordid brown in age, staining yellow and finally brown when bruised, about 2 per mm.

Stipe 7-12 cm long, 10-20 mm at apex, equal to slightly enlarged downward, white within, cut surface soon vinaceous-avellaneous, dark wood-brown finally around larval tunnels; surface white, at apex whitescabrous, streaked or spotted brownish to blackish lower down from colored ornamentation, somewhat longitudinally striate at maturity, ground color whitish overall.

Spores 12-16.5 × 4.5-6 μ, smooth, with a minute apical hyaline spot, wall about 0.2 μ thick, in face view fusoid to subfusoid, in profile somewhat narrowly inequilateral, the suprahilar depression rather deep but not sharply defined at the edges, in KOH dingy ochraceous, slightly darker in Melzer's or finally some weakly dextrinoid.

Basidia 4-spored, 25-30 × 9-11 μ, clavate, hyaline in KOH, scarcely yellow in Melzer's. Pleurocystidia 36-52 × 8-12 μ, mostly near the tube edges, thin-walled, smooth, content ochraceous-brown. Cheilocystidia similar to pleurocystidia. Caulocystidia small and large (25-36 × 7-10 μ and 40-60 × 12-20 μ), both types intermixed, walls thin and smooth, content smoky ochraceous.

Tube trama obscurely bilateral, the *Boletus* subtype. Pileus cuticle a tangled layer of hyphae 6-10 μ wide, the terminal cells 6-10 μ wide, tapered to a subacute apex, with yellow-brown pigment in KOH, walls smooth or slightly roughened, rarely some end-cells bullet-shaped (on the broader hyphae), no inflated or sphaerocyst-like cells observed in the elements—most of the cells elongate and tubular, but these cells readily disarticulating (fig. 18). Context hyphae with content not distinctively colored in Melzer's. Clamp connections absent.

Habit, habitat, and distribution.—Scattered under birch and aspen, Carp Lake Bog, Carp Lake, Cheboygan County, July 13, 1961, Smith 63555.

Observations.—In revived material the hyphae adjacent to the subcutis are not orange to red as revived in Melzer's. This distinguishes dried specimens from *L. proximum.* Both stain gray when the context is cut. Although the cuticle is composed of narrow hyphae the cells disarticulate readily. This is a rare feature in this genus. The broad hyphae are usually the ones in which the cells separate.

Subsection SCABRA

In this subsection are grouped the species with yellow-brown to fuscous-brown or blackish pilei in which the sectioned context in the stipe apex does not stain gray to any appreciable extent. However, nearly all species show a change to blue or greenish in some part of the stipe and some will in addition show a change to pink or red or yellow or various combinations of these.

Type species: *Leccinum scabrum.*

KEY

1. Odor when fresh strongly raphanoid; pores staining rusty cinnamon when bruised . . . . *L. parvulum*
1. Not as above 2

    2. Spores 10-15 X 3.5-4.5 μ; stipe becoming yellow over a considerable portion especially in drying . . . . *L. flavostipitatum*
    2. Spores 5-7 μ wide 3

3. Stipe apex staining pink or redder when cut . . . . (see *L. snellii* also) 4
3. Stipe apex unchanging when cut or slowly merely brownish 6

    4. Pileus coffee-brown to dark rusty brown; pileus cutis of hyphae 4-6 μ wide and with numerous pigment globules when mounted in Melzer's . . . . *L. coffeatum*
    4. Not as above. 5

5. Pileus dull cinnamon young and darker cinnamon when mature; stipe ornamentation fine and long remaining pallid . . . .*L. pallidistipes*
5. Pileus pale pinkish buff, becoming grayer; stipe ornamentation coarse . . . . *L. rimulosum*

    6. Caulocystidia often proliferated at apex . . . .*L. singeri*
    6. Caulocystidia not proliferated 7

7. Stipe ornamentation fine, pallid becoming orange to cinnabar but finally darkening—especially in drying . . . . *L. subpulchripes*
7. Stipe ornamentation over lower half coarse and soon blackish. . . . . *L. scabrum* f. *scabrum*

97. Leccinum parvulum Smith, Thiers, & Watling

Mich. Bot. 6:135. 1967

Pileus 2.5-3.5 cm broad, obtuse, expanding to broadly umbonate, viscid, pallid when very young, soon becoming pale tan (dull "pinkish cinnamon" to "pinkish buff"), in some specimens pallid beneath minute pinkish buff areolae and hence giving a somewhat mottled appearance but some pilei remaining perfectly even. Context firm but soon soft, white, unchanging when cut in young material, in old ones slowly changing to near wood-brown, slowly and weakly olive with FeSO$_4$, with KOH slowly brownish; odor strong of radish but taste mild.

Tubes 1 cm deep, dull cinnamon when mature ("Sayal-brown"), adnate at first, becoming nearly free from the stipe; pores minute, dull cinnamon but staining rusty cinnamon where injured.

Stipe 6-8 cm long, 7-9 mm thick at apex, up to 12 mm at base, with a napiform bulb (pointed), ornamented with dingy cinnamon scabrosities which do not darken much in age, solid, white within but slowly becoming dingy wood-brown when cut.

Spore deposit dull cinnamon ("Sayal-brown"); spores 14-17 X 4.5-5.5 (6) μ, smooth, pale tawny in KOH, duller in Melzer's, narrowly suboblong to obscurely fusoid in face view, in profile obscurely inequilateral from a shallow broad suprahilar depression, wall only slightly thickened and no apical differentiation evident.

Basidia 4-spored, 18-23 X 8-9 μ, clavate, very soon gelatinizing. Pleurocystidia scattered to rare, 36-45 X 9-15 μ, fusoid-ventricose, apex subacute, hyaline in KOH and in Melzer's. Caulocystidia smooth, clavate and up to 40 X 18 μ, broadly fusoid-ventricose and 40-60 X 15-22 μ, or elongate and up to 70 X 15 μ, content mostly hyaline in KOH but in some pale ochraceous.

Pileus trichodermium, if one is present at first, collapsing and forming a cutis of appressed tangled smooth hyphae 5-11 μ wide, with homogeneous yellow-brown content in KOH (near "argus-brown"), often only the terminal cell colored, some terminal cells bullet-shaped and with granular to somewhat stringy content in Melzer's but lacking pigment globules. Subcutis of hyphae which are hayline in KOH, interwoven, subgelatinous, and orange-ochraceous in Melzer's. Hyphae of context with homogeneous reddish orange content when revived in Melzer's. Clamp connections absent.

Habit, habitat, and distribution.—Scattered under oak and pine, Douglas Lake, June, rare.

Observations.—This species is outstanding because of its dull cinnamon spores, tubes dull cinnamon, the rusty cinnamon stains on the pores and the dull cinnamon stipe ornamentation along with the strong raphanoid odor. This variant appears to be isolated in its combination of characters and is placed here tentatively. As wood-brown stains are very slow to develop, the species is placed here rather than in subsection *Fumosa*. It will also be noted that in the whitish species of this section the color change occurs to a greater or lesser degree in some of them.

98. Leccinum flavostipitatum Dick & Snell

Mycologia 57:456. 1965

Pileus 6-10 cm broad, very minutely reticulately fibrillose, the disc later subglabrous to glabrous, gray, dark gray, brownish, or blackish gray, occasionally with tinges of greenish. Context dingy white changing very slowly to pale salmon-pink or to bright blue-green quickly when wet.

Tubes depressed to free, very dingy white or dirty yellowish white, quite long; pores concolorous with sides at first.

Stipe up to 11 cm long, 1-2 cm thick, enlarged at apex and base, faintly black subscabrous or faintly concolorous-reticulate here and there, but mostly reticulately torn or "brown- to black-fibrillose-sub-reticulate-scabrous" usually pale yellow, occasionally more grayish or dingy whitish, drying bright yellow, often with spots or areas of blue-green at the base, dingy whitish to dingy tan within, the cortex variously coloring (some pinkish, pale yellow, or blue-green), the base deeper dingy tan with greenish blue areas especially when wet.

Spore deposit dark olive-buff; spores 13-17 X 3.8-5 μ, smooth, lacking an apical pore, weakly yellow in KOH, dingy ochraceous in Melzer's, in face view narrowly fusoid to subfusoid, in profile narrowly inequilateral, wall thin (about 0.2 μ thick).

Basidia 20-26 X 7-11 μ, clavate, hyaline, 4-spored. Pleurocystidia not found. Cheilocystidia like small basidioles and discolored brownish. Caulocystidia in fascicles, 40-80 X 12-20 μ, thin-walled, smooth, empty, clavate to fusoid-ventricose, many smaller cystidia and basidioles also present; cortex of stipe of hyphae lemon-yellow in KOH.

Pileus cuticle a layer of interwoven hyphae 4-10 μ wide with thin smooth walls and bister content in KOH and Melzer's, becoming granular in the latter medium but not forming pigment globules, very few short scarcely inflated cells present, terminal cell cylindric to narrowly cystidioid. Subcutis of hyaline subgelatinous hyphae 4-12 μ wide and content hyaline in KOH. Context hyphae scarcely yellow in Melzer's. Clamp connections not present.

Habit, habitat, and distribution.—The type locality for this species is under spruce in Nova Scotia, but on several occasions we thought we had found it here in Michigan. An examination of the type proved this to be erroneous, but we expect that it will eventually turn up in our area.

99. Leccinum coffeatum Smith & Thiers, sp. nov.

Illus. Pl. 86.

Pileus 5-10 cm latus, convexus demum expansus, glaber, demum viscidus, coffeatus. Contextus albidus tactu pallide vinaceus. Tubuli pallidi; pori pallidi tactu luteo-cinnamomei. Stipes 6-10 cm longus, 1-2 cm crassus, brunnei punctatus. Sporae 15-19 X 5-7 μ. Specimen typicum in Herb. Univ. Mich. conservatum est; prope Brimley, Chippewa County, July 11, 1968, legit H. D. Thiers (Smith 75536).

Pileus 5-10 cm broad, convex becoming plane or finally the margin uplifted; surface glabrous and at maturity viscid, color dark coffee-brown (near "russet"), evenly colored; margin even and fertile. Context white and when cut developing a faint flush of vinaceous but the change not progressing to gray and no gray showing on dried basidiocarps, borders of larval tunnels merely pallid-brown, becoming olive-yellow in the base of the stipe in age and yellow persisting to some extent in dried material, $FeSO_4$ bluish; NH_4OH on flesh no reaction.

Tubes 1-1.5 cm long, depressed around the stipe, pallid when young, becoming dingy bister; pores pallid at first, when bruised staining dingy yellowish cinnamon, small (1-2 per mm), round.

Stipe 6-10 cm long, 1-2 cm thick at apex, solid, color changes as described above for context, surface ornamented by fine brown points and the ground color more or less grayish avellaneous shading to pallid at apex and base, base with white mycelium.

Spore deposit dark cinnamon-brown; spores 15-19 (20) X 5-7 μ, smooth, with a minute apical thin spot, more or less fusoid in face view, in profile somewhat inequilateral, wall 0.5 μ thick as revived in Melzer's; color in KOH ochraceous to yellowish cinnamon, in Melzer's pale tawny.

Basidia 4-spored, hyaline in KOH, clavate. Pleurocystidia—none seen except near the pores, 20-36 X 8-12 μ, content hyaline to ochraceous, fusoid-ventricose with subacute apex, smooth, thin-walled. Cheilocystidia clavate or similar to pleurocystidia. Caulocystidia 35-50 (60) X 10-18 μ, thin-walled, clavate to ovate-pedicellate or elliptic-pedicellate, more rarely somewhat fusoid or mucronate, content mostly pale dull cinnamon as revived in KOH, walls thin and even.

Tube trama of weakly divergent hyphae, hyphae tubular, thin-walled and 4-6 μ wide, some large laticiferous elements with dull cinnamon amorphous content also present; in Melzer's numerous nearly hyaline globules present in the mount. Pileus cuticle of appressed narrow (3-6 μ wide) hyphae with septa widely spaced, the cells not disarticulating, content cinnamon in KOH and slowly forming indistinct globules in Melzer's nearly all hyphae with cinnamon pigment globules 2-6 μ in diameter. Hyphae of subcutis and the adjacent context hyphae hyaline to yellowish in Melzer's. Clamp connections absent.

Habit, habitat, and distribution.—Scattered in aspen-birch woods near Brimley, Chippewa County, July 11, 1968, H. D. Thiers (Smith 75536).

Observations.—This is one of the most striking and distinct species of *Leccinum* known. The strong tendency for pigment globules to form in Melzer's in mounts of the cuticular hyphae, the narrowness of these hyphae, the lack of color in the subcutis and the adjacent context hyphae—also as revived in Melzer's—the color change to vinaceous in fresh basidiocarps and the fact that it does not continue on to become fuscous, the coffee-brown to russet pilei, the fine, brown, stipe ornamentation, clavate caulocystidia, and large spores are an unusual set of characters.

100. Leccinum rimulosum Smith & Thiers, sp. nov.

Illus. Pl. 87.

Pileus 3-10 cm latus, convexus, siccus demum subviscidus, pallide argillaceus. Contextus pallidus, tactu tarde vinaceus. Tubuli 1.5 cm longi, pallidi; pori tactu griseobrunnei. Stipes 4-9 cm longus, 1-2 cm crassus, clavatus vel fusiformis, sparse punctatus; punctae pallidae, tarde brunneae. Sporae 15-18 X 5-6 μ. Specimen typicum in Herb. Univ. Mich. conservatum est; prope Paradise, July 8, 1968, legerunt Wells et Patrick (Smith 75513).

Pileus 3-10 cm broad, convex becoming plane or nearly so; margin even; surface unpolished at first and often with minute spotlike squamules over the disc, subviscid in age, color a dingy pale pinkish buff with an overlay of darker grayish brown squamules or fibrils in a matted-down layer, minutely rimulose over the marginal area at times. Context pallid when cut but slowly changing to vinaceous and in places finally dingy brownish (but not going to gray or fuscous), in the base of the stipe yellowish and then pale gray but this fading in drying, gray around the larval tunnels.

Tubes up to 1.5 cm deep, depressed around the stipe, pallid then dingy gray-brown; pores readily staining a dingy gray-brown.

Stipe 4-9 cm long, 1-2 cm thick at apex, conspicuously clavate to fusiform when young, nearly equal in age, white within at first, slowly pinkish and then weakly brownish on standing after being cut; surface with pallid coarse ornamentation which becomes brownish and finally darker but very sparse, ground color showing prominently.

Spores 15-18 X 5-6 μ, smooth, with a thin spot at apex, fusiform in face view, in profile somewhat inequilateral, wall about 0.2 μ thick, ochraceous to pale tawny in KOH, not much different in Melzer's.

Basidia 4-spored, clavate, hyaline in KOH, 10-12 μ broad. Pleurocystidia—none observed. Cheilocystidia 18-27 X 4-9 μ, mostly clavate and soon gelatinizing, content typically ochraceous revived in KOH. Caulocystidia in fascicles, 30-50 X 9-20 μ, clavate to ovate-pedicellate or mucronate, thin-walled, content dingy ochraceous in KOH; some of the subiculum hyphae with ochraceous content in age.

Tube trama weakly divergent; laticiferous elements widely scattered and not conspicuously differentiated, content ochraceous in KOH. Pileus cutis of appressed hyphae 5-9 μ wide, content ochraceous to yellow-brown in KOH, homogeneous, cells elongate (more than 4 times as long as wide), apical cell tapered to a subacute apex, midportion not inflated, wall smooth or nearly so, in Melzer's the cell content near "argus-brown" (a bright yellow-brown), homogeneous to granular; hyphae of subcutis and adjacent context orange in Melzer's. Clamp connections absent.

Habit, habitat, and distribution.—Under birch, Lower Falls, Tahquamenon Falls State Park, Chippewa County, July 8, 1968, Mary Wells and W. Patrick (Smith 75513).

Observations.—The outstanding features of this species are the pale pileus which becomes minutely squamulose to rimulose, the very fine sparse stipe ornamentation, the slow change to pink in the apex of the stipe as well as in the pileus when a basidiocarp is sectioned, the lack of pigment globules in mounts of cuticular hyphae in Melzer's, and the orange reaction of the subcuticular hyphae and those adjacent to it in the context. It is at once distinguished from *L. scabrum* in the field by the change to pink on the cut surfaces as mentioned, and the fine, sparse, ornamentation of the stipe. It is intermediate between subsections *Scabra* and *Pallida,* but is placed in the former because as dried the pileus is pale avellaneous. The ground color of the stipe remains pallid throughout in drying.

101. Leccinum pallidistipes Smith & Thiers, sp. nov.

Pileus 5-10 cm latus, convexus, siccus demum viscidus, glaber, sordide cinnamomeus. Contextus pallidus, tactu pallide vinaceus. Tubuli 1-2 cm longi pallidi; pori cinerei-pallidi tactu luteobrunnei. Stipes 9-14 cm longus, 1-2 cm crassus, sparse punctatus; punctae pallidae, tarde griseae. Sporae 15-18 X 5-6 μ. Specimen typicum in Herb. Univ. Mich. conservatum est; prope Brimley, Chippewa County, July 13, 1968, legerunt H. D. Thiers et W. Patrick (Smith 75572).

Pileus 5-10 cm broad, convex to broadly convex, the margin even; surface dry at first, becoming viscid later, glabrous, color near "Sayal-brown" (dull cinnamon) on the disc, the margin pallid at first, in age dull cinnamon ("Verona-brown") over the disc and with pallid blotches toward the margin, pale gray-brown as dried, some obscurely fibrillose-spotted near the margin. Context thick, pallid, when cut staining slightly vinaceous-brown (reddish by a Mazda bulb) $FeSO_4$ on context bluish; odor and taste not distinctive.

Tubes 1-2 cm deep, depressed around the stipe, pallid becoming wood-brown; pores grayish pallid staining dingy yellow-brown and in age the stains dark cinnamon-brown.

Stipe 9-14 cm long, 1-2 cm thick at apex, equal or nearly so, solid, staining as in pileus, merely dull brown around larval tunnels; surface pallid or at base yellowish where decay had started; ornamentation fine, sparse, pallid until late maturity, slowly becoming gray or slightly darker, apex remaining pallid and pruinose.

Spores 15-18 X 5-6 μ, smooth, with a minute apical thin spot, in face view fusoid, in profile somewhat inequilateral, color in KOH dingy ochraceous-tawny, slightly darker in Melzer's, wall about 0.3 μ thick.

Basidia 4-spored, about 30 X 9 μ, clavate hyaline in KOH and yellowish in Melzer's. Pleurocystidia not found. Cheilocystidia clavate to subfusoid or fusoid-ventricose, 26-42 X 8-12 μ, neck narrow and short, apex subacute, content ochraceous in KOH, walls thin and smooth. Caulocystidia fusoid to fusoid-ventricose, 36-62 X 8-11 μ, walls thin and smooth, content mostly hyaline in KOH. Caulobasidia with dingy ochraceous content.

Tube trama of somewhat divergent hyphae, the latter tubular and with refractive cross walls, hyaline in KOH and not obviously gelatinous, laticiferous elements rare. Pileus cutis of appressed narrow (4-8 μ), tubular hyphae with yellow-brown content in KOH and Melzer's, not rounding into globules in Melzer's, end-cells tubular to narrowly clavate or weakly cystidioid. Hyphae of subcutis and adjacent context yellow in Melzer's. Clamp connections absent.

Habit, habitat, and distribution.—Gregarious under birch and aspen, Brimley, Chippewa County, July 13, 1968, H. D. Thiers and W. Patrick (Smith 75572).

Observations.—This species is worth placing on record since it is the only one known to us with predominantly hyaline caulocystidia but with the other elements of the caulohymenium colored. The vinaceous-brown stains of the cut flesh, fine ornamentation of the stipe, and the narrow cuticular hyphae are also distinctive.

102. Leccinum scabrum (Fries) S. F. Gray

Nat. Arr. Brit. Pls. 1:647. 1821.

f. scabrum

Illus. Pl. 88.

Pileus 4-10 cm broad, pulvinate to broadly convex, margin even and scarcely projecting beyond the tubes, moist to dry or when wet fairly viscid, glabrous and often with depressions in age, grayish brown to dull yellow-brown and often flushed olivaceous in age. Context white, when cut not staining or slowly becoming slightly brownish (near "pinkish buff"); odor and taste mild.

Tubes 8-15 mm deep, pallid, slowly becoming wood-brown as spores mature, deeply depressed around the stipe; pores small, pallid then concolorous with the sides and either not staining when bruised or staining yellow.

Stipe 7-12 (15) cm long, 7-12 (16) mm thick above, evenly enlarged downward, with dark brown to blackish punctate to nearly reticulate or squamulose decorations, the pallid ground color visible, solid, white within, slowly staining pinkish buff in the cortex when sectioned, also developing both blue and red stains in restricted areas lower down if wet.

Spores 15-19 X 5-7 μ, smooth, subfusiform in face view, in profile ventricose-inequilateral to elongate-inequilateral, often with a pronounced suprahilar depression, pale tawny singly, in groups darker (both as revived in KOH), pale tawny in Melzer's, wall thickened to about 0.5 μ as revived in KOH.

Basidia 4-spored, 11-13 μ broad, hyaline in KOH and yellowish in Melzer's. Pleurocystidia—none observed. Caulocystidia mostly clavate, 32-46 X 9-17 μ, with smoky ochraceous content in KOH, some with an apical protrusion causing the cell to be clavate-mucronate and a few varying to fusoid-ventricose.

Pileus cuticle a tangled layer of floccose hyphae 6-12 (15) μ wide, above a subgelatinous subcutis, their content smoky ochraceous in KOH and with a grayer shadow in Melzer's but content remaining granular to homogeneous, wall at first seen to be enveloped in an outer gelatinous matrix as observed revived both in KOH or in Melzer's, the layer or matrix of irregular extent and thickness but at times enveloping two adjacent hyphae, this material slowly dissolving in KOH, at times a few short slightly inflated cells observed. Clamp connections absent.

Habit, habitat, and distribution.—Not uncommon throughout the state during the summer and fall under birch.

Observations.—The dingy yellow-brown to grayish brown pileus, the failure of sectioned basidiocarps to stain gray, and the broad spores are characteristic of this, the type form, of the *L. scabrum* complex. On the basis of the species as presently defined it has been one of the most frequently misidentified species in North America. It is not possible at this time to present a complete study on this subsection either for

Michigan or any other area. For this reason we have not given synony-my nor are we discussing variation in the complex. A comprehensive treatment will be included in a monograph on *Leccinum* now in prepara-tion by Smith, Thiers, and Watling.

103. Leccinum singeri Smith & Thiers, sp. nov.

Pileus 3-8 cm latus, demum late convexus, glaber, viscidus, brunneogriseus vel subspadiceus. Contextus pallidus tactu incarnato-argillaceus. Tubuli pallidi; pori avellanei tactu lutei, tarde lignobrunnei. Stipes 6-12 cm longus, 10-15 mm crassus; contextus tactu sordide cinna-barinus; deorsum atrosquamulosus, sursum brunneo-punctatus. Sporae 15-20 X 5-6.5 μ. Caulocystidia 30-35 X 10-20 μ demum 100-200 X 6-10 (10-20) μ, proliferata. Specimen typicum in Herb. Univ. Mich. conservatum est; prope Mackinaw City, Emmet County, July 23, 1965, legit Smith 71844.

Pileus 3-8 cm broad, convex becoming broadly convex; surface gla-brous, viscid, pale grayish brown to bister, margin even and not appendi-culate. Context soft and soon collapsing, pallid when cut but soon staining dingy pinkish tan at least near the cutis, with $FeSO_4$ quickly bluish; taste acid, odor mild.

Tubes pallid becoming wood-brown or (when bruised) darker and near "warm sepia," depressed, seceding and free in age, ventricose, 2 cm deep in widest place; pores small, round, avellaneous, staining yellowish and finally dark wood-brown.

Stipe 6-12 cm long, 10-15 mm thick at apex, solid, when cut cinnamon reddish to cinnabar, in the base dark red, at times yellow below and in old specimens olivaceous near apex; surface pallid to brownish; ornamentation weakly brownish at first but finally blackish, typically fine above and coarser below, blackish at maturity, that over the upper part usually remaining brownish.

Spores dark yellow-brown ("snuff-brown") in deposit; spores 15-20 X 5-6.6 μ, smooth, ochraceous to clay color singly in KOH, yellow-fulvous in groups, in Melzer's about the color of the mounting medium (yellowish) and the content finely granular; apex with a small hyaline spot.

Basidia 4-spored, 18-25 X 10-15 μ, hyaline in KOH, with a large hyaline globule in many. Pleurocystidia—none found. Cheilocystidia re-sembling basidioles but varying to subfusoid, ochraceous in KOH. Caulo-cystidia mostly ventricose-mucronate and 30-35 X 10-20 μ, tapered to a mucro abruptly but in age the mucro often developing into a hyphal-like prolongation 23-200 μ long, small cells (basidioles ?) often when ventricose part 6-9 μ wide also proliferating to 100 μ or more long, clavate cells with 2 sterigmata also present, and 10-15 μ broad but no spores seen attached, cell content bister in KOH, some broadly rounded cells that are ventricose in midportion are also present (utriform types, as found in *Psathyrella* of the Coprinaceae), the latter usually rare.

Pileus cutis a tangle of hyphae, the hyphae 4-12 (15) μ wide, mostly tubular but at times some cells disarticulating, content clay colored in KOH, duller brown and granular to homogeneous in Melzer's, the end-cells on narrower hyphae narrowly fusoid and up to 100 μ or more long, those on broad hyphae often bullet-shaped, rarely ellipsoid, short cells present in some hyphae. Clamp connections none.

Habit, habitat, and distribution.—Scattered on humus near Mackinaw City, but in Emmet County, July 23, 1965, Smith 71844.

Observations.—In view of the fact that *L. scabrum* is now regarded as essentially a species lacking pronounced color changes other than to blue at times in the lower half of the stipe, it is logical to recognize the present variant as a species, particularly in view of the proliferated caulocystidia. We cannot be sure it is identical with Singer's *L. scabrum* f. *coloratipes.* In any event, it is a species parallel in its aggregate of characters with the others as recognized in this work, and we take pleasure in dedicating it to Dr. Rolf Singer who at least called attention to a similar if not identical fungus.

104. Leccinum subpulchripes Smith & Thiers, sp. nov.

Pileus 3-4.5 cm latus, convexus, glaber, viscidus, pallide argillaceus. Contextus pallidus, immutabilis. Tubuli 14 mm longi, lignobrunnei; pori minuti, grisei tactu luteobrunnei. Stipes 3-5.5 cm longus, 5-9 mm crassus, deorsum tactu aurantio-cinnabarinus, furfuraceus; squamulae pallide subochraceae demum aurantiacae vel cinnabarinae. Sporae 14-18 X 5-6.5 μ. Specimen typicum in Herb. Univ. Mich. conservatum est; prope Culhane Lake, Luce County, August 2, 1965, legit Smith 71933.

Pileus 3-4.5 cm broad, convex, scarcely expanding, margin even; surface glabrous, viscid, "pale pinkish buff" to near "pinkish cinnamon" (dingy pale tan to pale cinnamon), becoming dingy dull clay color in age or with the marginal area more or less olivaceous. Context pallid (olive-buff when water-soaked), unchanging when cut; taste acid, odor none, with FeSO₄ instantly blue to grayish blue.

Tubes wood-brown (all mature), ventricose, 14 mm deep, unchanging when cut, free to deeply depressed around the stipe; pores minute, round, grayish, staining yellow-brown when bruised.

with $FeSO_4$ instantly blue to grayish blue.

Stipe 3-5.5 cm long, 5-9 mm thick at apex, slightly enlarged downward, solid, interior watery yellowish, when cut with cinnabar-orange stains in lower part or merely ochraceous there, at very base merely watery buff; surface minutely furfuraceous-scabrous from fine particles of ornamentation which are buffy pallid at first but become yellowish to orange or finally cinnabar, base whitish.

Spore deposit about cinnamon-brown; spores 14-18 $\times$ 5-6.5 μ, smooth, shape in face view narrowly oval, in profile obscurely inequilateral to more or less narrowly oval (not typically "boletoid"), near buckthorn-brown in KOH singly, in groups tawny, in Melzer's paler; wall only slightly thickened.

Basidia 4-spored, 26-32 $\times$ 9-12 μ, clavate, with large hyaline globules as revived in KOH. Pleurocystidia—none found. Caulocystidia 32-46 $\times$ 7-11 μ, and 50-80 $\times$ 9-16 μ, fusoid in shape more or less but apex obtuse.

Pileus cutis a tangled mass of appressed hyphae 4-9 μ wide (when fresh appearing as a collapsing trichodermium in a gelatinous matrix) and with yellowish, granular content in water mounts when fresh but more or less homogeneous in KOH, in Melzer's the hyphal content granular and dingy brownish, the hyphae tubular, the cells readily disarticulating and the end-cells narrowly cystidioid (tapered gradually to the apex). Hyphae of context adjacent to the subcutis hyaline to weakly yellow when revived in Melzer's. Clamp connections none.

Habit, habitat, and distribution.—Scattered under pine with alder near by, Culhane Lake, Luce County, August 2, 1965, Smith 71933.

Observations.—The color changes of the stipe ornamentation are unique in this subsection and do not progress to dark brown or blackish in the manner typical of the genus. The species, however, is so obviously related to others in this section that it would be folly to place it elsewhere. It should be carefully compared with *L. singeri.* If one disregards the color change of the stipe ornamentation, the latter species still differs from *L. subpulchripes* in having more short cells in the epicuticular hyphae, and in lacking proliferated caulocystidia.

Section LUTEOSCABRA Singer

Amer. Midl. Nat. 37:112. 1947

Pileus cuticle usually a trichodermium with at least a fair number of short inflated cells in the trichodermial hyphae, or the layer appearing cellular because the inflated cells are so compactly arranged; the hymenophore pallid to yellow when young.

Type species: *Leccinum crocipodium.*

KEY TO SUBSECTIONS

1. Context and/or hymenophore yellow in some degree when young or on drying; stipe ornamentation unchanging or darkening but darker as dried . . . . Subsect. *Luteoscabra*
1. Young hymenophore pallid to whitish or if yellowish at some stage then the stipe ornamentation darkening and pileus context never wholly yellow . . 2

  2. Pileus trichodermium forming a more or less distinct cellular epithelium; pileus often glabrous and viscid . . . . Subsect. *Albella*
  2. Pileus trichodermium of hyphae with constituent cells particularly toward the end, inflated sometimes to sphaerocyst proportions but the enlarged cells not numerous enough to form an epithelium . . . .
  . . . . Subsect. *Pseudoscabra*

Subsection PSEUDOSCABRA Smith, Thiers, & Watling

Mich. Bot. 6:119. 1967

The hyphae of the pileus cutis frequently have one or more cells in a filament inflated sufficiently to cause the cells concerned to be elliptic, subglobose, or globose.

Type species: *Leccinum snellii.*

KEY TO SPECIES

1. When the stipe-apex is cut there is a distinct color change to pink or reddish on exposed surface 2
1. Not staining as indicated above 3

  2. Pileus streaked with dark fibrils or nearly black from them; many caulocystidia with a septum distal to the ventricose portion . . . . . *L. snellii*
  2. Pileus white but discoloring to buff or pale crust-brown; caulocystidia not as above . . . . *L. oxydabile*

3. Pileus granulose-squamulose to areolate; caulocystidia often with the neck greatly proliferated . . . . *L. subgranulosum*
3. Pileus merely slightly appressed-fibrillose; caulocystidia not having greatly proliferated necks . . . . *L. aberrans*

105. **Leccinum snellii** Smith, Thiers, & Watling

Mich. Bot. 6:120. 1967

Illus. Figs. 84-86, 88; Pl. 89.

Pileus 3-9 cm broad, obtuse to convex, margin sterile at times but not lobed, expanding to plane or with a low umbo, dry but becoming subviscid in age, minutely fibrillose with fuscous-brown fibrils, pallid ground color showing near margin or (rarely) almost overall, disc usually blackish, in age at times with minute spotlike squamules, toward the margin where injured staining yellowish brown. Context thick, firm but very soon soft, blue with $FeSO_4$ and yellow on cutis with KOH, but no reaction with NH_4OH; odor slight, taste slightly acid.

Tubes depressed, 10-15 mm deep, becoming ventricose, white becoming wood-brown by maturity; pores small, white becoming grayish by maturity and staining dingy ochraceous as bruised.

Stipe 4-11 cm long, 1-2 cm thick near apex, evenly enlarged downward or equal, ornamented with drab-gray to blackish punctae and squamules to the pallid apex, white ground color showing but obscured over midportion, greenish to yellow to yellowish in some degree over basal area as dried, solid, pallid within, when cut staining pink to dull red, in the lower half caerulean blue in some places, rarely yellowish near the blue stains except where treated with alkali.

Spore deposit between cinnamon-brown and bister when moist, dull cinnamon on drying; spores (15) 16-22 X 5.5-7.5 μ, smooth, in face view definitely fusoid, in profile inequilateral but the suprahilar depression often shallow, clay color and tawny singly and in groups respectively as revived in KOH, in Melzer's tan or slowly becoming reddish tawny, wall slightly thickened, apex with a minute hyaline spot but not truncate.

Basidia 22-30 X 10-13 μ, 4-spored, clavate, hyaline in KOH, with 1-2 large oil globules, merely yellowish in Melzer's. Pleurocystidia observed neither in young nor old basidiocarps. Cheilocystidia 28-42 X 7-12 μ, fusoid-ventricose, hyaline to dingy brownish ochraceous within. Caulocystidia distinctive (60) 75-150 (250) X 10-20 μ, mostly ventricose but with a long filamentose often flexuous neck 4-6 μ in diameter, ending in a subacute apex, often with a septum just distal to the ventricose portion, if septate as indicated the filamentose part beyond the septa hyaline, ventricose part with bister content in KOH, these cells occurring in large fascicles with few caulobasidia or caulobasidioles present.

Pileus cutis of appressed fibrils, the cells of the hyphae varying from as long as wide to greatly elongated, the inflated cells 9-25 μ wide, the end-cell clavate to cystidioid and often almost globose, with homogeneous smoky brown content in KOH, near bister and remaining homogeneous or in some cells becoming slightly stringy as mounted in Melzer's, hyphal cells appreciably disarticulating by maturity. Subcutis of interwoven subgelatinous hyphae 4-6 μ wide and merely yellowish in Melzer's, numerous laticiferous elements present which are dark brown to blackish brown in Melzer's. Clamp connections absent.

Habit, habitat, and distribution.—Scattered to gregarious along roads in slashings through hardwood forests with scattered yellow birch in them but with heavy yellow birch reproduction along the roads, common, late summer and fall in the Upper Peninsula.

Observations.—The distinguishing features of this *Leccinum* are the caulocystidia as described, the yellow stains which develop over the pileus margin and the bruised pores, the broad spores, and the almost moniliform appearance of many of the epicuticular hyphae on young pilei. The inflated cells of the cuticular hyphae, the broader spores, and change to red when bruised or sectioned, distinguish the species from *L. flavostipitatum. L. oxydabile* Singer is a paler species with different caulocystidia (see Watling, 1968).

106. Leccinum oxydabile (Singer) Singer

Amer. Midl. Nat. 37:123. 1947

Krombholtzia oxydabilis Singer, Schwiz. Zeitschr. Pilzk. 16:136. 1938.

Pileus 2.5-3 cm broad, convex, the margin fertile and even, white when fresh and young but soon becoming somewhat yellowish to buff to pinkish buff or pale crust-brown from handling or on aging, appressed-fibrillose becoming glabrous but dry to the touch. Context white, when sectioned staining pinkish buff under the subcutis and pale vinaceous in the region of the stipe apex, pea-green to bluish green with $FeSO_4$, odor and taste mild.

Tubes ventricose, about 10 mm deep, depressed to practically free around the stipe, snow-white when young, staining cream-buff after cutting; pores white, minute, staining ochraceous when bruised.

Stipe 6-7 cm long, 8-10 mm at apex, 12-13 mm at base, with very fine ornamentation which is white at first, darkening over the lower two-thirds to avellaneous and on drying yellowish over the basal third, base blue where handled, solid, watery-streaked on the cut surface, pallid but slowly staining vinaceous (not becoming gray, however).

Spores 15-21 X 5-6.5 μ, smooth, in face view bluntly fusiform, in profile elongate-inequilateral, suprahilar depression shallow and indistinct, dingy ochraceous and pale tawny singly or in groups respectively as revived in KOH, not changing significantly in Melzer's; wall somewhat thickened and apex with a pallid minute spot but not a true pore.

Basidia 4-spored, 26-32 X 10-13 μ, hyaline in KOH, with globules of various sizes in the interior, hyaline in Melzer's. Pleurocystidia scattered, 48-75 X 9-16 μ, fusoid-ventricose with prolonged narrow neck and subacute apex. Cheilocystidia 18-32 X 6-10 μ, subfusoid to sub-cylindric, smooth, thin-walled, hyaline in KOH. Caulocystidia in the fas-cicles with basidia, 40-75 X 7-15 μ and narrowly fusoid-ventricose neck up to 20-30 μ long, also some large clavate to mucronate cells present which measure up to 20 μ broad, content ochraceous-brown in KOH.

Pileus cuticle of hyphae 7-15 μ wide, the cells more or less in-flated, the end-cells clavate to subglobose, with hyaline to brownish content in KOH and Melzer's. Subcutis of hyaline (in KOH), interwoven, subgelatinous (revived in KOH) hyphae also hyaline in Melzer's. Context hyphae 9-17 μ wide and hyaline to yellowish as revived in Melzer's. Clamp connections absent.

Habit, habitat, and distribution. – Along logging roads through hard-wood forests, with young growth of *Betula lutea* in the vicinity, late summer, Upper Peninsula, rare.

Observations. – This is the species as based on Singer's original de-scription. Watling (1968) has discussed this problem and given an ac-count based on specimens borrowed from the Komarov Bot. Institute (LE) identified by Singer. In more recent accounts Singer (1967) de-scribed the pileus as very dark brown to grayish brown or blackish.

107. Leccinum subgranulosum Smith & Thiers, sp. nov.

Pileus 4-10 cm latus, late convexus, subsquamulosus vel areolatus, pallide spadiceus. Contextus pallidus tactu subluteus. Stipes 7-11 cm longus, 8-16 mm crassus, atrosquamulosus. Sporae 15-21 X 5.5-7.5 μ. Specimen typicum in Herb. Univ. Mich. conservatum est; prope Pellston, July 2, 1967, legit Smith 74398.

Pileus 4-10 cm broad, obtuse to convex, in age broadly convex to nearly plane, margin even to very obscurely crenate and not extended significantly beyond the tubes; surface granulose-squamulose, in old pilei minutely areolate as dried; a dull "snuff-brown" (dingy yellow-brown) and approximately this color as dried. Context pallid, when cut slowly becoming brownish, yellow just above the tubes, brownish in the apex of the stipe.

Tubes whitish becoming bister (very dark yellow-brown) 1-1.5 cm deep, nearly free from the stipe; pores small, pallid, staining ochraceous and then brownish as bruised.

Stipe 7-11 cm long, 8-16 mm thick at apex, evenly and slightly enlarged downward, solid, pallid within, brownish slowly in apex when cut; surface covered by a fine blackish ornamentation above, squamules coarser lower down; ground color and base of stipe whitish.

Spores 15-21 × 5.5-7.5 μ, smooth, wall ~0.5 μ thick, apex with a minute pore; shape in face view elongate-fusiform, in profile elongate-inequilateral; color near "buckthorn-brown" in KOH, more or less dextrinoid in Melzer's.

Caulocystidia 40-80 (120) × (8) 10-16 (20) μ, fusoid to fusoid-ventricose, the neck often greatly proliferated but no secondary cross walls present, apex subacute, and in many with a hyaline refractive deposit against the wall at times involving the whole apex, content mostly smoky brown in KOH. Pileus cutis of hyphae 5-15 μ broad with cells of broader hyphae short and inflated up to 2 μ at times, the cells disarticulating and often floating free in the mount; end-cells cystidioid to bullet-shaped and often greatly reduced in size, walls smooth to ornamented; content smoky brown in KOH and duller as well as granular in Melzer's, no distinct globules present in any of the cuticular hyphae. Context hyphae adjacent to the cuticle hyaline to weakly yellowish in Melzer's. Subcutis of gelatinizing hyphae. Clamp connections absent.

Habit, habitat, and distribution.—Under *Acer saccharum* and *Betula lutea*, Pellston, July 2, 1967, Smith 74398.

Observations.—The distinctive features of this species are the dark yellow-brown granulose-squamulose to fibrillose pilei, lack of a color change to red or gray in the stipe apex, the large spores, the proliferated caulocystidia many with a deposit of highly refractive material in the apex, and the inflated cells in the pileus cutis. It differs from *L. oxydabile* in lacking a change to red on the cut flesh and in having proliferated caulocystidia.

108. Leccinum aberrans Smith & Thiers, sp. nov.

Pileus circa 7 cm latus, convexus, siccus, griseo-brunneus. Contextus pallidus tactu tarde brunneolus. Stipes 5 cm longus, 1 cm crassus, griseus, squamulosus. Sporae in cumulo vinaceo-brunneae, 16-22 × 5-7 μ. Cellulae terminales 15-30 μ crassae. Specimen typicum in Herb. Univ. Mich. conservatum est; prope Vermillion, July 29, 1965, legit Smith 71899.

Pileus about 7 cm broad, convex, margin fertile and even; surface dry, under a lens subtomentose with dark grayish brown tomentum but pallid ground color showing under the tomentum (but specimens not areolate). Context white, soft, taste strongly acid at first but fading, odor none, when cut or injured slowly changing to pinkish tan in the pileus and in the apex of the stipe.

Tubes dark wood-brown, 1.5 cm deep, depressed around the stipe; pores minute, gray where undamaged, where injured staining dark cinnamon-brown.

Stipe 5 cm long, 1 cm thick at apex, solid, white within, very little color change (merely pinkish tan) when sectioned; surface dark gray to apex from closely set punctations, white beneath but ground color hardly showing.

Spore deposit about wood-brown fresh (with a vinaceous tone); spores 16-22 X 5-7 μ, smooth, yellow-brown in KOH singly, rusty brown in groups, in Melzer's dingy yellow-brown or a few dextrinoid; in profile elongate-inequilateral, in face view subfusiform to nearly elongate-oval, with a hyaline apical spot.

Basidia 20-25 X 8-10 μ, 4-spored, clavate, hyaline in KOH. Pleurocystidia 32-46 X 8-12 μ, fusoid-ventricose, the neck often flexuous, apex subacute, hyaline in KOH. Cheilocystidia clavate to fusoid-ventricose, 15-28 X 5-10 μ, pale dingy ochraceous in KOH to hyaline. Caulocystidia clavate to subfusoid or submucronate (not typically fusoid-ventricose), 35-50 X 8-20 μ, many clavate cells also present, caulohymenial elements as well as many ground hyphae with smoky ochraceous content in KOH.

Cuticle of pileus with epicuticular hyphae 5-10 μ wide, with end-cells inflated to 15-25 (30) μ, walls smooth, hyphal cells not disarticulating, in KOH the content dingy ochraceous to dull brown, in Melzer's the hyphae of the subcutis with homogeneous orange-red content and the epicuticular hyphae with dull brown content aggregating in part to form globules and particles 5-10 μ (or more) in diameter and colored smoky brown, end-cells subellipsoid to cystidioid and almost uniformly wider (usually much wider) than the cell from which they were differentiated. Clamp connections absent.

Habit, habitat, and distribution.—Solitary, Vermillion, July 29, 1965, Smith 71899.

Observations.—The distinguishing features of this species are the greatly enlarged end-cells of the epicuticular hyphae, their dull brown pigment masses (globules) as seen in Melzer's, and the mostly clavate to ellipsoid or submucronate caulocystidia. The spore-print color is somewhat problematic as under the microscope the spores are the same color as in the rest of this subsection. The ornamentation of the stipe reminds one of *L. murinaceo-stipitatum,* but the cuticular hyphae of the pileus distinguish them at once.

Subsection ALBELLA   Smith, Thiers, & Watling

Mich. Bot. 6:116. 1967

Pileus epicutis in the form of a trichodermium of elements more or less fused into an epithelium by the 2-6 distal cells of the individual hyphae becoming so closely packed as to obscure the hyphal structure of the layer; the young tubes never truly yellow.

Type species: *Leccinum albellum.*

KEY TO SPECIES

1. Pileus yellow at first. . . . . see *L. luteum*
1. Pileus yellow-brown to fuscous or whitish 2

    2. Context in stipe apex not staining when cut . . . . *L. albellum* and variants
    2. Context staining gray and finally blackish. . . . .*L. griseum*

109. Leccinum albellum (Peck) Singer

Mycologia 37:799. 1945

f. albellum

Boletus albellus Peck, Rept. N. Y. State Mus. 41:149. 1889.
Ceriomyces albellus (Peck) Murrill, Mycologia 1:145. 1909.

Pileus 3-6 cm broad, at maturity convex to obtuse, becoming broadly convex to plano-convex or remaining pulvinate, moist to dry, apparently not becoming viscid or subviscid when wet or old, smooth, occasionally appearing pitted with age, glabrous to subvelutinous to somewhat pruinose at times when young, appearing glabrous in age, sometimes remaining subvelutinous, frequently becoming rimose areolate with age, whitish to pale pinkish buff to pale olive-buff to pallid, sometimes yellowish during all stages of development, margin incurved but not appendiculate, decurved or finally spreading, entire. Context 1-1.5 cm thick, white, unchanging when injured, odor and taste not distinctive.

Tubes broadly and typically deeply depressed when young, becoming more broadly depressed in age, white when young, becoming more or less olive-buff in age, up to 10 mm deep; pores angular, less than 1 mm broad, pallid, unchanging or darkening only slightly in age.

Stipe 5-8 cm long, 7-11 mm broad at apex, subequal to tapering toward the apex, dry, white to pale olive-buff during all stages of development, when young appearing subtomentose to appressed-fibrillose, typically becoming scabrous but often not strongly or conspicuously so in age, scales white when young, typically becoming darker with age or when handled but still pale colored when compared with the genus as a whole, solid, white within, unchanging.

Spores (10) 15-20 (24) X 4-6 μ, smooth, in face view cylindric to subfusiform, in profile somewhat elongate-inequilateral, pale yellow revived in KOH, with a moderately thick wall (-0.5 μ).

Basidia 4-spored, 21-26 X 9-12 μ, clavate, hyaline in KOH. Pleurocystidia scattered to numerous, 30-48 X 7-12 μ, more abundant near the pores, often difficult to locate in old basidiocarps, hyaline, thin-walled, versiform–aciculate, clavate, with elongated, tapered, distal portion, subcylindric or fusoid-ventricose.

Hymenophoral trama obscurely divergent to subparallel, hyaline in KOH and Melzer's. Pileus cuticle more or less of upright hyphae forming a trichodermium but composed of globose to pyriform cells, these disarticulating with age and disappearing leaving a glabrous pileus. Cuticle of stipe differentiated as a layer of loosely interwoven hyphae, the filaments with occasionally upright tips with inflated cells similar to those of the pileus trichodermium, hyaline in KOH or Melzer's. Clamp connections absent.

Habit, habitat, and distribution.—This is a common species of the Gulf Region not yet with an authentic record from Michigan.

Observations.—One reason for including this species here is to get records and complete studies of it and other variants since many of the Gulf Coast fungi occur rarely in Michigan. Singer (1947) described 2 forms, f. *reticulatum* having an alveolate-reticulate pileus which becomes slightly viscid, and f. *epiphaeum* having an olive-gray pileus becoming rimose-tessellate in age.

110. Leccinum griseum (Quélet) Singer

Röhrlinge II. In Pilze Mitteleuropas p. 89. 1967

Gyroporus griseus Quélet, Assoc. Fr. Avanc. Sci. (1901):496. 1902.

Illus. Pl. 90.

Pileus 3-9 cm broad, obtuse to convex becoming broadly convex or with a low broad umbo, at times nearly plane, margin not extending beyond the tubes, glabrous at first and often conspicuously rugulose-pitted, eventually areolate (particularly when dried) as in *Boletus chrysenteron* and then the areolae appearing subtomentose, pallid context showing in the cracks, dingy yellow-brown ("snuff-brown" to "tawny-olive"), in age often olivaceous or pea-green at least in patches. Context very thick and soft, pallid but gradually avellaneous when cut.

Tubes 10-20 mm deep, free or nearly so at maturity, pallid, slowly becoming wood-brown but when bruised staining purplish brown; pores pallid-avellaneous but staining greenish when bruised, the stains often slowly becoming dingy yellowish.

Stipe 4-12 cm long, 8-15 mm thick, pallid to avellaneous and densely furfuraceous to scabrous, the ornamentation avellaneous to wood-brown, base of stipe with white appressed mycelium which frequently is stained greenish, solid, pallid within but slowly staining gray when sectioned, drying whitish over the cut surfaces.

Spore deposit near cinnamon-brown; spores 11-15 X 4.5-6 μ, smooth, fusiform to subfusiform in face view, in profile inequilateral, the suprahilar depression broad and shallow to very distinct, ochraceous to dingy ochraceous in KOH, in Melzer's mostly ochraceous but many very dark rusty brown, wall slightly thickened.

Basidia 4-spored, 26-32 X 9-12 μ, hyaline in KOH, yellowish in Melzer's. Pleurocystidia 30-42 X 9-14 μ, fusoid to fusoid-ventricose, with obtuse apex, thin-walled, smooth, content hyaline in KOH and Melzer's. Pileus trichodermium with elements having the upper 2-4 cells inflated to nearly isodiametric and enlarged to 15-35 μ, in older stages the layer appearing pseudoparenchymatous; with a dull brown cellular content when fresh in water mounts, nearly hyaline as revived in KOH or Melzer's. Clamp connections absent.

Habit, habitat, and distribution.—Gregarious to scattered in old-growth stands of hardwoods, southeastern part of the state during late summer and early fall after heavy rains. Under the right conditions it is abundant.

Observations.—This is basically a glabrous species tending to become rugose-reticulate, but it becomes very conspicuously areolate in age or on drying, a feature not present in anything like the same degree in other species of this group. No pseudocystidia were found. It differs from *L. carpini* in the flesh not distinctly turning vinaceous when basidiocarps are sectioned and in the tubes showing no yellow flush when young.

Subsection LUTEOSCABRA

For the diagnostic features see key to subsections.
Type species: Same as for the section.

KEY TO SPECIES

1. Spores 6-8 μ in diameter . . . . *L. crocipodium*
1. Spores 3.5-6.5 μ wide 2

   2. Pileus typically uneven to pitted; stipe reticulate and furfuraceous-punctate as well as resinous to touch . . . . *L. rugosiceps*
   2. Pileus smooth to slightly rugulose or if pitted then the stipe ornamentation dry to the touch and never a reticulum 3

      3. Stipe ornamentation pallid becoming blackish in age; young hymenophore pallid and pileus subviscid . . . . *L. luteum*
      3. Stipe ornamentation very fine and pallid to yellowish, rarely reddish and scarcely changing in drying 4

         4. Pileus cutis a cellular layer appearing to be 2-4 cells deep . . . .
         . . . . see *Boletus subglabripes*
         4. Pileus cuticle a disarranged hymeniform layer . . . .
         . . . . see *L. brunneo-olivaceum* (not treated here)

111. Leccinum rugosiceps (Peck) Singer

Mycologia 37:799. 1945

Boletus rugosiceps Peck, Bull. N. Y. State Mus. 94:20. 1904.
Krombholzia rugosiceps (Peck) Singer, Ann. Mycol. 40:34. 1942.

Pileus 5-15 cm broad, convex, unpolished, dry to moist but not viscid, uneven to rugulose or pitted, becoming areolate-rimulose in age and then showing the white context in the cracks, orange-yellow to yellow-ocher (duller) and finally some shade of crust-brown. Context thick but thinner than the depth of the tubes, pallid yellowish but slowly changing to dull reddish when cut or when bruised in at least some specimens.

Tubes pale bright yellow at first, dingy ochraceous in age, 10 mm or more deep at maturity, depressed around the stipe to free, not changing color when bruised; pores small and round, less than 1 mm in diameter.

Stipe 8-10 cm long, 2-3 cm thick, fleshy-fibrous, solid, surface reticulate overall and furfuraceous-punctate from a resinous (to the touch) covering, apex pale yellow, pallid-orange-yellow over the remainder from the punctate elements which in age become blackish on the yellow ground color, yellow within, not staining blue anywhere.

Spores (14) 16-21 X 5-5.5 μ, smooth, in face view fusiform, in profile elongate-inequilateral, ochraceous revived in KOH, in Melzer's pale tawny but some finally dark reddish brown overall or over the basal half, the wall slightly thickened.

Basidia 4-spored, 24-30 × 10-12 μ, clavate, yellowish in both KOH and Melzer's, mostly with a homogeneous content. Pleurocystidia 36-48 × 9-13 μ, fusoid-ventricose with more or less obtuse apex, thin-walled, hyaline, smooth, content yellow at first but the pigment diffusing into the mount, not distinctive in either KOH or Melzer's. Cheilocystidia similar to pleurocystidia or with apex more acute, many yellowish in KOH. Caulocystidia 35-70 × 7-15 μ, fusoid-ventricose to mucronate, thin-walled, smooth, content hyaline to yellow-brown as revived in KOH; brachybasidioles also present and intergrading with the caulocystidia.

Hymenophoral trama of the bilateral *Boletus* subtype, the hyphae with yellow content when revived in KOH but pigment diffusing into the mounting medium; subhymenium cellular to filamentose. Pileus trichodermium more or less cellular and consisting of enlarged hyphal cells with the upper 3-6 cells of the filament subglobose and the largest 10-25 (30) μ in diameter, the ultimate and penultimate cells often greatly reduced in size and cystidioid in shape (10-30 × 7-14 μ) or at times almost reduced to a papilla on the end of the trichodermial element, the third and fourth cells back usually the largest, all of them thin-walled, smooth, yellow in KOH but soon fading and in Melzer's with reddish orange to orange-brown content. Context of inflated thin-walled hyphae "empty" as revived in KOH and Melzer's or a few of the hyphae with ochraceous to orange content; laticiferous elements occasional, (4) 6-12 (15) μ wide. Clamp connections absent.

Habit, habitat, and distribution.—Solitary to gregarious in thin grassy oak woods during wet weather in July, August, and early September, rare in Michigan and to be expected in the lower 2 to 3 tiers of counties in the Lower Peninsula. It is much more common southward.

Observations.—This species is most closely related to *L. crocipodium,* but the latter differs in wider spores and a less cellular cuticle. From *B. subglabripes* it is readily distinguished by the resinous feel of the stipe and the darkening stipe ornamentation.

112. Leccinum.crocipodium (Letellier) Watling

Trans. Bot. Soc. Edinb. 39:200. 1961

Boletus crokipodius Letellier, Champ. Pl. 666. 1838.

Pileus 4-7.5 cm broad, obtuse expanding to convex or the margin finally flaring, the margin obtuse when young, not viscid but the surface with a resinous feel, soon becoming irregularly rimose, very rugulose to pitted, the prominences soon blackish brown, the valleys mottled with pallid where the context shows, on drying the surface blackish with an undertone of yellow. Context pallid to yellowish pallid, soon stained vinaceous-gray where cut, finally distinctly vinaceous near apex of stipe, green with $FeSO_4$ and distinctly orange-ochraceous with KOH; odor and taste not distinctive.

Tubes about 10 mm deep, depressed around the stipe, grayish brown in sections; pores about 2 per mm, pale olive-ochraceous and staining brown where injured, yellow when dried.

Stipe 5-7 cm long, 10-15 (17) mm thick at apex, equal to narrowed at base at maturity, solid, olive-buff and brownish resinous-furfuraceous from discolored glandular papillae, more or less squamulose below and at base scarlet to dull red in at least some American specimens, interior pallid to olive-buff, brown around the larval tunnels, staining vinaceous when cut, cortex yellowish.

Spore deposit honey-yellow; spores 14-20 X 6-9 μ, smooth, in face view broadly ventricose-fusiform to fusiform, in profile broadly inequilateral, dingy yellow-brown in KOH, more rusty tan in Melzer's but not dextrinoid, wall about 0.5 μ thick in Melzer's but appearing thicker in KOH, apex with a minute pore in some spores.

Basidia 4-spored or some 1-spored, (18) 20-28 X 9-13 μ, short-clavate with a broad base or more elongate-clavate and with a narrow (3-4 μ) base, hyaline in KOH, yellowish in Melzer's but with no distinctive content in either medium. Pleurocystidia rare, 38-62 X 9-15 μ, thin-walled or wall slightly thickened and distinctly refractive, smooth, hyaline to yellowish in either KOH or Melzer's. Cheilocystidia 18-36 X 6-12 μ, fusoid-ventricose to clavate-mucronate, content dingy yellow-brown in KOH, walls thin to slightly thickened (0.3μ), and slightly refractive as in the pleurocystidia. Caulocystidia 30-66 X 7-15 (20) μ, narrowly fusoid-ventricose with flexuous neck and often with a short narrow subacute to rounded apical extension, or broadly fusoid-ventricose to clavate-mucronate, content hyaline to yellowish and walls slightly thickened and refractive or not in KOH.

Tube trama with a thin central scarcely gelatinous strand of thin-walled hyphae from it on either side a slight divergence of the hyphae giving an obscurely bilateral effect; subhymenium cellular to filamentose. Pileus trichoderm of tangled hyphae 8-15 μ wide, the cells of various lengths but some inflated to 20-30 X 14-20 μ, not forming a chain of subglobose cells at the distal end as in *L. rugosiceps* although the end-cell is often reduced in size as in that species, content ochraceous-brown in KOH and dark dingy orange-brown in Melzer's. Context hyphae beneath the trichodermium with dark brownish orange content as revived in Melzer's. Clamp connections absent.

Habit, habitat, and distribution.—This species is apparently very rare in North America. Our description was taken from material found at the Ypsilanti golf course, Washtenaw County, September 15, 1961. It grew under oak on rather hard-packed soil.

113. Leccinum luteum Smith, Thiers, & Watling

Mich. Bot. 6:114. 1967

Illus. Pl. 91.

Pileus 3-6.5 cm broad, convex to obtuse when young, remaining merely obtuse in age, glabrous, subviscid, alveolate, or smooth and even, becoming areolate-rimose, pale yellow, becoming olive-brown in age, on drying becoming nearly date-brown, with KOH orange-cinnamon to pale red. Context thin, soft, pallid or flushed yellow near the trichodermium, staining vinaceous-gray to avellaneous when sectioned; slowly olive with $FeSO_4$, odor none, taste acidulous or mild.

Tubes ventricose, up to 15 mm deep, deeply depressed around the stipe to nearly free, pallid to pale gray to fuscous; pores on old pilei about 2-3 per mm (small), pallid when young, slowly becoming olive-buff and when injured avellaneous or wood-brown.

Stipe 6-13 cm long, 8-10 mm thick at apex, 10-15 (17) mm at base, nearly equal, pallid and solid, cut surface in lower portion staining pale green in local areas and fuscous overall slowly, dark blue with guaiac; surface pallid above, pale yellow below, ornamentation present as sparse fine blackish points and squamules.

Spore deposit olive-brown; spores 17-20 X 5-6.5 μ, smooth, no apical differentiation visible or merely as a thin spot, in face view sub-fusoid, in profile elongate-inequilateral, dull yellow-brown to dingy clay color in KOH, nearer ochraceous-tawny in Melzer's but not dextrinoid, wall about 0.5 μ thick as measured in Melzer's, slightly thicker revived in KOH.

Basidia 18-20 X 10-14 μ, near tube mouths 25-30 X 10-15 μ near pileus context, clavate, content homogeneous in KOH or with a large hyaline globule, merely yellowish in Melzer's but a few with dark yellow-brown granules. Pleurocystidia scattered, 37-52 (75) X 6-10 (12) μ, fusoid to subfusoid with acute to subacute apex, smooth, thin-walled, not incrusted, content when revived in KOH homogeneous and hyaline to pale ochraceous to ochraceous-brown, dark yellow-brown (bister) in many when revived in Melzer's. Cheilocystidia similar to the pleurocystidia but varying to subcylindric to narrowly clavate and mostly with colored content. Caulocystidia mostly fusoid-ventricose with proliferated neck, 50-120 X 10-20 μ, flexuous and many branching once but lacking a cross wall at the base of the neck, many appearing to be flexuous hyphal end-cells with only a slight basal or subbasal inflation, both types intergrading, smooth, thin-walled, hyaline to bister when revived in KOH and globules tending to form in some. Caulobasidioles rare and intergrading with caulocystidia. Caulobasidia mostly 2-spored, usually with yellow-brown content when revived in KOH.

Tube trama of hyaline thin-walled, smooth hyphae with a subfloccose central strand and a diverging subgelatinous lateral band of hyphae extending into the subhymenium which is cellular near the tube mouths but hyphal near the pileus trama. Pileus trichodermium very compact, composed of elements in which the 2-4 distal cells in each filament are inflated to 10-30 μ and are subglobose, walls thin and smooth, content yellow in KOH and paler bister in Melzer's but no pigment globules forming. Context of thin-walled hyphae dingy yellowish revived in KOH or in Melzer's, cells inflated to 10-20 μ. Clamp connections absent.

Habit, habitat, and distribution.—Scattered under blue beech (*Carpinus caroliniana*) on low ground; known from the southeastern part of the state.

Observations.—The yellow pileus, the cellular cuticle, the KOH reaction of the cells of the trichodermium when fresh, the yellow tint over most of the stipe in age, the color changes in the stipe apex when cut, and the proliferating caulocystidia as a combination distinguish this species.

BOLETUS Fries

Syst. Mycol. 1:385. 1821

Hymenophore tubular, the pores wide and angular to minute and round; context distinctly fleshy and soon riddled by insect larvae or decaying; basidiocarp consisting of a pileus, hymenophore, and stipe but stipe not always central; spores more or less elongate-inequilateral in most species and smooth to ornamented (but species with longitudinally striate spores are segregated in *Boletellus*); spore deposit dull yellow-brown to amber-brown, cinnamon-brown, bister, olive-brown, olive, or earth-brown (in air-dried deposits); stipe lacking ornamentation of squamules which become gray to dark brown or black at maturity or in age (or in rare instances are colored from the beginning against a paler ground color but which do not darken in aging).

Type species: *Boletus edulis.*

KEY TO SECTIONS

1. Spores ornamented but not by longitudinal lines or ridges; stipe more or less lacerate-reticulate . . . . Sect. *Allospori*
1. Spores not ornamented (though apex may be truncate) 2

   2. Spores flattened (truncate) or notched at apex . . . . Sect. *Truncati*
   2. Spore apex unmodified 3

3. Hymenophore in age tending to become pinkish red throughout or in part; taste acrid in some species; pileus subviscid or merely soft to the touch . . . . Sect. *Piperati*
3. Not as above 4

   4. Pores some shade of orange, red, dark brown to bay-brown when young . . . . Subsect. *Luridi* of Sect. *Boletus*
   4. Pores not colored as above when young 5

5. Stipe typically reticulate at least near apex; pores usually stuffed when young . . . . Sect. *Boletus*
5. Stipe furfuraceous to pruinose or glabrous, at times coarsely ridged or with a coarse wide-meshed reticulum but this not constant 6

   6. Pileus unpolished to velvety or subtomentose. . . . . Sect. *Subtomentosi*
   6. Pileus glabrous and moist or viscid 7

7. Stipe furfuraceous to punctate but ornamentation not darkening as in *Leccinum* . . . . Sect. *Pseudoleccinum*
7. Stipe more or less pruinose to naked . . . . Sect. *Pseudoboleti*

Section SUBTOMENTOSI Fries

Epicr. Syst. Mycol. p. 412. 1838

The diagnostic characters are given in the key to sections.
Type species: *Boletus subtomentosus.*

The majority of the species included here have been placed in *Xercomus* by modern authors, and we also have tried to recognize this group as distinct from *Boletus* at times but have finally given it up. Watling (1968) has also recognized that on a world basis the genus is untenable. In the past various characters have been used in attempts to circumscribe it such as: Do the tubes separate from each other when the pileus is broken downward through the tubes, or do the tubes split in such a way that you see the hymenial surface as you view the tube layer? We have found this distinction unusable and so cannot consider it of importance in the recognition of a genus. As most boletes have the pileus cuticle in the form of a trichodermium which often collapses to an interwoven tangled layer of hyphae, this feature is of little help at the generic level although the rather coarse development of a trichodermium is a feature of the type species of *Xerocomus* and its closest relatives. It is now known that the trichodermium and the ixotrichodermium are common to most of the genera of the Boletaceae. In fact it is one of the basic types for the family.

Our approach is to divide the section as here defined into numerous smaller groups of obviously closely similar species in order to contrast species sufficiently to allow them to be fairly readily recognizable. In this section, however, it is always an open question as to whether one is dealing with relationships as compared to accidental parallelisms. The major problem, in spite of all that has been done in this group, is to clearly define the species.

KEY TO SUBSECTIONS

1. Basidiocarps attached to basidiocarps of the gastromycete *Scleroderma*
. . . . Subsect. *Parasitici*
1. Habitat terrestrial or lignicolous 2

    2. Spores often over 20 μ long; pileus granulose-squamulose at maturity
    . . . . .Subsect.*Mirabiles*
    2. Not as above 3

3. Spores short (7-9 × 3-3.5 μ); habitat on decayed wood . . . . . Subsect. *Sulphurei*
3. Spores longer than in above choice or habitat terrestrial 4

    4. At least some spores notched or truncate at apex . . . . . see Sect. *Truncati*
    4. Spores lacking above-mentioned differentiation at their apex. 5

5. Pileus dark to bright red when young; tubes yellow and pores staining blue (often slowly) when bruised (but see *B. carminipes* also) . . Subsect. *Fraterni*
5. Pileus differently colored when young 6

   6. Pores staining greenish to blue when injured (often slowly) . . . . Subsect. *Subtomentosi*
   6. Pores staining yellow to brownish or unchanging when bruised . . . . Subsect. *Versicolores*

Subsection SULPHUREI (Singer), comb. & stat. nov.

Phlebopus section *Sulphurei* Singer

Amer. Midl. Nat. 37:2. 1947

Stirps SPHAEROCEPHALUS

114. Boletus sphaerocephalus Barla

Champ. Nice p. 72. 1859

Ixocomus sphaerocephalus (Barla) Quélet, Flore Mycol. Fr. p. 419. 1888.

Illus. Pl. 92.

Pileus 5-10 cm broad, convex with a curved-in irregular margin, expanding to broadly convex or nearly plane; surface soft and subviscid to tough but no gelatinous layer present, unpolished to subtomentose, in age at times areolate; color evenly sulphur-yellow overall but on aging slowly fading to pallid; veil lacking. Context thick, sulphur-yellow under the cuticle, paler inward, turning blue when cut, with $FeSO_4$ no reaction, taste mild to bitterish, odor slight.

Tubes adnate becoming depressed around the stipe, pale sulphur-yellow when young, when injured staining blue, 10-15 mm deep; pores 2-3 per mm when young, broader at maturity, staining blue and then slowly brownish where injured.

Stipe 6-10 cm long, 1-2.5 cm thick (or larger in giant basidiocarps), central to eccentric, narrowed to a point below, solid, yellow within, surface yellow and unpolished (appearing somewhat matted-fibrillose), yellow tomentum and mycelium usually holding much debris around the base.

Spore deposit olive brownish or tinged ochraceous in the olive-brown range; spores 7-9 $\times$ 3-3.5 μ, oblong to ellipsoid, walls slightly thickened, smooth, lacking an apical pore, dull ochraceous in Melzer's, bright yellow in KOH.

Basidia 4-spored, 27-36 × 7-9 μ, narrowly to abruptly clavate, ochraceous in Melzer's, in KOH yellow fading to hyaline. Pleurocystidia 33-50 × 9-15 μ, mostly fusoid-ventricose, thin-walled and with ochraceous to rusty ochraceous content as revived in KOH, none seen with incrustations.

Tube trama of long-cylindric yellow to hyaline (revived in KOH) filaments 5-8 μ in diameter, subparallel in arrangement with diverging elements to the subhymenium. Cuticle of pileus of interwoven hyphae dull orange-ochraceous in Melzer's, lacking incrustations in KOH and yellow fading to hyaline, the epicuticular zone compactly interwoven, hyphal· end-cells with blunt to narrowed apex and some inflated to 8-12 μ in the midportion. Clamp connections absent.

Habit, habitat, and distribution.—Cespitose on sawdust, near Betsy Lake, Luce County, August 9, 1963, H. D. Thiers; rare in Michigan.

Observations.—The name *Boletus sulphureus* is not available for this species (Watling, in press). No veil was present in our collection. Barla's original plate shows an extension of the margin of the pileus somewhat like that found in Section *Leccinum* of *Leccinum.* The habitat on wood, the colors, and the short, ellipsoid to oblong spores form a distinctive combination of features.

Subsection MIRABILES (Singer), stat. et comb. nov.

Boletellus section *Mirabiles,* Singer

Farlowia 2:128. 1945

KEY TO SPECIES

1. Growing on or beside decaying conifer wood; pileus granulose-roughened
   to coarsely rough-tomentose . . . . *B. mirabilis*
1. Terrestrial; surface of young pileus subtomentose . . . .*B. projectellus*

115. Boletus mirabilis Murrill

Mycologia 4:217. 1912

Ceriomyces mirabilis Murrill, Mycologia 4:98. 1912.
Xerocomus mirabilis (Murrill) Singer, Rev. de Mycol. 5:6. 1940.
Boletellus mirabilis (Murrill) Singer, Farlowia 2:129. 1945.

Illus. Pls. 93-94.

Pileus 5-16 cm broad, convex when young, the margin inrolled, becoming broadly convex in age, sterile margin distinct and either continuous or broken segments, hence somewhat crenate, surface of the buttons with a lubricous feel when wet, otherwise dry and at maturity coarsely rough tomentose to somewhat areolate or rimose-scaly, almost granular-scaly in some, color dark dull red to brownish red to grayish brown ("chocolate," "burnt-umber," "dark vinaceous-fawn," or duller). Context thick, firm, pallid, or tinged vinaceous near the cuticle, unchanging when bruised, yellow when dried, odor and taste not distinctive.

Tubes depressed around the stipe, about 2 cm long, pallid at first but very soon pale yellow and finally greenish yellow ("olive-ocher"), when bruised staining mustard-yellow rather readily; pores round or nearly so, about 1 mm in diameter, staining yellow when bruised.

Stipe 8-12 (20) cm long, 1-3.5 cm thick at apex, clavate, 3-7 cm at base, solid, cortex sordid vinaceous near apex, inward vinaceous especially around the larval tunnels, becoming yellow in drying; surface dull sordid reddish to reddish brown, at times with a tinge of purple (more or less concolorous with pileus), often reticulate near apex, cuticle often splitting and becoming rough and unpolished, with a thin pruinose coating at first, surface uneven to shallowly pitted at times, the base usually becoming a darker reddish brown where handled.

Spore deposit olive-brown; spores 19-24 X 7-9 μ, smooth, shape in profile somewhat inequilateral, in face view subventricose or varying to narrowly elliptic, color pale ochraceous in KOH, revived in Melzer's many spores clouded grayish and only a few clearly dextrinoid, lacking an apical pore.

Basidia 4-spored, 36-42 X 10-12 μ, hyaline in KOH, yellow in Melzer's. Pleurocystidia 52-74 X 10-15 μ, fusoid-ventricose with apex obtuse, thin-walled, hyaline in KOH, fairly abundant.

Tube trama gelatinous, the hyphae divergent from a central floccose strand. Cuticle of pileus a turf of filamentose septate hyphae the cells comparatively short (26-80 X 10-15 μ), the pigment rapidly disintegrating in KOH, groups of hyphae pale yellowish, walls smooth in KOH but with amorphous debris present in the fascicles. Clamp connections absent.

Habit, habitat, and distribution.—Solitary to subcespitose on decayed conifer logs, mostly on hemlock (*Tsuga*). It is rarely collected in the Upper Peninsula (Luce County, Smith 38627) in regions where hemlock grows, but is common in the Pacific Northwest. It is an excellent edible species.

Observations.—The large spores, colors, rough pileus, and habitat on conifer logs make it a species anyone can learn to recognize accurately. The reticulation of the stipe apex may be quite coarse and conspicuous (Pl. 88) or it may be absent. We have seen specimens with pileus dull cinnamon ("Sayal-brown") which is much duller than the colors as given in the description. We have never seen a veil of any sort on young basidiocarps though hundreds have been examined. In Michigan the closely related *B. projectellus* is much more frequent. It is terrestrial, usually growing on sandy soil under pine.

116. Boletus projectellus Murrill

Mycologia 30:525. 1938

Ceriomyces projectellus Murrill, Mycologia 30:524. 1938.
Boletellus projectellus (Murrill) Singer, Farlowia 2:129. 1945.

Illus. Pl. 95.

Pileus 4-12 cm broad, convex becoming plane, margin slightly overlapping the tube layer; surface dry, subtomentose, soon areolate-squamulose, dry at all times; color pale to dark dingy cinnamon varying to olive tinged in some areas or more or less reddish and at times bay-red. Context pallid, when cut becoming flushed vinaceous near the cuticle but elsewhere slowly changing to yellow-brown, scarcely colored around larval tunnels, taste acid, odor not distinctive.

Tubes pale olivaceous, 1-1.5 cm long, depressed around the stipe, not staining when cut; pores 1-1.5 (2) mm broad, round or nearly so, pale olive fresh, staining lemon-yellow bruised.

Stipe 7-12 cm long, 1-2 cm thick, equal or slightly enlarged downward, solid, pallid within, base with a white mycelial outer layer, then a dingy olive-brown zone outlining the cortex, interior pale pinkish buff with some olive-brown stains from the context having been cut; surface dingy vinaceous-buff to cinnamon-buff (or reddish below), unpolished and both coarsely and shallowly reticulate at times nearly to base.

Spore deposit olivaceous; spores $18\text{-}33 \times 7.5\text{-}10$ (12) μ, ovate to boat-shaped in face view or at times slightly angular, in profile inequilateral and with a prominent suprahilar depression, smooth, wall thickened slightly (less than 1 μ thick), when dried hymenophoral tissue is crushed in Melzer's the tissue becomes bluish black and under transmitted light many spores seen to be amyloid but on standing a few minutes many become dextrinoid at least in the distal half or on one side or, more rarely, overall; lacking a distinct apical pore, in KOH golden yellow and wall measuring 1-1.5 μ thick.

Basidia (3) $40\text{-}50 \times 9\text{-}12$ μ, clavate, yellowish in KOH fading to hyaline, 2- and 4-spored. Pleurocystidia subcylindric to fusoid-ventricose, $55\text{-}80 \times 9\text{-}18$ μ, smooth, thin-walled, hyaline, readily collapsing, some also present on tube edges as cheilocystidia.

Tube trama as revived in KOH parallel to subparallel, the outer hyphae diverging slightly to the subhymenium. Pileus cuticle a collapsed trichodermium of hyphae 4-8 μ wide, thin-walled, smooth and at first (as revived in KOH) with greenish yellow content, the elements branched to some extent; hyphae of context 8-15 μ wide, thin-walled and hyaline in KOH. Clamp connections none. Tissues of pileus and hymenophore nonamyloid under the microscope.

Habit, habitat, and distribution.—Under pine (either red or white), and to be expected throughout the pine region of the state, but to date found most abundantly along the south shore of Lake Superior. Mrs. Bartelli has collected it in quantity in the Marquette region.

Observations.—Mrs. Bartelli has collected a variant in which the margin of the pileus is fertile, the sterile band—for which the species was named—being absent. The Melzer's reaction on the spores is much like that for *B. amylosporus,* only more pronounced.

Material examined.—Luce: Smith 72030, 72842. Marquette: Bartelli 2156, 2157, 2284, 2425, 2517, 2525, 2529, 2540; Smith 67677.

Subsection PARASITICI (Singer), comb. et stat. nov.

Xerocomus section *Parasitici* Singer

Ann. Mycol. 40:43. 1942

Type species: *Boletus parasiticus.*

117. Boletus parasiticus Fries

Syst. Mycol. 1:389. 1821

Xerocomus parasiticus (Fries) Quélet, Fl. Myc. Fr. p. 418. 1888.

Illus. Pl. 96.

Pileus 2-8 cm broad, convex becoming broadly convex; surface dry and unpolished, appearing matted-fibrillose under a lens; color evenly "tawny-olive" (dingy yellow-brown), or varying more toward olive (buffy brown), with a narrow sterile margin incurved at first. Context pale lemon-yellow, unchanging when cut, instantly orange-ochraceous in KOH, with $FeSO_4$ pale olive, KOH on cuticle of pileus dark orange-brown.

Tubes 3-5 mm deep or in age up to 1 cm, adnate to decurrent by lines, or in age depressed around the stipe but still with decurrent lines, honey-yellow to a more greenish yellow; pores angular, honey-yellow, 1-2 mm broad, slowly staining ochraceous where bruised.

Stipe 3-6 cm long, 8-13 mm thick, equal, solid, pale yellow within, surface under a lens of appressed, matted fibrils to fibrillose-squamulose with yellow-brown fibrils ("bister" to "tawny-olive"), instantly cinnabar-orange in KOH, at base matted-fibrillose and pallid yellowish from the fibrils.

Spore deposit dark olive. Spores 12-18.5 X 3.5-5 μ, smooth, in face view boat-shaped, in profile narrowly inequilateral, color bright yellow revived in KOH, dingy yellowish tan with a gray shadow in Melzer's, walls slightly thickened.

Basidia 4-spored, 35-40 X 8-10 μ, clavate, yellow in KOH, hymenium bluish black crushed out in Melzer's but no amyloid elements showing under the microscope. Pleurocystidia abundant, hyaline to yellow in KOH, 35-50 (60) X 8-14 μ, fusoid-ventricose, tapered to an acute apex, rarely obtuse. Cheilocystidia similar to pleurocystidia but shorter and with shorter necks and more obtuse apices.

Trichodermium of pileus of hyphae with hyaline incrustations on many which disappear in Melzer's and more slowly in KOH, the hyphae 4-9 μ wide, scattered in the layer are cells inflated to 15-20 μ and yellowish to hyaline under the microscope (in KOH), terminal cells tubular with apex obtuse. Context bluish black in Melzer's but no amyloid hyphae present, in KOH the pileus tissue canary-yellow at first, soon fading. Clamp connections absent.

Habit, habitat, and distribution.—Solitary to cespitose on basidiocarps of *Scleroderma vulgare*, rare in Michigan but known from Tahquamenon Falls State Park, Pellston, and Otis Lake in Barry County. It is more common in the southeastern states.

Observations.—This is a readily recognized species by virtue of its parasitic habit of growing on basidiocarps of *Scleroderma*, the dark fibrils on the stipe, and the brilliant colors when tested with KOH.

Subsection VERSICOLORES, subsect. nov.

Tubuli tactu lutei vel brunnei sed non caerulei.

In this group there is no change to blue on the injured surface of the hymenophore. The pilei are variously colored depending on the species, and many species have short, inflated cells in the trichodermial hyphae of the pileus.

Type species: *Boletus illudens.*

KEY TO STIRPES

1. Spore deposit bright yellow-brown (about amber-brown or brighter)
. . . . Stirps *Affinis*
1. Spore deposit duller than in above choice, usually olive to olive-brown
or dark yellow-brown. 2

   2. Stipe narrowed at base and firmly attached to substratum by a mat
of mycelium. . . . . .Stirps *Tenax*
   2. Not as above 3

3. Pileus appearing glabrous, the cuticle a rather well-defined epithelium of
inflated cells; stipe ornamentation (if present) not darkening with aging
. . . . see Sect. *Pseudoleccinum*
3. Not as above. 4

   4. Spores 8-11 × 3-5 μ . . . . Stirps *Roxanae*
   4. Spores typically 10 μ or more long 5

5. Pileus trichodermium containing elements having some short cells which
are inflated, at times many of them . . . . Stirps *Illudens*
5. Pileus trichodermium of hyphae with essentially tubular cells . . . .
. . . . (see Stirps *Subtomentosus* also) Stirps *Subpalustris*

Stirps AFFINIS

118. Boletus affinis Peck

Ann. Rept. N. Y. State Mus. 25:81. 1873

Suillus affinis (Peck) Kuntze, Rev. Gen. Pl. 3(2):535. 1898.
Ceriomyces affinis (Peck) Murrill, Mycologia 1:149. 1909.
Xanthoconium affine (Peck) Singer, Amer. Midl. Nat. 37:88. 1947.

Var. affinis

Illus. Pls. 97, 101 (above).

Pileus 5-10 cm broad, convex to broadly convex; surface dry and decidedly rugulose overall at first, irregularities not so pronounced in age, vinaceous-brown ("pecan-brown") overall at first, gradually changing to "clay color" or "cinnamon-buff" (pale dingy tan), surface micaceous under a lens, margin fertile. Context firm, white, unchanging, taste mild to faintly disagreeable, odor not distinctive, with $FeSO_4$ slowly olive-gray; NH_4OH on cuticle rusty tan.

Tubes 10-12 mm long, becoming depressed around the stipe, white at first, becoming "cinnamon-buff" or a stronger yellowish tan; pores round, 1-2 per mm, staining yellowish slightly when injured.

Stipe 8-12 cm long, 10-15 mm thick, slightly enlarged downward, often pointed at base, solid, flesh white and unchanging, apex faintly reticulate at times, "fawn color" (vinaceous-tan) except for a whitish base and pallid apex, more or less streaked with whitish areas, evenly pruinose under a lens.

Spore deposit near "buckthorn-brown" fresh (rather bright yellow-brown); spores 12-16 X 3-3.5 μ, narrowly ventricose to more or less cylindric in face view, narrowly inequilateral in profile view, smooth, hyaline sheath indistinct, color in KOH pale dingy yellow, pale tawny in Melzer's.

Basidia 4-spored, 30-36 X 7-8 μ, hyaline in KOH and Melzer's. Pleurocystidia 36-48 X 7-12 μ, fusoid-ventricose and hyaline, thin-walled, apex subacute.

Tube trama parallel to obscurely divergent, subgelatinous. Pileus cuticle not distinctively colored in KOH, in the form of a trichodermium with cells short and near the distal end isodiametric and compacted into an epithelium from mutual pressure, the terminal cells clavate, 24-36 X 10-20 μ, walls pale brownish yellow in KOH. Context hyphae interwoven, floccose, all hyphae nonamyloid. Clamp connections absent.

Habit, habitat, and distribution.—The typical variety of this species is rare in Michigan; we have it from Haven Hill in Oakland County.

Observations.—Singer (1947) commented on the type variety being common in the south but the spotted variety not so; with us it is the opposite.

118a. Boletus affinis Peck var. maculosus Peck

Rept. N. Y. State Mus. 32:57. 1879

Illus. Figs. 108, 111.

Pileus 3.5-10 cm broad, convex expanding to broadly convex to nearly plane, often highly irregular during all stages of development, margin strongly incurved becoming slightly to broadly decurved with age, entire, often becoming irregularly crenulate and incised with age; surface dry, subpruinose to minutely velvety-tomentose, smooth to rugulose or pitted, often markedly rimose in old basidiocarps in dry weather, colored dark seal-brown at first, becoming dark yellow-brown and finally strongly ochraceous ("dark seal-brown," "raw umber" or "bister" to "mummy-brown" or lighter to "Brussels-brown" to "Dresden-brown" or "buckthorn-brown"), more or less spotted with paler yellow to pallid rounded irregular spots. Context white, compact, firm, unchanging or rarely becoming yellow when exposed, frequently pale yellow around wormholes, up to 2 cm thick; taste mild, odor faintly disagreeable, KOH orange to cinnabar on tubes of mature pileus, NH_4OH slightly reddish tan; $FeSO_4$ scarcely any change.

Tubes adnate to short-decurrent when young, becoming depressed with age, typically white to "pale pinkish buff" when young, changing to yellow ("Mars-yellow," "primuline-yellow" or "ivory-yellow") with age, occasionally some more tawny than yellow; pores 1-2 per mm, round, "olive-ocher" when bruised, or in age not changing, not stuffed when young.

Stipe 3-6 (10) cm long, 1-2 cm thick at apex, more or less equal or tapering toward the base or rarely ventricose, frequently with an enlarged base, solid, flesh white, surface dry, glabrous to pruinose, typically distinctly reticulate at apex but occasionally smooth, colored white, brownish or pale yellow, sometimes flushed vinaceous, often with reddish tan streaks, occasionally darkening when bruised.

Spore deposit "buckthorn-brown" fresh; spores 9-16 $\times$ 3-5 μ, smooth, nearly oblong in face view or varying to slightly ventricose, in profile narrowly inequilateral, with an inconspicuous hyaline sheath, pale golden ochraceous in KOH, pale brownish in Melzer's.

Basidia 4-spored, 17-22 $\times$ 9-10 μ, hyaline in KOH and readily disintegrating. Pleurocystidia scattered, 30-40 $\times$ 9-12 μ, ventricose with aciculate neck and an acute apex, hyaline or ochraceous-tawny in KOH, thin-walled. Cheilocystidia golden-yellow in KOH from the colored content, some of them hyaline, fusoid-ventricose with tapered neck and subacute apex.

Tube trama hyaline and gelatinous, more or less divergent. Cuticle of pileus golden yellow to golden brown in KOH, the trichodermium at the upper level of the elements compacted into an epithelium of cystidioid and clavate end-cells, their content golden yellow in KOH and nonamyloid, cystidioid cells 40-80 $\times$ 10-20 μ, apex often acute. Context hyphae of hyaline thin-walled, nonamyloid hyphae. Caulobasidia abundant. Caulocystidia scattered to numerous, similar to cheilocystidia.

Habit, habitat, and distribution.—Scattered to cespitose in hardwood forests especially stands which have been opened by selective logging, common during wet summers in northern Michigan.

Observations.—A feature frequently noticed in Michigan material is the way the margin of the pileus becomes intergrown with the stipe, reminding one of the condition frequently encountered in *Gastroboletus.* We venture to predict that a *Gastroboletus* of the *B. affinis* relationship will eventually be found.

The type collections of the type variety and var. *maculosus* were studied. Smith felt that the material was no longer reliable for chemical features, but some details were verified.

B. affinis (type): Spores 10-13 X 3-3.5 μ, smooth, lacking an apical pore; shape in face view narrowly subfusiform, in profile obscurely to somewhat inequilateral, color in KOH yellowish to nearly hyaline, in Melzer's weakly dextrinoid to ochraceous, wall thin (-0.2 μ).

Basidia 4-spored, 6-8 μ in diameter, clavate, hyaline in KOH, readily gelatinizing. Pleurocystidia 33-50 X 8-13 μ, clavate to fusoid-ventricose, content ochraceous in KOH, darker (dark red) in Melzer's. Cheilocystidia smaller than pleurocystidia, content ochraceous in KOH. Cuticle of pileus apparently cellular but not reviving well, no distinctive Melzer's reactions evident but dark debris present—possibly of foreign origin. No distinctive reactions in Melzer's were evident in the context or subcutis. Clamp connections not found.

B. affinis var. *maculosus*. The spores measure 10-13 X 3.5-4.5 μ and are slightly more ventricose in profile than in the type variety, but in face view they are suboblong. The pileus cuticle is a trichodermium of enlarged cells 10-25 μ in diameter in chains with the terminal cell cystidioid to clavate or vesiculose. In Melzer's the material is ochraceous but no distinctive content present either in this layer or in the hyphae beneath. No clamp connections were found.

From all that could be determined from the types of both taxa they are the same as the Michigan collections. However, we have a variant in Michigan (Smith 75796) in which the stipe is white and smooth or rarely faintly reticulate at the apex. Also the stipe dries ochraceous, but it is not vinaceous-brown as in the type variant. The laticiferous elements as revived in KOH have a greenish yellow amorphous content and this is evident in the pleurocystidia also. In Melzer's the cuticular elements are flooded with copious amounts of dull tawny incrusting debris (with much of it free from the hyphae themselves). The pilei were not spotted when fresh but dried distinctly rimulose with the pale context showing. This species complex deserves further study in the state. Our material examined is merely identified to species without any attempt to list each variant separately.

Material examined.—Cheboygan: Charlton G-223, G-274; Smith 36851, 36984, 37327, 42815, 58255; Thiers 1103, 3611, 3758, 4020. Chippewa: Shaffer 2204; Smith 72357, 72379, 72438. Crawford: Thiers 3395. Emmet: Kauffman 1905; Smith 25855; Thiers 3583, 3713, 4238, 4428. Gratiot: Potter 8337, 10949, 11735, 11765, 11775, 11777, 12869, 13151. Mackinac: Smith 37362. Marquette: Bartelli 2527,

2536, W-288..Oakland: Smith 18537, 64255. Luce: Beach 138; Shaffer 1580; Smith 37080, 37170, 37257, 37441, 37796, 39280, 42296, 42469, 42624; Thiers 3027, 3137, 3855, 3868, 4059, 4100, 4353. Washtenaw: Kauffman 7-29-27; Thiers 4557.

Stirps ROXANAE

KEY

1. Pleurocystidia 38-70 (100) × 9-16 μ, content dark brown in Melzer's, yellow in water mounts or in KOH. . . . . *B. auriporus*
1. Pleurocystidia smaller, not dark brown or red in Melzer's 2

    2. Pileus ferruginous (rusty red). . . . . *B. roxanae*
    2. Pileus olive to pale tan . . . . , ... 3

3. Pileus olive to honey-yellow . . . . (see *B. minutiporus* also) *B. mariae*
3. Pileus pale tan ("cinnamon-buff") . . . . *B. subilludens*

119. Boletus auriporus Peck

Rept. N. Y. State Cab. 23:133. 1872

Suillus auriporus (Peck) Kuntze, Rev. Gen. Pl. 3(2):595. 1898.
Ceriomyces auriporus (Peck) Murrill, Mycologia 1:147. 1909.
Xerocomus auriporus (Peck) Singer, Rev. de Mycol. 5:6. 1940.
Pulveroboletus auriporus (Peck) Singer, Amer. Midl. Nat. 37:13. 1947.

Pileus 3-7 cm broad, convex becoming plane or the margin up-lifted in age; surface dry and subpruinose at first, matted-fibrillose under a lens, viscid when wet, color varying from "tawny-olive" to dull reddish brown, when young often flushed ochraceous, surface sometimes areo-late in age, when dried the pileus pale drab to dull cinnamon. Context pale yellow, 5-8 mm deep, unchanging when cut; odor none; taste slightly acid to mild.

Tubes 3-5 mm deep, (10 mm—Coker and Beers), adnate then de-pressed or with decurrent lines, "lemon-chrome" and remaining a bright yellow in drying, unchanging when cut or bruised; pores 2-3 per mm at first but widening in large caps to 1-2.5 mm wide near stipe in age.

Stipe 3-6 cm long, 1-1.2 (1.5) cm thick, equal or flared either above or below (or both), solid, pale yellow becoming brownish and darker below, staining near bister (dingy yellow-brown) when handled, somewhat viscid when wet; surface thinly coated at first with a pale lemon colored pruinosity, not at all reticulate (rarely so—Coker and Beers).

Spores 8-11 × 3.5-4.5 μ, smooth, apex lacking a pore; in face view subelliptic, in profile suboblong to obscurely inequilateral, pale yellow in KOH, slightly yellower in Melzer's, wall thin.

Basidia 4-spored, 8-12 μ broad, yellow becoming hyaline in KOH. Pleurocystidia abundant, 38-70 (100) × 9-16 μ, fusoid-ventricose, smooth, thin-walled, content yellow in water, in KOH yellow but soon dissolving into the medium. Cheilocystidia similar to pleurocystidia or smaller. Caulocystidia of elongate-clavate cells with curved pedicels, 36-70 × 9-10 μ, thin-walled, soon hyaline in KOH mounts.

Tube trama of the *Boletus* subtype, yellow pigment diffusing out when the tissue is revived in KOH. Pileus cutis a layer of matted-down, tangled hyphae at maturity, the hyphae 6-12 (20) μ wide and some of the cells slightly inflated, cell content yellow in KOH but soon hyaline, in Melzer's orange-ochraceous and many of the wide hyphae with conspicuous dextrinoid patches of material adhering variously, these less evident in KOH because of being nearly hyaline, end-cells tubular to somewhat cystidioid. Hyphae of subcutis and adjacent context orange-ochraceous in Melzer's. Clamp connections absent.

Collected in the Waterloo Recreation Area, Washtenaw County, August 13, 1968, by Mrs. Florence Hoseney 1081.

Observations.—The descriptive data included here are from Smith 10913 collected in the Great Smoky Mountains National Park in 1938. The species is very rare in Michigan according to our experience. Mrs. Hoseney's collection is the only one to come to our attention to date.

The stipe is moist at first but becomes sticky if handled very much; it is not viscid from a gelatinous layer. The large cystidia, small spores pale in color, and the brilliant yellow hymenophore are distinctive.

120. Boletus roxanae Frost

Bull. Buff. Soc. Nat. Sci. 2:104. 1874

Xerocomus roxanae (Frost) Snell, Mycologia 37:383. 1945.

Illus. Figs. 59-62; Pl. 98.

Pileus 3-8 cm broad, obtuse to convex, becoming expanded-umbonate to broadly convex; surface dry, granulose-roughened by minute tufts of epicuticular elements but in age merely glabrous and unpolished; color near ferruginous when young and fresh ("hazel"), but soon more ochraceous toward the margin which is even and fertile; surface when dried ochraceous to tan. Context firm, thick, buffy white, taste acid-disagreeable, odor of cut specimens slightly pungent, cut or bruised context unchanging; with $FeSO_4$ slowly grayish; with KOH brownish.

Tubes very pale dingy yellowish, 5-8 mm deep, depressed around the stipe; pores "colonial-buff" (yellowish), round, 1-2 per mm, scarcely any change on bruising.

Stipe 4-8 cm long, 8-12 mm at apex and 15-20 mm at the base, evenly enlarged downward, solid, cortex lemon-yellow, stained brown around the larval tunnels; surface "light orange-yellow" darkening to "raw sienna" downward, longitudinally striate and obscurely pruinose but not reticulate, KOH brownish on surface, $FeSO_4$ no change.

Spore deposit dark olive-brown; spores 8-11 X 3-4.5 μ, suboblong in face view, in profile obscurely inequilateral and with a slight suprahilar depression, smooth, hyaline to yellow in KOH, in Melzer's merely yellow to pale brown but immature spores bluish gray in many instances (weakly amyloid).

Basidia 4-spored, 24-32 X 8-9 μ, clavate, yellow in KOH and Melzer's. Pleurocystidia scattered to rare, 33-46 X 8-12 μ, fusoid-ventricose, apex subacute, thin-walled, smooth, content hyaline to weakly yellowish in KOH or Melzer's.

Tube trama with a central interwoven strand and from this rather strongly divergent hyphae curve to the subhymenium, gelatinous, hyaline, all hyphae thin-walled and smooth. Pileus trichodermium of thin-walled hyphae 5-12 μ wide, with some cells 10-30 μ wide, yellow in KOH at first and content somewhat "colloidal," soon becoming hyaline, apical cell obtuse to cylindric or cystidioid to clavate, content in Melzer's (of all cells of the trichodermial elements) weakly yellowish, some amyloid debris present in some mounts. Context hyphae interwoven and many with orange-brown homogeneous content as revived in Melzer's. Clamp connections absent.

Habit, habitat, and distribution.—Solitary to scattered in sandy oak woods, Pinckney Recreation Area, Washtenaw County, September, rare.

Observations.—This is a striking species because of the ferruginous red pileus and orange-yellow stipe. It is most closely related to *B. illudens* outside of the stirps *Roxanae*. *B. illudens* differs in wider pores and more yellow in the pileus. In *B. roxanae* the elements of the pileus trichodermium are merely weakly yellow in Melzer's instead of orange-brown as in *B. illudens*. In the hyphae of the context the reaction is also reversed. In *B. illudens* the hyphae show very little orange-brown color in Melzer's but in *B. roxanae* most of them are rather dark orange-brown. In both, the elements of the trichodermium show many short cells which become inflated in age.

A comparison of the Frost collection at Albany with Smith's material clearly indicates that they are the same species. The most important possible difference is in the ferruginous color of the pileus at first for Michigan specimens; they faded, however, to the colors usually given for the species. In order to present as complete an account of the species as possible the original description is quoted below and Smith's study of the microscopic features of the Frost specimen at Albany are added: "Pileus flat, convex, yellowish brown, fasiculated red pilose, sub-tomentose young. Tubes at first whitish, then light yellow, not large, falling away around the stem, or arcuate adnate. Stem light cinnamon or weak gamboge color, striate at apex, thickened downwards, and sub-tuberous. Flesh yellowish white, just tinged."

Spores 8-11 (12) X 3.5-5 μ, smooth, lacking an apical pore, ovate to subfusoid in face view, in profile obscurely to somewhat inequilateral, nearly hyaline in KOH, yellowish in Melzer's, wall thin (-0.2 μ).

Basidia 4-spored, 8-11 μ, broad, clavate, hyaline in KOH. Pleuro-cystidia fusoid-ventricose, 28-42 X 7-12 μ, thin-walled, smooth, hyaline in KOH, content not distinctive in Melzer's (mostly appearing "empty" but they were poorly revived). Cheilocystidia smaller than pleuro-cystidia. Caulocystidia clavate, 9-16 μ broad, length variable, wall thin and smooth, content hyaline ("empty").

Tube trama of weakly divergent hyphae inamyloid in all respects, hyaline and smooth in KOH. Pileus cuticle a trichodermium of hyphae 6-15 μ wide, the hyphae tubular or with cells inflated and some dis-articulating, walls thin and smooth to minutely roughened, content hyaline to yellowish in KOH and yellowish and homogeneous in Melzer's, both short and long cells present in varying amounts, end-cells bullet-shaped to cystidioid, or more rarely tubular to narrowly clavate. Hyphae of subcutis yellowish in Melzer's. Clamp connections none.

121. Boletus mariae Smith & Thiers, sp. nov.

Pileus 4-9 cm latus, convexus demum late convexus, siccus, ve-lutinus, olivaceus vel melleus. Contextus luteolus, immutabilis. Tubuli 1 cm longi olivaceolutei; pori olivaceolutei, immutabiles. Stipes 5-9 cm longus, 8-15 mm crassus, intus luteolus, extus reticulatus, reticulum latum, deorsum pruinosus. Sporae 9-12 (13) X 3.5-4.5 μ. Specimen typicum in Herb. Univ. Mich. conservatum est; prope Cheboygan, July 4, 1968, legit Mary Wells (Smith 75480).

Pileus 4-9 cm broad, convex with an incurved margin, becoming broadly convex with the margin decurved, surface dry and velvety to subtomentose, dried pilei rimulose near margin in large ones and un-polished in all; olive to honey-yellow, drying ochraceous to clay color, NH_4OH brown on cuticle (no green or violet flash). Context firm, yellowish, unchanging, odor and taste not distinctive.

Tubes up to 1 cm deep, olive-yellow, adnate; pores olive-yellow, 0.5-1.5 mm wide, unchanging when bruised.

Stipe 5-9 cm long, 8-15 mm at apex, equal or slightly enlarged downward, solid, yellowish within, unchanging, buff color in dried material; surface reticulate above with a wide-meshed reticulum having poorly defined ridges, densely pruinose to subtomentose downward, the ornamentation brownish to dull cinnamon against an ochraceous ground color, these colors persisting in dried material, base with a yellow my-celium.

Spores 9-12 (13) $\times$ 3.5-4.5 μ, smooth, lacking a distinct pore or thin spot at apex, shape in face view narrowly elliptic to suboblong, in profile obscurely inequilateral, the suprahilar area scarcely depressed in many, color weakly olive-yellow in KOH, in Melzer's about the color of the mounting medium, wall relatively thin.

Basidia 4-spored, 8-9 μ broad, clavate, yellowish hyaline in KOH. Pleurocystidia scattered, 36-48 $\times$ 9-12 μ, fusoid-ventricose, obtuse to subacute, thin-walled, smooth, hyaline in KOH and Melzer's. Cheilo-cystidia mostly small and more or less basidiole-like. Caulocystidia abundant, 36-60 $\times$ 10-24 μ, clavate to clavate-mucronate, thin-walled, smooth, weakly yellow in KOH.

Tube trama of tubular hyphae 5-11 μ wide, scarcely divergent as revived in KOH and scarcely gelatinous, thin-walled. Pileus cuticle a trichodermium of elements 9-16 (18) μ wide, the cells often short and inflated, end-cell often bullet-like, content colloidal in KOH, walls smooth to slightly roughened, in Melzer's the cuticular hyphae appearing nearly or quite smooth (lacking dextrinoid incrustations) and with ochraceous content. Hyphae of subcutis and adjacent hyphae of the context dingy orange-ochraceous in Melzer's and lacking dextrinoid in-crustations. Clamp connections absent.

Habit, habitat, and distribution.—Gregarious in mixed woods, Mud Lake Bog, Cheboygan County, July 4, 1968, collected by Mary Wells (Smith 75480).

Observations.—In this species there was no change to blue when bruised, there are no dextrinoid incrustations on the cuticular or sub-cuticular hyphae of the pileus, the spores are small, and many of the cells in the trichodermial hyphae are inflated.

122. Boletus subilludens Smith & Thiers, sp. nov.

Pileus 4-9 cm latus, convexus, glaber, siccus, demum areolatus, pallide alutaceus. Contextus pallide luteus, immutabilis. Tubuli 1 cm longi, laete lutei, immutabiles; pori lati (circa 1 mm diameter), immutabiles. Stipes 3-8 cm longus, 6-12 mm crassus, intus luteus, extus sordide luteo-brunneus, glaber. Sporae 8-11 X 4-5 μ. Specimen typicum in Herb. Univ. Mich. conservatum est; prope Otis Lake, Barry County, August 19, 1966, legit Smith 73096.

Pileus 4-9 cm broad, broadly convex to plane; surface glabrous, dry, dull (unpolished), finally somewhat areolate with yellow context showing in the cracks, evenly pinkish cinnamon to cinnamon-buff, diffracted squamulose in age. Context pale yellow, unchanging when cut, taste acid, odor none; slowly becoming pinkish to vinaceous around the wormholes, with $FeSO_4$ olive, with guaiac no reaction.

Tubes bright yellow becoming bright olive-ochraceous as dried, about 1 cm deep, depressed at the stipe, unchanging when injured; pores about 1 mm in diameter, unchanging when bruised.

Stipe 3-8 cm long, 6-12 mm at apex, solid, yellow within and brighter as dried, bright yellow in base when fresh; surface drying dingy yellow-brown, uneven but neither pruinose nor reticulate.

Spores 8-11 X 4-5 μ, smooth, apical pore not evident, walls -0.2 μ thick, nearly hyaline as revived in KOH and merely yellowish in Melzer's, no fleeting-amyloid reaction seen; in face view elliptic to subfusoid, in profile somewhat inequilateral to subelliptic, the suprahilar depression very shallow.

Basidia 23-32 X 7-10 μ, short-clavate, 4-spored, yellowish to hyaline in KOH and Melzer's. Pleurocystidia scattered, 34-56 X 6-11 μ, subcylindric, to narrowly fusoid, obtuse at apex, smooth, thin-walled, hyaline and homogeneous in KOH and Melzer's. Cheilocystidia—none found. Edges of dissepiments mostly of small ochraceous basidioles.

Tube trama of hyaline parallel hyphae 3-6 μ wide, subgelatinous, no appreciable divergence noted toward the subhymenium (the *Xerocomus* subtype). Pileus cutis a tangle of hyphae 4-9 wide, with tubular end-cells or the latter tapered somewhat to apex, content yellowish in Melzer's, soon hyaline in KOH, walls thin, smooth and hyaline (no true trichodermium present), hyphae with cells often disarticulating. Clamp connections absent.

Habit, habitat, and distribution.—Gregarious under hardwoods, Otis Lake, Barry County, August 19, 1966, Smith 73096.

Observations.—This species is similar to *B. roxanae* in its short spores, but the pileus is much paler, the FeSO$_4$ reaction is olive, and the cells of the cuticular hyphae often disarticulate.

Stirps SUBPALUSTRIS.

KEY

1. Pileus olive-ocher darkening to dark yellow-brown; stipe not staining when cut . . . . *B. subpalustris*
1. Pileus dingy pale cinnamon to tan and slowly darkening; stipe staining vinaceous-buff when cut . . . . *B. alutaceus*

123. Boletus subpalustris Smith & Thiers, sp. nov.

Illus. Pl. 99.

Pileus 3-5 cm latus, convexus, siccus, velutinus, olivaceoluteus demum subspadiceus. Contextus luteolus, immutabilis. Tubuli 1 cm longi, sordide lutei; pori 1-2 per mm, lutei, immutabiles. Stipes 3-6 cm longus, 5-10 mm crassus, luteus, pruinosus; pruina cinnamomea. Sporae 10-13 × 3.5-4.2 μ. Specimen typicum in Herb. Univ. Mich. conservatum est; prope Douglas Lake, Cheboygan County, June 23, 1968, legit Smith 75298.

Pileus 3-5 cm broad, convex expanding to broadly convex or plane; surface dry and velvety to subtomentose, becoming areolate in age, color olive-yellow to dingy yellow-brown ("olive-ocher" to "Dresden-brown"); NH$_4$OH on cuticle dull olive. Context firm, yellowish, unchanging when cut or injured, with FeSO$_4$ no reaction, guaiac no reaction; odor and taste not distinctive.

Tubes up to 1 cm deep, nearly free from the stipe at maturity, dull yellow, not changing to blue when bruised; pores dull yellow to olive-yellow (when old), unchanging when injured, 1-2 per mm, round to angular, yellow, unchanging when bruised.

Stipe 3-6 cm long, 5-10 mm thick, equal, yellow within, unchanging; with a dull cinnamon scurfiness or pruinosity over a dull yellow ground color, yellowish at base, in some longitudinally ridged near the apex.

Spores 10-13 (14) × 3.5-4.2 μ, smooth, lacking an apical pore or thin spot, shape in face view narrowly subfusoid, in profile somewhat inequilateral, pale olive-yellow in KOH, with a bluish cast in Melzer's (weakly amyloid), wall thin (about 0.2 μ).

Basidia 4-spored, 26-34 × 8-11 μ, hyaline to yellow in KOH, clavate. Pleurocystidia fusoid-ventricose, 32-47 × 8-13 μ, apex subacute to obtuse, smooth, thin-walled, content hyaline to yellowish in KOH. Tube edges lined almost entirely with basidioles. Caulocystidia in fascicles, 28-42 X 8-15 μ, clavate, thin-walled, smooth, hyaline and "empty" in KOH.

Tube trama of the *Xerocomus* subtype, yellowish in KOH but fading as pigment diffuses into the mount. Pileus cuticle a trichodermium of tubular elements 6-9 (11) μ wide, content yellow in Melzer's, walls smooth to roughened, end-cells tubular and rounded at apex, scattered, short, slightly inflated cells present in some of the elements. Hyphae of subcutis with conspicuous dextrinoid incrustations, the hyphae merely with yellowish content in Melzer's. Clamp connections absent.

Habit, habitat, and distribution.—In a wet pasture under aspen near a rotten log. Douglas Lake, Cheboygan County, June 23, 1968, Smith 75298.

Observations.—The distinguishing features of this species are the dull olive NH$_4$OH reaction on the pileus cuticle, lack of blue stains when injured, small narrow spores, scurfy to pruinose stipe, elements of pileus cuticle mostly of tubular cells, and the generally small size of the basidiocarps.

124. Boletus alutaceus Morgan in Peck

Bull N. Y. State Mus. 2 (8):109. 1889

Xerocomus alutaceus (Morgan in Peck) Dick & Snell, Mycologia 53:228. 1961.

Pileus 5-10 cm broad, pulvinate becoming broadly convex to nearly plane; surface dry and subpruinose (unpolished), even, tacky to the touch in wet weather, cuticle slightly separable (but not gelatinous); "pinkish cinnamon" to "cinnamon-buff" (pale tan with a reddish tint), at first a duller yellow-brown ("Saccardo's umber" to "tawny-olive"). Context white with a reflection of pink near the cutis, 7-16 mm thick at the stipe, unchanging when injured; odor and taste not distinctive.

Tubes about 6-15 mm long, whitish young becoming "pale olive-buff" or finally more olivaceous, depressed around the stipe, not changing to blue when injured; pores 1-1.5 mm broad, not stuffed when young, olivaceous pallid but slowly brownish in age, stipe with thin extensions of tube dissepiments down it for a short distance.

Stipe 6-8 cm long, 1-2 cm thick, straight or curved slightly, subequal, solid, pallid inside and out but context in older specimens changing to vinaceous-buff when cut (pale avellaneous); surface smooth, even or at times with lines or faint reticulations from extensions of the tubes, generally becoming dingy brownish from handling.

Spores 9-12 (13) X 4-5 μ, suboblong in face view, slightly inequilateral in profile, smooth, wall slightly thickened, nearly lemon-yellow in KOH, in Melzer's pale dingy yellow-brown (not dextrinoid).

Basidia 26-32 X 8-9 μ, 4-spored, clavate, yellow in KOH and Melzer's. Pleurocystidia scattered, 33-45 X 9-14 μ, clavate to fusoid-ventricose with obtuse apex, content often dingy yellow as revived in KOH, in Melzer's merely dingy yellow-brown.

Tube trama of hyaline gelatinous hyphae somewhat divergent to the subhymenium. Pileus cuticle a tangled layer of hyphae 3-6 μ wide, the end-cells tubular or nearly so and blunt at apex. Context hyphae 8-15 μ wide, hyaline, smooth, thin-walled, hyaline to yellowish in KOH, in Melzer's orange-brown. Clamp connections extremely rare. No amyloid debris present or no fleeting-amyloid reactions evident.

Habit, habitat, and distribution. —On swampy ground, Saline, Washtenaw County, July 18, 1922, C. H. Kauffman. Apparently, the species is very rare in Michigan.

Observations. —The tube color as indicated in Kauffman's notes was more olive than yellow, but a strong yellow cast is present in the dried specimens. This is a nondescript bolete which in mature stages or when dried might be confused with *B. pallidus,* but it does not stain blue and is not whitish when young. The hyphae of the context react in Melzer's similarly to those of *B. roxanae.*

Stirps ILLUDENS

B. spadiceus, which at times lacks a change to blue on bruising, may be sought for here also, but see subsection *Subtomentosi.*

KEY

1. Spores (13) 14-17 x 4-5.5 μ; pileus vinaceous-brown or orange-cinnamon
   . . . . *B. hoseneae*
1. Spores 10-14 μ long and pileus differently colored. 2
   2. Pores 1-2 per mm at maturity. 3
   2. Pores 1-2 mm wide at maturity. 4
3. Terminal cells of trichodermial elements often globose to dumbbell-shaped. . . . . *B. brunneocitrinus*
3. Terminal cells of trichodermial elements mostly bullet-shaped to cystidioid
   . . . . see *B. minutiporus*
   4. NH$_4$OH on pileus cuticle Paris-green then brown . . . . *B. illudens*
   4. NH$_4$OH on pileus cuticle purple then darkening . . . . *B. nancyae*

125. Boletus illudens Peck

Ann. Rept. N. Y. State Mus. 50:108. 1897

Ceriomyces illudens (Peck) Murrill, North Amer. Fl. 9:145. 1910.
Xerocomus illudens (Peck) Singer, Farlowia 2:293. 1945.

Illus. Figs. 64, 66, 71; Pl. 100.

Pileus 3-9 cm broad, convex to broadly convex expanding to plane, rarely finally with a depressed disc; surface moist but not viscid, unpolished to velvety but not becoming conspicuously areolate in age; pale pinkish cinnamon to pinkish cinnamon, flushed lemon-yellow near margin and drying with a strong yellow tone, margin even to irregular; NH$_4$OH on surface Paris-green, slowly changing to fuscous, with KOH dark brown. Context pallid when young, soon bright yellowish, unchanging when cut, odor and taste not distinctive; with guaiac no reaction, with FeSO$_4$ greenish gray going to olive-black.

Tubes adnate to subdecurrent, 1-1.5 cm long, honey-yellow to olive-yellow; pores large and angular (1-2 mm wide or more in radial direction), lemon-yellow and unchanging when bruised or only slightly yellower but in age bruised portions often brownish; tubes splitting rather than separating when pileus is broken downward.

Stipe 3-9 cm long, (4) 6-13 mm wide at apex, narrowed downward, mustard-yellow in the cortex and base, paler in the interior; surface yellowish to mustard-yellow around the base, pallid to brownish upward but pale lemon-yellow near apex, usually marked by coarse ridges and anastomosing lines but not finely reticulate.

Spore deposit olive becoming olive-brown to buffy brown; spores 10-14 X 4-5 μ, shape in face view subelliptic to subfusoid, in profile somewhat inequilateral, surface smooth, suprahilar depression often distinct, yellow in KOH, bluish gray slowly becoming brownish in Melzer's (fleeting amyloid), rarely slightly dextrinoid, no apical pore evident.

Basidia 4-spored, 24-32 X 7-9 μ. Pleurocystidia narrowly and obtusely fusoid, varying to narrowly fusoid-ventricose, 28-42 X 7-12 μ, thin-walled, hyaline to yellowish in KOH and in Melzer's, content homogeneous as revived. Cheilocystidia smaller than pleurocystidia (20-35 X 5-8 μ), more frequently varying to clavate and with more highly colored (brownish) content as revived.

Tube trama of subparallel hyphae only slightly divergent near the subhymenium, in KOH in some mounts long, hyaline, needle-like crystals form, no crystals seen in mounts in Melzer's. Pileus trichodermium of tangled hyphae 4-12 μ wide, the cells often short and some inflated to 12-17 μ and globose or nearly so, the terminal cell of an element often reduced in size or in the form of an obtuse flexuous hyphal tip; walls thin and with some adhering dextrinoid debris occasionally, content in Melzer's ochraceous-orange-brown and homogeneous, in KOH the content "colloidal." Hyphae of context with some orange content in Melzer's but often mostly hyaline especially in mature basidiocarps, interwoven, thin-walled. Clamp connections absent.

Habit, habitat, and distribution.—Under oak and in mixed oak-pine forests, late summer and fall during wet seasons, common at times.

Observations.—The distinguishing features are the lack of blue stains anywhere, the inflated cells in the pileus trichodermium and their orange-brown content in Melzer's, the fleeting-amyloid reaction of the spores, the green then fuscous NH_4OH reaction of the fresh pileus cuticle, and the strong $FeSO_4$ reaction. In the material we have studied from Michigan the formation of the needle-like crystals was not a constant feature. However, at times it is very pronounced. This species has the stature of *Boletus fraternus,* which is strictly in accord with Peck's original description. The type has been studied and a complete description of it follows:

"Pileus convex, dry, subglabrous, yellowish brown or grayish brown, sometimes tinged with red especially in the center, flesh pallid or yellowish; tubes bright yellow, plane or somewhat convex when old, adnate, their mouths angular to subrotund, often larger near the stem; stem nearly equal, sometimes abruptly pointed at the base, glabrous, pallid or yellowish, coarsely reticulated either wholly or at the top only; spores oblong or subfusiform, yellowish-brown tinged with green, .00045-.0005 in. long, .00016-.0002 broad.

"Pileus 1.5-3 in. broad; stem 1.5-2.5 in. long, 3 to 5 lines thick."

Microscopic characters from the type (by Smith): Spores 11-13 X 4-4.5 μ, smooth, apex lacking a pore or a thin spot, color in KOH ochraceous to pallid yellowish, in Melzer's merely slightly tan; shape in face view narrowly subovate to subfusoid, in profile obscurely to somewhat inequilateral, wall about 0.3 μ thick.

Basidia 4-spored, mostly gelatinized. Pleurocystidia if present gelatinized and content not homogeneous. Caulocystidia not studied.

Tube trama of the *Xerocomus* subtype (weakly divergent), completely inamyloid, large wormlike laticiferous elements present with yellow amorphous content in KOH or Melzer's.

Pileus cuticle a collapsed trichodermium of wide hyphae (8-15 μ) with many short often inflated cells and with bright yellow content in KOH, the pigment soon fading, walls thin and smooth, end-cell elongate bullet-shaped to short bulletlike or subglobose (rarely the latter), the cells showing a distinct tendency to disarticulate, walls smooth in KOH, in Melzer's the cuticular elements in the zone near the subcutis and the subcuticular hyphae showing some tawny-brown incrusting material in Melzer's but the cell content yellowish and homogeneous. Hyphae of the subcutis and context yellow in Melzer's.

Notes: The iodine reactions of the type are not quite the same as those of the Michigan material, but here we do not have a basis for critical evaluation since we do not know what the specimens have been subjected to by way of treatment or the effect of time on the chemicals giving the reactions with iodine. The positive characters connecting the Michigan material to the type are the yellow tones of the dried pileus, the inflated short cells in the pileus cuticle, the weakly dextrinoid incrustations on the hyphae of the subcutis, the relatively small spores, wide pores, and lack of any change to blue on the pores when fresh.

The type is apparently a mixture of 2 species: We designate as lectotype the material designated in packets 1, 2, and 3 in the box housing the type collection at Albany. These represent the concept of a thin-stiped species, which is what Peck described. Loose in the box are certain basidiocarps which as dried have stipes about 2 cm thick at the apex and pilei about 8 cm broad. These have a distinct fine reticulum like *B. tenacipes,* which is not the kind of reticulum Peck described in his original description.

126. **Boletus nancyae** Smith & Thiers, sp. nov.

Pileus 4-10 cm latus, convexus, siccus, fibrillose squamulosus, subspadiceus. Contextus pallide luteus, immutabilis. Tubuli 0.9-1.3 cm longi, ochraceolutei; pori 1-1.5 mm lati, ochracei, immutabiles. Stipes 4-9 cm longus, 8-15 mm crassus, reticulatus, deorsum luteo-myceliosus. Sporae 11-13.5 × 3.8-4.5 μ. Specimen typicum in Herb. Univ. Mich. conservatum est; prope Burt Lake, Cheboygan County, July 8, 1967, legit N. J. Smith (A. H. Smith 74469).

Pileus 4-10 cm broad, convex becoming broadly convex; surface dry and tomentose-squamulose, in dried specimens the ornamentation appearing as appressed-fibrillose squamules, near snuff-brown to bister when fresh (dingy yellow-brown), when dried with a strong ochraceous ground color showing beneath or between the squamules. Context yellowish and drying pale yellow, odor of dried basidiocarps rather strong (but difficult to define), unchanging when injured, NH_4OH on cuticle when fresh purplish before darkening.

Tubes 9-13 mm deep, adnate to subdecurrent, dull yellow to ocher-yellow and unchanging when bruised; pores 1-1.5 mm broad, ochraceous and not staining when bruised.

Stipe 4-9 cm long, 8-15 mm thick, equal or nearly so, yellow within, unchanging when injured, surface coarsely reticulate and pruinose from a reddish brown pruinose coating; base coated with pale yellow mycelium.

Spores 11-13.5 × 3.8-4.5 μ, smooth, elongate-inequilateral in profile, subfusoid in face view, color in KOH pale ochraceous, in Melzer's pale brownish ochraceous, wall about 0.2 μ thick, no amyloid reaction observed.

Basidia 4-spored, 28-33 × 8-10 μ, yellowish in KOH becoming hyaline. Pleurocystidia 34-55 (63) × 8-13 μ, elongate-fusoid-ventricose to subfusoid, thin-walled, hyaline to yellowish from a diffusing yellow pigment in KOH, in Melzer's yellowish hyaline and content homogeneous. Caulocystidia clavate, 30-52 × 8-16 μ, thin-walled, smooth, with some interior thickened patches in the wall of the enlarged part, pale yellow in KOH and subiculum hyphae also pale sulphur-yellow.

Pileus cuticle a trichodermium the elements of which are divided into long and short cells, the latter often terminal and sphaerocyst-like or oval, the cells often disarticulating, in Melzer's the hyphae of the subcutis with some incrusting patches of dextrinoid pigment, yellow pigment diffusing into the mount when made in KOH. Clamp connections absent.

Habit, habitat, and distribution.—On soil at edge of a bog, Burt Lake, Cheboygan County, July 8, 1967, N. J. Smith collector (A. H. Smith 74469).

Observations.—This species is similar to *B. illudens* in many respects, but is distinguished by its scaly pileus and instantaneous purple reaction on the fresh pileus cuticle with ammonia. The hyphae of the context have some dextrinoid incrustations near the subcutis and these hyphae have a dull orange homogeneous content in Melzer's. The trichodermial elements have smooth thin pale yellow walls as revived in KOH and are very boldly defined.

127. Boletus hoseneae Smith & Thiers, sp. nov.

Pileus 5-6 cm latus, udus, subhygrophanus, aurantio-cinnamomeus. Contextus luteus, immutabilis; pori olivaceolutei, immutabiles, lati. Stipes 6-7 cm longus, 10-12 mm crassus, sursum laete luteus, deorsum sordide vinaceus. Sporae (13) 14-17 × 4-5.5 μ. Specimen typicum in

Herb. Univ. Mich. conservatum est; prope Waterloo, Washtenaw County, September 5, 1965, legit Florence Hoseney (Smith 72503).

Pileus 5-6 cm broad, plane; surface moist and subhygrophanous becoming slightly rimose in age, color "mikado-brown" to "orange-cinnamon" (bright orange-brown to orange-cinnamon), when dried a dingy olivaceous-brown. Context slightly acid to the taste, yellow near tubes and with an olive line just above the tubes, dingy vinaceous in the central part and near the cuticle as well as in apex of stipe, pale lemon-yellow in base of stipe, not changing to blue when cut or bruised; with $FeSO_4$ no reaction on the pileus context, on base of stipe bright olive to olive-fuscous.

Tubes up to 1 cm long, greenish yellow and drying olivaceous, slightly depressed but decurrent down the apex of the stipe; pores about 1 mm wide, greenish yellow and unchanging when bruised.

Stipe 6-7 long, 10-12 mm at apex, tapered downward, solid (for color of context see above); surface even, apex bright yellow, with lines from tubes extending down a short distance but not reticulate, base dull vinaceous-red and unpolished.

Spore deposit dull olivaceous; spores (13) 14-17 X 4-5.5 μ, smooth, no apical pore evident, shape in face view boat-shaped to fusoid, in profile narrowly somewhat inequilateral but suprahilar depression shallow, ochraceous in KOH, ochraceous-tan in Melzer's, amyloid when fresh but soon fading.

Basidia 4-spored, 23-30 X 7-11 μ, short- to long-clavate, hyaline in KOH, pale yellowish in Melzer's. Pleurocystidia and cheilocystidia—none found. Tube trama of the bilateral type but hyphal divergence only moderately prominent, all hyphae hyaline and smooth.

Pileus cutis a poorly formed layer of clavate to versiform cells in a weak palisade which collapses in age, some cells up to 20 μ wide, some globose and of various sizes, all thin-walled, smooth, and with ochraceous content in KOH, merely yellowish in Melzer's. Context hyphae merely yellowish in Melzer's; many large hyaline to yellowish droplets present in mounts in Melzer's. Clamp connections absent.

Habit, habitat, and distribution.—Gregarious in sandy soil, Waterloo Recreation Area, Washtenaw County, September 5, 1965, collected by Florence Hoseney (Smith 72503).

Observations.—Many hyphal cells in the pileus are over 20 μ wide and collapse readily. To this extent the structure of the pileus differs from that of other species in the group. The tubes separate when the pileus is broken downward, but in spite of this the species appears closely related to *B. illudens.* It differs in having a more olivaceous pileus as dried, differently shaped cells in the hyphae of the pileus cuticle, and in having spores averaging slightly over 15 μ long.

128. Boletus brunneocitrinus Smith & Thiers, sp. nov.

Pileus 3.5-5 cm latus, late convexus, siccus, subtomentosus, badius demum olivaceus. Contextus luteus, immutabilis. Tubuli 15 mm longi, sordide lutei; pori 1-2 per mm, sordide lutei, immutabiles. Stipes 5-8 cm longus, 5-8 mm crassus, deorsum laete luteo-myceliosus, sursum pruinosus; pruina sordide luteobrunnea; nonreticulatus. Sporae 10-14 X 3.5-4.5 (5) μ. Specimen typicum in Herb. Univ. Mich. conservatum est; prope Douglas Lake, Cheboygan County, July 1, 1968, legit Smith 75435.

Pileus 3.5-5 cm broad, broadly convex and remaining so, margin at first incurved but not sterile; surface subtomentose to unpolished, continuous, dull reddish brown when wet, olivaceous after moisture has escaped ("Vandyke-brown" wet, "dark olive-buff" to "citrine-drab" on loss of moisture). Context yellow, not staining blue, pale buff as dried, odor and taste not distinctive, NH_4OH on pileus dull brown immediately.

Tubes ventricose and deeply depressed around the stipe (up to 15 mm deep), dull yellow and unchanging; pores at maturity 1-2 per mm, angular to round, dull yellow, not staining blue.

Stipe 5-8 cm long, 5-8 mm thick, slender, solid, equal, yellowish within becoming pallid, no color change when cut; base covered with bright yellow mycelium, pruinose above the lower third with bister pruina on a yellowish to brownish ground color, not at all reticulate.

Spores 10-14 X 3.5-4.5 (5) μ, smooth, apex lacking a pore, shape in face view subfusoid, in profile more or less inequilateral, color in KOH pale bright yellow, in Melzer's yellow with a greenish gray shadow (fleeting-amyloid reaction strong).

Basidia 4-spored, 24-35 X 8-11 μ, yellow to hyaline in KOH or Melzer's. Pleurocystidia 35-52 X 8-14 μ, fusoid-ventricose with obtuse to subacute apex, thin-walled, hyaline or yellow in KOH but fading if yellow at first. Cheilocystidia similar but smaller. Caulocystidia 28-42 X 9-13 μ, subcylindric to clavate, thin-walled, smooth, hyaline to yellow in KOH, more rarely fusoid-ventricose.

Tube trama of the *Xerocomus* subtype, the hyphae hyaline in KOH, 5-9 μ broad, walls not obviously gelatinous, fleeting-amyloid reaction strong (sections violet-black in Melzer's but pigment not localized). Pileus cuticle a trichodermium of wide (9-15 μ) hyphae of globose to elongate (and then often dumbbell-shaped) cells hyaline in KOH and with minutely ornamented walls, terminal cells globose to elongate or dumbbell-shaped, all thin-walled; hyphae beneath the trichodermium with dextrinoid incrusting material in addition to patches of purplish vinaceous debris between the cells. Hyphae of upper context merely yellow in Melzer's. Clamp connections absent.

Habit, habitat, and distribution.—Gregarious on an old conifer log, Douglas Lake, Cheboygan County, July 1, 1968, Smith 75435.

Observations.—This species has the stature of *Tylopilus gracilis* but belongs near *B. subtomentosus.* From the variety *perplexus* of that species it differs in the dull yellow-brown pruina on the stipe, lack of a change to blue when injured, having purple-vinaceous debris between the cells of the trichodermium in mounts in Melzer's, and in having bright yellow mycelium at the base of the stipe. The small pores are the best field character, but one must be careful to compare mature specimens.

129. Boletus minutiporus Smith & Thiers, sp. nov.

Illus. Pl. 101 (lower).

Pileus 4-6 cm latus, late convexus, siccus, subtomentosus, sordide olivaceus, demum ochraceus. Contextus pallide luteus tactu tarde sub-caeruleus. Tubuli 5-7 mm longi, adnati vel subdecurrentes, pallide lutei; pori pallide lutei. Stipes 3.5-6 cm longus, 9-14 mm crassus, demum cavus, deorsum luteo-myceliosus, sursum subluteus. Specimen typicum in Herb. Univ. Mich. conservatum est; prope Mackinaw City, Emmet County, July 4, 1968, legit H. D. Thiers (Smith 75467).

Pileus 4-6 cm broad, convex becoming broadly convex, surface dry and subtomentose, dull olive-buff ("deep olive-buff") becoming ochraceous in drying, margin even. Context pallid to very pale yellow, with a brownish line under the cuticle, when injured slowly staining greenish blue, NH_4OH on cuticle dingy olive then quickly brownish; $FeSO_4$ no reaction on context.

Tubes rather shallow (5-7 mm deep), adnate to short-decurrent, pale wax-yellow and unchanging when cut; pores small 1.5-2.5 per mm, round; "Naples-yellow" (pale yellow) and unchanging when bruised.

Stipe 3.5-6 cm long, 9-14 mm thick, nearly equal, pallid within, becoming hollow in midportion; surface yellow below, pale dingy cinnamon above from a faint pruinosity, basal mycelium pale sulphur-yellow, not discoloring when handled.

Spore deposit olive-brown; spores 10-13 X 4-5 μ, smooth, apical pore not evident, shape in face view somewhat fusiform, in profile somewhat inequilateral, color as revived in KOH pale olive-yellow, duller in Melzer's as though weakly amyloid, wall 0.2 μ thick.

Basidia 4-spored, narrowly clavate, 28-36 X 6-9 μ, yellow in KOH fading to. hyaline, yellowish in Melzer's. Pleurocystidia scattered, 34-55 X 6-11 μ, narrowly fusoid-ventricose, neck in some flexuous, hyaline to yellow in KOH, wall thin and smooth. Cheilocystidia smaller than pleurocystidia and necks shorter. Caulocystidia 35-60 X 9-15 μ, clavate, thin-walled, hyaline in KOH, wall thin and smooth, fasciculate.

Tube trama of the *Xerocomus* subtype (weakly divergent), the hyphae tubular, 7-12 μ wide and septa refractive in KOH, no fleeting-amyloid reaction present in the trama (though possibly a weak reaction on the spores). Pileus cuticle a trichodermium of hyphae 6-12 (15) μ in diameter, hyaline in KOH and in Melzer's, end-cells often short and bullet-like, both short and long cells present and a tendency for them to disarticulate is evident, walls smooth or nearly so (either in KOH or in Melzer's). Hyphae of subcutis and adjacent context with orange-cinnabar content as revived in Melzer's but slowly fading to orange-ochraceous, walls smooth, hyphae more or less tubular and 5-12 μ wide. Clamp connections absent.

Habit, habitat, and distribution.—Mackinaw City area, Emmet County, July 4, 1968, H. D. Thiers (Smith 75467).

Observations.—This species is unusual because the hyphae of the subcutis as revived in Melzer's have such brightly colored contents, whereas the elements of the trichodermium show relatively little color. In addition, the small pores are a good field character for distinguishing it from *B. illudens,* to which it seems related. It is also close to *B. spadiceus* var. *gracilis,* but has smaller pores.

Stirps TENAX

130. Boletus tenax Smith & Thiers, sp. nov.

Illus. Pls. 102-3 (upper).

Pileus 4-10 cm latus, convexus demum planus vel late depressus, siccus, velutinus, olivaceo-brunneus demum lateritius vel obscure roseus. Contextus pallidus tactu tarde subvinaceus. Tubuli circa 12 mm longi, sordide lutei, immutabiles; pori lati (1-3 mm), laete lutei tactu leviter aurantio-cinnamomei. Stipes 3-9 cm longus, 1-3 cm crassus, reticulatus, subroseus, sursum demum subluteus, adbasin tenax. Sporae 9-12 × 4.5-5.5 (6) μ. Specimen typicum in Herb. Univ. Mich. conservatum est; prope Ypsilanti, Washtenaw County, September 11, 1965, legit Smith 72556.

Pileus 4-10 cm broad, convex with an incurved to inrolled margin, becoming plane to shallowly depressed, margin at times uplifted and irregular to lobed; surface plushlike, even at first but surface developing depressions (as the flesh softens); color when young with an olive sheen ("buffy brown") overall, with a dull brick-red ground color, in age "Pompeian-red" to "madder-brown" (almost rose-russet), margin obtuse and even in age. Context soon soft, pallid, with a watery line next to the tubes, bright pale yellow as dried, becoming vinaceous-brown around the larval tunnels when fresh, taste acid, odor not distinctive, with NH_4OH on cuticle green then blue, finally fuscous; with KOH fuscous; cut surface slowly flushing vinaceous (in 15 minutes); KOH and $FeSO_4$ neither giving an appreciable reaction.

Tubes up to 12 mm deep, long-decurrent by lines, dull yellow with a faint reflection when young, soon bright yellow, not staining blue when injured, splitting if pileus is broken downward, not readily separable from the context; pores broad (1-3 mm radially or near stipe more elongated), isodiametric near pileus margin, bright yellow becoming dingy yellow in age, when bruised turning slightly reddish cinnamon to orange (in age).

Stipe 3-9 cm long, 1-3 cm thick at apex, narrowed to 8-12 mm at base, solid, pallid within, near base the cortex often tinged vinaceous, basal area covered by yellowish to pallid mycelium; surface reticulate overall or only over the upper half with a distinct wide-meshed fine reticulum, dingy rose to pinkish cinnamon over most of its length but becoming decidedly yellow in age or on drying, consistency very hard in the lower part and the base tenaciously attached to the substratum by binding mycelium.

Spores 9-12 × 4.5-5.5 (6) μ, smooth, no apical pore apparent, in profile somewhat inequilateral, in face view subfusoid to ovate, yellow in KOH, in Melzer's amyloid when fresh, very weakly so as revived, wall relatively thin.

Basidia 4-spored, 24-30 × 7-10 μ, clavate, hyaline in KOH, merely yellowish in Melzer's. Pleurocystidia scattered, 38-60 × 7-11 μ, narrowly clavate to cylindric with subacute apex to narrowly fusoid, thin-walled, smooth as revived in KOH, in Melzer's with fine hyaline granules adhering to distal portion, content yellow in KOH in fresh material, hyaline as revived, content not distinctive in Melzer's.

Tube trama parallel but slightly divergent near the subhymenium. Pileus cutis a tangled trichodermium the elements of which are 5-10 μ wide, a fair number of the cells short and these often disarticulating, walls thin and smooth, cell content yellowish to hyaline in KOH, in Melzer's the layer rusty brown from patches of incrusting material adhering variously to the cells back from the terminal cell (similar to the material in the cuticle of *B. spadiceus*). Hyphae of the context not distinctive in either KOH or Melzer's. Clamp connections absent.

Habit, habitat, and distribution. —Gregarious among poison ivy under oak, Washtenaw County, September, rare. It seems to fruit in the early fall if the weather is exceptionally wet.

Observations. —The outstanding features of this species are its poroid hymenophore, conspicuously reticulate stipe, the tenacious hold it has on the substratum, and the rusty brown incrustations on the cells toward the base of the trichodermial hyphae as revived in Melzer's. This material is somewhat dextrinoid.

Subsection SUBTOMENTOSI

This grouping features species that stain blue slightly on the pores when bruised, but see also *B. minutiporus* of the previous subsection in which the context stains slightly bluish. The pores typically are angular at maturity, and the pileus is velvety when young. Singer (1965) has, in our estimation, given a good account of this group, one of the most puzzling of the boletes, and we have adopted his treatment with such new features added as the hyphal incrustations, as seen in mounts revived in Melzer's, and the amyloid reactions.

KEY TO SPECIES

1. Cuticular hyphae (elements of the pileus trichodermium) with bister walls in Melzer's and some incrustations present in KOH mounts, the incrustations dark yellow-brown in KOH; pileus surface brownish when NH$_4$OH is applied directly to fresh material . . . .
. . . . (see *B. subparvulus* also) *B. chrysenteron*
1. Not with above combination of characters 2
   2. NH$_4$OH on pileus giving a fleeting green to blue-green reaction . . . . .
   . . . .(see *B. minutiporus* also) *B. spadiceus*
   2. NH$_4$OH giving a brown to mahogany-red reaction 3

3. Hyphae of pileus subcutis lacking dextrinoid incrustations (rusty brown) in Melzer's . . . . *B. tomentosulus*
3. Dextrinoid incrustations present 4

    4. Elements of pileus trichodermium with red content as viewed in Melzer's (using dried material) . . . . *B. subparvulus*
    4. Not as above 5

5. Pileus cuticle with hyphae having many short cells (15-30 μ long); pileus 2-6 cm wide . . . . *B. subtomentosus* var. *perplexus*
5. Pileus cuticle lacking a significant number of short cells in the elements of the trichodermium; pileus 4-12 cm or more wide. . . . .
. . . . *B. subtomentosus* var. *subtomentosus*

131. Boletus tomentosulus Smith & Thiers, sp. nov.

Pileus circa 12 cm latus, late convexus, ochraceo-luteus, siccus, granuloso-rimulosus. Contextus albus, tactu caeruleus. Tubuli 8-12 mm longi, ochraceo-lutei tactu caeruleus; pori 1-1.5 mm lati. Stipes circa 6 cm longus, 17 mm crassus, costatus, subluteus. Sporae 11-14 X 3.8-4.8 μ. Specimen typicum in Herb. Univ. Mich. conservatum est; prope Harbor Springs, Emmet County, July 1, 1967, legit Smith 74383.

Pileus about 12 cm broad, broadly convex; surface dry and granulose-rimulose, yellow-ocher overall and drying with a strong ochraceous tone, margin even. Context white, becoming weakly yellowish in drying, staining blue near the tubes when cut, the fresh cutis merely brownish with NH_4OH and the same reaction with KOH.

Tubes shallow (8 mm deep in a 12 cm pileus), decurrent, evenly about ocher-yellow (almost concolorous with the pileus), staining blue when cut; pores wide, 1-1.5 mm, yellow, staining blue when injured.

Stipe 6 cm long, 17 mm thick, solid, narrowed below, yellow within down to the basal area which is dingy cinnamon (but it dries an even color throughout); surface coarsely reticulate to longitudinally ribbed, neither distinctly fibrillose nor pruinose, yellow except for pinkish base which is partly covered by yellow mycelium.

Spores 11-14 X 3.8-4.8 μ, smooth, ochraceous in KOH, duller in Melzer's and with a distinct "fleeting-amyloid" reaction; in profile elongate-inequilateral, in face view subfusoid, wall about 0.2 μ thick.

Basidia 4-spored. Pleurocystidia scattered, 33-47 X 8-13 μ, subfusoid, hyaline, smooth, thin-walled. Pileus cutis a trichodermium of hyphae 6-15 μ wide and with inflated cells in some of the hyphae up to 20 μ broad, the cells not infrequently short-ellipsoid to subglobose, walls smooth, all contents colloidal and yellowish in KOH but soon fading, yellowish in Melzer's; subcutis of interwoven hyphae yellow in Melzer's

and lacking dextrinoid incrustations. Context of pileus and hymenophore crushed out in Melzer's exhibiting areas of blue-green color but the
pigment not localized under the microscope. Clamp connections none.

Habit, habitat, and distribution.—On a road bank, Harbor Springs,
Emmet County, July 1, 1967, Smith 74383.

Observations.—This species lacks the dextrinoid incrustations on
the hyphae of the subcutis characteristic of the variants around *B. subtomentosus.* Its distinguishing features are the ochraceous colors overall,
the rimulose-squamulose pileus, the short inflated cells in the hyphae of
the pileus trichoderm, the lack of pruina on the stipe, and presence of
blue staining on the hymenophore and to some extent on the context of
the pileus. Specimens of *B. subtomentosus* from Josserand in France did
not show short inflated cells in the cuticle of the pileus, and they did
show weakly dextrinoid incrustations on the lower cells of the trichodermial hyphae.

132. Boletus spadiceus Fries

Epicr. Syst. Mycol. p. 415. 1838

Suillus spadiceus (Fries) Kuntze, Rev. Gen. Pl. 3(2):536. 1898.
Xerocomus spadiceus (Fries) Quélet, Fl. Myc. p. 417. 1888.
Xerocomus coniferarum Singer, Farlowia 2:297. 1945.

Pileus 5-11 (18) cm broad, convex becoming broadly convex,
sometimes nearly plane; surface dry and subtomentose to velvety, remaining so throughout its life or in dry seasons somewhat areolate, in
age often appearing somewhat matted-fibrillose under a lens, color "dark
olive-buff" to near "Isabella color" (olive to olive yellowish) but a reddish tinge gradually developing as an undertone and in age sometimes
reddish brown when wet, often merely some shade of dingy yellow-
brown or dull reddish along the margin, margin not exceeding the tubes.
Context thick, soft, pallid-yellow with a reddish line beneath the cuticle,
bright yellow and finally pinkish around larval tunnels, rarely slightly
blue when bruised, odor none or very faintly pungent, taste mild;
NH_4OH on cutis green to bluish green then changing to fuscous then to
reddish brown; $FeSO_4$ no reaction on the cuticle, context or stipe.

Tubes becoming depressed around the stipe, 1-2 cm long, yellow-
ocher; pores large and angular (1-2 mm wide), pale yellow young, more
olive-yellow in age, staining bluish slightly when bruised.

Stipe 4-10 (12) cm long, 1-2.5 (3) cm thick, often narrowed
downward, with yellow mycelium around the base, pallid over apical

region at first, pruinose with brownish to reddish pruina, sometimes with coarse longitudinal ridges or obscurely widely reticulated.

Spore deposit olive to olive-brown; spores (9) 11-14 × 4.5-5 μ, smooth, oblong to ventricose in face view, narrowly inequilateral with a broad suprahilar depression in profile view, dingy ochraceous in KOH, greenish gray to yellowish in Melzer's, finally weakly dextrinoid, no apical pore evident.

Basidia 4-spored, 36-40 × 10-12.5 μ, clavate, pedicel narrow. Pleurocystidia 38-50 × 9-14 μ, fusoid-ventricose to mucronate, content homogeneous and hyaline, walls thin and smooth. Cheilocystidia clavate to subfusoid, yellow in mass from pigment in the wall, 36-48 × 9-14 μ. Caulocystidia fasciculate, clavate, thin-walled, 32-56 × 8-16 μ, yellow in KOH but content homogeneous. Tube trama parallel as revived in KOH, the hyphae long, straight, and 7-12 μ wide.

Pileus trichodermium of long tangled but more or less upright hyphae 6-12 μ wide, the walls smooth and hyaline in KOH and the content colloidal, content ochraceous or orange-brown in Melzer's, the cells near the tips shorter than the remainder but 2-4 times longer than broad, no inflated cells seen, the tips agglutinating into a layer (as revived in KOH or Melzer's). Context hyphae near the trichodermium with orange-brown homogeneous content in Melzer's, smooth or with some incrusting debris in the region of the subcutis as well as the lower cells of the elements of the trichodermium but this disappearing in KOH. Clamp connections absent.

Habit, habitat, and distribution.—Gregarious especially along wet road banks, its abundance in the state has not been accurately ascertained.

132a. Boletus spadiceus var. gracilis var. nov.

A typo differt: Pileus 3-8 cm latus; stipes 3-6 cm longus, 5-10 mm crassus; sporae 10-12 (13) × 4.5-5.5 μ. Typus Smith 75911 (MICH).

Pileus 3-8 cm broad, expanding to broadly convex or nearly plane, margin fertile and even; surface dry and velvety-subtomentose, in age or when wet appearing matted-fibrillose, color dingy yellow-brown ("Saccardo's umber" to "Prout's brown"), when wet with a dull red tinge, when dried with a subgranulose texture. Context whitish in midportion and bright yellow near the tubes, when mature with a brown to reddish brown line under the cuticle; odor and taste not distinctive; $FeSO_4$ no reaction, NH_4OH on cuticle giving a green flash and then turning reddish brown; staining blue when bruised near the pileus margin.

Tubes 6-10 mm deep, dull yellow becoming dull olive-yellow, readily splitting when pileus is broken downward; pores large (about 1 per mm), near stipe sublamellate, slowly staining cinnamon when bruised.

Stipe 3-6 cm long, 5-10 mm thick, equal to a pinched off base, solid, yellow within, with bright yellow mycelium around the base, surface yellowish with a thin overlay of cinnamon pruina, not reticulate.

Spore deposit olive when fresh; spores 9-12 (13) X 3.5-4.5 μ, smooth, lacking an apical pore, shape in face view suboblong or tapered slightly to apex, in profile somewhat inequilateral, color in KOH yellowish hyaline to greenish yellow, duller in Melzer's; wall about 0.2 μ thick.

Basidia 4-spored, 24-36 X 7-10 μ, narrowly clavate, pale yellow in KOH, a fleeting-amyloid reaction present and very strong. Pleurocystidia 32-50 X 7-12 μ, narrowly fusoid, thin-walled, hyaline in KOH, content homogeneous. Cheilocystidia similar to pleurocystidia but smaller and neck shorter. Caulocystidia 30-50 X 10-16 μ, short- to long-clavate, thin-walled, smooth, in KOH content hyaline to yellowish, occurring in fascicles from the subiculum.

Tube trama of wide (8-12 μ) scarcely gelatinous and only obscurely divergent hyphae, with a strong fleeting-amyloid reaction (sections at first violet-black in Melzer's). Pileus cuticle a trichodermium, the elements incrusted with dextrinoid patches of amorphous material, the cells 20-60 X 9-12 μ, tubular or nearly so and the apical cell merely blunt. Hyphae below cuticle merely yellow in Melzer's. Clamp connections absent.

Habit, habitat, and distribution.—Scattered under oak-pine, Topinabee, Cheboygan County, August 10, 1968, W. Patrick and Smith 75911.

Observations.—This variety differs from var. *spadiceus* in the slightly smaller spores, and narrower stipe as well as in the strong fleeting-amyloid reaction of the hymenophoral trama. The stipe was not ridged near the apex. This is a taxon somewhat intermediate between the *B. illudens* group and the *B. subtomentosus* group, in fact it is a bridging variant.

133. Boletus subtomentosus Fries

Syst. Mycol. p. 389. 1821

Leccinum subtomentosum (Fries) S. F. Gray, Nat. Arr. Brit. Pls. 1:647. 1821.
Rostkovites subtomentosus (Fries) P. Karsten, Rev. Mycol. 3:16. 1881.
Versipellis subtomentosa (Fries) Quélet, Enchir. Fung. p. 158. 1886.

Xerocomus subtomentosus (Fries) Quélet, Flor. Myc. Fr. p. 418. 1888.
Ceriomyces subtomentosus (Fries) Murrill, Mycologia 1:153. 1909.

var. subtomentosus

Illus. Pl. 104.

Pileus 5-18 (20) cm broad, convex becoming broadly convex, rarely plane, margin at first incurved against the tubes; surface dry and subtomentose to velvety, gradually rimose and in age often conspicuously areolate; color dull olivaceous to olive-yellowish or finally yellow-brown, with a dull reddish line under the cuticle which finally may give a reddish tone to the pileus after the collapse of the epicutis (dull "olive" to "Isabella color," "argus-brown," "antique brown," or in age redder). Context thick and firm, whitish slowly becoming pale yellow, not changing color when cut or bruised, odor and taste not distinctive. Pileus surface mahogany-red with ammonia (NH$_4$OH).

Tubes 1-2.5 cm long, adnate becoming depressed or decurrent on stipe by lines, yellow, when cut slowly very slightly greenish blue; pores 1-2.5 mm broad at maturity, somewhat angular or more rarely sublamellate near stipe, yellow, staining slightly greenish and finally brownish where injured.

Stipe 4-10 cm long, 1-2 (3) cm thick, equal downward to a narrowed base, solid, whitish in center, yellow in cortex, yellow around larval tunnels; surface pruinose to scabrous, obscurely reticulate to widely ridged-reticulate over upper part, around the base sulphur-yellow, pallid to ocher-yellow above or overall but in age with brown areas or near "orange-cinnamon" finally, at times reddish brown from handling.

Spore deposit olive to olive-brown; spores 10-13 (15) X 3.5-5 μ, in face view subfusoid to oblong, in profile narrowly inequilateral, smooth, walls slightly thickened, in KOH pale lemon-yellow, in Melzer's dingy pale olive yellowish (as if with a weak fleeting-amyloid reaction).

Basidia clavate, 4-spored, 25-34 X 6.5-9 μ, hyaline to yellow in KOH and Melzer's. Pleurocystidia scattered, lanceolate, 47-72 X 8-12 μ, hyaline in KOH, yellowish in Melzer's, thin-walled, smooth.

Tube trama of gelatinous hyaline hyphae in KOH, some divergence to the subhymenium but central strand subparallel. Cuticle of pileus a trichodermium of smooth thin-walled hyphae 8-14 μ wide and with the terminal cell merely obtuse, rarely slightly inflated at the midportion. Clamp connections absent.

Habit, habitat, and distribution.—Common in the summer and fall especially on road banks, but seldom do large numbers of basidiocarps occur in one small area.

Observations.—European specimens examined have the dextrinoid incrustations on the hyphae of the subcutis, but they are often limited to certain areas. In American specimens they are much more numerous. The NH_4OH reaction to green will distinguish *B. spadiceus* and *B. illudens* from *B. subtomentosus.*

133a. Boletus subtomentosus var. perplexus Smith & Thiers, var. nov.

Illus. Pl. 103 (lower).

Pileus 3-6 cm latus, demum planus, siccus, subtomentosus, olivaceus vel olivaceo-ochraceus, tarde olivaceo-brunneus. Contextus pallide luteus, immutabilis. Tubuli olivaceo-lutei; pori 1-2 per mm, lutei tactu olivaceo-caerulei. Stipes 4-6 cm longus, 8-12 mm crassus, pruinosus; pruina roseo-cinnamomea. Sporae 11-14 X 4-5 μ. Typus Smith 74476 (MICH).

Pileus 3-6 cm broad, convex, expanding to plane; surface dry and tomentose to plushlike, dry, olive to olive-ochraceous ("Isabella color") or dingy olive-brown finally, margin even. Context weakly yellow-buff to ivory-yellow (yellowish white), lacking a distinct line at the tubes, not staining blue when injured, with a dark olive-brown line just under the cutis; odor and taste mild.

Tubes 8-12 mm deep, broadly adnate to decurrent, "Isabella color" (olive yellowish) or more olivaceous, unchanging when cut; pores 1-2 per mm, yellow becoming olive-yellow and when bruised readily staining olive-blue.

Stipe 4-6 cm long, 8-12 mm thick, equal, solid, greenish pallid within including the base, brownish around wormholes, surface covered more or less overall with a reddish cinnamon pruina, base with a thin coating of pallid mycelium.

Spore deposit olivaceous; spores 11-14 X 4-5 μ, smooth, wall -0.5 μ thick, apex lacking a pore; shape in profile somewhat inequilateral, in face view narrowly subelliptic, dull yellow-brown in KOH, about the same in Melzer's. Caulocystidia clavate, 26-38 X 9-15 μ, thin-walled, smooth, with a yellow content diffusing into the mount as revived in KOH, hardly distinct from basidioles.

Pileus cuticle a trichodermium of nongelatinous hyphae 5-15 μ wide, tubular or nearly so, with many short cells in the elements that are 15-30 μ long, walls smooth to roughened slightly, with a yellow content diffusing into the mounts as revived in KOH, as mounted in

Melzer's the content granular and the color ochraceous to orange. Subcutis of hyphae with dextrinoid patches of pigment on the walls causing the layer to appear spotted under the microscope. Clamp connections absent.

Habit, habitat, and distribution.—Solitary to scattered on mossy hummocks under hardwoods, Berry Creek, Cheboygan County, July 10, 1967, Smith 74476.

Observations.—This variant is constant in its characters and was observed frequently in Cheboygan County. The trichodermial elements contain more short cells than do those of var. *subtomentosus,* the basidiocarps are usually solitary, and the mycelium at the base of the stipe is pallid, not yellow.

134. Boletus subparvulus Smith & Thiers, sp. nov.

Pileus 4-5 cm latus, convexus, glaber, atrobrunneus. Contextus pallidus tactu subcaeruleus. Tubuli sordide lutei, tactu subcaerulei. Stipes circa 6 cm longus, 10 mm crassus, deorsum ferrugineo-roseus, sursum luteus, roseo-pruinosus. Sporae 10-13 X 4-4.5 μ. Specimen typicum in Herb. Univ. Mich. conservatum est; prope Emerson, Chippewa County, July 27, 1967, legit Smith 74678.

Pileus 4-5 cm broad, convex; surface moist, glabrous, uneven, color near "mummy-brown" to dark bister, margin even and slightly inrolled. Context pallid, yellowish later, reddish around the wormholes and staining weakly blue when cut, $FeSO_4$ slowly bluish gray.

Tubes dull yellow, staining bluish green when cut; pores weakly dull yellow, when bruised weakly bluish.

Stipe 6 cm long, 10 mm thick, narrowed below, weakly yellow within in upper half, rusty rose in base; surface reddish pruinose on a yellowish ground color above, and rusty rose near the base and the buried portion weakly yellowish.

Spores 10-13 X 4-4.5 μ, smooth, wall thin, no visible pore at apex; shape in profile elongate-inequilateral to almost oblong; subfusiform to suboblong in face view; ochraceous in KOH, duller ochraceous in Melzer's.

Basidia 4-spored. Pleurocystidia scattered, 32-40 X 8-13 μ, fusoid to fusoid-ventricose, content dingy ochraceous in KOH, content not distinctively colored in Melzer's. Caulocystidia basidiole-like or slightly larger.

Pileus cuticle a trichodermium of heavily incrusted hyphae with the apical cells clavate and up to 15 μ broad, the incrustations rusty brown in KOH, content granular and in Melzer's dull red, nearly hyaline in KOH. Clamp connections absent.

Habit, habitat, and distribution.—Solitary on hummocks, hardwood slashings, Emerson, Chippewa County, July 27, 1967, Smith 74678.

Observations.—This is an autonomous species by virtue of the color of the pileus (date-brown), the heavily incrusted elements of the pileus trichodermium, and their red content in Melzer's. It is one of the small species which occurs singly and is easily overlooked when the collecting is good.

135. Boletus chrysenteron Fries

Epicr. Syst. Myc. p. 415. 1838

Xerocomus chrysenteron (Fries) Quélet, Fl. Myc. Fr. p. 418. 1888.

var. chrysenteron

Illus. Pl. 105.

Pileus 3-8 cm broad, convex becoming broadly convex or finally nearly plane; surface dry and evenly velutinous to subtomentose when young, the epicutis areolate in age exposing the context which soon reddens; color when young dark olive to olive-brown, paler olive to olive-gray in age and then often with a red marginal zone, rarely becoming reddish overall. Context whitish at very first but soon yellowish, slowly staining bluish when cut, becoming lemon-yellow in $FeSO_4$, in KOH buff colored, very soft and punky, odor and taste mild; on the cutis blackish with KOH, no reaction with $FeSO_4$ and brownish with NH_4OH.

Tubes up to 5-10 mm deep, depressed around the stipe, bright yellow, slowly becoming "Isabella color" in age, blue when wounded; pores about 1 mm in diameter, irregular in outline, bright yellow young, in age reddish over some areas, mostly splitting lengthwise when the pileus is broken downward.

Stipe 4-6 cm long, 5-10 mm thick, equal or narrowed downward, solid, yellowish to pallid at apex, purplish red in the base, staining bluish green where injured; surface pallid becoming bright yellow, pinkish red below, scurfy to pruinose overall and at times with longitudinal lines or ridges but not reticulate, basal mycelium white.

Spore deposit olive to olive-brown; spores 9-13 X 3.5-4.5 (5) μ, smooth, oblong to boat-shaped in face view, narrowly inequilateral in profile, bright yellow in KOH, dull grayish yellow in Melzer's, smooth, lacking an apical pore.

Basidia 24-30 X 8-10 μ, clavate, 4-spored, content yellow in KOH. Pleurocystidia scattered, 28-42 X 9-14 μ, fusoid-ventricose, smooth, thin-walled, yellow in KOH and in Melzer's, apex obtuse to subacute, walls thin and smooth. Cheilocystidia clavate to subfusoid, 9-12 μ wide, with bright yellow content in KOH. Caulocystidia numerous, 24-38 X 9-15 μ, clavate, mucronate or fusoid-ventricose with mostly obtuse apices, often with a wall thickening in the apex, yellow to hyaline in KOH.

Tube trama of gelatinous hyphae somewhat divergent to the subhymenium. Pileus cutis a trichodermium of hyphae with bister walls in Melzer's and some incrusting material present, the cells of the hyphae up to 30 μ wide and keg-shaped to globose, terminal cells somewhat cystidioid to globose, in KOH dark yellow-brown and often heavily incrusted. Clamp connections absent.

Habit, habitat, and distribution.—Scattered to gregarious in the summer and early fall on earth exposed for 3-5 years, such as around uprooted trees, along roadsides, on mossy banks, etc., in hardwood and mixed forests, common during some seasons, rare during others, occurring throughout the state.

Observations.—It is doubtful if any species of bolete has been more frequently misidentified than this one. In North America *B. truncatus* has been most frequently confused with it, but is readily distinct by having spores with a truncated apex. As *B. chrysenteron* is presently delimited, it grades into subsection *Fraterni*, and we key it there also to allow for old pilei which have become entirely red. In Smith 77892 the trichodermial hyphae contained numerous globose cells.

135a. Boletus chrysenteron var. subnudipes Smith & Thiers, var. nov.

Pileus 3-6 cm latus, convexus demum late convexus, siccus, velutinus vel subtomentoseus, olivaceus demum olivaceo-brunneus, areolatus, ad marginem demum roseus. Contextus luteus. Tubuli lutei demum olivaceo-lutei; pori 1-2 per mm, lutei tactu caerulei. Stipes 4-7 cm longus, 3.5-6 mm crassus, glaber, sursum luteus, deorsum roseus. Sporae 11-14 (15) X 4-5 μ. Specimen typicum in Herb. Univ. Mich. conservatum est; prope Manchester, August 24, 1966, legit Smith 73150.

Pileus 3-6 cm broad, convex to pulvinate but in age broadly convex to nearly plane; surface dry, dull and subtomentose, olive to olive-brown young, soon flushed dull rose-red near margin and finally overall, irregularly rimulose showing the yellow in the cracks but not as regularly areolate as in the type variety, margin even and fertile. Context yellow, staining slightly greenish blue; odor and taste not distinctive.

Tubes yellow when young (near flavous), more olive yellowish when mature, adnate to subdecurrent, 5-10 mm deep; pores yellow and staining bluish slightly, 1-2 per mm (or broader in age), almost bole-tinoid near the stipe.

Stipe 4-7 cm long, 3.5-6 mm thick, equal, solid, yellow with red streaks lower down; surface naked to slightly pruinose, apex yellow, remainder rose-red and drying this color, base coated with pale yellow to flavous mycelium.

Spores 11-14 (15) $\times$ 4-5 μ, smooth, no apical pore evident, color in KOH near yellow-ocher after standing a few minutes, in Melzer's amyloid at first, slowly becoming dingy yellow-brown; shape in face view subfusoid to narrowly elliptic, in profile somewhat inequilateral, wall about 0.2 μ thick.

Basidia 4-spored, 24-33 $\times$ 8-11 μ, clavate, yellow in KOH, more brownish in Melzer's but content not distinctive. Pleurocystidia scattered, 38-60 $\times$ 8-13 μ, fusoid-ventricose or neck elongated and somewhat flexuous, apex acute to subacute, walls smooth, thin and hyaline, content not distinctive in either KOH or Melzer's. Cheilocystidia similar to pleurocystidia. Caulocystidia present throughout a rather well-developed caulohymenium (hence the stipe surface not appearing scurfy as in the type variety, but the layer becomes separated into patches in drying), the cells clavate to fusoid-ventricose, and in some collections frequently with a wall thickening in the apex as revived in KOH, 27-40 $\times$ 10-16 μ.

Tube trama of slightly divergent hyphae lacking distinctive content in either KOH or Melzer's. Trichodermium of pileus of hyphae 5-12 μ wide, the end-cells with a filamentous flexuous prolongation 3-5 μ wide and up to 20 μ long arising from some of them and often differentiated by a basal septum, the ventricose part and the cells below the terminal element having a granular hyaline content in KOH which in Melzer's is ochraceous to orange, the walls hyaline in KOH and Melzer's and becoming ornamented from the outer layer breaking into small thin hyaline plates and flaking off as in the cuticular hyphae of *Leccinum insigne*. Hyphae of the context hyaline and the cells "empty" as revived in KOH or Melzer's, their walls thin and smooth. Clamp connections absent.

Habit, habitat, and distribution.—Gregarious in low wet woods, near Manchester, August 24, 1966, Smith 73150.

Observations.—The distinguishing features of this variety are centered in the hyphae of the trichodermium of the pileus rather than on the naked stipe as the latter is viewed in the fresh condition. The content of the trichodermial cells, which is somewhat granular-colloidal in appearance and colors ochraceous-orange in Melzer's, is distinctive, and the hyphal proliferation from an occasional terminal cell reminds one of *B. hortonii.* The colors are generally redder than in the type variety so that the basidiocarps are frequently mistaken for *B. truncatus* when collected. The type variety differs in the hyphae of the trichodermium having more or less dark yellow-brown walls as revived in KOH. The name *B. versicolor* is not available for this variant at the species level, and we are not sure the European bolete to which the name is often attached is identical with our variety.

Subsection FRATERNI Smith & Thiers, subsect. nov.

Pileus flammeus vel ruber, tactu caeruleus.
Typus: *Boletus fraternus* Peck.

Pileus red at first or if olive to olive-brown very soon becoming red; context yellow or yellowish and staining blue when injured; tubes yellow staining blue to greenish blue on injury; stipe naked to pruinose-scurfy rarely faintly reticulate at apex.

KEY TO STIRPES

1. Stipe 1-3 cm thick near apex . . . . Stirps *Sensibilis*
1. Not as above, usually 3-10 or rarely 13 mm thick . . . . Stirps *Fraternus*

Stirps FRATERNUS

KEY

1. At least a fair number of the cells of the pilear trichodermium inflated and with amyloid inclusions present in them or such inclusions present in some of the noninflated cells also . . . . *B. flavorubellus*
1. Not as above 2

   2. Spores 5-7 μ wide . . . . (see *B. subfraternus* also) 3
   2. Spores 3-4.5 (5) μ wide 4

3. Pores large (1 mm or more) and usually angular, KOH not causing a bright yellow color in the terminal cells of the pilear cuticle. . . . . *B. fraternus*
3. Pores 1-2 per mm, cuticular cells lemon-yellow in KOH . . . . . . *B. campestris*

   4. Cuticle of pileus an epithelium but the cells not compacted into a rigid layer. . . . . *B. harrisonii*
   4. Not as above 5

5. Terminal cells of pilear trichodermium dingy brown in KOH . . . .
. . . . see *B. chrysenteron* var. *chrysenteron*
5. Not as above 6
    6. Spores 10-12.5 × 3-4 μ . . . . *B. rubeus*
    6. Spores 4-5.5 μ wide 7
7. An occasional cell of the trichodermial hyphae having an apical proliferation . . . . see *B. chrysenteron* var. *subnudipes*
7. Not as above . . . . (but see *B. subfraternus* also) 8
    8. Pileus dark dull red . . . . *B. rubellus* var. *rubellus*
    8. Pileus flame-red . . . . *B. rubellus* var. *flammeus*

136. Boletus flavorubellus Thiers & Smith

Mich. Bot. 5:117. 1966

Illus. Pl. 125 (upper).

Pileus (2.5) 5-6 (8) cm broad, convex to pulvinate, becoming broadly convex; surface dry and subtomentose, soon deeply areolate exposing the yellow context; the margin fertile, color deep red ("brick-red" to "Hay's russet"), gradually paler and in age at times chrome-yellow. Context bright rich yellow (flavous), slowly changing to greenish blue when cut, with KOH ochraceous-tan, with $FeSO_4$ grayish; taste mild, odor none.

Tubes sharply depressed at the stipe, 4-5 mm deep, flavous staining greenish blue bruised; pores small (2-3 per mm), angular to round, flavous staining greenish when bruised.

Stipe 3-7 cm long, 5-11 mm thick, equal, solid, flavous within, streaked reddish around larval tunnels, slowly staining greenish where cut; surface above flavous beneath the reddish pruinose coating, flavous below, tinged red in the midportion, base with yellow-ocher mycelium.

Spore deposit dark olive; spores compressed to some degree, 10-13 × 4-5 × 4.5-7 μ, smooth, ovate to ventricose-ovate in face view, in profile subelliptic to inequilateral or ventricose-inequilateral, dingy ochraceous in KOH, in Melzer's a dull yellowish tan with a slight greenish gray shadow when first mounted, wall slightly thickened, with a minute apical thin spot.

Basidia 4-spored, 20-24 × 8-10 μ, clavate, yellowish in KOH. Pleurocystidia scattered, 30-40 × 9-12 μ, fusoid-ventricose, apex obtuse, nearly buried in the hymenium and with content merely yellowish in KOH and Melzer's. Tube trama somewhat gelatinous, hyphae slightly divergent from a central area to nearly parallel. Pileus trichodermium of short-celled hyphae 12-20 μ broad and 2-3 times as long, but some almost globose, thin-walled, yellow in KOH and dingy yellowish in Melzer's, in Melzer's often observed to have aggregations of dark violet rods and particles of an amorphous nature but these not present in every cell. Clamp connections absent.

Habit, habitat, and distribution.—Scattered on wet earth, Ann Arbor late summer, during wet weather.

Observations.—The inflated to sphaerocyst-like cells in the trichodermium of the pileus remind one of *Leccinum* section *Luteoscabra,* and the flaking off of thin plates of wall material is similar to the condition found in *L. insigne.*

Material examined.—Washtenaw: Smith 62872, 72948, 73138, 73239.

137. Boletus fraternus Peck

Torrey Club Bull. 24:144. 1897

Illus. Figs. 76-78.

The original description is quoted as follows:

"Pileus convex, becoming plane or depressed, slightly tomentose, deep red when young, becoming dull red with age, flesh yellow, slowly changing to greenish blue where wounded; tubes rather long, becoming ventricose, slightly depressed about the stem, their walls sometimes slightly decurrent, the mouths large, angular or irregular, sometimes compound, bright yellow, quickly changing to blue when wounded; stem short, cespitose, often irregular, solid, subtomentose, slightly velvety at the base, pale reddish yellow, paler above and below, yellow within, quickly changing to dark green where wounded; spores .0005 in. long, .00025 in. broad. Pileus 1-1.5 in. broad; stem 1-1.5 in. long, 3-6 lines thick."

Peck stated further that when the pileus cracked the chinks became yellow as in *B. subtomentosus.* Our microscopic data from the type follows:

Spores 12-16 X 5-6.5 μ, smooth, lacking an apical pore, shape in face view rather broadly subfusoid, in profile inequilateral with a markedly distinct suprahilar depression, color in KOH bright ochraceous, in Melzer's dull ochraceous, wall 0.3-0.5 μ thick (some elongate narrow spores also present having the wall 1 μ thick); spores with a subapical lateral protuberance rare.

Basidia 4-spored, 9-13 μ broad, hyaline in KOH. Pleurocystidia none observed. Hymenophoral trama giving a strong positive fleeting-amyloid reaction. Pileus cuticle a weak trichodermium of hyphae 7-15 μ wide, the elements mostly of short cells often inflated, the walls thin and essentially smooth, cells readily collapsing, end-cells bullet-shaped or beaked, rarely the penultimate cell with a subapical beak, content hyaline in KOH and weakly yellow in Melzer's. Hyphae of subcutis and adjacent context strongly ochraceous in Melzer's. Clamp connections absent.

Observations.—This species differs sharply in the features of the cuticular hyphae from *B. campestris*. *B. fraternus* is a southern species which apparently lacks a coating of bright yellow mycelium over the base. We have not yet recognized it in Michigan, but include it here in order to clarify the concept as based on a type study and to complete our presentation of this very difficult group.

138. Boletus rubeus Frost

Bull. Buff. Soc. Nat. Sci. 2:102. 1874

Illus. Figs. 79-80.

"Pileus flat convex, rather thin edge, at first inflexed, extended, turning up in age, bright brick red when young, afterwards mottled with red and yellow, very finely apressed subtomentose, yellow under cuticle. Tubes bright lemon yellow when young, stuffed, afterwards yellow, and sometimes with red mouths, generally adnate but sometimes with a slight depression. Stem small, often flexuous, brick red or matted as pileus, white tomentose at base. Flesh yellow, pale in pileus and tinged reddish in the stem, changing to blue.

"Spores .00095-.0042 mm. In deep woods. Rare. August."

We have quoted the original description above.

Our data from a Frost specimen in the Peck collections follows:

Spores 10-12.5 X 3-4 μ, smooth, lacking a thin spot or pore at apex, shape in face view narrowly oblong or nearly so, in profile suboblong to obscurely inequilateral, suprahilar depression scarcely differentiated, color in KOH weakly ochraceous, in Melzer's ochraceous; wall thin (−.2 μ thick).

Basidia 4-spored, 7-9 μ broad, hyaline in KOH when well revived. Pleurocystidia 28-38 X 5-11 μ, small, fusoid-ventricose, neck 3-5 μ wide and apex obtuse to subacute, content hyaline in KOH, merely yellowish in Melzer's. Hymenophore tissue not amyloid in any way when crushed in Melzer's. Laticiferous hyphae 8-14 μ wide prominent in the hymenophoral tissue, their content yellow in Melzer's.

Pileus cuticle of a tangle of tubular hyphae 4-9 μ wide, the cells mostly long but a few short ones present, walls thin and often with a gelatinous halo (in KOH), content ochraceous in KOH, strongly ochraceous in Melzer's and homogeneous, end-cells mostly tubular. Hyphae of subcutis orange-ochraceous to yellowish in Melzer's. Clamp connections absent.

Observations.—Our account here is the original description supplemented with data obtained from the type. We designate the Frost specimen at Albany as lectotype. Smith noted that this material was not the same as *B. bicolor* in general appearance as dried, and this point is further emphasized by the wide angular pores, as is evident on the dried specimen. The species does appear to be closely related to *B. miniato-olivaceus,* but again the size and shape of the pores distinguish it. Although the species is clearly in the stirps *Fraternus* of this work, it is marginal in the sense that the pileus cuticle is of the type which becomes at least subviscid in age or when wet and does not become areolate. We have not as yet recognized the species in the Michigan flora.

139. Boletus campestris Smith & Thiers, sp. nov.

Illus. Pl. 106 (lower).

Pileus 3-4 cm latus, siccus, velutinus, roseus, tarde luteus. Contextus luteus tactu caeruleus. Pori 1-2 per mm, lutei tactu caerulei. Stipes 4-5 cm longus, 5-10 mm crassus, flavus, pruinosus. Sporae 11-14 (15) X 4.5-6 (7) μ. Specimen typicum in Herb. Univ. Mich. conservatum est; Ann Arbor, July 16, 1966, legit Rup Chand (Smith 72961).

Pileus 3-4 cm broad, convex becoming broadly convex; surface dry and velvety to unpolished, rose-red or deeper when young, paler pinkish red when older, slightly rimose and a yellow flush showing in the cracks in age, dull ferruginous as dried. Context thick, yellow (nearly flavous), staining blue to greenish blue when bruised, odor and taste not distinctive.

Tubes 6-8 mm deep, sharply depressed at the stipe, flavous when young, greenish yellow old, greenish blue when injured; pores 1-2 per mm, round to angular, bright yellow young, greenish yellow in age, bluish when injured.

Stipe 4-5 cm long, 5-10 mm thick, equal or nearly so, solid and bright yellow within, staining greenish blue, surface flavous to pale yellow at apex, pruinose and concolorous with pileus downward, base coated with flavous mycelium.

Spores 11-14 (15) X 4.5-6 (7) μ, smooth, not truncate at apex, soon dull yellow-ocher in KOH, in Melzer's with a fleeting-amyloid reaction, soon pale dingy yellow-brown, elliptic to subfusoid in face view, in profile inequilateral with a sharply defined suprahilar depression, apex obtuse, wall about 0.5 μ thick.

Basidia 4-spored, 26-38 X 10-14 μ, clavate, yellow in KOH and Melzer's. Pleurocystidia scattered, 34-56 X 9-13 μ, ventricose with a narrow wavy neck (in some) and subacute apex, content yellowish to hyaline in KOH and Melzer's. Cheilocystidia similar to pleurocystidia but content more highly colored.

Tube trama slightly divergent from a central area of more or less parallel hyaline thin-walled tubular hyphae; laticiferous, elements present. Pileus trichodermium of long hyphae 5-12 μ wide, the cells tubular or nearly so to the terminal cells which gradually taper to an obtuse to subacute apex, walls smooth or roughened slightly as in *Leccinum insigne,* content lemon-yellow in KOH and this diffusing into the mount, lacking amyloid particles. Hyphae of context thin-walled, smooth, hyaline in KOH or Melzer's except for yellowish walls. Clamp connections absent.

Habit, habitat, and distribution. –In a lawn, gregarious, Ann Arbor, July 16, 1966, collected by Rup Chand (Smith 72961).

Observations. –The more or less noninflated, tapered terminal cells in the hyphae of the trichoderm, the intense yellow pigment in KOH (which diffuses into the mount readily), the relatively broad spores and bright yellow basal mycelium features this variant among the species with the trichodermal hyphae having roughened walls in the manner of *Leccinum insigne.* The basal mycelium is like that of *B. flavorubellus,* but no amyloid particles were found in the trichodermial cells. Greatly inflated cells in the cuticular elements are also lacking.

140. **Boletus harrisonii** Smith & Thiers, sp. nov.

Pileus 2.5-5 (7) cm latus, convexus, lateritius, areolatus, siccus. Contextus luteus; lamellae lutei tactu caeruleae; stipes 4-6 cm longus, 4-10 mm crassus, sursum luteus, deorsum roseus, ad basin luteo-brunneus. Sporae 9-12 (13-18-22) × 4.5-5.5 (6-7.5) μ. Specimen typicum in Herb. Univ. Mich. conservatum est; Ann Arbor, July 11, 1967, legit K. A. Harrison 9504.

Pileus 2.5-5 (7) cm broad, irregularly convex, margin incurved; surface finely rimose-areolate, brick-red to ocher-red, dry. Context yellow, soft, odor pungent but fading. Tubes yellow, staining blue slightly; pores medium broad. Stipe 4-6 cm long, 4-10 mm thick, flexuous, expanding at apex, solid, yellowish streaked with brown within, brittle; surface yellow above, midportion red and basal area dull yellow-brown.

Spores 9-12 (13-18-22) × 4.5-5.5 (6-7.5) μ, smooth, with a minute thin spot at apex, wall up to 1.5 μ thick in larger spores, mostly the wall 0.5 μ thick; shape in profile inequilateral, in face view subelliptic to fusoid, color in KOH yellow-brown, pale dingy tawny in Melzer's (fleeting-amyloid reaction slight).

Basidia 4-spored or 1-, 2-, and 3-spored also. Pleurocystidia scattered, 36-63 X 9-18 μ, fusoid to ventricose-subcapitate, content when revived in Melzer's orange brownish and amorphous, fading on standing, walls smooth and thin. Caulocystidia in fascicles or scattered, 18-32 X 9-15 μ, content ochraceous in KOH and the same for the cortical hyphae of the stipe.

Pileus cutis a tangled trichoderm of hyphae 6-12 (15) μ wide, with the last 3-4 cells short and becoming inflated to elliptic or subglobose, the end-cell cystidioid to subglobose, content ochraceous in KOH, paler in Melzer's, no amyloid material present, walls thin and smooth or nearly so. Subcuticular and tramal hyphae finally nearly hyaline in Melzer's but fleeting-amyloid reaction strong and relatively persistent, no dextrinoid incrustations seen on any of the hyphae. Context hyphae often with "colloidal" content in Melzer's but their color merely yellowish.

Habit, habitat, and distribution.—Gregarious in grass near a spruce tree, Ann Arbor, July 11, 1967, K. A. Harrison 9504.

Observations.—Although described from a single collection the positive features of this species are sufficient to justify describing it. The inflated cells in the pileus trichodermium are so numerous that in sections of mature pilei one is reminded of an epithelium with the cells loosely arranged. The ultimate cell in a filament is often "aborted"— possibly formed secondarily by a cross wall in the original apical cell at the point where the taper begins. The pleurocystidia are large and conspicuous in mounts revived in Melzer's. The spores vary greatly. In some basidiocarps small ellipsoid spores 8-10 X 4-5 μ are present along with the typical range (9-12 X 4.5-5.5 μ), and in all caps a varying number of larger spores was present. The species is close to *B. fraternus,* but the almost epithelial structure of the cuticle of the pileus distinguishes it from all in this group.

141. Boletus rubellus Krombholz

Nat. Abb. Sche. 5:12. 1836

var. rubellus

Illus. Fig. 81.

Pileus 3.5-8 cm broad, broadly convex expanding to plane or margin finally uplifted; surface dry and plush-like when young, becoming irregularly rimulose or in age distinctly areolate, color dull brick-red with an olivaceous overtone, when dried distinctly brick-red in marginal area and pale olive-brown over disc; margin even. Context pale flavous, staining blue slightly when cut.

Tubes adnate to decurrent, 7-10 mm deep, dingy yellow; pores greenish yellow staining bluish, irregular in outline, up to 1 mm or more wide in age.

Stipe 4-8 cm long, 5-10 mm thick, equal or enlarged above, solid, yellow within, streaked reddish in the base; surface bright yellow at apex shading to dull rose-red toward base and reddish pruinose, with a coating of yellow mycelium around the base.

Spores 10-13 × 4.5-5 μ, smooth, wall -0.5 μ thick, apex rounded to obscurely truncate in a few; shape in profile ventricose-inequilateral to merely inequilateral, in face view subelliptic to subfusoid; color strongly ochraceous in KOH, in Melzer's dulled by a bluish cast (amyloid when first revived).

Basidia 4-spored. Pleurocystidia—none found. Pileus cutis a trichodermium with the hyphae 5-11 μ wide, and with smooth walls or the latter only very minutely roughened; terminal cell cylindric to an obtuse apex or narrowed to apex, rarely subcapitate, content dingy ochraceous in Melzer's to ochraceous orange-brown, walls thin, cells for the most part remaining agglutinated to each other. Hyphae of subcutis and tramal hyphae orange-red to dull ochraceous in Melzer's; all hyphae tubular or cells only slightly inflated, no amyloid material present in any of the cells.

Habit, habitat, and distribution.—Collected in Cheboygan County, August 21, 1967, in the vicinity of Douglas Lake by a member of the American Mycological Association.

Observations.—The name *B. rubellus* as used here excludes the synonymy given by Singer (1967). The species is very close to *B. rubeus* Frost, but the latter has white mycelium over the base of the stipe and narrower spores.

141a. Boletus rubellus var. flammeus var. nov.

Illus. Fig. 82; Pl. 107 (lower).

Pileus 2-4 cm latus, convexus, siccus, aurantio-ruber, demum luteus. Contextus luteus. Tubuli 6-10 mm longi; pori lutei tactu caerulei. Stipes 4-8 cm longus, 4-8 mm crassus, deorsum sordide fulvus, sursum olivaceo-luteus, pruinosus. Sporae 10-13 × 4-5 μ. Specimen typicum in Herb. Univ. Mich. conservatum est; prope Ann Arbor, August 2, 1966, legit S. Mazzer (Smith 73026).

Pileus 2-4 cm broad, pulvinate becoming broadly convex; surface unpolished, color "flame-scarlet" with a yellowish margin, soon decoloring to pale reddish on disc and the remaining area yellowish, surface velvety at first, becoming rimose-areolate, margin even. Context pallid-yellow but soon blue when cut, taste soapy, odor slightly pungent, with $FeSO_4$ quickly olive on the cuticle, with guaiac no reaction.

Tubes canary-yellow young, becoming olive-yellow, staining blue-green, 6-10 mm long, depressed around the stipe; pores bright yellow, dingy with age or flushed brownish, staining blue when bruised.

Stipe 4-8 cm long, 4-8 mm thick, equal, solid, greenish yellow within except for flavous basal area, surface rusty to clay color below and base surrounded by flavous mycelium, pinkish pruinose on a yellow ground to the bright greenish yellow apex.

Spore deposit olive-brown; spores 10-13 X 4-5 μ, smooth, apex lacking a distinct pore, in KOH ochraceous to clay color, in Melzer's dingy yellowish brown but first with a grayish shadow (fleeting amyloid), in face view subelliptic, in profile obscurely inequilateral, wall about 0.2 μ thick.

Basidia 4-spored, (28) 30-45 X 8-10 μ, clavate, content yellow in KOH and in Melzer's. Pleurocystidia 34-52 X 8-14 μ, rostrate to fusoid-ventricose, thin-walled, smooth, content hyaline to yellow in KOH and Melzer's. Cheilocystidia similar but usually smaller and more highly colored than the pleurocystidia. Tube trama of hyaline subparallel hyphae diverging somewhat in the area next to the subhymenium, the hyphae tubular, thin-walled. Laticiferous elements yellow, 4-10 μ wide, rare to scattered.

Pileus trichodermium with end-cells inflated cystidioid, clavate or subsphaeroid, 8-25 μ wide, penultimate and other cells of the elements not greatly inflated but wall roughened by hyaline plates of a broken outer wall, content yellow to yellowish or hyaline in KOH, lacking amyloid granules. Hyphae of context as revived in Melzer's with a strong fleeting-amyloid reaction but not distinctive when seen under a microscope, in fresh material some hyphae seen with greenish granular material in the cells. Clamp connections absent.

Habit, habitat, and distribution.—Scattered on humus in mixed forest of oak and pine, Saginaw Forest, Ann Arbor, August 2, 1966, Samuel Mazzer (Smith 73026).

Observations.—The distinctive features of this variety are the flame-scarlet pileus when young and the inflated terminal cells of the trichodermial hyphae correlated with a somewhat soapy taste and a slight pungent odor. It is close to *B. flavorubellus,* but lacks amyloid particles in the cuticle cells. *B. harrisonii* has an ocher-red to brick-red pileus and its cuticle is almost an epithelium.

Stirps SENSIBILIS

KEY

1. Pleurocystidia 26-38 × 12-18 μ; vesiculose to utriform, rarely fusoid-
   ventricose . . . . *B. miniato-olivaceus*
1. Pleurocystidia, if present, narrow and with apex subacute to obtuse 2

     2. Spores (11) 12-16 (17) × (3) 4-5.5μ 3
     2. Spores smaller 4

3. Spores distinctly blue in Melzer's as revived; pileus rich vinaceous-red,
   fading slowly . . . . *B. bicoloroides*
3. Spores pale ochraceous tawny in Melzer's; pileus soon fading to yellow or
   orange-yellow . . . . *B. miniato-pallescens*

     4. Pores wide and boletinoid toward the stipe. .(see *B. tenax* also) *B. subfraternus*
     4. Pores round and 1-2 per mm 5

5. Base of stipe persistently carmine-red . . . . *B. carminipes*
5. Base of stipe rusty-rose to purplish red 6

     6. Many wide (15-20 μ) short cells in elements of pilear cuticle *B. pseudosensibilis*
     6. Wide short cells lacking in cuticular hyphae (the latter tubular) 7

7. Pileus and stipe dark apple-red; tubes 5-10 mm long . . . . *B. bicolor*
7. Pileus and stipe with more orange or yellow than in above; tubes ± 1.5
   cm long . . . . *B. sensibilis*

142. **Boletus miniato-olivaceus** Frost

Bull. Buff. Soc. Nat. Sci. 1:101. 1874

"Pileus at first vermillion color, then disappearing and becoming olivaceous, pulvinate, smooth, rather soft and spongy, margin at first incurved, then applanate, 2-6 inches broad. Tubes bright lemon yellow, partly adnate, then slightly decurrent. Stem light yellow, generally not always lurid at base, very smooth, enlarges as it enters the pileus, about ¼ to ½ of an inch thick. Flesh yellow, changing to blue, the pileus less yellow than in stem. Spores .0125-.0063 mm.

"Borders of woods. July and August."

The above is the original description by Frost. He sent a specimen to Peck, which is now on deposit in the Peck collections at Albany. The following data are taken from it:

Spores 11-14 (14.5-16) × 3.5-5 (5.5-7) μ, smooth, apex lacking a pore or thin spot; color in KOH ochraceous, in Melzer's ochraceous slowly darkening slightly but not dextrinoid; shape in face view sub-clavate to subelliptic, in profile typically somewhat inequilateral but varying to subclavate; wall about 0.2 μ thick.

Basidia 4-spored, 25-32 × 7-10 μ, narrowly clavate, hyaline in KOH when well revived, with yellowish walls in Melzer's. Pleurocystidia abundant, 26-38 X 12-18 μ, vesiculose to utriform or obtusely fusoid-ventricose (the latter rare), content weakly ochraceous in KOH, reddish tan and homogeneous in Melzer's, walls thin and smooth. Cheilocystidia small (18-28 X 5-10 μ), fusoid-ventricose, clavate, or pyriform. Caulocystidia not studied.

Tube trama of the *Boletus* subtype, nonamyloid in all parts, pieces of hymenophore mounted in Melzer's merely brownish, laticiferous elements not conspicuous (very few seen).

Pileus cuticle a tangled layer of hyphae 6-10 μ wide with some cells inflated to 15 μ or more, hyphae either tubular or with enlarged cells, cells mostly long but short ones (2-4 times as long as broad) also present, walls thin and smooth or some with a slight gelatinous halo; end-cells cylindric to slightly cystidioid or subglobose-pedicellate, the head up to 18 μ wide at times, contents yellow to orange-ochraceous in Melzer's, merely ochraceous in KOH. Hyphae of subcutis with homogeneous content. Clamp connections absent. Pileus context of very compactly interwoven hyphae yellowish to reddish in Melzer's, texture very friable in dried condition—almost crumbly.

Observations.—The pedicellate-clavate to pedicellate-globose end-cells of the cuticular hyphae and the curious voluminous pleurocystidia mark this as one of the most distinct species in the genus in spite of the fact it has been one of those most frequently misidentified. The concept given in the above description is drawn entirely from Frost's specimen and description and furnishes good characters to "anchor" the concept. We designate the Frost specimen at Albany as lectotype, since Frost apparently did not label a specimen as the type.

143. Boletus miniato-pallescens Smith & Thiers, sp. nov.

Pileus 8-20 cm latus, convexus, demum late convexus vel irregularis, siccus, impolitus, roseus vel lateritius demum armeniacus. Contextus luteolus, tactu caeruleus. Tubuli adnati vel subdecurrentes, 1-2 . cm longi, ceraceo-lutei tactu caerulei; pori lati tactu caerulei. Stipes 6-14 cm longus, 1-4.5 cm crassus, sursum laeti luteus, deorsum subbadius, incarnato-pruinosus. Sporae (11) 12-16 X (3) 4-5 μ. Specimen typicum in Herb. Univ. Mich. conservatum est; prope Chelsea, Washtenaw County, August 1, 1935, legit Smith 1935.

Pileus 8-20 cm broad, convex to hemispheric when young, becoming broadly convex to plano-convex to occasionally shallowly depressed or highly irregular when older, margin entire, incurved becoming decurved or sometimes flared, glabrous but appearing minutely fibrillose-granular under a lens, dry, unpolished, almost velvety at maturity or in age, often becoming areolate or rimose with age, colored "Morocco-red" at very first, fading to "Acajou-red" to "brick-red" and in age "apricot-buff" or more or less orange-yellow. Context up to 3 cm thick, pale yellowish, becoming blue when exposed, odor and taste not distinctive.

Tubes adnate to subdecurrent becoming depressed around the stipe, 1-2 cm deep, "deep chrome" to "wax-yellow" (bright yellow) at first, turning greenish blue where bruised; pores small (1-2 per mm), round, yellow at first, greenish blue when bruised but in age often orange to reddish especially after being bruised, but the color never as pronounced as in the *Luridi*—and never present when the basidiocarps are young.

Stipe 6-14 cm long, 1-4.5 cm thick, equal or nearly so, occasionally flared at the apex, often crooked, solid, yellow within, surface dry, ridged with dots but not reticulate, bright yellow above, orange to reddish brown lower down, or when young yellow all over but with a pinkish bloom, somewhat longitudinally striate at first, pruinosity more or less persistent, colors dingy in age.

Spore deposit brownish olive; spores (11) 12-16 (17) X (3) 4-5 μ, inequilateral in profile view, in face view narrowly fusiform, bright yellow in KOH, very pale ochraceous-tawny in Melzer's, smooth, walls slightly thickened, in Melzer's at first with a slight fleeting-amyloid reaction.

Basidia 4-spored, 20-24 X 7-9 μ, (rarely 2-spored), yellowish in KOH and in Melzer's. Pleurocystidia scattered to abundant, almost imbedded, 28-40 X 6-12 μ, fusoid-ventricose with acute apex, hyaline in KOH, thin-walled, smooth, some clavate elements (30-35 X 12-15 μ) also present scattered through the hymenium. Cheilocystidia often crowded, mostly clavate and with ochraceous to orange content revived in KOH.

Tube trama of gelatinous divergent hyphae from a central strand. Cuticle of pileus a more or less tangled hymeniform layer of clavate to subfusoid cells 30-60 X 9-14 μ and with rich ochraceous content as revived in KOH, the walls of the pedicel often up to 1 μ thick and with some pale yellow to hyaline incrusting material, some elements of the cuticle typical of the trichodermial type of cutis. Context hyphae floccose, smooth, hyaline, interwoven. Clamp connections absent.

Habit, habitat, and distribution.—Gregarious and rather common in open thin oak woods during wet weather in midsummer to early fall. Rather common in the southeastern part of the state.

Observations.—The large spores distinguish this and *B. bicoloroides* in the stirps. There is some gelatinization of the cuticular elements to the extent that the pedicels of the terminal cells often appear thicker-walled than the other parts of the cell or adjacent cells. *B. sensibilis* is the most closely related species but has the stipe apex reticulate, smaller spores, the terminal cells of the cuticular hyphae are tubular to only slightly cystidioid.

144. Boletus subfraternus Coker & Beers

Bol. North Carol. p. 61. 1943

Illus. Pl. 108.

Pileus 3-4.5 cm broad, broadly convex becoming nearly plane, margin even; surface unpolished, dry and velvety, dull rose (red) in buttons, becoming flame-scarlet and colors fading further to orange-buff and in dried pilei more or less olive-buff with a reflection of red remaining along the margin, margin undulating in age at times. Context buffy pallid and staining blue when cut, yellowish as dried, soft and fragile, ammonia on cuticle no reaction.

Tubes up to 1 cm deep, adnate to nearly free from the stipe, dull yellow but soon olive-yellow and stained greenish blue slightly where injured; pores almost boletinoid—lamellate toward the stipe, poroid near pileus margin, olive-yellow and staining blue slightly on bruising.

Stipe 2.5-4 cm long, 1-1.5 cm thick, solid, narrowed slightly toward base, yellow (flavous) in base, paler above, staining blue in apex when cut; surface above ochraceous, downward pale scarlet over an ocher ground color, the red pigment located mostly in a faint pruinosity.

Spores 9-13 (14) $\times$ 4-5 (6) μ, smooth, apex lacking a pore, shape in face view subfusiform, in profile obscurely inequilateral (the suprahilar depression broad and shallow); color in KOH bright ochraceous, in Melzer's dull yellow-brown (as if with a gray shadow indicating a fleeting-amyloid reaction), wall about 0.5 μ thick.

Basidia 4-spored, clavate, 8-11 μ broad, hyaline to pale yellow in KOH. Pleurocystidia scattered, 28-42 $\times$ 9-14 μ, fusoid-ventricose with subacute to obtuse apex, thin-walled, hyaline in KOH and Melzer's. Cheilocystidia smaller than pleurocystidia and often with yellow content revived in KOH. Caulocystidia clavate to fusoid-ventricose or some subcylindric, yellow in KOH but soon hyaline as color dissolves in the mounting fluid, thin-walled, variable in size but not gigantic (18-40 $\times$ 8-15 μ and clavate, 25-38 $\times$ 9-15 μ and fusoid-ventricose, and 40-56 $\times$ 8-12 μ and subcylindric), thin-walled, smooth.

Tube trama as revived in KOH of hyphae 5-11 μ wide, the walls thin and hyaline, not clearly divergent, inamyloid in all parts. Pileus cuticle a trichodermium, its elements 2-5 cells long, the apical cell 20-50 X 10-18 μ or at times becoming secondarily septate with the tip cell subglobose and 8-10 μ wide, apical cells generally cylindric, ovate, or cystidioid with the apex obtuse, the wall roughened with thin plates of wall material which are hyaline in KOH or in Melzer's, the remainder of the cells similarly ornamented, content as revived in KOH yellow to hyaline (soluble in KOH). Hyphae of subcutis and of adjacent context not colored distinctively in Melzer's. Clamp connections absent.

Habit, habitat, and distribution.—Three basidiocarps gregarious along a woods road on Colonial Point, Burt Lake, Cheboygan County, July 24, 1968, were found, Smith 75673.

Observations.—The distinctive features of this species are the flame-colored pileus at maturity—which then dries olive-buff—the smooth spores, wide boletinoid pores, weak change to blue when injured, the thick stipe, and the very fragile context along with the details of the trichodermial elements of the pileus. We did not note colored (reddish) pores on the collection cited.

145. Boletus bicolor Peck

Ann. Rept. N. Y. State Mus. 24:78. 1872

var. bicolor

Suillus bicolor (Peck) Kuntze, Rev. Gen. Pl. 3(2):535. 1898.
Ceriomyces bicolor (Peck) Murrill, Mycologia 1:152. 1909.
Boletus rubellus ssp *bicolor* (Peck) Singer, Amer. Midl. Nat. 37:53. 1947.

Illus. Figs. 70, 73-75; Pl. 109.

Pileus 5-15 cm broad, convex becoming plane to irregular; surface dry, unpolished, under a lens appearing pruinose-granulose to occasionally glabrous at first, in age becoming subtomentose and areolate, margin entire, with a narrow sterile band at maturity, deep apple-red ("Acajou-red" to "Pompeian-red," "old rose," or "Vandyke-red") and gradually becoming pinkish red in age, or toward the margin yellowish ("warm buff"), in age yellowish from exposed flesh but areolae remaining reddish. Context thick and firm (up to 1.5 cm), punky in age, pale yellow, slowly turning blue when cut, odor none, taste mild.

Tubes about 1 cm long, when mature, adnate becoming shallowly or sharply depressed around the stipe, yellow ("honey-yellow" to "lemon-yellow"), becoming somewhat paler with age, slowly staining blue when bruised; pores closed and appearing as white dots at first, at maturity small (1-2 per mm) and somewhat angular, yellow but occasionally reddish in places in age.

Stipe 5-10 cm long, 1-3 cm thick, clavate to subequal to tapering slightly toward the base, solid and yellow within, slowly or not at all turning blue when cut, surface dry, dull, smooth, appearing pruinose under a lens, yellow ("straw-yellow") at apex, yellow ground color showing toward the base but this overlaid with apple-red on lower two-thirds or more, the color line sometimes sharp.

Spore deposit dull olive-brown; spores 8-11 (12) X 3.5-4.5 (5) μ, oblong to slightly ventricose in face view, oblong to obscurely inequilateral in profile, with a shallow suprahilar depression, smooth, nearly hyaline to pale dingy ochraceous in KOH, slightly more ochraceous to tawny in Melzer's.

Basidia 4-spored, 26-34 X 7-8 μ, hyaline in KOH, yellowish in Melzer's. Pleurocystidia scattered to abundant, (20) 35-50 X 7-12 μ, fusoid-ventricose, apex subacute, hyaline or with yellow content, thin-walled, smooth, nonamyloid. Cheilocystidia abundant, often forming a sterile edge, similar to pleurocystidia but clavate cells also present.

Tube trama gelatinous and divergent from a central strand, not amyloid, often staining yellow-brown in KOH. Pileus cuticle a tangled mass of hyphae 4-7 μ wide with bright red content when fresh (in water mounts), hyaline or yellowish in KOH, filaments sparingly septate, originating as a trichodermium possibly, but if so the elements soon disarranged. The terminal cells tubular to inflated (clavate, oval, or cystidioid). Context hyphae hyaline, thin-walled, 8-15 μ wide, those adjacent to the cuticle having some orange to red content in Melzer's. Clamp connections absent.

Habit, habitat, and distribution. —Solitary to scattered to gregarious under oak and oak-pine stands, rather common in the southern half of the Lower Peninsula.

Observations. —This is an easily recognized species close to *B. sensibilis* and *B. miniato-olivaceus,* but the three are readily distinguished by the characters given in the key. The group around *B. fraternus* is sharply distinguished by the details of the trichodermium of the pileus. The following data were recorded from a study of the type:

Spores 10-13 X 3.5-5 μ, smooth, lacking a pore or thin spot at apex, shape in face view narrowly elliptic to oblong, in profile suboblong to obscurely inequilateral, suprahilar depression scarcely visible in most, color in KOH ochraceous to nearly hyaline, in Melzer's ochraceous, wall thin (-0.2 μ).

Basidia 4-spored, 8-10 μ broad, clavate, hyaline in KOH. Pleuro-cystidia 30-40 $\times$ 8-12 μ, mucronate to fusoid-ventricose, walls thin and smooth, content not distinctive in either KOH or Melzer's. Cheilo-cystidia smaller than pleurocystidia, pore edge soon gelatinized. No fleet-ing-amyloid reaction on tissue of hymenophore when dried material is crushed in Melzer's.

Tube trama of the *Boletus* subtype, hyaline in KOH, nonamyloid throughout. Pileus cutis a tangle of more or less decumbent tubular hyphae 4-8 μ wide with thin walls, often with a gelatinous halo as revived in KOH, no pigment incrustations seen, cells long and with ochraceous content in both Melzer's and KOH, rarely a short cell seen and very few of these inflated, end-cells tubular and blunt, rarely some-what cystidioid, content of hyphae not forming globules in Melzer's. Hyphae of subcutis strongly ochraceous in Melzer's, content homo-geneous. Clamp connections absent.

Material examined.—Barry: Mazzer 4270; Smith 73159. Cheboy-gan: Smith 37529. Eaton: Smith 1665. Gratiot: Potter 10052, 11754. Jackson: Smith 62805. Livingston: Smith 64202. Montcalm: Potter 13207. Oakland: Smith 6670, 6926, 7284, 64844, 73263. Washtenaw: Kauffman 1912; Shaffer 2656, 2658; Smith 18483, 72526, 723282, 8-8-32, 8-15-32, 7-15-35; Thiers 4564, 4592.

145a. Boletus bicolor var. **borealis** Smith & Thiers, var. nov.

A typo differt: Tubuli circa 6 mm longi; pori aurantio-ruber tactu caerulei. Typus: Smith 75623 (MICH).

Pileus 12 cm broad, convex with an incurved margin; surface dry and unpolished, glabrous but appearing to be appressed-fibrillose; margin pruinose-furfuraceous, even; color evenly "brick-red" to "madder-brown" (dark rusty rose). Context thick and soft, lemon-yellow above the tubes, paler and duller near the cuticle, not staining blue when cut; $FeSO_4$ no reaction.

Tubes 6 mm long, adnate-depressed, greenish yellow staining blue when bruised; pores dull orange from evenly distributed pigment or shading to orange-red, the color covering the entire hymenophoral sur-face.

Stipe 10 cm long, 22 mm thick, tapered downward, lemon-yellow within but slowly blue in apex when cut, in the base dingy vinaceous-brown; surface evenly dark "madder-brown" (dark rusty rose), apical region yellow, surface dull and unpolished (but not velvety or tomentose anywhere), context around larval tunnels dull rusty rose.

Spores 11-13 (15) X 4-4.5 (5) μ, smooth, lacking an apical spot or pore, shape in face view narrowly suboblong to subfusoid, in profile obscurely to somewhat inequilateral, color pale yellow to nearly hyaline revived in KOH, in Melzer's weakly tawny, wall too thin to measure.

Basidia 4-spored, clavate, 28-37 X 9-12 μ, tapered to a narrow base in many, yellowish in Melzer's. Pleurocystidia 32-47 X 8-14 μ, scattered, fusoid-ventricose to ventricose-subrostrate, content yellow in KOH but slowly becoming hyaline as pigment disolves, in Melzer's the content homogeneous and yellowish. Cheilocystidia similar to pleurocystidia but smaller and neck not as elongate. Caulocystidia mostly clavate and up to 15 μ broad, content yellow in KOH, more rarely ventricose-subcapitate.

Tube trama of the *Boletus* subtype, with a weak fleeting-amyloid reaction. Pileus cutis of interwoven hyphae resulting from the collapsing of a trichodermium (?), the hyphae smooth, 6-12 μ wide, the cells not infrequently shorter than four times as long as broad and some oval to subglobose cells present, wall thin, content yellow in KOH, pigment readily dissolving in the mount, in Melzer's the content homogeneous and dull ochraceous to orange-brown, end-cells ovate in optical section, tubular to clavate or weakly cystidioid. Hyphae of subcutis and of adjacent context hyaline to merely yellowish in Melzer's. Clamp connections absent.

Habit, habitat, and distribution.—Solitary under aspen, near Pellston, Cheboygan County, July 22, 1968, H. D. Thiers (Smith 75623).

Observations.—Additional basidiocarps were found but were too old to save. In overall aspect the specimen is a typical *B. bicolor,* but the occasional oval to subglobose cells in the pileus cuticle and the typical pores of the *Luridi* make it worthy of placing on record.

145b. Boletus bicolor var. subreticulatus Smith & Thiers
var. nov.

A typo differt: stipes sursum reticulatus.   Typus: Smith 76042 (MICH).

Pileus 5-10 cm broad, convex; surface dry and unpolished to velvety or subtomentose, in age the cuticle pulling apart to form patches revealing the yellow context but surface not becoming deeply areolate, in age merely scurfy at times, color evenly dull rose when young and when dried rusty rose to dingy cinnamon on the cuticle. Context pale lemon-yellow, when cut weakly blue, NH_4OH on cuticle no reaction, $FeSO_4$ on context no reaction, odor and taste fungoid.

Tubes 5-9 mm deep, adnate to subdecurrent, dull ocher-yellow becoming greenish yellow; pores small (about 2 per mm), dull ocher-yellow, if injured staining blue readily.

Stipe 5-10 cm long, 1-2 cm thick, equal to narrowed at the base, solid, lemon-yellow above and flavous below, rusty rose around the wormholes, base with reddish to brownish tomentum, over most of length the surface dull rose, ground color yellow, apex finely reticulate for about 2 cm, below this the surface pruinose.

Spores 9-12 X 3.5-4.5 μ, smooth, lacking an apical pore, in face view elliptic to oblong, rarely subovate, in profile suboblong to obscurely inequilateral, ochraceous in KOH, in Melzer's bluish but soon fading to dingy ochraceous or very weakly dextrinoid.

Basidia 4-spored, narrowly clavate, 8-10 μ broad. Pleurocystidia scattered, 32-40 X 8-12 μ, embedded in the hymenium or slightly projecting, subcylindric to fusoid-ventricose, apex obtuse, content bright yellow in KOH, walls smooth and thin. Caulocystidia mostly clavate varying to fusoid-ventricose with obtuse apex, 26-38 X 8-16 μ, content dull ochraceous in KOH.

Cuticle of pileus a trichodermium of hyphae 5-10 μ wide, tubular to the cylindric, clavate, oval, fusoid-ventricose or subglobose end-cell which is yellow to ochraceous in KOH and Melzer's, content often metallic in appearance, final differentiation of the apical cell from the basic tubular shape often delayed until the basidiocarp is at late maturity. Hyphae of the subcutis and context with some yellow to tan content in Melzer's but no dextrinoid incrustations present and no significant pigment dissolved in the cells. No clamp connections observed.

Habit, habitat, and distribution.—Gregarious under oak, George Reserve, Pinckney, Livingston County, August 20, 1968, Ruth Zehner (Smith 76042-type, and 76043).

Observations.—Collection 76042 has retained its red color in drying and very few of the terminal cells of the trichodermium vary from the basic tubular shape. Since it has the colors of *B. bicolor* as dried, it is designated the type. In 76043 the color has faded to a reddish cinnamon and more of the terminal cells are enlarged. Both, however, have the reticulum at the apex of the stipe.

146. Boletus bicoloroides Smith & Thiers, sp. nov.

Pileus 6-15 cm latus, convexus, glaber, vinaceosanguineus tarde obscure ochraceus. Contextus luteus. Tubuli lutei, 1.5 cm longi; pori 1-2 per mm, lutei tactu caerulei. Stipes 6-12 cm longus, 1-2 cm crassus, vinaceosanguineus, pruinosus, non-reticulatus. Sporae 14-17 X 4-5.5 μ, amyloideae. Specimen typicum in Herb. Univ. Mich. conservatum est; prope Waterloo, August 13, 1968, legit Ruth Zehner 349.

Pileus 6-15 cm broad, broadly convex, glabrous, surface even (not rimose or areolate in old caps), color deep vinaceous-red, slowly fading to dull ochraceous with red splashes, margin even; context thick, yellow, drying dingy pale straw color.

Tubes yellow, up to 1.5 cm deep, slightly bluish bruised, depressed at the stipe; pores small (1-2 per mm), yellow, bluish if injured.

Stipe 6-12 cm long, 1-2 cm thick, equal or nearly so, solid, yellow within, surface dark red like the pileus to apex, pruinose, not reticulate, base coated with ochraceous mycelium (as dried).

Spores 14-17 X 4-5.5 μ, smooth, in face view subfusoid, in profile somewhat inequilateral, color in KOH pale ochraceous to brownish ochraceous, in Melzer's dingy blue as revived, apex with a minute pore, wall about 0.3 μ thick.

Basidia 4-spored, clavate, 8-11 μ wide at apex. Pleurocystidia scattered, 38-56 X 9-16 μ, fusoid-ventricose, hyaline or yellow in KOH, neck 10-30 μ long, apex subacute, wall thin and smooth. Cheilocystidia yellow in KOH, mostly basidiole-like or narrower. Caulocystidia very numerous, basidiole-like to fusoid-ventricose, yellow in KOH, mostly 6-11 μ wide when fusoid, having scarcely any neck, thin-walled and lacking incrustations.

Cuticle of pileus a short trichodermium of hyphae 9-15 μ wide with end-cells cystidioid to clavate and 14-25 μ wide, their walls showing a slime coating in KOH, content yellow in KOH and red in Melzer's, secondary septa present in some cutting off a short, tapered apical cell, the cells of the hyphae back from the 1-2 tip cells 6-12 μ wide. Hyphae of context with deep red content in Melzer's. Clamp connections apparently absent.

Habit, habitat, and distribution. —Known from the Waterloo Recreation Area, Washtenaw County, type collected by Ruth Zehner (349), August 13, 1968.

Observations. —This species was mistaken in the field for *B. bicolor,* which it resembles perfectly. Since the collector knew *B. bicolor* well, including its tendency to stain blue, and since the cut surfaces of the pilei and stipes of the basidiocarps of *B. bicoloroides* dried as if blue staining had taken place, we have assumed a change to blue in the fresh basidiocarps did take place even though no mention was made of it on the label—the collection was simply identified as *B. bicolor* at the time of collection. *B. bicoloroides* differs from the latter markedly in the almost hymeniform pileus cuticle of broad cystidium-like cells, the deep red color of the cap trama in Melzer's, the strongly amyloid spores (for a bolete), their larger size, and the much deeper tubes. It is likely that the pileus would show some degree of viscidity when wet.

147. Boletus pseudosensibilis Smith & Thiers, sp. nov.

Illus. Pls. 100-111.

Pileus 6-14 cm latus, late convexus, ferrugineus, siccus, impolitus. Contextus luteus, tactu caeruleus. Tubuli 5-8 mm longi, lutei, tactu caerulei. Stipes 8-16 cm longus, 1.5-3 cm crassus, sursum luteus, deorsum sordide ferruginus, non-reticulatus. Sporae 9-12 × 3-4 μ. Specimen typicum in Herb. Univ. Mich. conservatum est; prope Gun Lake, Barry County, August 22, 1966, legit Smith 73128.

Pileus 6-14 cm broad, convex expanding to plane, surface glabrous and unpolished, not velvety at any stage, dry, color dull ferruginous but the red soon fading and pileus more or less dingy yellow-brown ("cinnamon-buff" to "Sayal-brown"), the yellow flesh conspicuous as pileus becomes areolate, when dried the pilei dingy yellow-brown to "Isabella color" (with yellow in the cracks and as a pervading undertone). Context bright yellow, quickly azure-blue when cut, thick, firm, odor and taste mild, NH_4OH blue on the cuticle then changing to dull purplish, KOH brownish on context, guaiac no reaction.

Tubes shallow (5-8 mm long), subdecurrent to adnate on stipe, bright ocher-yellow staining blue and then changing to brown when injured; pores small (−1-3 per mm), staining blue then brown at maturity, if undamaged slowly becoming olive-ocher, lemon-yellow when young.

Stipe 8-16 cm long, 1.5-3 cm at the flared apex, often equal or nearly so, solid yellow within above, vinaceous-red in the base and around larval tunnels; surface naked and glabrous, not reticulate anywhere, apex pale lemon and duller yellow as dried, base rusty rose and purplish red as dried, ochraceous mixed with rose in midportion.

Spore deposit olive, becoming olive-brown; spores 9-12 × 3-4 μ, smooth, apex lacking a pore, wall about 0.2 μ thick, shape in face view subfusiform to suboblong, in profile narrowly inequilateral to suboblong, the suprahilar depression shallow to almost absent; yellow in KOH, in Melzer's dingy yellowish.

Basidia 4-spored, 22-28 × 7-9 μ, clavate, yellowish to hyaline in KOH or Melzer's. Pleurocystidia 36-58 × 8-14 μ, fusoid-ventricose with narrow often flexuous neck, smooth, thin-walled, content long remaining ochraceous in KOH and Melzer's. Cheilocystidia similar to pleurocystidia but smaller. Caulocystidia in patches, 25-35 X 8-12 μ, clavate, thin-walled, rarely some fusoid-ventricose.

Tube trama of the *Boletus* subtype, all hyphae smooth and with thin yellowish to hyaline walls revived in KOH. Pileus cutis a layer of intricately entangled hyphae 5-15 μ wide, with many short cells (some up to 20 μ or more wide), hyphae very clearly defined in either KOH or Melzer's, walls thin and smooth, content orange-brown in Melzer's, yellow to ochraceous-orange in KOH, no pigment globules present in the mounts; terminal cells cystidioid to clavate. Clamp connections absent.

Habit, habitat, and distribution.—Gregarious in an oak woods, Gun Lake, Barry County, August 22, 1966, Smith 73128.

Observations.—This species is in the *B. miniato-olivaceus* group but the red of the pileus fades quickly though the purplish red of the stipe is persistent. The cutis of the pileus in this species is never a tricho-dermium—ample button stages were available for examination. The spores are smaller than for *B. miniato-olivaceus* and the characteristic pleurocystidia of the latter were not present. Also, in Peck's species the tubes are deeper. The NH_4OH reaction of the pileus cuticle is unusual in this group.

Material examined.—Barry: Mazzer 4224, 4230; Smith 73128 (type), 73200. Washtenaw: Smith 72538.

148. **Boletus carminipes** Smith & Thiers, sp. nov.

Pileus 5-10 cm latus, convexus, siccus, subtomentosus demum impolitus, subcinnamomeus vel sordide alutaceus. Contextus luteus tactu caeruleus. Tubuli 4-6 mm longi, adnatae vel subdecurrentes, lutei tactu caerulei; pori circa 1 mm lati, lutei. Stipes 5-10 cm longus, 1-3 cm crassus, clavatus, sursum luteus, deorsum carmineus, pruinosus. Sporae 9-12 X 3-3.5 μ. Specimen typicum in Herb. Univ. Mich. conservatum est; prope Detroit, September 2, 1965, legit E. Kidd (Smith 72473).

Pileus 5-10 cm broad, convex, becoming broadly convex, surface dry, plushlike at first but becoming unpolished, color dingy "Sayal-brown" to near "snuff-brown" or "bister" (dull cinnamon or redder when young, becoming dingy yellow-brown in age, becoming slightly rimulose but not areolate. Context pale yellow drying buffy pallid, almost instantly blue when broken, taste and odor not distinctive, $FeSO_4$ slowly bluish, KOH on blue areas pale ochraceous-tan, on cuticle dark yellow-brown.

Tubes short, 4-6 mm deep when context is 1-2 cm deep, adnate to subdecurrent in a few finally slightly depressed, yellow but quickly blue if bruised; pores finally depressed, yellow but quickly blue if bruised; pores 1 mm wide at maturity, the injured areas finally orange to brown.

Stipe solid, 5-10 cm long, 1-3 cm thick at apex, enlarged downward to 4-5 cm in the bulb, base short-rooting or merely pinched off, the surface over lower part carmine-red or splashed this color and retaining the color in drying, yellow above, in age some areas dingy brown (from staining ?), coated with a very fine pruina, not at all reticulate.

Spore deposit olive becoming olive-brown; spores 9-12 X 3-3.5 μ, smooth, narrowly oblong in face view, in profile narrowly and obscurely inequilateral to oblong, color olive-ochraceous in KOH (from print), more ochraceous when taken from the tube tissue, slowly somewhat dextrinoid, wall -0.2 μ thick.

Basidia 4-spored, clavate, 8-11 μ wide at apex. Pleurocystidia abundant, 42-64 X 9-14 μ, fusoid-ventricose, neck up to 30 μ long and tapered to an acute apex, content ocher-yellow in KOH. Cheilocystidia narrowly clavate-mucronate to subfusoid, 24-32 X 5-9 μ. Caulocystidia clavate, ovate-pointed, mucronate or fusoid-ventricose, the latter 32-50 X 9-15 μ.

Cuticle of pileus at first a short trichodermium but this soon collapsing to form a cutis of interwoven hyphae, the hyphae 5-10 (12) μ wide, thin-walled, smooth, nongelatinous, mostly tubular but often flexuous and with bright yellow content (all in KOH), in Melzer's the content of the cuticular hyphae and those of adjacent context bright red, terminal cells mostly tubular, rarely clavate, cystidioid, globose or with short protuberances. Clamp connections none.

Habit, habitat, and distribution. – In a low open wood lot near Detroit, September 2, 1965, Edward Kidd (Smith 72473).

Observations. – This is a most interesting bolete obviously related to *B. bicolor* but differing in the color of the pileus and stipe as well as in the microscopic features such as spore size, size of pleurocystidia, and Melzer's reaction of the cuticle. The tubes are short as in *B. bicolor*.

149. Boletus sensibilis Peck

Ann. Rept. N. Y. State Mus. 32:33. 1879

Boletus miniato-olivaceus var. *sensibilis* (Peck)

Bull. N. Y. State Mus. 2:107. 1889

var. sensibilis

Illus. Pl. 112.

Pileus 6-15 (30) cm broad, convex, becoming broadly convex to nearly plane, margin even to slightly lobed, surface dry and pruinose to unpolished, cuticle tending to remain intact throughout life of the basidiocarp (not areolate or only slightly so in dry weather); color dark brick-red to pale brick-red ("Etruscan-red"), slowly fading to pale dingy rose and at times finally dull cinnamon, quickly yellow with a drop of KOH. Context thick, pale yellow, instantly blue when cut, $FeSO_4$ yellow-brown in pileus and stipe, walls of larval tunnels not appreciably discolored, odor and taste mild.

Tubes 1-1.5 cm long, adnate becoming only slightly depressed, bright yellow and instantly blue when injured; pores 1-2 per mm, round, bright yellow and instantly blue when bruised and these places slowly brownish to reddish.

Stipe 8-12 cm long, 1-3 (4) cm thick at apex, solid, equal or enlarged downward, punky in the base, bright yellow within but soon blue when cut but then soon fading to yellow again, surface brilliant yellow above or overall or dull red below, apex finely and only slightly reticulate.

Spores 10-13 X 3.5-4.5 μ, suboblong to slightly ventricose in face view, in profile slightly inequilateral from a shallow suprahilar depression, smooth, wall slightly thickened, ochraceous in KOH, greenish gray fading to dingy yellow-brown as revived in Melzer's (fleeting-amyloid reaction present).

Basidia 4-spored, 20-26 X 7-10 μ, dingy yellow in Melzer's, hyaline to yellow in KOH. Pleurocystidia scattered, 36-50 X 9-13 μ, ventricose at base and the neck long and narrow (3-5 μ wide), content hyaline to dingy ochraceous in KOH or Melzer's.

The tube trama divergent from a central strand. Pileus cutis a compactly tangled layer of hyphae 3-7 μ wide, with bright yellow content in KOH and the content orange-brown in Melzer's, some hyphae more or less coated with hyaline incrustations, the end-cell scarcely wider than the remainder of the hypha and often narrower. Clamp connections absent.

Habit, habitat, and distribution. —Scattered to gregarious in sandy open woods of maple, aspen, and birch with scattered beech, during wet seasons common throughout the state, fruiting from early summer on into the fall.

Observations. —This is the largest species of the stirps. The reticulations at the apex of the stipe are very faint and best seen with a hand lens. The instant change to blue on the cut flesh will aid in distinguishing it from *B. miniato-olivaceus* in the field. *B. bicolor* is a different shade of red. The blue stain in *B. sensibilis* is evident on well-dried material moistened with alcohol and then water. In Melzer's the hymenial tissue is bluish black when first revived. It is possible that the fleeting-amyloid reaction of this species is merely associated with the change to blue on moistening of the dried hymenophoral tissue in pure chloral hydrate. The following description is based on Peck's original and Smith's study of the type:

"Pileus at first firm, convex, pruinose-tomentose, brick red, then expanded, paler or ochraceous-red, glabrous, soft; tubes at first plane or concave, bright yellow, then tinged with green, finally sordid-yellow, small, subrotund; stem firm, smooth, lemon-yellow, narrowed at the top when young, and sometimes slightly fibrose from the decurrent walls of the tubes, often stained with red or rhubarb color; spores greenish brown, .0005' long, .00016' broad; flesh of the pileus pale yellow, of the stem brighter colored and marbled, both flesh and tubes quickly changing to blue when wounded.

"Plant scattered to caespitose, 4'-6' high, pileus 3'-8' broad, stem 6'-12' thick."

Spores 10-13 (15) X 3-4.5 μ, smooth, lacking a pore or thin-spot at apex, in shape suboblong to subfusiform in face view, in profile obscurely inequilateral to suboblong, color in KOH ochraceous, in Melzer's ochraceous to pale clay color, wall about 0.2 μ thick.

Basidia 4-spored, 25-33 X 7-9 μ, narrowly clavate, hyaline in KOH. Pleurocystidia 33-45 X 9-13 μ, fusoid-ventricose, the neck 3-4 μ wide and apex subacute, walls thin, smooth and hyaline, content yellowish to hyaline in either KOH or Melzer's. Cheilocystidia smaller than pleurocystidia and varying to basidiole-like in shape. Hymenial tissue mounted in Melzer's merely ochraceous to brownish.

Tube trama of the *Boletus* subtype, hyphae tubular and 4-8 μ wide, lacking any amyloid reaction, broad (up to 12 μ), laticiferous elements present and ochraceous in KOH or Melzer's. Pileus cutis a tangle of more or less appressed hyphae 4-8 μ wide, the walls smooth and thin, the content ochraceous in KOH, and in Melzer's orange-ochraceous, pigment globules not forming; cells tubular and mostly long, end-cells tubular to slightly clavate or cystidioid. Hyphae of subcutis and context only weakly ochraceous in Melzer's. Clamp connections none.

Observations.—Both *B. sensibilis* and *B. miniato-olivaceus* have very friable context—it almost crumbles when one attempts to section it. We assume that the fleeting-amyloid reaction is lost after basidiocarps have stood for years in an herbarium. This needs to be verified, but only time will tell. Previous to our study this species and *B. miniato-olivaceus* were hopelessly confused—starting from the time of Peck himself.

Material examined.—Cheboygan: Smith 63754, 63759, 63779, 63854.

149a. Boletus sensibilis var. **subviscidus** Smith & Thiers, var. nov.

Pileus viscidus, luteus ad centrum laete roseus. Stipes deorsum demum cadmio-aurantiacus. Sporae 9-13 × 3.5-4.2 μ. Specimen typicum in Herb. Univ. Mich. conservatum est; prope Waterloo, September 5, 1965, legit Florence Hoseney.

Pileus 5-11 cm broad, broadly convex to plane, in age margin uplifted, margin even, surface glabrous and somewhat viscid; color near "Naples-yellow" (pale yellow) on margin and this forming more or less of a ground color for the entire pileus, disc flushed "Carnelian-red" to "rufous" (rose to rusty red, and some areas blood-red where in contact with leaves), surface bluish fuscous where injured. Context thick, soft, yellow, instantly turning dark caerulean blue when injured, KOH orange on the blue-stained areas, $FeSO_4$ tan, odor and taste not distinctive.

Tubes shallow (5-7 mm deep), yellow but instantly blue when injured, broadly adnate to slightly depressed around the stipe; pores 1-2 per mm, round or nearly so, pale bright yellow staining blue when injured, the blue-stained areas slowly becoming orange.

Stipe 7-10 cm long, 1-2 cm at apex, equal or nearly so, solid, yellow within but instantly blue when injured, with KOH orange to reddish brown; surface smooth or nearly so, apex scarcely reticulate (with faint lines of the tubes), bright yellow, soon orange to bitter-sweet-orange at base or lower part, with fine white mycelium surrounding the base.

Spore deposit olive; spores 9-12 (13) × 3.5-4.2 μ, smooth, lacking an apical pore, wall -0.2 μ thick, shape in face view suboblong to obscurely fusiform, in profile suboblong to obscurely inequilateral, suprahilar depression broad and shallow but nearly lacking in many, yellow in KOH, greenish yellow (weakly amyloid) in Melzer's.

Basidia 4-spored, 23-30 × 7-11 μ, clavate, yellow to hyaline in KOH or Melzer's. Pleurocystidia scattered, 32-54 × 7-12 μ, fusoid-ventricose with flexuous neck 3-4 μ wide and subacute apex, thin-walled, smooth, content yellow to yellow-brown in KOH, fading, hyaline to yellowish in Melzer's.

Tube trama of the *Boletus* subtype, all hyphae thin-walled and smooth, 4-9 μ wide, large (15-50 μ) hyaline to yellow globules present in mounts in Melzer's. Pileus cuticle a tangled layer of hyphae with free ascending tips, the hyphae with a slime-coating as seen revived in KOH or Melzer's but in KOH this material slowly dissolves, the terminal cells 20-75 μ long and 5-9 μ broad, tapered slightly to the apex or inflated to

8-10 μ near the apex and then tapered, in Melzer's the content orange-ochraceous. Hyphae of context 4-15 μ wide, interwoven, content yellowish to hyaline in Melzer's, no amyloid material seen. Clamp connections absent.

Habit, habitat, and distribution.—Scattered under oak on thin soil, Waterloo Recreation Area, Waterloo, September 5, 1965, Florence Hoseney.

Observations.—The diagnostic features are the viscid pileus with the cuticular hyphae showing a slime-coating under the microscope, the basically shallow adnate (to stipe) tube layer, and the fact that the pilei dry with tints of rose readily visible. Nothing is known as yet about the edibility of this variety, which is reason enough for not recommending it. The stirps is to be regarded as dangerous.

Section TRUNCATI Smith & Thiers, sect. nov.

Sporae truncatae vel subtruncatae.

Typus: *Boletus truncatus.*

In veiw of the importance attributed to spore ornamentation in the boletes by previous investigators, and to the presence of an apical pore in spores with somewhat thickened walls throughout the fleshy fungi generally, we believe this group should be given more recognition than it has received in the past.

KEY TO SPECIES

1. Spores 9-12 × 4-5 μ. 2
1. Spores 11-15 μ long or 5-7 μ wide . . . . (see *B. betula* also) 3

    2. Stipe rhubarb-red from pruina . . . . see *Boletellus intermedius*
    2. Stipe pruina yellowish brown. . . . . *B. subdepauperatus*

3. Taste bitter to disagreeable, pileus seldom areolate . . . . *B. patriciae*
3. Taste mild, pileus typically areolate at maturity 4

    4. Spores 4.5-6.5 (7) μ wide, pileus soon with conspicuous red tones
    . . . . *B. truncatus*
    4. Spores 4-5 μ wide; pileus lacking pronounced red tones . . . .
    . . . . *B. porosporus* var. *americanus*

150. Boletus truncatus (Singer, Snell, & Dick) Pouzar

Česká Mycol. 20:2. 1966

Xerocomus truncatus, Singer, Snell, & Dick, in Snell, Singer, & Dick, Mycologia 51:573. 1959.

Illus. Figs. 67-69; Pl. 113.

Pileus 3-8 (11) cm broad, convex becoming broadly convex to nearly plane; the margin finally often crenulated; surface velvety and evenly dark olive to olive-brown, very soon red to reddish along the margin or overall and the cutis becoming areolate with red usually showing in the cracks. Context whitish but rose-red under the cutis, slowly becoming pale yellowish, usually red around the larval tunnels, staining blue when cut; odor slight, taste mild; $FeSO_4$ greenish gray on cut surface.

Tubes yellowish olive ("Isabella color") when mature, pale yellow young, becoming depressed around the stipe, 7-15 mm deep; pores pale yellow young, when mature 1-2 per mm or in age up to 2 mm wide, irregular in outline but not boletinoid, staining greenish when injured; some splitting and some separating as the pileus is broken downward.

Stipe 4-8 cm long, 4-12 mm thick, equal or nearly so, solid, pallid yellowish within above, soon rose-red from base up (dark brown within with KOH); surface inconspicuously pruinose to naked, with dingy ochraceous mycelium around the base.

Spore deposit olive-brown; spores 10-15 × 4.5-6.5 (7) μ, smooth, many with a truncate apex or apex in many appearing slightly notched, with a small apical pore, shape in profile with a broad shallow suprahilar depression, ovate to oblong in face view or some slightly ventricose, wall slightly thickened, many with a weak fleeting-amyloid stage, in KOH dingy ochraceous to pale yellowish brown.

Basidia 4-spored, clavate, 28-36 × 9-12 μ, mostly yellow in KOH. Pleurocystidia 50-70 × 10-16 μ, ventricose with a long neck and acute apex, smooth, thin-walled, readily collapsing, scattered to rare, content not distinctive.

Pileus cuticle a trichodermium of hyphae 8-14 μ wide, with dull rusty brown (in KOH) patches or plates of pigment adhering, the terminal cells tapered to an obtuse apex at times. Context hyphae hyaline in Melzer's next to the cuticle. Clamp connections none.

Habit, habitat, and distribution.—Gregarious to scattered or solitary around old stumps on sandy soil, in hardwood slashings where soil has been disturbed, along woods, borders of roads, etc. Common throughout the state in the summer and fall, but not usually found in quantity in any one place.

Observations.—This species has usually passed under the name *Boletus chrysenteron* in North America (see Coker and Beers, 1943). It is certainly common in Michigan. Also, it is very easy to get mixed collections of the two since they resemble each other so closely macroscopically and fruit in the same habitat at the same time of the year.

Material examined.—Barry: Mazzer 4060. Chippewa: Smith 67089, 67184, 67113, 67236, 73030. Cheboygan: Smith 57297, 67004. Emmet: Smith 67031; Thiers 3533, 3595. Gratiot: Potter 3723, 7763. Livingston: Smith 72597. Luce: Smith 72376. Montmorency: Smith 66996. Oakland: Smith 67327, 73280. Washtenaw: Smith 62671, 64056, 73110.

151. Boletus patriciae Smith & Thiers, sp. nov.

Illus. Pl. 106 (upper).

Pileus 4-5 cm latus, convexus, udus demum impolitus, olivaceo-brunneus demum olivaceo-griseus. Contextus pallide luteus tactu caeruleus. Sapor amarus vel subamarus. Tubuli lutei; pori 2 per mm, pallide lutei tactu tarde caerulei. Stipes 4-5 cm longus, 9-12 mm crassus, pallide luteus, derosum roseus. Sporae 11-15 X 5-6.5 μ, truncatae. Specimen typicum in Herb. Univ. Mich. conservatum est; prope Highlands, Oakland County, August 26, 1966, legit Smith 73236.

Pileus 4-5 cm broad, plano-convex, surface moist but dull and unpolished, evenly dark olive-brown fresh and moist, slowly paler olive-brown and finally olive-gray along the margin, no red showing. Context yellowish pallid young, becoming pale olive-buff to pallid, when cut soon staining blue but soon flushed rose and slowly rose-red around larval tunnels, odor unpleasant, taste disagreeable and bitterish, $FeSO_4$ slightly olive on context.

Tubes dull pale yellow aging to olive-buff, up to 1 cm long, depressed around the stipe; pores 2 per mm, pale yellow becoming yellowish olive, slowly staining dingy blue where bruised and these areas changing to brown.

Stipe 4-5 cm long, 9-12 mm thick, solid, yellowish pallid within near apex and dull rose in lower half or just in the base; surface rose-red and finely pruinose to apex, coated with olive-buff mycelium over the base.

Spores 11-15 X 5-6.5 μ, smooth, apex truncate but not notched, in Melzer's showing a fleeting-amyloid reaction when fresh, as revived showing none or a very faint one, in KOH pale bister (dingy yellow-brown), in face view suboblong to somewhat boat-shaped, or ventricose in largest spores, in profile inequilateral-elongate and with a pronounced suprahilar depression, on larger spores and those most highly colored often ventricose-inequilateral, wall about 0.5 μ thick.

Basidia 4-spored, 28-45 × 10-15 μ, yellowish in KOH or Melzer's, clavate. Pleurocystidia scattered, 36-55 × 9-15 μ, fusoid-ventricose and with a tapering neck to a subacute apex, thin-walled, readily collapsing, content yellowish to hyaline in either Melzer's or KOH. Cheilocystidia similar to pleurocystidia varying to merely obtusely fusoid-ventricose and smaller. Caulocystidia in patches, clavate to clavate-mucronate, fusoid-ventricose to lanceolate, 32-60 × 8-17 μ, also some flexuous filaments 4-6 μ in diameter in the patches of caulohymenium; content of all types either hyaline or yellowish to brownish in KOH.

Tube trama of subgelatinous hyphae nearly parallel in arrangement but slightly diverging to subhymenium; laticiferous elements 5-9 μ wide also present. Pileus trichodermium of hyphae with heavily brown-incrusted walls from plates, zones or spirals of incrusting material; the hyphae 8-15 μ wide with the 2-4 cells back from the end-cell inflated and 14-20 μ wide but not globose, end-cell short and cystidioid, some divided by a secondary septum into an inflated cell and a tapered tip-cell, cell content not distinctive in either KOH or Melzer's. Hyphae of the subcutis also heavily incrusted, 4-10 μ wide, lacking distinctive content in either KOH or Melzer's. Context hyphae smooth, thin-walled, content not distinctive in Melzer's; laticiferous elements numerous, yellowish in Melzer's. Clamp connections none found.

Habit, habitat, and distribution.—Gregarious under hardwoods, Highlands Recreation Area, Oakland County, August 26, 1966, collected by Patricia Mazzer, for whom it is named (Smith 73236-type).

Observations.—This species differs from *B. truncatus* in the olive mycelium around the base of the stipe, the disagreeable taste, and in the distinctly cystidioid heavily incrusted terminal cells of the trichodermial elements. The pileus (and some basidiocarps were old) did not become areolate as in *B. truncatus,* which is almost always areolate by maturity. We have not noted a change to rose on the context of *B. truncatus* after the blue fades, but in view of the general tendency of that species to develop red we do not emphasize the change noted for *B. patriciae. B. porosporus* Imler is perhaps the most similar bolete. It apparently lacks a bitter taste, the mycelium around the base of the stipe is slightly grayer, and the cap becomes areolate.

Material examined.—Cheboygan: Smith 74871, 74624. Emmet: Smith 74497. Chippewa: Smith 74679a, 74817. Otsego: Smith 74776. Washtenaw: Hoseney 600. Oakland: Smith 73236.

152. Boletus porosporus (Imler) Watling

Notes from the Royal Bot. Garden, Edinb. 28(3):305. 1968

var. **americanus** Smith & Thiers, var. nov.

Xerocomus porosporus Imler, Bull. Soc. Mycol. Fr. 71:21. 1955.

Pileus 4-6.5 cm broad, convex to nearly plane, the margin even, surface dry and plushlike from youth to maturity, gray-brown to dingy yellow-brown, dark grayish brown as dried. Context yellowish pallid, in apex of stipe reddish, unchanging when cut, odor faintly fragrant, taste mild, $FeSO_4$ weakly brownish.

Tubes nearly free, 1-1.5 cm deep, dull yellow staining dingy blue when cut; pores round, 1-2 per mm, yellowish gray staining blue if injured, some near the stipe tinged reddish orange.

Stipe 5-6 cm long, 8-12 mm thick, 1.5 cm toward base, solid, brownish gray within but stained reddish when cut, yellowish in the base and with yellow mycelium over the surface, upper area scabrous-roughened (like a finely ornamented *Leccinum*), the ornamentation dull brown on a reddish to brown ground color.

Spore deposit dark olive. Spores 13-16 X 4-5 μ, smooth, yellow in KOH, bister in Melzer's (dingy brown), amyloid at first but quickly changing, apex truncate or notched; shape in profile obscurely inequilateral, in face view elongate-suboblong, wall about 0.2 μ thick.

Basidia 4-spored, clavate, 28-35 X 7-10 μ, yellowish to hyaline in KOH. Pleurocystidia scattered, 37-58 X 9-14 μ, fusoid-ventricose, apex subacute, walls thin and smooth, content hyaline (all in KOH). Cheilocystidia scattered, more or less similar to pleurocystidia but smaller and content often yellow in KOH, numerous basidioles 10-12 μ broad on edge also. Caulocystidia of 2 types: (1) narrowly fusoid and with elongated (to proliferated) neck 3-5 μ wide, these thin-walled and smooth, hyaline in KOH; (2) clavate to mucronate and 34-50 X 8-15 μ, hyaline to pale cinnamon in KOH and wall often faintly thickened. Hyphae of subiculum on stipe with finely roughened walls in part at least and especially in the area near base of caulohymenium.

Cuticle of pileus a compact trichodermium reviving poorly in KOH or Melzer's, the cells of the clavate, inflated or keglike type and with coarse plates of tawny-brown to dull cinnamon incrusting material or the adhering material not highly colored. Subcuticular hyphae also with conspicuous incrustations but these not dextrinoid. Hyphae of context not distinctively colored (weakly yellow) in Melzer's. Clamp connections not found in Michigan material.

Observations.—The *B. porosporus* complex needs further concentrated study. Although we have used the European name for our material at the species level, the narrower spores and the elongated caulocystidia clearly indicate that it is distinct. The collection described is from Binarch Creek, Priest Lake, Idaho, September 9, 1968, Smith 76096. The Michigan specimens have been single basidiocarps which did not survive the drying process.

153. Boletus subdepauperatus Smith & Thiers, sp. nov.

Illus. Pl. 107 (upper fig.).

Pileus 3-5 cm latus, convexus, ad marginem rimulosus, olivaceobrunneus demum pallide spadiceus, siccus, subvelutinus. Tubuli tactu caerulei. Stipes 4-6 cm longus, 6-8 mm crassus, subfurfuraceus, sursum pallide olivaceus. Sporae 9-12 × 4.5-5 μ. Specimen typicum in Herb. Univ. Mich. conservatum est; prope Berry Creek, Cheboygan County, July 28, 1961, legit Smith 63736.

Pileus 3-5 cm broad, broadly convex becoming plane and finally the margin uplifted; surface dry and velvety to unpolished, irregularly rimulose toward the margin, color dingy olive brownish young, in age dingy yellow-brown and drying exactly "snuff-brown" (dingy yellow-brown). Context in the cracks pale dull buff and buff as dried (not bright yellow), staining blue as in *B. chrysenteron.*

Tubes depressed around the stipe, about 10 mm deep, dingy yellow fresh, staining blue; pores large, about 1 mm in diameter, irregular, dingy olive yellowish when mature and staining bluish.

Stipe 4-6 cm long, 6-8 mm thick, equal, when dried almost concolor with the pileus, pruinose-furfuraceous, mycelium at base yellowish gray as dried.

Spores 9-12 × 4.5-5 μ, smooth, apex truncate to slightly notched, short-inequilateral in profile, subelliptic in face view, ochraceous in KOH, somewhat tawny in Melzer's. Pileus cuticle a trichodermium of hyphae with cells inflated to 25-80 × 30 μ, the terminal cells tapered to an obtuse apex or fusoid-ventricose, all rather heavily incrusted by relatively coarse plates of pigment, cuticular cells and cells of contextual hyphae all lacking distinctively colored contents in Melzer's. Clamp connections absent.

Habit, habitat, and distribution.—Scattered on barren soil, Berry Creek Woods, Wolverine, July 28, 1961, Smith 63736.

Observations.—A collection of this species from a wet aspen log was made July 15, 1968, at the University of Michigan Biological Station, Cheboygan County. In it the spores measure 9-12 X 4-5 μ and are truncate. The stipe was rhubarb-red in the cortex, and this color was evident near the base on the surface. The surface of the stipe was pruinose-scabrous on the order of a species of *Leccinum* with moderately fine ornamentation. The small spores of this species at once distinguish it from *B. truncatus,* but there are in addition other differences such as lack of red or purple in the cracks in the pileus, at least in material observed to date, and the olivaceous-pallid stipe apex when young as contrasted to yellow. Singer, Snell, and Dick described the spores of *Xerocomus truncatus* as 12-17 (26) X (4.2) 4.5-6.5 μ, but Coker and Beers for their *B. chrysenteron* gave the spores as 3.8-5 X 9.5-12.5 μ, the size which distinguishes *B. subdepauperatus* from *B. truncatus.*

Section PIPERATI (Singer), comb. nov.

Ixocomus sect. *Piperati* Singer

Rev. de Mycol. 3:38. 1938

For the distinguishing characters see the key to sections.
Type species: *Boletus piperatus.*

KEY TO SPECIES

1. Hymenophore almost lamellate . . . . see a variant under *B. piperatus*
1. Hymenophore definitely tubulose, pores small to large 2

  2. Pores staining blue if injured . . . . *B. piperatoides*
  2. Not changing to blue anywhere if injured 3

3. Taste mild 4
3. Taste peppery to acrid. 6

  4. Spores 12-15 μ long; pileus subtomentose and red . . . . *B. rubinellus*
  4. Not with above combination of features 5

5. Spores bright olive to olive-yellow in KOH; pleurocystidia absent to rare
  and inconspicuous. . . . . *B. pseudorubinellus*
5. Spores hyaline to dingy yellowish in KOH; pleurocystidia scattered to
  abundant . . . . *B. rubritubifera*

  6. Taste sharply acrid; pores mostly 1-2 per mm. . . . . *B. piperatus*
  6. Taste merely peppery; pores in age 2-3 mm wide. . . . . *B. amarellus*

154. Boletus piperatus Fries

Syst. Mycol. 1:388. 1821

Viscipellis piperata (Fries) Quélet, Enchir. Fung. p. 157. 1886.
Ixocomus piperatus (Fries) Quélet, Flor. Myc. Fr. p. 414. 1888.
Chalciporus piperatus (Fries) Bataille, Bolets, p. 19. 1908.

Illus. Pls. 114-15.

Pileus (1.5) 2-5 (9) cm broad, obtuse to convex, becoming nearly plane to plano-convex to occasionally subumbonate or margin turned up slightly in age, margin entire, incurved when young, surface dry becoming subviscid, mostly appearing unpolished, pellicle typically somewhat separable, glabrous to somewhat fibrillose-streaked or appressed-squamulose, often glabrous with age, at times areolate; clay color to some shade of yellow-brown or orange-cinnamon ("clay color," "cinnamon-buff," "buckthorn-brown," "ochraceous-tawny," "tawny," or "orange-cinnamon"), often darker on the disc to rusty cinnamon. Context up to 1 cm thick, firm, pale yellow ("warm buff"), in age often with a slight rose tint or distinctly vinaceous above the tubes, becoming dingy vinaceous-buff in age, soft and subgelatinous when old, taste very sharply and distinctly acrid, odor not distinctive, with $FeSO_4$ the line above the tubes is grayish, in KOH no reaction but on the cutis of the pileus staining it dark red-brown, NH_4OH violaceous-fuscous on context. The pileus cutis stains waxed-paper red.

Tubes adnate to subdecurrent or only faintly depressed, short (3-10 mm deep), dingy ochraceous but slowly becoming reddish in age ("ocher-yellow" to "ochraceous-tawny" to "tawny-olive" and finally with vinaceous-red tones); pores angular, unequal, averaging 2 per mm, dissepiments thin, dull yellow when young, soon more or less cinnamon and in age finally distinctly red to brick-red, not staining appreciably when injured (merely dark brown).

Stipe (2) 4-10 (12) cm long, (3) 4-10 (15) mm thick, solid, lemon-yellow within, unchanging when injured; surface reddish cinnamon from pruina, ground color yellow, in age with lines from the tubes extending down the apical portion, base bright yellow from a coating of bright yellow mycelium, negative when tested with $FeSO_4$.

Spore deposit "Sayal-brown" (dull cinnamon); spores (8.5) 9-12 X 4-5 μ, smooth, narrowly fusiform in face view, obscurely inequilateral in profile, with a hyaline sheath, olivaceous-hyaline to dingy ochraceous in KOH, rusty brown in Melzer's.

Basidia 4-spored, 24-30 X 7.5-8.5 μ, hyaline in KOH, yellow in Melzer's. Pleurocystidia 43-60 X 8-13 μ, scattered, conspicuous, fusoid-ventricose to subcylindric, thin-walled, smooth, neck tapered to a subacute or slightly enlarged apex, content hyaline and homogeneous in KOH or in Melzer's. Cheilocystidia more or less similar to pleurocystidia, sometimes with a slight amount of adhering debris. Tube trama gelatinous and divergent, nonamyloid.

Cuticle of pileus a tangled trichodermium of broad (10-17 μ) hyphae with fusoid to elliptic or cylindric end-cells, hyaline in KOH and yellowish in Melzer's, the layer not at all gelatinous. Context of compactly interwoven floccose nonamyloid hyphae. Clamp connections none found.

Habit, habitat, and distribution.—Solitary to gregarious, mostly scattered under conifers, less common in pure stands of hardwoods; common throughout the state in its favorite habitats.

Observations.—This species is placed in *Suillus* by some authors, and these same authors recognize *Xerocomus* as a genus featured by the tubes splitting rather than separating when the pileus is broken downward. The tubes in *B. piperatus* split rather than separate under the above condition, but we do not consider the species closely related to *B. chrysenteron* or *B. subtomentosus* as relationships in the boletes go. However, there is a very obvious related group of boletes including *B. piperatus, Suillus castanellus,* and "*Boletinus squarrosoides.*" The last two are extralimital to this work. The limits of the group and the relationships of the species are under study at the present time. The characters involved are the color of the pores before full maturity—they are yellow-brown in most of our collections—the degree of reticulation at the apex of the stipe, the size of the pores, and staining reactions. The collections featuring a blue stain when injured we have here described as an autonomous species. A variant known from a single collection in which the hymenophore is more lamellate than poroid is not named, but our notes on it follow:

Lamellate variant.—Pileus about 6 cm broad, when expanded broadly depressed and the margin wavy and lobed (reminding one of *Cantharellus cibarius*), surface unpolished, both when fresh and when dried resembling *B. piperatus.* Context buff colored and punky as dried, taste of dried material slightly acid.

Hymenophore lamellate to poroid variously, the lamellae narrow, crowded, decurrent, rich cinnamon as dried.

Stipe about 6 cm long, 10 mm thick (estimated from dried material), surface unpolished (much like that of pileus), glabrous to pruinose, base (as dried) coated with bright ocher-yellow mycelial coating, this layer when mounted in KOH exuding a bright yellow pigment in the mount, the hyphae lacking clamp connections and variable in width (4-18 μ).

Spores 8-11 X 3-4.5 μ, smooth, ocher-yellow in KOH singly, mostly weakly to moderately dextrinoid in Melzer's, in face view obscurely fusoid to subellipsoid, in profile somewhat inequilateral, suprahilar depression shallow, no apical differentiation observed.

Basidia 20-26 X 7-9 μ, 4-spored, yellowish hyaline in KOH, ochraceous-orange in sections of hymenium. Pleurocystidia 45-75 X 7-12 μ, narrowly fusoid-ventricose to subcylindric, with obtuse to subacute apex, smooth, thin-walled, uncolored in KOH or in Melzer's. Cheilocystidia similar to pleurocystidia but smaller. Tube trama slightly divergent from a central strand but scattered through it are large hyaline oleiferous hyphae 8-12 μ wide.

Pileus cuticle a trichodermium of hyphae 4-19 μ wide, with cylindric to narrowly clavate terminal cells 12-15 μ wide, the walls all thin and smooth and content of cells hyaline in KOH, in Melzer's varying to slightly yellowish. Context of interwoven hyphae 6-15 μ wide, cell content not colored distinctively in KOH or Melzer's. Clamp connections absent.

Habit, habitat, and distribution.—Solitary beside Crowdon's Road near Marquette, July 23, 1963, Ingrid Bartelli 1169. When Mrs. Bartelli annotated the collection she wrote: "A freak *Boletus piperatus.*" In some respects the collection recalls *Phylloporus lariceti* Singer.

Material examined.—Barry: Smith 73198. Cheboygan: Charleton G-224; Harding 102; Smith 57421, 58082, 58254, 61413, 63760, 74627 (red tubes), 74873 (stipe apex reticulate); Thiers 1230, 2457, 3995, 4433. Chippewa: Smith 72164, 74996. Emmet: Smith 57797, 58127, 63555, 63716, 67030, 74410, 74607; Thiers 3074, 4711. Gratiot: Potter 3707, 6119, 6217, 6309, 8423, 8581, 7766, 7841, 9032, 9676, 10128, 11825, 12295, 12441, 12561, 12839, 12980. Houghton: Pennington 1906. Keweenaw: Pennington 1906. Livingston: Smith 66368, 75217. Luce: Smith 41797. Mackinac: Thiers 3363. Marquette: Bartelli 230, 252. Montcalm: Potter 8755. Montmorency: Smith 66956, 75128. Oakland: Smith 65611. Ogemaw: Smith 50601, 67526. Washtenaw: Mazzer 4202; Kauffman 1925; P. M. Rea 746; Shaffer 2747, 2815; Smith 32-656, 1459, 6794, 14927, 18838, 62892, 72462, 73208; Thiers 4555.

155. Boletus amarellus Quélet

Assoc. Fr. Av. Sc. 1882:398

Illus. Pl. 115 (upper).

Pileus 3-10 cm broad, convex, becoming broadly convex, margin even; surface dry but soft to the touch and tacky when wet (subviscid), dull cinnamon to reddish cinnamon or finally vinaceous-red ("vinaceous-rufous" to "Dragon's blood-red"), drying dull ochraceous-brown to dull cinnamon. Context yellow, developing a vinaceous-red zone near the tubes, taste slowly peppery, with $FeSO_4$ negative, odor slightly acid.

Tubes up to 1 cm deep in large caps, dull ochraceous to buckthorn-brown, not staining blue if injured; pores at first 1-2 per mm but in age up to 2-3 mm broad, angular, yellow-brown young but becoming reddish cinnamon or redder in age; not staining blue if injured.

Stipe (3) 4-8 cm long, 5-10 mm thick, equal, interior lemonchrome in lower half, surface. streaked with reddish cinnamon fine fibrils, base coated with bright yellow mycelium, apex ornamented with decurrent lines from the tubes.

Spore deposit pale dull cinnamon; spores 9-11 X 3-4 μ, smooth, apical pore not distinct, shape in profile obscurely inequilateral, the suprahilar area flattened, in face view subfusiform, color in KOH dingy olivaceous becoming bright olive-yellow, in Melzer's many becoming distinctly dextrinoid, wall about 0.2 μ thick.

Basidia 4-spored, 22-30 X 7-9 μ, yellowish in KOH and more ochraceous in Melzer's. Pleurocystidia 36-68 X 6-9 μ, subcylindric and usually curved near the apex, smooth, content homogeneous and ochraceous in either KOH or Melzer's, wall thin and smooth. Cheilocystidia 32-60 X 4-7 μ, otherwise like the pleurocystidia. Caulocystidia versiform: (a) elongate-clavate and 38-65 X 3-4 X 5-7 μ; (b) fusoid-ventricose and 28-42 X 7-12 μ, apex obtuse; (c) some subcylindric like the cheilocystidia; all types smooth, thin-walled, and content not distinctive in KOH or Melzer's.

Cuticle of pileus basically a tangle of hyphae 6-12 (15) μ wide, cells mostly elongate but some short ones present, walls thin, smooth, hyaline, content yellowish, cells tubular to slightly inflated, in large caps some hyphae 30 μ or more wide and the cells inflated as well as having the cells disarticulating to a distinct degree leaving ellipsoid cells free in the mount; end-cells clavate to fusoid or cylindric, 10-20 (30) μ wide and up to 200 μ long. Context hyphae with orange-red content in some as viewed in Melzer's. Cuticular hyphae merely ochraceous in Melzer's— no evidence of emulsification of the content. No clamp connections observed.

Habit, habitat, and distribution.—Scattered under conifers, especially spruce, Cheboygan and Chippewa counties, August and September, Smith 77697 is the collection described.

Observations.—We have identified our material with Quélet's species on the basis of the peppery (not acrid) taste, wide pores at maturity, and their red color. On the basis of the cuticular hyphae in the larger specimens, the species as we have described it is amply distinct from *B. piperatus*. In younger pilei of *B. amarellus* we found the cuticular hyphae to be 8-15 μ wide, and no disarticulation of cells was noted. Our study has shown that there is considerable more diversity of characters in the group than was previously supposed.

156. Boletus rubinellus Peck

Rept. N. Y. State Mus. 32:33. 1879

"Pileus at first broadly conical or subconvex, then expanded, subtomentose, red, becoming paler with age; tubes convex, adnate or somewhat depressed around the stem, rather large, subrotund, pinkish red then sordid yellow; stem equal, smooth, yellow with reddish stains; spores oblong, .0005'-.0006' long, .00016' broad; flesh of both pileus and stem bright yellow.

"Plant about 2' high, pileus 1-2' broad, stem 1-2' thick.

"Ground in woods. Gansevoort, Aug."

The species belongs in the section Subtomentosi and is apparently related to *B. rubinus*.

Data from the type:

Spores 12-15 × 3.3-4.5 (5) μ, smooth, apical pore not visible, color in KOH pale dull ochraceous, in Melzer's a large number soon rather strongly dextrinoid, shape in profile somewhat inequilateral-elongate, in face view subfusoid, wall about 0.3 μ thick.

Basidia 24-32 × 8-10 μ, 4-spored, clavate, content not distinctive in either KOH or Melzer's. Pleurocystidia scattered to abundant, 34-62 × 7-12 μ, fusoid-ventricose with neck elongated and apex obtuse to subacute, wall thin and smooth, content not distinctive in either KOH or Melzer's. Cheilocystidia and caulocystidia not studied.

Cuticle of pileus a tangled mass of hyphae 3-6 μ wide, hyphae tubular, thin-walled, smooth, end-cells tubular or nearly so, no short cells evident, in KOH ochraceous from colored homogeneous content, in Melzer's content homogeneous and dull orange-buff. Context hyphae with orange-buff content in Melzer's. No clamps observed.

Observations.—We have not encountered this species in the state. Its diagnostic features are the subtomentose pileus, strongly dextrinoid spores many of which reach a length of 15 μ, the abundant pleurocystidia with elongated necks, and the red pileus and tubes. The portion of the type studied did not show caulocystidia but more than likely they are (when present) similar to the end-cells of the cuticular hyphae of the pileus. The spores lack any distinctive color in KOH. It may be of some significance to note that Peck did not mention taste in his original account. Hence we do not know if the type was mild in taste.

157. Boletus rubritubifera Kauffman

Bull. N. Y. State Mus. 179:88. 1915

Data from original description:

Pileus 2-5 cm broad, fleshy, convex, obtuse, glabrous or obscurely subtomentose, dry, even, cinnamon-rufous (Ridgway), slightly variegated with yellowish. Context whitish tinged yellow, unchanged, very thick, odor and taste mild.

Tubes Pompeian-red (Ridgway) throughout; pores red, depressed around the stem, convex, 5-8 mm long, pores 2 per mm, subangular, dissepiments rather thick.

Stem 5-6 cm long, 1-2 cm thick above, tapering downward, dingy apricot-yellow (Ridgway), concolor within, even, glabrous, solid.

Data from type:

Spores 9-12 (13.5) X 3-4 μ, smooth, lacking an apical pore, yellowish hyaline in KOH (not olive), remaining uncolored in Melzer's or some slowly pale tan (weakly dextrinoid), shape in profile narrowly and somewhat to obscurely inequilateral, a few "sway-backed," in face view narrowly subfusoid to suboblong, wall about 0.2 μ thick, rarely wavy-angular in optical section.

Basidia 4-spored, clavate, up to 35 X 11 μ, not distinctively colored in either KOH or Melzer's. Pleurocystidia 36-68 X 7-14 μ, narrowly fusoid-ventricose varying to almost subcylindric, walls thin and smooth, content not distinctive in either KOH or Melzer's. Cheilocystidia similar to but smaller than the pleurocystidia. Caulocystidia 35-70 X 4-7 μ, some almost seta-like but thin-walled, mostly subcylindric to flexuous with obtuse apex (resembling end-cells of the cuticular hyphae of the pileus; color not distinctive in Melzer's.

Cuticle of pileus of interwoven matted-down hyphae 4-7 μ wide, with end-cells appressed or ascending, the filaments tubular and the cells long, walls smooth, thin, a few with a thin gelatinous halo, content homogeneous and ochraceous in KOH, more orange-ochraceous in Melzer's but remaining homogeneous. Context hyphae with some ochraceous-orange content but not truly distinctive. In mounts in Melzer's scattered yellowish globules 2-4 μ in diameter occur throughout the mount. No clamps observed.

Observations.—This species has been placed in synonymy with *B. rubinellus* by some, but we are not prepared to accept this at present. *B. rubinellus* has a subtomentose pileus which is red, has bright yellow flesh, more strongly dextrinoid spores which are slightly longer than in Kauffman's species and possibly differs in the color of the stipe.

158. Boletus pseudorubinellus Smith & Thiers, sp. nov.

Pileus 1.5-5 cm latus, obtuse conicus vel convexus, glaber vel subsquamulosus, pallide luteus demum rufo-cinnamomeus vel testaceus. Contextus pallide luteus. Tubuli rosei. Stipes 2-6 cm longus, 4-12 mm crassus, intus luteus vel subroseus, extus pruinosus, roseus. Sporae 9-13 X 3-4 μ. Pleurocystidia infrequentibus. Specimen typicum in Herb. Univ. Mich. conservatum est; prope Burt Lake, legit Smith 77910.

Pileus 1.5-5 cm broad, obtusely conic to convex, expanding to umbonate or broadly convex, glabrous to minutely appressed-squamulose, surface viscid to subviscid, when young "cream-buff" (yellowish), slowly becoming pinkish cinnamon and finally testaceous, sometimes squamulose, reddish on an ochraceous ground color. Context pale yellow, unchanging when cut, $FeSO_4$ no reaction, NH_4OH on cuticle of pileus no reaction, taste acid but not acrid.

Tubes 8-10 mm deep, bright rose-red overall including the pores; depressed to decurrent; pores 1-2 mm wide, unchanging when bruised.

Stipe 2-6 cm long, 4-12 mm at apex, equal to clavate, solid, yellow to pinkish within, unchanging when cut, surface finely pruinose and more or less rose-pink except for yellowish region near base, base with yellow mycelium over it.

Spores 9-13 X 3-4 μ, smooth, no apical pore visible, in KOH distinctly olivaceous but on standing bright olive-yellow, in Melzer's dull tan (as if shaded by a weak-amyloid reaction), in profile somewhat inequilateral and narrow, in face view narrowly subfusoid, wall thin (about 0.2 μ).

Basidia 4-spored, 18-26 X 8-10 μ, clavate, hyaline to yellow in KOH, not distinctive in Melzer's. Pleurocystidia very rare and mostly near pores, 32-45 X 7-12 μ, hyaline, thin-walled, smooth, not distinctive in Melzer's. Cheilocystidia rare to scattered, resembling pleurocystidia but smaller. Caulocystidia—some resembling pleurocystidia but most of them resembling the end-cells of the trichodermial hyphae of the pileus. The cells 40-70 X 5-8 µ and often flexuous, walls smooth and thin, content not distinctive in KOH or in Melzer's.

Cuticle of pileus at first a trichodermium of upright hyphae but this very soon collapsing, the hyphae 4-8 µ wide, the cuticle in age a tangled mass of hyphae, their cells tubular, the walls thin and smooth and the content orange-brown to orange-buff but fading on standing (all in mounts in Melzer's), in KOH the content ochraceous. Context hyphae (in Melzer's) orange to orange-red, in KOH hyaline and granular ("coloidal"). No clamps observed.

Habit, habitat, and distribution.—Gregarious near spruce, Lower Falls of the Tahquamenon River, Chippewa County, August to September, Smith 75830, 77910(type), 78050. The type was from Cheboygan County.

Observations.—This species differs from *B. rubritubifera* and *B. rubinellus* in the olive tone which the spores assume when revived in KOH and in the very rare pleurocystidia. The color of the pileus generally has less red in it than the other species—especially in the immature stages.

159. Boletus piperatoides Smith & Thiers, sp. nov.

B. piperatus similis sed pori tactu caerulei; sapor tarde subacris. Specimen typicum in Herb. Univ. Mich. conservatum est; prope Otis Lake, Barry County, August 19, 1966, legit S. Mazzer (Smith 73092).

Pileus 3-6 cm broad, obtuse becoming plane or finally the margin uplifted, surface soft (like kid gloves), subviscid, color tan to dull orange-cinnamon, margin near cinnamon-buff. Context thick, pale pinkish buff, when cut slowly staining dull blue, taste slowly peppery (not as "hot" as in *B. piperatus*) with guaiac slowly blue, with $FeSO_4$ bluish gray, with KOH a yellow pigment dissolves into the mount, the reaction on the pileus cutis soon going to brownish.

Tubes up to 1 cm long, decurrent, dull ochraceous, staining dingy inky blue when bruised; pores near "snuff-brown" (yellow-brown), staining bluish, 2-3 per mm, splitting when the pileus is broken downward.

Stipe 4-6 cm long, 4-6 mm thick, solid, honey-color above, lemon-yellow in the base; base coated with lemon-yellow mycelium, surface above naked or nearly so and dingy yellowish brown, basal portion lemon-yellow as dried both in the interior and over the surface including the mycelium.

Spores 7-9 (10) X 3-3.5 μ, smooth, pale clay color in KOH but soon fading to bright yellow singly, more brownish in groups, in Melzer's "fleeting-amyloid" fresh and as revived pale bister to dingy pale tan, in face view somewhat boat-shaped, in profile somewhat inequilateral, suprahilar depression fairly distinct, no apical differentiation observed.

Basidia 4-spored, 23-28 X 5.5-7.5 μ, clavate, hyaline to yellowish in KOH, yellowish in Melzer's. Pleurocystidia 36-57 X 7-12 μ, narrowly fusoid-ventricose to subcylindric, hyaline to yellow in KOH and Melzer's, content not otherwise distinctive in either medium. Cheilocystidia similar to pleurocystidia only smaller and content yellow to yellow-brown. Caulohymenium in patches near apex and with basidia and cystidia resembling the cheilohymenial elements.

Tube trama bilateral, the divergent hyphae gelatinous, sections mounted in Melzer's and crushed becoming dark greenish to bluish green as seen under the microscope (but no localized amyloid reaction present). Cutis of pileus if interwoven hyphae 4-8 μ wide, yellowish in KOH and with end-cells tubular to narrowly clavate, the walls showing a slight tendency to become roughened as in *Leccinum insigne*.

Context hyphae hyaline in KOH, thin-walled, much-branched, interwoven, yellowish to yellowish hyaline in KOH and a dingier more yellowish brown in Melzer's but content not distinctive as single cells are viewed. Clamp connections absent.

Habit, habitat, and distribution.—Gregarious in an oak woods, Otis Lake, Barry County, August 19, 1966, collection S. Mazzer (Smith 73092).

Observations.—This species differs from *B. piperatus* in the fleeting-amyloid reaction of the spores when fresh, the "fleeting-amyloid" (unlocalized) reaction noted in the revived tube tissue, in the slowly only somewhat peppery taste, the blue staining of freshly injured basidiocarp tissue, and the dull-colored pileus context.

Section PSEUDOBOLETI (Singer), comb. nov.

Xerocomus section *Pseudoboleti* Singer

Farlowia 2:299. 1945

In addition to the species keyed here see *B. subilludens* and *B. alutaceus.*

KEY TO SPECIES

1. Context of pileus when cut or broken changing to blue instantly 2
1. Context changing more slowly or not changing. 3

    2. Pileus lemon-yellow at first; pores 1-2 per mm . . . . . *B. pseudosulphureus*
    2. Pileus dark yellow-brown to blackish brown; pores 1-2 mm in
    diameter . . . . *B. pulverulentus*

3. Taste of pellicle bitter; spores 8-10 × 4-5 μ . . . . *B. calvinii*
3. Not as above 4

    4. Pileus whitish when young, becoming slowly leather-brown; NH₄OH
    on cuticle no reaction . . . . *B. pallidus*
    4. Pileus yellow-brown, gray-brown, smoky buff, or cinnamon to
    bay-red 5

5. Pileus dry, bay-red; spores merely yellowish as revived in Melzer's
. . . . *B. albocarneus*
5. Not as above 6

    6. Stipe clavate, 1-3 cm thick above, up to 5 cm at base before ex-
    panding; pileus dingy cinnamon . . . . *B. huronensis*

7. Pileus bay-red, NH₄OH olive to green around the spot of application (on
fresh caps); spores greenish as revived in Melzer's . . . . *B. badius*
7. Pileus alutaceous to smoky buff; NH₄OH not giving an olive to green reaction on pileus (?) . . . . *B. glabellus*

160. Boletus calvinii Smith & Thiers, sp. nov.

Pileus 5-10 cm latus, convexus, viscidus, ferruginus. Contextus albidus, immutabilis. Tubuli lutei tactu subviridis. Stipes 4-6 cm longus, 6-15 mm crassus, sursum rufus, deorsum sordide brunneus, non-reticulatus. Sporae 8-10 × 4-5 μ. Specimen typicum in Herb. Univ. Mich. conservatum est; prope Ann Arbor, August 14, 1915, legit C. H. Kauffman.

Pileus 5-10 cm broad, convex, with a viscid pellicle, glabrous, even, "chestnut" to "hazel" (ferruginous), surface bitter to the tongue. Context white, scarcely changing color when bruised, 5-12 mm thick.

Tubes adnate, at length depressed, 10 mm deep, yellow ("Martius-yellow") and at length greenish (almost "olive-lake"); pores angular about 1 mm broad, concolor with sides, surfaces slowly greening where bruised.

Stipe 4-6 cm long, 6-15 mm thick, tapering downward or sub-equal, rufous to near apex, at length smoky brown toward the base, solid, *not* reticulate, apex pallid, faintly pruinose, subpunctate toward the base.

Spores 8-9 (10) X 4-4.5 (5) μ, smooth, wall thin, no apical differentiation, in face view suboblong or narrowed slightly to distal end, in
profile obscurely inequilateral, suprahilar depression typically shallow;
color in KOH weakly yellow, in Melzer's merely tinged fulvous; no
fleeting-amyloid reaction present on revived material.

Basidia 18-22 X 6-7 μ, 4-spored, hyaline. Pleurocystidia scattered,
38-52 X 10-14 μ, subfusoid, apex subacute, thin-walled, content when
revived in Melzer's or KOH yellow, often originating from broad yellow
laticiferous hyphae. Caulocystidia in patches near apex, 16-22 X 8-11 μ,
short-clavate, hyaline in KOH, thin-walled, content "empty."

Cuticle of pileus a collapsing trichodermium the hyphae tubular
and 4-6 μ wide, in a gelatinous matrix, hyaline or nearly so (in Melzer's
or KOH), apical cells tubular with blunt apex. Context hyphae next to
cuticle not colored in Melzer's. Clamp connections none.

Habit, habitat, and distribution.—On a hillside among grasses under
oak, Ann Arbor, August 14, 1915, C. H. Kauffman.

Observations.—Kauffman had tentatively regarded this as a small-
spored *B. badius.* In many respects—especially spore characters—it is
similar to *Suillus castanellus* but has a differently colored hymenophore
with the pores staining blue slightly when bruised. The colors are more
ferruginous than in *B. badius,* and the very small spores at once distinguish it from the variants of that species. It is named in honor of C. H.
Kauffman, the collector.

161. Boletus pallidus Frost

Bull. Buff. Soc. Nat. Sci. 2:105. 1874

Suillus pallidus (Frost) Kuntze, Rev. Gen. Pl. 3(2):536. 1898.
Ceriomyces pallidus (Frost) Murrill, Mycologia 1:152. 1909.

Illus. Pls. 116-17 (above).

Pileus 4.5-15 cm broad, hemispheric to convex young, becoming
broadly convex to nearly plane to slightly depressed, surface dry and
unpolished, smooth but in age often strongly areolate, glabrous to obscurely tomentose to subtomentose, tacky in age when wet, pallid to
cinereous to "pale pinkish buff" or near avellaneous, finally dingy
leather-brown, occasionally with faint rose tints, margin sterile and often
crenated. Context up to 2.5 cm thick, soft, nearly white or near the
tubes pale yellow, often with a trace of pink where bruised, when young
and firm with a slight tendency to turn blue when injured; taste mild to

slightly bitter, odor not distinctive, NH_4OH on cuticle of pileus giving no color change; cut context finally greenish gray.

Tubes about 1-2 cm deep, adnate to subdecurrent becoming depressed, pallid when very young but soon pale yellow and finally olivaceous-yellow; pores small to medium (1-2 per mm), round, becoming somewhat angular, staining blue to greenish blue and then brownish where injured.

Stipe 5-12 cm long, 8-30 mm thick at apex, enlarged downward, finally nearly equal, solid, white within, reddish around wormholes, surface dry, smooth to slightly reticulated near apex, white when young, often yellow ("Chalcedony-yellow") at the apex, white mycelium at the base, finally becoming streaked brownish gray downward, often with reddish flushes, especially toward the base.

Spore deposit near "Saccardo's olive" and "brownish olive" (Singer), dingy olive-brown (Smith). Spores 9-15.4 X 3-4.5 μ, narrowly ovoid to subfusoid in a face view, in profile inequilateral, smooth, with a hyaline sheath, greenish hyaline in KOH, yellowish to pale tawny in Melzer's.

Basidia 4-spored, 20-26 X 9-10 μ, yellowish in KOH and Melzer's. Pleurocystidia 35-44 X 8-13 μ, fusoid-ventricose with a subacute apex, hyaline to yellow in KOH, thin-walled, smooth. Cheilocystidia more or less similar to pleurocystidia but many clavate cells also present.

Tube trama gelatinous and divergent, nonamyloid. Pileus cuticle a tangled turf about 300 μ deep of hyphae 4-8 μ in diameter, the terminal cells tubular and obtuse, hyaline in KOH, thin-walled, nonamyloid. Clamp connections absent.

Habit, habitat, and distribution.—Solitary to gregarious or cespitose in open hardwoods especially of oak, common in the Lower Peninsula in midsummer if the weather is wet. Its favorite habitat appears to be mossy areas near the edges of woodland pools, but on wet years it is common in the oak-pine barrens around Topinabee.

Observations.—Although almost lacking in pigments, this is one of our most attractive boletes when discovered in prime condition. The pallid to glaucous pileus, pale stipe, and weakly yellowish tubes distinguish it in the field. It is frequently parasitized by a white mold. Smith 73112 represents a slender miniature variant of this species with a stipe 3-4 mm thick at the apex. For a discussion of variability of this species in New England see Dick (1960).

Material examined.—Barry: Mazzer 4262, 4327, Aug. 1966; Smith 51178, 73122, 73160, 73184. Cheboygan: Charlton G 176; Shaffer 1924a; Smith 57330, 58147, 72404; Thiers 4383. Gratiot: Potter 3706,

3812, 7730, 7762, 7849, 7983, 8134, 8153, 8243, 8371, 8445, 10043, 10157, 10214, 11207, 11887, 12090, 12220, 12683, 13977. Livingston: Smith 7035, 18468, 18623. Oakland: Smith 6778, 7067, 7215, 18534, 73290. Ogemaw: Shaffer 2692, 2710. Washtenaw: Kauffman 7-26-14, 8-19-21; Homola 1629; Shaffer 2649, 2661, 2753, 2791, 2823; N. J. Smith 92A; A. H. Smith 1529, 1685, 62647, 64153, 72468, 73118; Thiers 4565, 4582, 4583, 4587, 4603.

162. Boletus huronensis Smith & Thiers, sp. nov.

Illus. Pls. 117 (lower), 118 (left).

Pileus 8-14 cm latus, late convexus, siccus, impolitus, sordide cinnamomeus vel luteo-brunneus. Contextus pallide luteus, tactu tarde viridi-caeruleus. Tubuli 1-1.5 cm longi, obscure lutei, pori lutei tactu caerulei. Stipes 7-10 cm longus, 1.8-4 cm crassus, sursum luteus, deorsum tactu brunneus. Sporae 12-15 X 3.5-4.5 μ. Specimen typicum in Herb. Univ. Mich. conservatum est; prope Canyon Lake, Marquette County, August 10, 1968, legit J. Ammirati 2188.

Pileus 8-14 cm broad, obtuse to convex, margin inrolled, expanding to plano-convex, margin often irregular at first and sometimes remaining so, surface dry and matted-fibrillose or obscurely velvety at first; color dull yellow-brown ("snuff-brown") to dull cinnamon ("Sayal-brown"), not areolate at any stage, when dried a grayish cinnamon-buff. Context thick (2-3.5 cm), pale yellow and slowly staining blue when bruised, with a reddish brown line under the cuticle, reddish tints developing on context after blue has faded; with $FeSO_4$ grayish green to bluish gray; odor and taste not distinctive.

Tubes 0.5-1.5 cm deep, nearly free from stipe, dull yellow, staining blue where injured; pores minute about 2 per mm, yellow to olive-yellow, when bruised staining greenish blue and finally reddish brown.

Stipe 7-10 cm long, 1.8-4 cm thick at apex, clavate and up to 5 cm thick at base, solid, firm, yellow above, paler to whitish below, brownish orange around larval holes, in the base watery-streaked and watery to grayish yellow; surface unpolished and evenly pale bright yellow ("baryta-yellow"), brownish where handled, not reticulate.

Spore deposit olive; spores 12-15 X 3.5-4.5 μ, smooth, apex lacking a pore, in face view suboblong to obscurely fusoid, in profile mostly obscurely inequilateral, color bright olive-yellow in KOH, in Melzer's grayish at first (weakly amyloid ?), wall about 0.2 μ thick.

Basidia 4-spored, 8-11 μ in diameter, clavate, hyaline in KOH, with a large oil droplet, yellowish in Melzer's. Pleurocystidia 33-57 X 9-15 μ, fusoid-ventricose with subacute apex, hyaline in KOH and Melzer's, smooth, thin-walled. Cheilocystidia 23-34 X 6-11 μ, subfusoid, content yellow in KOH, edge of tube composed mainly of clavate basidioles with yellow content. Caulocystidia 33-56 X 9-15 μ, clavate, ovate-pedicellate or fusoid-ventricose, thin-walled, either hyaline or with bright yellow content as revived in KOH.

Tube trama typical for the genus: the hyphae somewhat gelatinous as revived in KOH. Pileus cutis of appressed-interwoven hyphae 5-9 μ wide, mostly tubular but apical cell tapered to a flexuous tip, walls thin and smooth to minutely ornamented, the roughness hyaline, content orange-ochraceous and homogeneous in Melzer's, hyaline to weakly yellowish in KOH. Hyphae of subcutis and adjacent context hyphae orange-ochraceous in Melzer's. Clamp connections absent (for all practical purposes—only 1 clamp seen).

Habit, habitat, and distribution.—Gregarious under hemlock, Canyon Lake, Huron Mountains, Marquette County, August 10-13, 1968, J. Ammirati 2188 and A. H. Smith 76005. Type, Ammirati 2188.

Observations.—This is a very robust species with a much more bulbous stipe and darker pileus than in *B. pallidus* to which it is otherwise related. Also it occurs under hemlock instead of oak or other hardwoods. *B. alutaceus* Morgan also appears closely related but is distinct in having smaller spores, and its context does not stain blue when cut.

163. Boletus badius Fries

Elench. Fung. p. 126. 1828

Xerocomus badius (Fries) Kuhner ex Gilbert, Bot. Ser. 17, fasc. i-iv:195. 1926.
Boletus castaneus badius Fries, Syst. Mycol. 1:392. 1821.

Illus. Pls. 119-20.

Pileus 3-10 cm broad, pulvinate to convex becoming broadly convex or nearly plane, in moist weather surface viscid to thinly glutinous but soon dry and unpolished, in age at times somewhat pruinose; yellow-brown to vinaceous-red ("terra-cotta") when young, becoming darker yellow-brown to chestnut or bay, sometimes slightly olive-tinted (color rather variable). Context whitish young, soon yellow near the tubes, pinkish under the cutis, staining weakly vinaceous when cut in the area above the stipe, rarely bluish when injured; odor slight, taste sour to mild, with $FeSO_4$ dull bluish green, with KOH on tubes golden brown; NH_4OH olive around the spot of application of the drop.

Tubes olive-yellow, adnate to depressed-adnate or somewhat decurrent, about 8-15 mm deep, pale yellow to greenish yellow; pores small at first but in age 1-2 per mm, isodiametric to angular, blue when injured and finally brownish.

Stipe 4-9 cm long, 10-22 mm thick, equal, solid, cortex yellowish near apex, pinkish red below, pallid, becoming brownish around larval tunnels; surface yellowish pallid at the very first at the apex, dull rose-red below, slowly brownish around the base, often with a pallid bloom at first, rarely slightly reticulate above from tube lines, surface when fresh somewhat pruinose.

Spore deposit olive-brown; spores 10-14 X 4-5 μ, in face view suboblong to slightly ventricose, in profile narrowly inequilateral from a slight suprahilar depression, yellow in KOH individually, greenish revived in Melzer's, smooth, wall very slightly thickened.

Basidia 4-spored, 23-30 X 7-9 μ, yellowish in KOH and in Melzer's. Pleurocystidia scattered, 30-55 X 10-14 μ, fusoid-ventricose, filled with dark yellow-brown pigment as revived in KOH or Melzer's. Pileus cutis a tangled mass of hyphae 5-10 μ wide, with bright golden yellow content in KOH, walls smooth and thin; end-cells not appreciably differentiated, no amyloid particles or areas of hyphae seen and clamp connections absent.

Habit, habitat, and distribution.—Solitary to scattered in beech-maple and hemlock-birch forests during the summer and early fall, common during most seasons. The best fruitings seen were near the Upper Falls in Tahquamenon Falls State Park.

Observations.—As recognized here this species is collective and deserves further study. We have spore sizes of 9-11 X 3-3.5 μ, 10-14 X 4-5 μ, and some 11-15 X 4-5.5 μ to go along with the sizes given by Singer (1945) who found them 11.5-18.5 (24) X 4-5 μ and (from Austria) 12.5-16.5 X 5-5.6 μ. In order to resolve this situation, however, we must have complete data on all collections in the fresh as well as dried condition and on chemical features from fresh basidiocarps. In the variant which gives the olive reaction with NH_4OH and stains weakly vinaceous on the context when fresh if injured, the spores are 11-14 X 4-5 μ.

Material examined.—Alger: Mains 32-232. Cheboygan: Smith 25921, 36507, 37553; Thiers 605, 649, 732. Chippewa: Smith 72066, 72440, 74493. Emmet: Shaffer 1614; Smith 25807, 57395, 62964, 74390, 74502, 74531. Houghton: Kauffman 1906. Luce: Shaffer 1962; N. J. Smith 158A; A. H. Smith 37085, 37179, 37229, 37434, 37468, 37822, 38145, 38341, 38517, 38658, 39285, 41651, 44110, 50238,

57387, 57901, 63817, 75001; Thiers 3138, 3864, 4069, 4341. Mackinac: Smith 44121. Marquette: Bartelli 2512; Pennington 1906; Smith
72710. Montmorency: Smith 74731, 74895, 75028. Oakland: Smith
7119. Washtenaw: Smith 18484, 64144.

164. Boletus glabellus Peck

Ann. Rept. N. Y. State Mus. Nat. Hist. 41:76. 1888

Illus. Figs. 129, 131-33.

"Pileus fleshy, thick, broadly convex or nearly plane, soft, dry,
subglabrous, smoky-buff, flesh white, both it and the tubes staining blue
when injured; tubes nearly plane, adnate, subrotund, ochraceous, tinged
with green; stem subequal, glabrous, even, reddish toward the base,
pallid above with a narrow reddish zone or circumscribing line at the
top, spores oblong, brownish ochraceous, with a tinge of green when
fresh; .004 to .005 in. long, .000016 in. broad. Pileus 3-5 in. broad; stem
1-3 in. long, 5-10 lines thick."

The above is the original description. A study of the type gave the
following data: Spores 10-13 × 4-5 μ (8-11 × 5-7 μ), smooth, lacking a
germ pore; shape in profile obscurely narrowly inequilateral to suboblong, in face view suboblong to obscurely fusiform, the abnormal
spores shorter and wider, color in KOH ochraceous at first and becoming paler, in Melzer's ochraceous to near clay color, wall -0.2 μ thick;
lacking any distinctive amyloid reaction.

Basidia 4-spored, 6-8 μ wide, clavate, yellowish hyaline in KOH.
Pleurocystidia 36-78 × 8-16 μ, fusiform to fusiform-ventricose or ventricose-rostrate, thin-walled, smooth, content dingy ochraceous in KOH but
homogeneous, in Melzer's weakly brownish and with homogeneous content, often with long pedicels.

Tube trama of the *Boletus* subtype, nonamyloid in all parts, very
few globules present in KOH mounts. Pileus cutis a tangled layer of
hyphae 5-11 μ wide, the end-cells cylindric to weakly clavate, smooth or
in KOH with a slight halo, content yellowish in KOH, orange to ochraceous in Melzer's, short cells in filaments not infrequent but these not
inflated. Hyphae of subcutis yellowish in Melzer's. Clamp connections
none.

Observations.—This species is very similar to specimens of *B. pallidus* as these appear in dry weather or when old and may have been
overlooked in Michigan on this account. The small number of abnormal
spores present in mounts of the type may be an important feature—the

same as they appear to be for *T. sordidus*—but more observations are required to establish this. It is close to *B. alutaceus,* which is exactly the position in which Peck placed it. The smoky buff pileus, reddish base of the stipe with a red line at the apex, and lack of reticulation are the main field characters. It may possibly have been confused with *B. badius* as a yellowish brown variant. We have had such a bolete in the state, and it does not react with NH_4OH in the same way as the red variant.

Smith 57612 from Tahquamenon Falls State Park appears to be this species. It was mistaken for a slender specimen of *B. badius.*

165. Boletus albocarneus (Peck) Peck

Bull. N. Y. State Mus. 150:65. 1911

Boletus chrysenteron var. *albocarneus* Peck, Bull. N. Y. State Mus. 54:185. 1901.

"Pileus 1-2.5 in. broad, convex, fleshy, dry, subglabrous, varying from brick-red to bay-red, flesh white, sometimes tinged with red near the surface, rarely rimulose. Tubes rather long, adnate or slightly depressed around the stem, greenish yellow, their mouths small, subrotund, becoming blue to greenish blue when bruised. Stipe 1-2 in. long, 2-4 lines thick, equal or nearly so, solid, subglabrous, colored like or a little paler than the pileus, white within. Taste pleasant."

Spores 12-15 X 4-5 μ, smooth, lacking an apical pore, shape in face view subfusoid to subelliptic, in profile obscurely to somewhat inequilateral, color in KOH yellowish hyaline, in Melzer's merely yellowish, wall about 0.2 μ thick.

Basidia 4-spored, 8-10 μ broad, clavate, hyaline in KOH. Pleurocystidia 30-45 X 9-15 μ, clavate to clavate-mucronate, content ochraceous in KOH and merely ochraceous-brown in Melzer's fading to ochraceous.

Pileus cutis a layer of interwoven hyphae 4-8 μ wide, the end-cells cylindric or weakly cystidioid (rarely over 11 μ broad), cells elongate, walls thin and smooth, content yellowish in KOH or Melzer's and not rounding into globules. Some amyloid debris present in the layer. Hyphae of the subcutis yellowish in Melzer's also. Clamp connections absent.

Observations.—The type specimen gives one the impression that it is the same as *Tylopilus subpunctipes,* but neither Peck's description nor the microscopic data support this conclusion. It is close to *B. badius* but smaller, and was described as having a dry pileus. We have had small specimens of *B. badius* that might key out here. However, we need to sharpen our concept of *B. badius* before placing American species in synonymy with it. *B. albocarneus* is not closely related to *B. chrysenteron*. Our account is based on the original description and Smith's study of the type; we have as yet not recognized the species in the Michigan flora.

166. Boletus pseudosulphureus Kallenbach

Zeitsch. f. Pilzk. 2:225-30. 1923

Illus. Pl. 121.

Pileus 4-9 cm broad, pulvinate to convex to broadly convex; margin incurved slightly and sterile for about 0.5-1 mm; surface dry and velvety when young, becoming merely unpolished and somewhat shining by maturity; bright lemon-yellow overall varying on the disc to bright yellow-brown ("Sudan-brown" to "Brussels-brown") and quickly blue to indigo-black when bruised. Context lemon-yellow to greenish yellow but instantly blue when cut, taste acidulous, odor none, with $FeSO_4$ quickly yellow on blue surfaces, KOH quickly orange on similar surfaces.

Tubes about 1 cm deep, depressed around the stipe, bright to dull yellow, instantly blue to greenish blue when injured; pores small (2-2.5 per mm), round to angular, bright lemon-yellow but instantly blue where bruised.

Stipe 8-12 cm long, 1-1.5 cm thick, equal, solid, yellow within, in age dark red in the base, quickly blue when cut; surface lemon-yellow above a reddish base, when mature sometimes lemon-yellow overall, smooth and merely pruinose, at times paler in lower third, soon greenish blue and then grayish from handling.

Spore deposit "olive-brown" (dark olivaceous); spores 10-14 X 4.5-5 X 5-6 μ, smooth, in face view obtusely fusoid, in profile narrowly inequilateral wall about 1 μ thick as seen in KOH mounts and golden ochraceous, in Melzer's pale dull tan (in groups with a slight grayish shadow).

Basidia 4-spored, 26-34 X 9-11 μ, clavate, hyaline to yellowish in KOH and brighter yellow in Melzer's. Pleurocystidia 44-65 X 8-12 μ, lanceolate to narrowly fusoid, often ventricose below, narrowed above then ventricose again and finally tapered to an acute apex, hyaline to yellowish in KOH and Melzer's, thin-walled, smooth.

Tube trama of very gelatinous divergent hyphae from a central strand. Pileus trichodermium of hyphae 3-6 μ in diameter, with the end-cells tubular and obtuse, content ochraceous in KOH and darker dingy yellow-brown in Melzer's, wall smooth, thin and only subgelatinous. Clamp connections absent.

Habit, habitat, and distribution.—Solitary to scattered in a woods of maple, yellow birch, hemlock, and aspen, or on or near a stream bank with species of ash in the adjacent lowland, West Branch of the Maple River, Emmet County, July.

Observations.—It seems that one or two basidiocarps can be collected in this one area every year. Watling's (1969) description covers the features of our collections remarkably well: the bright yellow color overall, the instant change to blue, the fairly wide spores and non-reticulated stipe together characterize it. The size variations noted by Kallenbach were not represented in our collections.

167. Boletus pulverulentus Opatowski

Wiegmaan's Archiv. Naturgesch. 2:27. 1836

Boletus mutabilis Morgan, Journ. Cincinnati Soc. Nat. Hist. 7:6. 1884.
Ceriomyces cyaneitinctus Murrill, Lloydia 6:225. 1943.

Illus. Pls. 122-23.

Pileus 4-8 (12) cm broad, pulvinate, convex to broadly convex; surface dry, dull to subtomentose, slowly becoming glabrous, often somewhat shiny and tacky to the touch, moist; color dark yellow-brown to blackish brown to dull cinnamon-brown ("mummy-brown" to "bister" to "Prouts-brown" or "Mars-brown"), at times with a dull reddish tinge, NH_4OH on cuticle giving a green flash. Context thick, soft and spongy, yellow but changing to blue so rapidly as to obscure original color; odor and taste not distinctive.

Tubes 6-10 mm deep, adnate to subdecurrent, not depressed, yellow but instantly blue when bruised and slowly sordid brownish; pores large and angular (about 1 mm or more in diameter at maturity), lemon-yellow fresh but instantly blue when bruised.

Stipe 4-8 cm long, (0.7) 1-2.5 (3) cm thick, equal or nearly so, solid, context yellow instantly turning blue when cut; surface bright yellow to orange-yellow above, reddish brown and pubescent below, apex at first minutely powdery, in age merely pruinose, not truly reticulate but often with some raised lines, base blackish brown from handling.

Spore deposit dark olivaceous to olive-brown; spores 11-14 (15) X 4.5-6 μ, ventricose in face view, inequilateral in profile, dingy ochraceous in KOH, dull tawny in Melzer's, no fleeting-amyloid reaction observed in most basidiocarps.

Basidia 4-spored, clavate, 22-28 X 6-8 (10) μ, yellowish in KOH to hyaline. Pleurocystidia abundant, 32-46 (60) X 8-13 (16) μ, fusoid-ventricose, neck 3-5 μ thick, apex subacute, thin-walled, smooth, in Melzer's with a dextrinoid (dark red) amorphous content. Caulocystidia (40) 60-120 X 8-15 μ, versiform, narrowly ventricose toward base and tapered to a long flexuous proliferated neck with an obtuse apex 5-9 μ broad, or clavate, or as hyphal segments shaped like a fusoid-ventricose cystidium; some elements with dextrinoid content in Melzer's as in the pleurocystidia.

Tube trama gelatinous and the hyphae somewhat divergent. Cutis of pileus a tangled layer of yellow (in KOH) hyphae 3-7 μ broad and the end-cells not enlarged, in KOH with adhering yellow-brown debris and particles, with rusty brown incrustations as seen in Melzer's. Context hyphae 6-10 μ broad, hyaline in KOH, content yellow to hyaline in Melzer's. Clamp connections absent. No amyloid hyphae observed but particles or aggregations of amyloid (?) material were seen in one collection (Smith 75888).

Habit, habitat, and distribution.—Solitary to scattered on moist soil in woods, often on banks or hillsides, summer and early fall. We have never found it in large quantities. Known throughout the state.

Observations.—The dark-colored content of the cystidia in Melzer's is unusual in this group. This feature in addition to the blackish brown pileus, the almost instantaneous change to blue and the relatively broad pores characterize the species. The septate caulocystidia, the green flash on the pileus with NH_4OH make it one of the most readily identified of all boletes. It is intermediate between *B. subtomentosus* and *B. badius* but with many features not present in either of these.

Section PSEUDOLECCINUM Smith & Thiers, sect. nov.

Pileus glaber, udus vel viscidus; stipes furfuraceo-squamulosus vel punctatus, squamulae vel punctae pallidae vel incarnatae vel luteae sed non brunnescens; sporae leves.

Type species: *Boletus rubropunctus* Peck.

KEY TO SPECIES

1. Spores 17-21 X 5.5-7.5 μ; stipe with red dots . . . . *B. rubropunctus*
1. Spores smaller if stipe has reddish dots. 2

    2. Pileus viscid; stipe ornamented with reddish dots in age . . . . *B. longicurvipes*
    2. Pileus moist to dry, stipe faintly pruinose to furfuraceous but the ornamentation weakly colored 3

3. Spores 14-18 × 5-6.5 μ; stipe pruinose. . . . . *B. sphaerocystis*
3. Spores 3-5 μ wide and less than 15 μ long 4

    4. Terminal cells (or many of them) in the form of a tubular extension
      from the parent cell, often up to 100 μ long . . . . *B. hortoni*
    4. Terminal cells of cuticular layer not elongated as in above choice
    . . . . *B. subglabripes*

168. Boletus longicurvipes Snell & Smith

Journ. Elisha Mitchell Sci. Soc. 56:325. 1940

Illus. Pl. 125 (lower).

Pileus 1.5-6 cm broad, obtuse to convex becoming broadly expanded; surface viscid from a tenacious separable pellicle, glabrous; color reddish orange to tawny or dingy ochraceous, often with a reddish brown reticulum, at times olivaceous in age. Context soft, white, becoming pallid yellowish, unchanging when injured, odor and taste mild; $FeSO_4$ no reaction.

Tubes 9-12 mm deep, adnate, becoming depressed, not changing when bruised; pores small (about 2 per mm), pale yellow but in age sordid greenish gray.

Stipe 5-9 cm long, 8-15 mm thick, equal or gradually enlarged downward, solid, white within becoming yellowish, unchanging; surface whitish above, pale pinkish brown below, furfuraceous to scabrous-punctate (or appearing subreticulate from breaking up of the fibrillose coating), the ornamentation becoming dull reddish in age or on drying.

Spores 13-17 × 4-5 μ, smooth, narrowly subfusiform to oblong in face view, inequilateral in profile, dingy honey color in KOH, in Melzer's near ochraceous-tawny, with a well-defined mucilaginous outer sheath, the wall 0.5 μ thick and with a thin spot at apex.

Basidia 4-spored, 20-26 × 9-12 μ, clavate, hyaline in KOH, yellowish in Melzer's. Pleurocystidia scattered, 33-50 × 9-14 μ, fusoid-ventricose, hyaline in KOH and yellowish in Melzer's, content not distinctively colored.

Pileus cuticle a turf of narrowly clavate hyphal ends 4-10 μ broad with a layer of slime above them, ochraceous in KOH. Hyphae of cutis yellowish in Melzer's; those of the subcutis the same color or paler. Clamp connections none.

Habit, habitat, and distribution.—Known in Michigan from a collection by Bill Isaacs at Muskegon, September 15, 1963, and another by Victor Potter from near Ithaca.

Observations.—The stipe ornamentation shows up as reddish dots in old or dried material, but it does not darken as in *Leccinum.* Singer mentioned (1947) that the end-cells of the cuticular hyphae may rarely be ellipsoid or even globose. We saw some cells approaching these shapes, but in ours the cuticle was in the form of a turf with the end-cells approaching cells as he described them. We did not obtain a positive $FeSO_4$ reaction on the white context of the pileus. Because this species has been considered a synonym of *B. rubropunctus* by some, the type of the latter was studied and the following data recorded (figs. 95, 97):

Spores 17-21 X 5.5-7 μ, smooth, with a minute thin spot at the apex, shape in face view elongate-subfusiform, in profile elongate and obscurely to somewhat inequilateral, color in KOH dingy ochraceous to brownish, in Melzer's pale to dark tawny, wall mostly 0.5 μ thick (in KOH mounts) but in some up to 1.5 μ thick, in Melzer's mostly 0.4-0.5 μ thick.

Basidia 4-spored (a few 2-spored observed), short-clavate, 26-32 X 10-15 μ, hyaline in KOH, not distinctive in Melzer's. Pleurocystidia scattered, 34-47 X 10-14 μ, fusoid-ventricose, apex subacute, walls thin, smooth and hyaline, content "empty" as mounted in KOH or Melzer's. Cheilocystidia numerous, 18-35 X 8-12 μ, clavate, clavate-mucronate or fusoid-ventricose, content yellowish in KOH, readily gelatinizing. Caulocystidia in patches, (28) 35-60 X (8) 10-18 μ, clavate, clavate-mucronate or fusoid-ventricose (the latter not numerous), thin-walled, smooth, hyaline in KOH and content "empty" (both in KOH and Melzer's).

Tube trama of the *Boletus* type, nonamyloid in all parts and laticiferous elements not conspicuous. Pileus cutis a thick layer of interwoven hyphae 4-10 μ wide, both long and short cells present in varying number and the cells tending to disarticulate, hyphae tubular or some cells slightly inflated, walls thin and smooth or some roughened slightly but lacking pigmented deposits; content yellow in KOH and Melzer's but fading, end-cells cylindric to narrowly clavate or cystidioid. Hyphae of subcutis and adjacent context 8-15 μ wide and content orange-cinnabar in Melzer's. Clamp connections doubtfully present (some hyphae of stipe have numerous false clamps some of which could have been complete but one could not be sure the clamp-branch had fused with the penultimate cell).

Notes: Several points of difference readily stand out. Most important, the spore sizes of *B. rubropunctus* and *B. longicurvipes* clearly indicate that 2 species are involved. It is doubtful if *B. rubropunctus* is viscid to anything like the degree shown by *B. longicurvipes.* Finally, in

B. rubropunctus the hyphae of the subcutis are orange-cinnabar as revived in Melzer's, and false clamps are numerous on the hyphae of the stipe.

169. Boletus subglabripes Peck

Bull. N. Y. State Mus. 8:112. 1889

Boletus flavipes Peck, Rept. N. Y. State Mus. 39:42. 1886 (non Berk. 1854).
Suillus subglabripes (Peck) Kuntze, Rev. Gen. Pl. 3(2):536. 1898.
Ceriomyces subglabripes (Peck) Murrill, Mycologia 1:153. 1909.
Leccinum subglabripes (Peck) Singer, Mycologia 37:799. 1945.
Krombholzia subscabripes Singer, Rev. de Mycol. 3:188. 1938, n. nud.

Illus. Pl. 124.

Pileus 4.5-10 cm broad, convex becoming broadly convex to plano-convex or broadly umbonate, glabrous, dry to moist, dull, often pitted or somewhat rugose, yellow to ochraceous to clay color or more rarely dull cinnamon ("warm buff" to "yellow-ocher," "clay color," or "tawny-olive" to "ochraceous-orange," "buckthorn-brown," or rarely "Sayal-brown"); margin flaring at times, entire, often decurved. Context thick (up to 2 cm), firm, whitish to pale yellow, occasionally irregular areas changing to olive-buff or "citron-yellow" in old specimens where damaged, typically unchanging when injured (rarely slightly bluish); odor not distinctive, taste mild to slightly acid, sometimes sweetish.

Tubes deeply and broadly depressed around the stipe, 10-15 mm deep, yellow ("barium-yellow") in sectioned pilei; pores yellow ("citron-yellow" to "sulphur-yellow"), in age wax-yellow to amber-yellow, unchanging when injured, not stuffed when young, small, round, about 2 per mm, many depressed areas present.

Stipe 5-10 cm long, 10-20 mm thick, equal to ventricose, tapering slightly toward the base or apex, solid, surface dry, furfuraceous to scabrous or fibrillose, never reticulate, whitish within at the base but pale bright yellow in the remainder, cortex sometimes staining reddish in old basidiocarps, surface pale to bright yellow, occasionally with reddish stains at the base.

Spore deposit olive to olive-ochraceous brown; spores 11-14 X 3-5 μ, smooth, in face view narrowly fusoid, in profile somewhat narrowly inequilateral, pale greenish yellow revived in KOH, merely yellowish in Melzer's, wall less than 0.2 μ thick and showing no evidence of an apical pore.

Basidia 4-spored, 18-26 X 8-10 μ clavate, hyaline to yellowish revived in KOH, yellowish in Melzer's. Pleurocystidia rare to scattered,

32-54 X 8-15 μ, fusoid-ventricose, thin-walled and smooth, apex acute, content hyaline to yellowish when revived in KOH. Cheilocystidia numerous, 20-32 X 8-12 μ, fusoid to fusoid-ventricose, hyaline to yellowish in KOH. Caulohymenium present in patches; caulocystidia clavate to subglobose, 35-60 X 10-30 μ, thin-walled, smooth, soon hyaline in KOH mounts.

Hymenophoral trama more or less divergent from a distinct central strand of nongelatinous hyphae, the divergent hyphae gelatinous, central strand hyaline to brownish in KOH; subhymenium subcellular to indistinct, of hyaline thin-walled cells.

Pileus cuticle a compact trichodermium the elements of which have the distal 2-5 cells inflated to sphaerocyst proportions and some so closely packed together as to produce a cellular layer, the cells (6) 10-24 μ in diameter, more or less isodiametric and with yellow walls as revived in KOH or in Melzer's. Hyphae of the context loosely interwoven, thin-walled, smooth, yellowish to brownish in Melzer's but content not distinctly colored. Clamp connections absent.

Habit, habitat, and distribution.—Scattered to gregarious under hardwoods in a mixed forest or in pure stands of hardwoods, later summer and early fall throughout the Great Lakes region, often abundant.

Observations.—This is reported as an edible species in the literature, but we have not tried it. It is abundant enough to warrant consideration as a worthwhile species if the flavor justifies it. Singer places this species in *Leccinum,* but Smith and Thiers (1968) pointed out that it lacks the major features of that genus. It lacks caulocystidia which darken in age and the hymenophore is yellow. A large number of boletes have caulocystidia so the mere presence of this type of cell cannot be argued as indicating *Leccinum;* also, very few species of *Leccinum* have a yellow hymenophore when young, so the present species is atypical to say the least as far as this feature goes. As amply pointed out in this work, many boletes in various genera have inflated cells in the elements of the pileus cutis.

170. Boletus sphaerocystis Smith & Thiers, sp. nov.

Pileus 5-7 cm latus, convexus, impolitus, cinnamomeus vel argillaceus. Contextus pallide luteus. Tubuli lutei, tactu subaurantiaci. Stipes 5-6 cm longus, 1-1.5 cm crassus, laete luteus, deorsum rufo-virgatus. Sporae 14-18 X 5-6 μ. Specimen typicum in Herb. Univ. Mich. conservatum est; prope Ann Arbor, September 15, 1961, legit Smith 64279.

Pileus 5-7 cm broad, convex becoming plane to broadly convex, surface dry and unpolished but matted down like felt under a lens (not velvety), surface very uneven; color dingy cinnamon to clay color and retaining a strong ochraceous tone on drying. Context yellowish, brighter yellow as dried, bright yellow near tubes when fresh, pallid near cuticle, no color change when basidiocarp is sectioned, taste acid, odor mild, KOH orange on tubes and flesh, $FeSO_4$ olive on tubes; in age reddish around larval tunnels.

Tubes dull olive, 1 cm long, depressed around the stipe, unchanging when injured; pores small (about 2 per mm), honey-yellow but often stained orange or finally brownish but not staining blue anywhere.

Stipe 5-6 cm long, 1-1.5 cm thick at base, solid, lemon-chrome within, red-brown around wormholes; surface somewhat pruinose, lemon-chrome at apex, toned reddish in lower portion or variegated yellow, red and brown, at apex subreticulate from decurrent tubes, rather uneven downward.

Spores 14-18 X 5-6.5 μ, smooth, with a thin spot at apex (appearing as a pore only in giant spores), smooth, wall thin (-0.2 μ thick), in face view fusoid, in profile elongate-inequilateral, the suprahilar depression shallow to elongate, color yellowish to olive-ochraceous revived in KOH, in Melzer's reddish tawny for the most part (moderately dextrinoid—often only in the distal half).

Basidia 4-spored, 23-31 X 7-10 μ, yellowish to hyaline revived in KOH, yellowish in Melzer's. Pleurocystidia and cheilocystidia—none found.

Tube trama of the bilateral *Boletus* type but hyphal divergence not pronounced, all hyphae having thin hyaline walls. Pileus cuticle a trichodermium with the elements in the form of chains of globose cells 10-55 μ in diameter or a few terminal cells clavate, the content yellowish to hyaline in KOH and orange-brown in Melzer's, walls thin. Context hyphae not distinctive in either KOH or Melzer's. Clamp connections absent.

Habit, habitat, and distribution.—Scattered under oaks, on a golf course near Ann Arbor, September 15, 1961, Smith 64279.

Observations.—This species is somewhat similar to *B. illudens*, but has a true trichodermium of upright hyphal elements composed mostly of globose to greatly inflated cells—the largest we have yet seen in a bolete cuticle. The pores are small as compared to *B. illudens*, and the spores are larger.

171. **Boletus hortonii**, nom. nov.

Boletus subglabripes var. *corrugis* Peck, Bull. N. Y. State Mus. 2(8):112. 1889.

Pileus 4-12 cm broad, obtusely umbonate becoming convex, surface dry and under a lens pubescent, uneven to corrugated or alveolate, colors as in *B. subglabripes* or merely dingy ochraceous with a reddish undertone, drying brownish with a slight Isabella color overtone. Context yellow, when cut very slowly and weakly staining bluish in some collections. Tubes about 8 mm deep, yellow, when cut slowly and weakly staining bluish; pores yellow and very minute, slowly bluish when bruised in some collections. Stipe 6-10 cm long, 10-20 mm thick, smooth, narrowly clavate, solid, yellow within; surface pale yellow, naked to faintly pruinose, around the clavate portion slightly lateritious, not rooting.

Spores 12-15 × 3.5-4.5 μ, smooth, narrowly inequilateral in profile, almost navicular in face view, yellow in KOH, yellow to pale tawny in Melzer's, wall less than 0.5 μ thick.

Basidia 4-spored, 26-30 × 7-11 μ when sporulating, yellow in KOH or in Melzer's. Pleurocystidia rare to scattered, 32-45 × 7-12 μ, fusoid-ventricose, apex obtuse, smooth, thin-walled. Cheilocystidia basidiole-like to fusoid-ventricose, 24-36 × 5-10 μ, yellowish hyaline in KOH.

Pileus cuticle a tangled trichodermium, the hyphae with the 2-4 cells back from the apical cell variously inflated (some up to 20 μ wide), the apical cell long or short but essentially tubular and 6-10 μ wide, up to 100 μ long, at times with secondary septa, slightly narrowed to apex, smooth, thin-walled, yellowish hyaline in KOH, content merely yellowish in Melzer's. Clamp connections none.

Habit, habitat, and distribution. —Solitary in moist hardwoods, Ann Arbor, Hoseney 1301. The Michigan collection consisted of young material.

Observations. —Peck's type has been studied and the long tubular extensions were found on the terminal inflated cells of the trichodermial elements; the spores were as given above. Singer's (1947, p. 116) comments on this variety should be disregarded. Smith studied Peck's type of variety *corrugis* and found the features as given in the above description. Singer commented on the type being lost, but Stanley Smith located it for A. H. Smith, who then made his study of it. Obviously, Singer did not see the type and his comment: "However, the original description of Peck's is all but convincing," is not relevant to the problem. The spore features alone distinguish *B. hortonii* from *L. rugosiceps,* to say nothing of the tubular cells at the extremities of many of the trichodermial elements. Singer undoubtedly studied additional specimens by Peck which had been misidentified.

It is interesting to note that there is a weak tendency in this variety to stain blue just as there apparently is in some collections of the type variety. It is also interesting that in the Michigan material, at least, the stipe was practically glabrous. Thus, when the close relatives of *B. subglabripes* are studied in detail, it is found that as a group they lack the diagnostic features of the genus *Leccinum.*

Section BOLETUS

Stipe finely to conspicuously reticulate at least at the apex; if stipe not reticulate then pores red to dark brown before maturity; basidiocarps often very robust (but see section *Subtomentosi* in which some species have a reticulate stipe at times, such as *B. affinis, B. sensibilis, B. tenax, B. pallidus, B. mirabilis,* etc.).

KEY TO SUBSECTIONS

1. Tubes yellow when young; taste usually bitter to disagreeable; stipe finely reticulate over apical portion at least . . . . Subsect. *Calopodes*
1. Not as above 2
    2. Pores some shade of red to dark yellow-brown *before* maturity
    . . . . Subsect. *Luridi*
    2. Pores not colored differently from the sides as indicated in above choice 3
3. Pleurocystidia with distinctly yellow content in KOH . . . . . Subsect. *Reticulati*
3. Cystidia not as above, content if colored brown to reddish, but mostly lacking a colored content . . . . Subsect. *Boleti*

Subsection CALOPODES (Fries), stat. nov.

Section *Calopodes* Fries

Epicr. Syst. Myc. p. 416. 1838

Type species: *Boletus calopus.*

KEY TO SPECIES

1. Pileus whitish to pale buff to dark olivaceous 2
1. Pileus red to reddish at least when young 3

2. Spores 10-13 μ long; pileus whitish when mature; stipe bright red,
    usually equal and slender . . . .*B. inedulis*
2. Spores 13-19 μ long; pileus dark olive to olive-brown; stipe typically
    red over basal area and clavate at first . . . .*B. calopus*

3. Pileus rich rose-red; pileus epicutis a trichodermium. . . . .*B. speciosus*
3. Pileus more ferruginous-red and fading readily—finally nearly tan; pileus
    cuticle a layer of appressed-interwoven hyphae . . . .*B. pseudopeckii*

172. Boletus calopus Fries

Syst. Mycol. 1:390. 1821

Illus. Pls. 126-27.

Pileus 10-26 (50) cm broad, convex with an incurved somewhat
cottony-fibrillose margin, broadly convex to flat in age, surface dry,
matted-fibrillose when young, in age unpolished or subtomentose and
usually areolate-rimose, "buffy brown" (olive-brown) to "deep olive-
buff" (paler), becoming "Saccardo's umber" (dark yellow-brown) in age,
often with a persistent sterile margin. Context yellow but gradually
shading to whitish, instantly turning bright blue when injured, thick,
firm at first but punky in age, taste bitter, odor slight or none, with
KOH brownish on context; with $FeSO_4$ olivaceous.

Tubes pale yellow when young and not stuffed, gradually more
greenish yellow in age, depressed-adnate and with decurrent lines on the
stipe, 15 mm or more deep; pores small and round at first but finally
angular and about 1 mm broad, yellow, bright blue when bruised, the
change rather rapid.

Stipe (6) 10-15 (20) cm long, (2) 3-7 cm thick, bulbous when
young in some to nearly equal in others, solid, yellow to pallid within
and instantly blue when cut, surface yellow to orange-yellow and unpolished below, base near ground line delicately pink, soon blue and
then blackish where handled; surface above finely but distinctly reticulate, pale yellow to pallid-yellow, rarely reddish in the upper portion.

Spores olive-brown in deposit; 13-19 X 5-6 μ, smooth, subventricose in face view, narrowly somewhat inequilateral in profile, the suprahilar depression broad, relatively thick-walled, very pale dingy ochraceous to tawny in Melzer's, merely pale ochraceous in KOH.

Basidia 4-spored, 24-30 X 9-12 μ, yellowish both in KOH and
Melzer's. Pleurocystidia 40-60 X 9-13 μ, fusoid-ventricose, thin-walled,
content yellowish in KOH, yellowish in Melzer's. Cheilocystidia clavate
to fusoid and strongly ochraceous in KOH and Melzer's.

Tube trama of more or less parallel hyphae diverging to subhymenium, hyaline to yellow in KOH and in Melzer's often with amyloid septa, subhymenium indistinct. Pileus trichodermium of hyphae with dull yellow-brown incrustations as seen in KOH mounts, the end-cells tubular to cystidioid and also with incrustations, the subcutis and context hyaline in KOH; in Melzer's the hyphae of the trichodermium incrusted and the hyphal walls mostly amyloid (the layer blackish in sections mounted in Melzer's), the hyphae 6-11 μ or more wide and the cell often somewhat enlarged but sphaerocysts not seen. Hyphae of the subcutis and context with pale to dark orange-brown content in Melzer's and the septa showing as amyloid rings (with a hole in the middle). Clamp connections absent.

Habit, habitat, and distribution.—Solitary to gregarious in the virgin conifer forests of the Pacific Northwest, where it is common from late August to October. In Michigan it is rare—it has been reported but as yet we have not located a correctly identified specimen.

Observations.—This species is one of the most easily identified of all boletes if one has a microscope and Melzer's reagent. We have noted in many boletes that the septa of the hyphae in the tube trama were more refractive than the side walls, but this is one of the few species that we know in which the septa are violet in iodine and hence stand out prominently in Melzer's mounts. Miller and Watling (1968) have recently discussed this reaction. In addition to amyloid septa, the trichodermial elements are heavily incrusted and amyloid. The color of the stipe varies a great deal from brilliant red over a large portion to no red at all, and in the Pacific Northwest there is a second species which regularly shows no red at all, *B. coniferarum* Dick & Snell. The basidiocarps develop slowly in the cool mountain climate, and the pattern of checking of the pileus surface varies accordingly. In Idaho one finds mostly coarsely areolate basidiocarps, but in the fall in the Cascades they occur frequently with the cuticle unbroken. The reticulum is usually pallid and one may have to look carefully to see it, but a few specimens with red reticulum have been encountered—all of them old.

173. Boletus inedulis (Murrill) Murrill

Mycologia 30:523. 1938

Ceriomyces inedulis Murrill, Mycologia 30:523. 1938.

Illus. Pls. 128-29.

Pileus 4-11 cm broad, convex expanding to broadly convex or nearly plane; surface dry, densely cottony tomentose at first, the layer breaking up into patches (areolae) in age, often conspicuously areolate with white flesh showing prominently; color when young whitish to pale cinereous to avellaneous, often with the areolae wood-brown. Context thick, soon soft and spongy, yellowish to white, changing to blue when cut or bruised, odor none, taste bitter, with $FeSO_4$ no reaction.

Tubes slightly depressed around the stipe, 1-1.5 cm long, pale greenish yellow becoming blue and finally orange brownish when injured; pores small and round, 1.5-2 per mm, pale yellow, blue when bruised.

Stipe 6-9 (12) cm long, 8-20 mm thick, equal to slightly enlarged downward, not rooting, yellow within, staining blue; surface reticulate over upper half or three-fourths to nearly smooth except at the apex, granular-pruinose overall at least when young; color yellow over apical region, lower down "grenadine-pink," base soon sordid reddish brown or blackish from handling.

Spore deposit olive-brown; spores 9-12 (14) X 3.3-4.5 μ, smooth, lacking an apical pore, in profile slightly sway-backed to inequilateral, suprahilar depression slight to prominent, ochraceous in KOH, in Melzer's pale ochraceous to dextrinoid.

Basidia 4-spored, 20-26 X 9-10 μ, short-clavate, yellowish in KOH and slightly more so in Melzer's. Pleurocystidia scattered, 30-42 X 8-12 μ, ventricose with a narrow (2-3 μ) flexuous neck, or fusoid-ventricose with a short neck and subacute apex, both types thin-walled, smooth, lacking distinctive contents in either KOH or Melzer's.

Pileus cutis a tangled trichodermium soon collapsing and hyphae 4-7 μ wide, with the cells tubular throughout, nongelatinous, thin-walled, smooth, or roughened (both occur on single sections), and with tubular end-cells obtuse at apex, in KOH pale cinnamon in epicuticular zone and with walls distinctly roughened by incrusting pigment as seen in Melzer's, content evenly orange-brown and darker in younger pilei, amyloid debris present in sections of old pilei, but no amyloid hyphal walls seen. Clamp connections none.

Habit, habitat, and distribution.—Solitary to gregarious around borders of woodland pools in oak-hickory forests of southern Michigan, but in upland woods during wet periods, July to early September. I have seen it most abundantly in Washtenaw County.

Observations.—This is truly a beautiful species with its rose-red stipe and pallid pileus. With us it is typically a slender-stiped species. It is readily distinguished from *B. calopus* by its smaller spores almost lacking a suprahilar depression, and lack of amyloid septa in the hyphae. It was confused with *B. calopus* previously in Michigan. The microscopic features of our collections check well with those from a specimen marked "part of type" and distributed by Murrill. Some use the name *B. radicans* for this species, but the stipe of our material is never attenuate-radicate and the pores are not at all like those of *B. subtomentosus*.

Material examined.—Jackson: Smith 18523. Livingston: Smith 64237. Washtenaw: Homola 1060; Hoseney 106, 473, 524, 579, 610; Smith 1677, 18545, 18593, 18647, 18702, 62630, 62859, 64141, 72989, 73003, 73006, 73153.

174. Boletus speciosus Frost

Bull. Buff. Soc. Nat. Sci. 2:101. 1874

Illus. Fig. 98; Pl. 130.

Pileus 8-15 cm broad, hemispheric to convex, becoming broadly convex, margin incurved and even, occasionally with pits and large irregularly shaped depressions; surface dry, unpolished, appearing glabrous at first, becoming matted-fibrillose in age, margin incurved, entire, bright red ("Corinthian-red") to rose-pink or finally duller and vinaceous ("vinaceous-russet") or brighter, or with patches or generally yellowish, yellowish where injured. Context thick, 15-20 mm, firm, becoming punky, pale yellow ("barium-yellow" to "naphthalene-yellow"), quickly turning blue when cut; odor and taste not distinctive; KOH orange, $FeSO_4$ grayish.

Tubes 9-15 mm long, adnate becoming depressed, with decurrent lines or reticulations on the stipe, bright yellow ("empire-yellow" to "pinard-yellow") often becoming "honey-yellow" to "olive-ocher" toward the stipe when older, immediately staining blue when bruised; pores small (2 per mm), lightly stuffed when young, staining blue and then brownish at maturity or at times becoming dingy reddish.

Stipe 5-13 cm long, 1.5-4 cm thick at the apex, equal to clavate to the narrowed pinched-off base, solid, flesh-yellow ("lemon-yellow") toward the base, becoming paler toward the apex; surface finely reticulated over upper half or overall, colored light yellow ("pinard-yellow" to "empire-yellow") above, reddish toward the base and in damaged areas, quickly turning blue when injured, in the base often chrome-yellow.

Spore deposit olive-brown; spores 11-15 × 3-4 μ, smooth, narrowly oblong to subfusoid with subacute ends in face view, narrowly inequilateral to subcylindric in profile view, pale ochraceous in KOH, yellow to pale tawny in Melzer's, lacking apical differentiation, wall thin (-0.2 μ).

Basidia 4-spored, 24-27 X 8-9 μ, clavate, yellow in KOH or Melzer's. Pleurocystidia 33-45 X 8-12 μ, narrowly ventricose to fusoid with acute apex, walls sometimes flexuous, hyaline to yellowish in KOH or Melzer's, content not distinctive, thin-walled, smooth. Cheilocystidia similar to pleurocystidia but smaller and varying more toward clavate and averaging broader.

Tube trama divergent and gelatinous, the hyphae hyaline, thin-walled and smooth with some dextrinoid debris along the hymenium. Pileus cuticle matted down into a layer of interwoven hyphae, hyphae 3-6 (8) μ wide, the hyphal cells tubular, the end-cells tubular and obtuse, at times with a slight amount of hyaline incrusting material or wall roughened by irregular adhesions, hyaline in KOH, but most hyphae smooth, in Melzer's smooth and the cell content ochraceous to orange brownish, amyloid debris present in the layer as well as some that is dextrinoid. Hyphae of context thin-walled and smooth, as seen in Melzer's having dull orange to ochraceous content. Laticiferous elements present, 4-8 μ wide and content ochraceous in either KOH or Melzer's. Clamp connections absent.

Habit, habitat, and distribution.—Solitary to scattered in open hardwoods and mixed forests, often near woodland pools or moist depressions in the woods, late summer or early fall, rare but most frequent in the Lower Peninsula.

Observations.—This is one of our most beautiful species. It is most closely related to *B. regius.* A critical comparison of these remains to be made (see Thiers, 1967).

In our Michigan collections the rose-pink is still evident, but in the type it has disappeared. The specimens described above agree with the type in having identical very narrow spores, a pileus cutis of appressed hyphae 3-7 μ wide, and some amyloid and some dextrinoid debris in the layer between the hyphae, in lacking cystidioid or otherwise distinctly differentiated end-cells on the cutis hyphae, in staining blue when injured, and in having rose tints lower down on the stipe.

The specimen studied is a Frost specimen sent to Peck and is in the Peck collections at Albany, New York, and is here designated as lectotype since Frost did not designate a specimen as the type. The account of its microscopic characters is that given in our description.

175. Boletus pseudopeckii Smith & Thiers, sp. nov.

Pileus 4-10 cm latus, convexus, demum subplanus, siccus, impolitus, roseus vel lateritius demum sordide luteobrunneus. Contextus crassus, pallide luteus, tactu tarde caerulescens, insipidus. Tubuli circa 10 mm longi, pallide lutei demum laete lutei, tactu caerulescens. Stipes 4-12 cm longus, 1-2 (3) cm crassus, intus luteus, tactu caerulescens, obscure reticulatus, extus luteus vel sursum roseus, saepe deorsum roseus vel purpureus. Sporae 10-14 × 3.5-4.5 μ. Cuticula pileorum hyphae appressi, 3-5 (7) μ diam., tubulosi. Specimen typicum in Herb. Univ. Mich. conservatum est; prope Tahquamenon Falls State Park, Luce County, July 2, 1965, legit Smith 71688.

Pileus 4-10 cm broad, convex becoming broadly convex or nearly plane; surface dry and unpolished, at times finely tomentose, with a narrow sterile margin; color some shade of rose or brick-red, but soon developing a grayish overtone from a very thin fibrillose coating, or in slowly developing basidiocarps "warm sepia" to yellowish brown in button stages. Context thick, pale yellow, slowly staining pale blue on injury, odor and taste not distinctive, with $FeSO_4$ bleaching the blue areas to yellow again.

Tubes about 10 mm deep, when mature depressed around the stipe, grayish yellow to yellow, finally bright yellow, staining blue when injured; pores nearly round and 2-3 per mm, yellow when young, often reddish near stipe in age in some specimens, staining blue where injured.

Stipe 4-12 cm long, 1-2 (3) cm thick, equal, solid, yellow within, finally red around larval tunnels, staining pale blue when cut; surface very finely reticulate over upper half or nearly to the base, yellow over-all at first but often with a red zone at the apex and often pinkish red to purplish red below, later on the base usually yellow, not distinctly rooting but sunken in the substratum at times.

Spore deposit olive-brown; spores 10-14 × 3.5-4.5 μ, narrowly ventricose (boat-shaped) in face view, in profile narrowly inequilateral, smooth, wall slightly thickened, bright yellow in KOH, in Melzer's dull yellowish or with a greenish gray shadow indicating a fleeting-amyloid reaction.

Basidia 4-spored, short-clavate, 20-27 × 9-11 μ, yellowish to hyaline in KOH and Melzer's (but blocks of tubes revived in Melzer's turn bluish to greenish black). Pleurocystidia scattered, 36-52 × 9-13 μ, fusoid-ventricose, neck tapered to a more or less acute apex, walls thin, smooth, content not distinctive in either KOH or Melzer's.

Tube trama of hyphae somewhat divergent from a central strand. Cuticle of pileus an interwoven layer of hyphae 3-5 (7) μ in diameter, ochraceous in KOH and with orange brownish content in Melzer's, the hyphae flexuous and intricately interwoven, smooth or at times with some hyaline adhering debris, walls thin and in KOH the outermost slightly gelatinous, end-cells merely tubular and obtuse. Context of hyaline interwoven hyphae with abundant oleiferous hyphae near the cutis, these ochraceous in KOH and Melzer's. Clamp connections absent. All hyphae inamyloid under the microscope.

Habit, habitat, and distribution.—Gregarious in beech-maple woods, summer and early fall mostly in the Upper Peninsula but known from Washtenaw County also, rather rare.

Observations.—For years this species has passed under the name *B. peckii* in Michigan. It has a strong resemblance to the European *B. appendiculatus,* as Watling noted upon seeing fresh material. Singer (1967) has given a detailed account of the European species in which the cuticle is described as "ein Trichoderm aus fädingen Hyphen, deren Endglieder oft keulig bis fast an der Spitze verdict sind, keine Pallisade bildend." Our species lacks upright hyphae in the cuticle of the pileus, in fact it has a cuticle of interwoven hyphae in which the end-cells are not enlarged. Our bolete also lacks a positive (green) $FeSO_4$ reaction. This combination of features we regard as sufficient to distinguish *B. pseudopeckii* from the European *B. appendiculatus.* From *B. peckii* it differs in its constantly mild taste, a feature tested over a period of twenty years. Also, Frost's description (in Peck, Rept. 29) indicates the stipe as strongly reticulate, a feature never observed in our species. In ours the reticulum is so fine that it takes a hand lens to discern it in freshly matured specimens. Frost also indicated the stipe as red with a yellow apex, but in ours the stipe is yellow at first and develops some reddish tints on maturing. *B. pseudopeckii* differs from *B. speciosus* in the less stable red pigment of the pileus. This is reflected in the brilliant red of the latter at maturity or in age.

176. Boletus peckii Frost in Peck

Rept. N. Y. State Mus. 29:45. 1878

Ceriomyces peckii (Peck) Murrill, Mycologia 1:151. 1909.

Illus. Figs. 63, 65.

The original description is quoted as follows:

"Pileus dry, firm minutely tomentulose, red, fading to buff-brown, the margin usually retaining its color longer than the disk; tubes nearly plane, adnate or slightly decurrent, yellow, turning blue when wounded; stem equal or subventricose, strongly reticulated, red, yellow at the top; spores ochraceous-brown, oblong, .00035'-.0004' long.

"Plant 3-4' high, pileus 2-3' broad, stem 3"-6".

"Found in deciduous woods. Sand Lake. August.

"The stem is generally brighter colored than the pileus and retains its color longer. This species should be referred to the Calopodes."

The following data are from the type (Sand Lake).

Spores 9-12 (13) X 3.5-5 μ, smooth, lacking a pore or thin spot at apex, color in KOH weakly ochraceous, in Melzer's merely yellowish, shape in face view suboblong, in profile suboblong to obscurely inequilateral, suprahilar area in many scarcely flattened; wall about 0.2 μ thick. Occasional large ellipsoid spores 13-14 X 6-7.5 μ present.

Basidia 4-spored, 26-35 X 8-10 μ, clavate, hyaline in KOH and yellowish in Melzer's. Pleurocystidia rare, 28-37 X 7-12 μ, fusoid-ventricose, content "empty" in KOH or Melzer's, thin-walled, smooth, hyaline. Cheilocystidia scattered among basidioles, 18-28 X 5-10 μ, similar in shape to pleurocystidia.

Tube trama of the *Boletus* subtype, inamyloid in all respects; laticiferous elements present but not conspicuous. Pileus cuticle a strong trichodermium, the elements 4-8 (10) μ wide and arranged in a distinct palisade, the cells long and often flexuous but essentially tubular, end-cells elongate and apex mostly obtuse to rounded, tubular to weakly cystidioid, content yellowish in KOH and brownish orange in Melzer's, the lower cells in the elements with pale ochraceous incrustations on the walls or seen revived in KOH but smooth in Melzer's. Hyphae of the subcutis and adjacent context with brownish orange content in Melzer's and walls smooth in that medium. Clamp connections none.

Observations.—The distinguishing features of this species are represented by the strong palisade of the elements of the pileus cuticle having elongate narrow terminal cells, the presence of an ochraceous to clay-colored incrusting material over the walls of many of the hyphae in the subcutis and lower cells of the trichodermial elements, the short spores, the lack of conspicuous pleurocystidia, the red pileus rapidly becoming buff-brown, the more or less persistent reddish colors over the lower part of the stipe and the bitter taste (the last as reported by Coker and Beers, 1943, and by Snell, personal communication).

Such a species as this has not as yet, to our knowledge, been reported from Michigan, but since the name has been used frequently for *B. pseudopeckii*, we include it here as part of the clarification of taxonomic concepts in this group.

Subsection RETICULATI, subsect. nov.

Stipes reticulatus; pleurocystidia cum "KOH" valde lutea.
Typus: *Boletus griseus.*

Stirps GRISEUS

KEY

1. Tubes yellow when young, staining orange-brown . . . . *B. ornatipes*
1. Tubes white when young; staining brown when injured . . . . *B. griseus*

177. Boletus ornatipes Peck

Ann. Rept. N. Y. State Mus. 29:67. 1878

Illus. Pl. 131.

Pileus 4-16 (20) cm broad, obtuse to convex, the margin bent in slightly or straight, expanding to broadly convex or nearly plane; surface dull and unpolished to velvety or minutely tomentose, when moist subviscid to the touch; color when young fuscous to violaceous-fuscous becoming dark olive-gray to olive-buff and later·paler as well as developing a strong lemon tint in the ground color, at times yellow-brown in age, at times appearing to have a slight yellowish pruina. Context thick, chrome-yellow, odor none, taste slightly bitterish, with $FeSO_4$ no reaction, with KOH bleaching the yellow to a paler tone or whitish but in the stipe causing a change to dingy brown.

Tubes lemon-yellow staining orange-yellow, adnate to subdecurrent, rarely depressed, 5-8 (10) mm deep; pores small about 2 per mm round, lemon-yellow staining orange-brown when injured.

Stipe 8-15 cm long, 1-2 (3.5) cm thick above, equal to narrowly clavate, solid, chrome-yellow throughout, becoming darker orange-chrome where injured (finally dingy yellow-brown); surface lacerate-reticulate to base or merely·sublacerate to reticulate, yellow overall. Veil absent.

Spore deposit olive-brown to near bister (dark yellow-brown); spores 9-13 X 3-4 μ, smooth, walls slightly thickened, dingy pallid yellow in KOH or a few ochraceous, in Melzer's pale orange-tan, in profile narrowly inequilateral with a broad suprahilar depression, in face view narrowly oblong to slightly ventricose with apex obtuse.

Basidia 4-spored, 22-30 X 6-8 μ, hyaline to yellowish in KOH. Pleurocystidia abundant, 38-65 X 9-15 μ, ventricose at base with a cylindric neck ending in a subacute apex (rostrate), thin-walled, surface smooth, hyaline in KOH when fresh, as revived in KOH with a strongly pigmented, ochraceous to bister, amorphous mass occupying most of the ventricose part and in Melzer's this body strongly dextrinoid (dark red-brown).

Tube trama gelatinous, somewhat divergent from a central strand. Pileus cutis a trichodermium with the end-cells cylindric and obtuse to somewhat cystidioid, the elements narrow (4-8 μ), hyaline in KOH, thin-walled, smooth or with a slight amount of hyaline adhering debris, in Melzer's the content dingy pale orange-ochraceous. Clamp connections absent. All hyphae nonamyloid.

Habit, habitat, and distribution.—Solitary to cespitose in sandy woods of second growth hardwoods, along roads, on banks, and along the borders of woods, often very abundant after heavy rains in late July or August. It occurs throughout the state.

Observations.—This is one of our most easily recognized boletes, but may be confused at times with yellowish forms of *B. griseus.* However, the latter has whitish tubes when young, and the degree of reticulation on the stipe is often relatively inconspicuous. In *B. griseus* the yellow tones mostly develop in age.

We have not followed the recent trend to consider *Boletus ornatipes* a synonym of *B. retipes.* The original description of *B. retipes* calls for a pileus "sicco, luteo-pulverulento"; a condition never seen in 30 years in Michigan collections. In Singer's (1947) account we do not know from what specimens he recorded his data on cystidia. He states: "Cystidia. . .moderately numerous, little projecting, hyaline to melleous, frequently incrusted at the apex." This information was probably obtained from fresh material. Certainly, this description does not apply to sections revived in KOH or Melzer's. *B. retipes,* as restricted to the type, could be a valid species in *Pulveroboletus,* that is, have at least the rudiments of a veil. Actually, Frost (see Peck, 29th Report) was undoubtedly closer to the truth than anyone else when he suggested a relationship of *B. griseus* to *B. ornatipes:* "Either this or a closely related form is regarded by my friend Mr. C. C. Frost, as a variety of *B. griseus* but the yellow flesh and the tubes, which are also yellow from the first, indicate to my mind a specific difference." We are inclined to accept Frost's idea of a close relationship but, with Peck, recognize both as distinct species. To us *Pulveroboletus* sensu Singer, as boletes go, seems to be a rather artificial assemblage of species.

Material examined.—Cheboygan: Shaffer 2662; Smith 38211, 63784, 64843, 72866, 72857, 74818; Thiers 3469, 3615, 3722, 4122,

4214, 4254, 4434. Emmet: Mains 1932; Smith 37628, 62997, 63010, 63654, 63855; Thiers 3717, 4425. Gratiot: Potter 3801, 10053, 11093, 11781, 12376, 12746, 12923. Houghton: Kauffman 1906; Pennington 1906. Luce: Smith 37253, 39292, 72318, 72380. Mackinac: Smith 26000. Oakland: Smith 72392.

178. Boletus griseus Frost in Peck

Ann. Rept. N. Y. State Mus. 29:45. 1878

Suillus griseus (Frost in Peck) Kuntze, Rev. Gen. Pl. 3(2):535. 1898.
Ceriomyces griseus (Frost in Peck) Murrill, Mycologia 1:145. 1909.
Xerocomus griseus (Frost in Peck) Singer, Ann. Mycol. 40:44. 1942.

Illus. Pl. 132.

Pileus 5-12 (17) cm broad, hemispheric to convex, becoming broadly convex to plane or shallowly depressed, in age margin at times recurved; surface dry, tomentose at first but soon felted-tomentose, often appearing fibrillose scaly to glabrous with age, rimose to areolate in age at times, with a narrow sterile margin; color pale gray overall with blackish appressed fibrils to give a darker effect, often fuscous to drab gray, at times with an undertone of ochraceous tones much as in *B. ornatipes.* Context pallid but soon dingy vinaceous where cut or scarcely changing to brownish, around larval tunnels very dark yellow-brown, no distinctive reaction in KOH, with $FeSO_4$ bluish gray on context and tubes; odor and taste not distinctive or rarely the odor slightly fragrant.

Tubes 8-15 mm long, adnate to decurrent or becoming depressed around the stipe, grayish to pallid young, near "wood-brown" (dark gray-brown) mature, staining brownish where bruised or unchanging; pores 1-2 per mm, round, unchanging or brownish where bruised or unchanging.

Stipe 4-11 cm long, 1-2.5 (4) cm thick, subequal or tapering to the base, often crooked lower down and base pointed, solid, pallid within above, greenish yellow in base when young, cortex soon yellow and in age at times yellow nearly to apex, dark yellow-brown around larval tunnels; surface pallid overall at first, soon yellow at the base and at times in age yellow nearly to apex, reticulate overall, reticulum pallid at first often yellowish to some extent and at times the ridges brownish to blackish finally, rarely showing no yellow at all in the stipe.

Spore deposit "dark olive-buff" when thick and fresh, when air-dried near bister (dark yellow-brown). Spores 9-12 (13) X 3.5-4 μ, oblong to narrowly subfusoid in face view, inequilateral in profile, smooth, dingy pale ochraceous in KOH, pale tawny in Melzer's (not truly dextrinoid).

Basidia 24-30 × 8-9 μ, clavate, 4-spored, hyaline in KOH, yellowish in Melzer's. Pleurocystidia and cheilocystidia similar and scattered to abundant, 38-50 × 8-13 μ, fusoid-ventricose, thin-walled, the ventricose portion in KOH filled with a smoky yellow content, hyaline in the neck and apex, apices subacute to acute, in Melzer's the content of the ventricose part bister.

Tube trama divergent and hyaline (gelatinous ?) in KOH, nonamyloid. Cuticle of pileus a well-formed trichodermium of hyphae 5-10 μ or more wide and at first with smoky yellow content in KOH, some hyaline incrustations present but not a conspicuous feature, cells slightly broader and shorter in upper part of the trichodermial element. Hyphae of context floccose, interwoven, nonamyloid, 6-12 μ wide. Clamp connections absent.

Habit, habitat, and distribution.—Solitary to scattered in grassy open oak woods or in open hardwoods with no grasses present, frequent during wet summers in the oak belt of Michigan.

Observations.—The color of the cystidia in KOH and in Melzer's is a valuable aid in the identification of herbarium specimens, but the same is true also for *B. ornatipes*. At times a strong ochraceous tone pervades the whole basidiocarp, and then the resemblance to *B. ornatipes* is very striking. Since a gradual progression from species with a white hymenophore when young to those with a strongly yellow one has been clearly demonstrated in *Leccinum* it will hardly do to argue that the white hymenophore in *B. griseus* precludes close relationship with *B. ornatipes*. In the summer of 1966 *B. griseus* was found by Smith in abundance at Highlands in Oakland County, and it was noted that the species actually intergrade in the disposition of the yellow pigment.

Subsection LURIDI (Fries) Smith & Thiers, comb. nov.

Section *Luridi* Fries, Epicr. Syst. Myc. p. 417. 1838.

For the features of this group see the key to subsections. The important point to keep in mind in the recognition of this group is that the pores are colored before the basidiocarp reaches maturity. In a number of species throughout the genus, especially in those involved with red pigments, the pores may become somwhat colored by old age.

Type species: *Boletus luridus*.

KEY TO STIRPES

1. Pores yellow-brown to dark reddish brown . . . . .Stirps *Vermiculosoides*
1. Pores orange to red (often dark red) 2

   2. Stipe reticulate from slight to prominent veins of tissue forming a netted pattern. . . . . .Stirps *Luridus*
   2. Stipe not reticulate though the pruina may be so arranged on some as to show the outline of a reticulum. . . . . .Stirps *Subvelutipes*

Stirps VERMICULOSOIDES

KEY

1. Stipe reticulate at least over the apical region . . . . *B. fagicola*
1. Stipe even to the apex but usually pruinose to furfuraceous 2

   2. Spores 9-12 × 3-3.5 μ. . . . . *B. vermiculosoides*
   2. Spores 10-15 μ long 3

3. Odor of dried specimens strong and disagreeable. . . . . *B. subgraveolens*
3. Odor not as above (merely fungoid) . . . .*B. vermiculosus*

179. Boletus vermiculosus Peck

Rept. N. Y. State Mus. 23:130. 1873

Illus. Fig. 103.

"Pileus broadly convex, dry, smooth, or at most minutely tomentose, grayish brown, tinged with red; tubes plane or slightly convex, free, small, round, yellow, with the mouths brownish orange, becoming almost black; stipe equal, solid, smooth paler than the pileus; flesh whitish, changing to blue, as also do the tubes, when cut or bruised.

"Height 3'-4'; breadth of pileus 3'-4', stipe 6" thick.

"Found in woods and open places. New Baltimore and Sandlake, July and August."

The original description is given above. Smith's study of the type (the New Baltimore collection) gave the following data:

Spores 11-15 × 4-5.5 (6) μ, smooth, apex lacking a pore or thin spot; shape in face view narrowly ovate to subfusoid, in profile obscurely to somewhat inequilateral, suprahilar depression shallow to fairly distinct, color in KOH about "buckthorn-brown" (dingy yellow-brown) fading to dull ochraceous, in Melzer's moderately dextrinoid; wall about 0.3 μ thick.

Basidia 4-spored, 24-32 X 7-9 μ, clavate, yellowish hyaline in KOH, yellowish in Melzer's. Pleurocystidia rare, 28-36 X 8-12 μ, fusoid-ventricose, content "empty" in KOH or Melzer's. Cheilocystidia scattered to numerous, 18-32 X 5-9 μ, subfusoid to fusoid-ventricose, content dingy yellowish in KOH or Melzer's. Caulocystidia numerous, 32-40 X 10-15 μ, mucronate to ventricose-rostrate, content when well revived "empty" in KOH or Melzer's but caulohymenium having many cystidia apparently with tawny amorphous content irregularly distributed but these generally poorly revived. All parts totally inamyloid under the microscope.

Tube trama of the *Boletus* subtype, all parts totally inamyloid under the microscope but a distinct fleeting-amyloid reaction present as trama is mounted in Melzer's. Pileus cuticle a distinct trichodermium of narrow hyphae (3-6 μ) with smooth thin walls, content ochraceous in KOH, sparingly septate and flexuous-tubular in outline, in Melzer's the cell content stringy to alveolate or homogeneous, dull orange-ochraceous; end-cells long, tubular and blunt at apex. Hyphae of subcutis ochraceous in Melzer's and lacking dextrinoid incrustations. Clamp connections none.

Observations.—The diagnostic features of this species are the brownish orange pores becoming almost black, the spores 4.5-5.5 (6) μ wide, grayish brown pileus tinged with red, whitish flesh, and ventricose-rostrate to mucronate caulocystidia. In Michigan, however, we have made the error, it now appears, of placing here a species with a distinctly yellow-brown pileus, yellow context, amber-brown (yellow-brown) pores, and spores 3.5-4.5 μ wide, and with larger caulocystidia. We also have now found, in Michigan, a species with reddish brown pores, spores 10-13 X 4-5 μ but the colors of the pileus and context do not "match" perfectly those given in the original description of *B. vermiculosus*. However, for the pileus they are close, and they are about the same as those of the type as dried. If we assume a bleaching effect of age on the flesh of the type (it was wormy when collected), we may disregard the discrepancy in color (whitish). The caulocystidia of our material are of two types, one of which very closely resembles those of the type of *B. vermiculosus*. We assume the second kind would also be found if the type were ample enough to justify extensive sampling. On the basis of our present information we conclude that *B. vermiculosus* as represented by the type of the species is present in the state but rare. The following is a description drawn from fresh specimens:

Pileus 4-12 cm broad, convex becoming broadly convex or nearly plane, the margin even or sometimes lobed, surface dry and velvety at first, in age unpolished, after rains somewhat viscid; color at first a dark brown tinged bay color (dull red), gradually becoming paler and then dull cinnamon ("Verona-brown") or dull reddish cinnamon, finally near "Sayal-brown" (dull cinnamon with little red evident). Context yellow when cut but almost instantly staining blue, odor slight, taste mild, no reaction in $FeSO_4$.

Tubes 1-1.3 cm deep, depressed around the stipe or at first adnate, greenish yellow becoming dull yellow and staining blue instantly when bruised; pores reddish brown, minute, readily staining bluish black.

Stipe 4-8 (9) cm long, 1-2 cm thick, equal or nearly so, solid, yellow within staining blue, drying pale buffy brown except for aging specimens in which the context gradually becomes rose-red; surface dull yellow beneath a brownish pruinose covering, soon staining blackish where handled, base in some slightly tomentose.

Spore deposit dark olive; spores 10-13.5 $\times$ 4-5 μ, smooth, apex lacking special differentiation, in face view suboblong to somewhat fusiform, in profile somewhat inequilateral, wall about 0.3 μ thick, color near tawny in mass (revived in KOH) or individually ochraceous with a tawny tinge, in Melzer's dingy yellow-brown, fleeting-amyloid reaction present.

Basidia 4-spored, 8-11 μ broad, short-clavate, yellowish to hyaline in KOH, yellowish in Melzer's. Pleurocystidia 34-52 $\times$ 9-16 μ, fusoid-ventricose, thin-walled, content ochraceous to hyaline (or rusty brown in Melzer's). Caulocystidia in patches (as pruina), 25-50 $\times$ (5) 7-12 (15) μ, flexuous-filamentose, varying to narrowly clavate, or ventricose mucronate to ventricose-rostrate, filled with yellow pigment readily soluble in KOH with the result that most cystidia are soon hyaline, walls thin and smooth, content not distinctive in well-revived individuals.

Tube trama typical for genus. Pileus cutis of appressed fascicles of hyphae 4-8 μ wide, tubular, thin-walled, smooth, content yellow in KOH and orange-buff to orange-yellow in Melzer's and in the latter medium in some hyphae rounding into pigment globules but mostly remaining granular to homogeneous, end-cells tubular and tips merely blunt, the cells long (4 times or more longer than broad) and not disarticulating. Hyphae of subcutis and adjacent context with yellow to orange-yellow content in Melzer's. Clamp connections absent.

Habit, habitat, and distribution.—Gregarious under beech, Pellston, the hills west of town, Emmet County, August 9, 1968, Smith 75865.

Notes: This species is distinguished by its reddish brown pores, lack of reticulations on the stipe, and the red tones of the fresh pileus. It differs from Peck's original account of *B. vermiculosus* in having yellow flesh and in not being readily infested by insect larvae—if the latter is significant. As Coker and Beers (1943) point out, Peck, in his later work did not think it was significant. It is likely that the original collection was old, as previously mentioned, and that the flesh had faded. It may be significant that all authors since Peck have described the flesh as yellow. *B. vermiculosus* differs from *B. firmus* in having more red in the pileus (it was described as gray for *B. firmus*), and there is no reticulation at the stipe apex.

180. Boletus subgraveolens Smith & Thiers, sp. nov.

Illus. Pl. 133.

Pileus 8-13 cm latus, convexus, demum late depressus, subviscidus, glaber, submaculatus, pallide spadiceus. Contextus alboluteus tactu caeruleus. Tubuli 6-12 mm longi pallide lutei tactu olivaceo-caerulei; pori circa 2 per mm, sordide lutei demum "Sudan-brown" (laete luteo-brunnei) tactu sordide caerulei. Stipes 6-9 cm longus, 2-3 cm crassus, sursum citrinus, deorsum badius. Sporae 10-13 × 3.5-4.5 μ. Specimen typicum in Herb. Univ. Mich. conservatum est; prope Crooked Lake, July 15, 1966, legit Mazzer (Smith 72955).

Pileus 8-13 cm broad, broadly convex to plano-depressed; surface slightly viscid, color evenly "snuff-brown" and drying much the same shade (dull yellow-brown), glabrous, slightly mottled. Context yellowish white ("cartridge-buff"), changing to dull blue when cut, around larval tunnels pale reddish cinnamon, with guaiac indigo-blue (but on stipe no reaction), no reaction with $FeSO_4$, taste mild, odor none or slight but when specimens have dried, the odor is very strong and disagreeable.

Tubes 6-12 mm long, depressed around the stipe, pale yellow, staining dingy bluish green when broken; pores about 2 per mm, dull yellow when very young, soon near "Sudan-brown" (bright yellow-brown) but paler in age, when dried near cinnamon-brown, staining inky blue where bruised with the spots discoloring to brown.

Stipe 6-9 cm long, 2-3 cm thick at apex, narrowed to the base, solid, reddish cinnamon in cortex, dull red in the base, surface lemon-yellow at apex, pallid and pruinose below this, dry, reddish at the base.

Spore deposit olive, drying olive-brown; spores 10-13 × 3.5-4.5 μ, smooth, apical pore not evident, wall –0.2 μ thick, weakly amyloid when fresh and soon fading, as revived nonamyloid and dingy yellowish brown, narrowly subfusiform in face view, narrowly elongate-inequi-lateral in profile, pale ochraceous revived in KOH.

Basidia 4-spored, 18-27 × 6-8 μ, clavate, yellowish to hyaline in KOH and Melzer's. Pleurocystidia 28-42 × (6) 7-11 μ, narrowly fusoid to fusoid-ventricose, often only the neck projecting, neck often tapered to an acute apex, content hyaline in KOH. Cheilocystidia similar to pleurocystidia but content more ochraceous-brown. Caulocystidia 36-50 × 12-20 μ, clavate to fusoid-ventricose or vesiculose, thin-walled, smooth.

Tube trama bilateral (the *Boletus* type), all hyphae thin-walled, smooth, and hyaline. Pileus cuticle a trichodermium (?) but soon collapsing, the hyphae 3-6 μ wide, sparingly branched, the cells tubular and elongate, hyaline in KOH fresh and ochraceous as revived, smooth or with a gelatinous sheath, the end-cells somewhat cystidioid—slightly enlarged near apex and tapered to the apex, in semidecumbent fascicles; subcutis of hyphae with orange to orange-brown content in Melzer's (the gelatinous layer paler). Clamp connections absent.

Habit, habitat, and distribution.—Under aspen, Crooked Lake, Barry County, July 15, 1966, Samuel Mazzer and A. H. Smith 72955.

Observations.—The terminal cells of the cuticular hyphae distinguish this species from *B. vermiculosus.* They are narrower and more cystidioid. The odor of the dried material was described by a biochemist as that of boiled urine. The viscid pileus is an added distinctive feature.

181. Boletus vermiculosoides Smith & Thiers, sp. nov.

Illus. Pls. 134-35.

Pileus 4-12 cm latus, late convexus demum late depressus, siccus, impolitus, sordide ochraceus demum pallide spadiceus. Contextus citrinus tactu caeruleus. Tubuli 1 cm longi, pallide olivacei tactu caerulei; pori 2-3 per mm, spadicei. Stipes 4-9 cm longus, 1-2 cm crassus, sursum olivaceo-pallidus, deorsum sordide brunneus. Sporae 9-12 × 3-3.5 μ. Specimen typicum in Herb. Univ. Mich. conservatum est; prope Hartland, August 30, 1966, legit Smith 73286.

Pileus 4-12 cm broad; convex, expanding to broadly convex or finally slightly depressed on the disc, sometimes nearly flat; surface dry, dull, unpolished, obscurely fibrillose at first in some, when perfectly fresh dull to bright yellow and soon becoming snuff-brown to bister, rather dingy brown in age and occasionally somewhat mottled, spotting darker brown where bruised, margin incurved at first, rarely with lobes produced by localized growth. Context thick, rigid and firm, lemon-yellow young fading to pallid yellow, staining blue to somewhat greenish in the stipe, with $FeSO_4$ no reaction in either place, with KOH on cutis very dark brown, on context yellow.

Tubes up to 1 cm deep, pallid-olivaceous, depressed to nearly free, dingy bluish when cut; pores minute (2-3 per mm), dark sepia to dark amber-brown when young and fresh, paler in age but never red (finally near cinnamon-buff), when bruised staining blue then orange.

Stipe 4-9 cm long, 1-2 cm thick, solid, watery olive-yellow within at first, pallid in age, dingy brown around larval tunnels; surface olivaceous-pallid to yellowish, obscurely pruinose, not at all reticulate, dingy brown on the base from handling.

Spores 9-12 X 3-3.5 μ, smooth, no apical pore evident, in profile view narrowly inequilateral, in face view elongate-subfusiform, yellow in KOH, duller and paler yellow in Melzer's, suprahilar depression distinct, wall -0.2 μ thick.

Basidia 4-spored, 20-26 X 7-9 μ, clavate, soon hyaline as revived in KOH, in Melzer's weakly yellowish. Pleurocystidia fusoid-ventricose, 38-57 X 7-12 μ, with a narrow flexuous neck and subacute apex, in KOH the content first ochraceous-brown then deep ochraceous, in Melzer's merely weakly yellowish. Cheilocystidia mostly with dark yellow-brown content and more or less resembling basidia.

Tube trama of the *Boletus* type, the hyphae 4-7 μ wide, thin-walled and gelatinous. Pileus cutis an interwoven cuticle of hyphae 2.5-5 μ wide, subgelatinous in KOH and with brownish ochraceous content, in Melzer's the content orange-brown or darker, cells all tubular and the end-cells merely obtuse, all hyphae smooth in both KOH and Melzer's. Clamp connections absent.

Habit, habitat, and distribution.—Gregarious under oak, Hartland, August 30, 1966, Smith 73286.

Observations.—This was thought to be *B. vermiculosus* when collected, but that species has spores 11-15 X 4.5-5.5 μ. There apparently is an important difference in the pileus cuticle also, since that of *B. vermiculosoides* is never a trichodermium.

182. Boletus fagicola Smith & Thiers, sp. nov.

Pileus 4-12 (18) cm latus, siccus, subtomentosus, ochraceus demum subspadiceus. Stipes 4-8 cm longus, 10-25 mm crassus, sursum reticulatus. Sporae (9) 11-14 (15) X 3.5-5 μ. Specimen typicum in Univ. Mich. Herb. conservatum est; prope Douglas Lake, Cheboygan County, August 7, 1961, legit Smith 63863.

Pileus 4-12 (18) cm broad, convex with the margin regular to irregular, becoming broadly convex to depressed-convex or nearly plane; surface dry and velvety to subtomentose, in age finally irregularly rimose to areolate, often merely unpolished at maturity; bright yellow on the young incurved margin, remainder dull yellow to yellow-brown ("ocher-yellow" to "buckthorn-brown"), at times darker when young and then "raw-umber" to "Dresden-brown," slowly becoming "buckthorn-brown," when dried retaining these colors. Context thick, firm, yellow but quickly changing to blue when cut, taste mild, odor none; with KOH orange, with $FeSO_4$ the yellow areas fading to pallid.

Tubes 1-2 cm deep, adnate becoming depressed around the stipe, pale yellow quickly staining blue when cut; pores dark yellow-brown ("argus-brown" to "Sudan-brown"), small (about 2 per mm), bluish bruised and staining darker brown with KOH.

Stipe 4-8 cm long, 10-25 mm thick, equal, solid, bright yellow within but soon blue when cut; surface reticulate in apical region to one-third the distance to the base, ground color ochraceous but this shaded from the grayish to brownish pruina, often reddish at the base and soon dingy brownish where handled (after the blue fades out), base at times with a velvety covering.

Spore deposit dull olivaceous; spores (9) 11-14 (15) X 3.5-5 μ, smooth, no apical pore evident, in face view nearly oblong to slightly ventricose, in profile slightly inequilateral with a shallow suprahilar depression, with an inconspicuous hyaline sheath, pale ochraceous to pale tawny in KOH and not appreciably darker in Melzer's, smooth, wall slightly thickened, no fleeting-amyloid reaction observed.

Basidia 4-spored, 17-25 X 8-10 μ, short-clavate, hyaline to dingy buff in KOH and in Melzer's. Pleurocystidia scattered, 35-55 X 9-14 μ, fusoid-ventricose, the neck often curved and apex subacute, thin-walled, hyaline in KOH or content in part yellowish, in Melzer's often with dark yellow-brown pigment in the neck. Cheilocystidia about like the pleuro-cystidia but usually narrowly fusoid-ventricose and 30-45 X 7-10 μ, hyaline to yellow-brown in KOH and often darker brown revived in Melzer's, the pigment apparently in the wall to some extent in some pilei but regularly in the cell content.

Tube trama of gelatinous divergent hyphae with thin walls and no incrustations (the *Boletus* subtype). Pileus cuticle a trichodermium of narrow (4-7 μ) hyphae with end-cells at times narrowly clavate, soon collapsing to form a cutis of interwoven hyphae, their content golden yellow in KOH and orange-brown in Melzer's. Clamp connections none. No amyloid cell inclusions seen but few scattered granules present between the hyphae in some mounts.

Habit, habitat, and distribution.—Gregarious to scattered in brushy places and in beech or oak woods or along woods borders, summer and early fall, fairly abundant during some seasons.

Observations.—This species passed as *B. firmus* in Michigan for many years, but Frost's original description rules against the identification. He described the pileus as gray, the pores as "tinged with red," and the stipe as finely "reticulated" (Frost, 1874, p. 103). In the present variant the stipe is only inconspicuously reticulated at the apex or for a narrow apical region rarely to one-third of the length, the pores are dark yellow-brown, and the pileus ocher-yellow to yellow-brown, becoming more or less date-brown in age. It differs from *B. vermiculosoides* in the reticulation on the stipe and in having a pileus cuticle in the form of a trichodermium.

Stirps LURIDUS

KEY

1. Not staining blue anywhere when injured 2
1. Staining blue on context or tubes or both if injured 3

   2. Pileus dark red; spores 11-14 × 5-7 μ . . . . *B. holoroseus*
   2. Pileus orange-brown to duller; spores 9-12 × 3.8-4.5 μ . . . . . *B. eberwhitei*

3. Spores 4-5 μ wide 4
3. Spores 5-7 μ wide 5

   4. Pileus glabrous and viscid . . . . *B. frostii*
   4. Pileus dry and appearing appressed-fibrillose at first . . . . *B. rubroflammeus*

5. Spores 12-17 × 4.5-7 μ . . . . *B. luridus*
5. Spores 9-12 × 5-6 μ . . . . *B. vinaceobasis*

183. **Boletus holoroseus** Smith & Thiers, sp. nov.

Illus. Figs. 109-10.

Pileus 3-10 cm latus, convexus, siccus, subtomentosus, testaceoroseus. Contextus vinaceus, tarde demum griseus. Tubuli 12 mm longi, pallidi, demum olivacei; pori testaceo-rosei. Stipes 5-11 cm longus, 1-2 cm crassus, testaceo-roseus, rubro-reticulatus. Sporae (9) 11-14 × 5-7 μ. Specimen typicum in Herb. Univ. Mich. conservatum est; prope Highlands, Oakland County, September 29, 1965, legit Smith 72730.

Pileus 3-10 cm broad, convex expanding to plane, margin incurved and even; surface dry and subtomentose, evenly vinaceous-red ("Prussian-red") but in oldest pilei becoming olive brownish with only a flush of red remaining, margin even and fertile, incurved at first. Context vinaceous-red when pileus was sectioned but not changing following cutting, the red slowly fading to grayish, taste mild, odor fungoid, with FeSO$_4$ olivaceous, with KOH yellow, with Melzer's violet-gray.

Tubes up to 12 mm long, adnate becoming depressed, pallid at first, becoming olivaceous and in age dull red ("Prussian-red") in places throughout their length; pores 1-2 per mm, round to slightly angular, dark dull red ("Prussian-red") in immature pilei, in age more reddish brown, unchanging when bruised.

Stipe 5-11 cm long, 1-2 cm thick at apex, equal or nearly so, solid, when cut olive-pallid in pith, cortex and lower half (especially the base) dark rusty rose-red, not staining blue; surface "Prussian-red" lower down and longitudinally ribbed to striate as well as with a fine dull red reticulum, ground color dingy yellowish pallid becoming dull red like the reticulum.

Spores (9) 11-14 × 5-7 μ, smooth, with a thin spot at apex but not a true pore, wall about 0.5 μ thick in Melzer's and 1 μ thick in KOH, ovate to subelliptic in face view, subfusiform to obscurely kidney-shaped in profile (as in many spores of species in *Inocybe dulcamera* group), cinnamon-brown individually in KOH, paler and near "snuff-brown" in Melzer's; some abnormal spores almost globose (fig. 110).

Basidia subelliptic in mature pilei, clavate when immature, 28-34 × 9-14 μ, 4-spored, yellowish in KOH and Melzer's. Pleurocystidia 32-48 × 5-10 μ, narrowly fusoid to fusoid-ventricose, smooth, thin-walled, in old pilei some found which were 30-40 × 10-15 μ and mucronate (giant basidia ?), all hyaline to yellowish in KOH or Melzer's. Cheilocystidia similar to pleurocystidia but smaller (20-30 × 5-9 μ), content yellowish brown in KOH.

Tube trama a central strand of more or less ochraceous-brown hyphae (in young tubes) with color in both content and wall, with hyaline, diverging hyphae extending to subhymenium, both types smooth. Pileus cuticle a turf of narrow hyphae, tubular and with elongated end-cells 4-6 μ wide, most of them seem to be surrounded by a mucilaginous sheath (revived in KOH), the content of the cells of the trichodermium amber-brown in Melzer's. Context hyphae near cuticle also with amber-brown content or hyaline. Clamp connections absent.

Habit, habitat, and distribution.—Gregarious by a trail in low hardwoods, Highlands Recreation Area, Oakland County, September 29, 1965, Smith 72730.

Observations.—The lack of any blue stains, the presence of deep rusty red colors overall, deeply colored spores scarcely inequilateral in profile view, and the reticulate stipe are distinctive. It is close to *B. rhodoxanthus* and in fact was mistaken for that species in the field, but it does not stain blue and has wider spores. The reddish color noted for the context may possibly be a matter of aging, not autooxydation on exposure to air.

184. Boletus eberwhitei Smith & Thiers, sp. nov.

Pileus circa 11 cm latus, convexus, siccus, aurantiobrunneus demum subalutaceus. Contextus laete luteus, immutabilis. Tubuli circa 1 cm longi, lutei; pori 1-2 per mm, ferrugineo-rosei. Stipes circa 10 cm longus, 2.5 cm crassus, sursum luteus et obscure reticulatus, deorsum subluteus. Sporae 9-12 × 3.8-4.5 μ. Specimen typicum in Herb. Univ. Mich. conservatum est; prope Ann Arbor, September 25, 1965, legit Florence Hoseney (Smith 72721).

Pileus about 11 cm broad, convex, the margin fertile but irregular; surface dry and even or slightly uneven and with obscure areolate mottling, dull orange-brown in color drying crust-brown; context pale bright yellow, unchanging when cut or slowly becoming salmon-buff (but not blue), $FeSO_4$ bluish gray, KOH pinkish, Melzer's on tubes dark blue.

Tubes up to 1 cm deep, adnate to slightly depressed, near ocher-yellow and unchanging when cut; pores 1-2 per mm, rusty rose near stipe, paler toward the cap margin, unchanging when bruised, in drying reddish stains show on sides of tubes in places.

Stipe 10 cm long, 2.5 cm thick, equal solid, cortex lemon-yellow, pith dingy yellow, base buff to grayish, surface yellow above, ochraceous at ground line, apical region reticulated by a fine inconspicuous reticulum.

Spores 9-12 × 3.8-4.5 μ, smooth, narrowly elliptic to oblong in face view, suboblong to obscurely inequilateral in profile, amyloid fresh, weakly so as revived; greenish ochraceous in KOH, finally weakly tan in Melzer's, wall about 0.2 μ thick, no apical pore evident.

Basidia 4-spored, 7-9 μ broad. Pleurocystidia scattered, 28-40 × 7-12 μ, clavate-rostrate with the neck flexuous and projecting only slightly beyond the hymenium, context hyaline in KOH or yellowish but soon fading. Caulocystidia clavate, mucronate, or fusoid-ventricose, 20-40 × 9-15 μ or some 40-60 × 15-22 μ, and subelliptic to clavate, thin-walled and not reviving readily.

Cuticle of pileus of interwoven hyphae 3-6 μ wide with bright yellow content in KOH and pigment leaching into the mount, hyphae tubular or nearly so, end-cells tubular or nearly so and not organized into any kind of layer, in Melzer's the cuticular hyphae dull ochraceous and context hyphae bright red. No clamp connections observed.

Habit, habitat, and distribution.—Solitary in low rich hardwoods, Ann Arbor, September 25, 1965, Florence Hoseney (Smith 72721).

Observations.—The Eber White Woods in Ann Arbor has been maintained as a nature study area. In Michigan the only other member of this stirps, which lacks a color change to blue on the injured context or tubes, is *B. holoroseus*. The latter has little in common with *B. eberwhitei*. The latter is close to *B. appendiculatus* and *B. impolitus*, but the combination of characters does not agree with either one.

185. **Boletus frostii** Russell in Frost

Bull. Buff. Soc. Nat. Sci. 2:102. 1874

Suillellus frostii (Russell) Murrill, Mycologia 1:17. 1909.
Boletus alveolatus Berkeley & Curtis in Frost, Bull. Buff. Soc. Nat. Sci. 2:102. 1874.
Suillus frostii (Russell) Kuntze, Rev. Gen. Pl. 3(2):535. 1898.

Illus. Pls. 136-37.

Pileus 5-13 (15) cm, hemispheric to convex becoming broadly convex to plane to shallowly depressed, or the margin turned up in age; margin incurved, sterile; surface hoary at first but soon shining, sometimes appearing finely areolate, viscid from a gelatinous pellicle, dark blood-red to blackish red at first ("garnet-red" to "Morocco-red" or "Brazil-red"), fading with age, typically blood-red with yellowish areas when old. Context up to 2.5 cm thick, pallid to pale yellow (lemon-yellow), turning blue immediately upon exposure; odor none, taste mild (cuticle acid to the tongue at times).

Tubes (5) 9-15 mm deep, adnate becoming depressed, with decurrent lines down the apex of the stipe, yellow to olivaceous yellow ("mustard-yellow"), dingy blue when bruised; pores small, 2-3 per mm, round (stuffed at first—Coker), deep red ("madder-brown" to "Morocco-red") until old age but finally a paler red, pore surface typically beaded with yellowish droplets when young and readily staining dingy blue when bruised.

Stipe 4-12 cm long, 1-2.5 cm thick at apex, equal to tapering toward apex to occasionally ventricose, solid, white within above, yellowish downward, surface dry, very coarsely reticulate to lacerate-reticulate overall and dark blood-red, often yellow at base or whitish, showing a slight change to blue when bruised (often none in the yellow base).

Spore deposit "olive-brown"; spores 11-15 × 4-5 μ, occasional giant spores up to 18 μ long also present; shape in face view narrowly boat-shaped, in profile narrowly inequilateral; pallid-ochraceous in KOH and in Melzer's hardly changing, smooth, no apical pore apparent.

Basidia 4-spored, 23-32 × 8-10 μ, yellowish to hyaline in KOH, nonamyloid or some areas of hymenium dingy bluish at first but soon fading (reaction not clearly localized). Pleurocystidia scattered, 24-38 × 8-12 μ, fusoid-ventricose, neck often curved, thin-walled, hyaline, nonamyloid, apex subacute to acute. Cheilocystidia abundant, with gelatinous pedicels, obfusoid above and in a dense palisade, often with elongated neck, content red when fresh.

Tube trama gelatinous and of divergent hyphae, nonamyloid though sections may appear dingy bluish at first when mounted. Cuticle of pileus a thick gelatinous tangled to matted-down layer of hyphae 3-6 μ wide, hyphae hyaline to yellowish in KOH. Context hyphae floccose-interwoven, hyaline in KOH. Nonamyloid in all parts under the microscope but a fleeting-amyloid reaction present on the tissue revived in Melzer's. Clamp connections absent.

Habit, habitat, and distribution.—Scattered to gregarious in open oak woods, common in July and August in southern Michigan during wet seasons.

Observations.—This species is supposed to be edible, but we advise against trying it in spite of the ease with which it can be identified. It is in a dangerous group. One of the best field characters is the fact that the hymenial surface, especially the pore surface, exudes droplets of yellowish liquid when rapidly developing. The coarsely reticulate stipe, dark apple-red viscid pileus and weak color change to blue are additional field characters.

Material examined.—Barry: Mazzer 4229, 4325; Smith 73168, 73206, 73209. Livingston: Homola 945, 1953. Oakland: Smith 6735. Washtenaw: Baxter 1920; Homola 1628; Hoseney 560; Smith 9615, 18436, 62625, 62848, 62898, 64154, 72519; Thiers 4507, 4556, 4596.

186. Boletus rubroflammeus, sp. nov.

Illus. Pl. 138.

Pileus 6-12 cm latus, convexus, siccus, adpresse fibrillosus demum glaber, vinaceo-ruber. Contextus luteus tactu caeruleus. Tubuli 1-2 cm longi, lutei tactu violacei; pori roseoflammii, stipe 6-8 cm longus, 1-3 cm crassus, roseo-reticulatus; sporae 10-14 × 4-5 μ. Specimen typicum in Herb. Univ. Mich. conservatum est; prope Ann Arbor, legit Smith 15352.

Pileus 6-12 cm broad, convex becoming broadly convex, 'the margin projecting slightly beyond the tubes; surface dry and appearing appressed-fibrillose or with matted grayish tomentum at first, soon appearing merely unpolished—the matted tomentum visible only along the margin, disc becoming slightly areolate in age, color evenly deep vinaceous-red ("Acajou-red" to "Vandyke-red"), this color remaining constant throughout the life of the basidiocarp. Context thick, soft, yellow, quickly turning blue when cut; taste mild, odor slight and hardly distinctive.

Tubes 1-2 cm deep, adnate but becoming free or nearly so, yellow, quickly turning blue; pores round, small (2 per mm), pore surface uneven or pitted, deep red ("Brazil-red"), but in age a more dingy red.

Stipe 6-8 cm long, 1-3 cm thick, solid, equal or clavate, yellowish within to reddish streaked and quickly blue when injured; surface coarsely reticulate with dark red reticulations, apex yellow beneath the reticulation, concolorous with pileus below.

Spore deposit not obtained. Spores 10-14 $\times$ 4-5 μ, smooth, suboblong to slightly ventricose in face view, in profile obscurely inequilateral. The suprahilar depression broad and shallow, walls only slightly thickened, yellowish hyaline in Melzer's and pale ochraceous in KOH, no sign of any amyloid reaction.

Basidia 30-40 $\times$ 8-9 μ, clavate with a long pedicel, 4-spored, many with ochraceous content as revived in KOH or Melzer's. Pleurocystidia rare to scattered, 28-37 $\times$ 9-14 μ, fusoid-ventricose, apex subacute, content hyaline in KOH. Cheilocystidia abundant, 18-35 $\times$ 5-9 μ, narrowly fusoid-ventricose to cylindric or narrowly clavate, content as revived in KOH dingy ochraceous, walls smooth and thin.

Tube trama bilateral, the central strand interwoven, floccose and ochraceous in KOH, the diverging hyphae continuing into the hymenium with the formation of a true subhymenium, hyphae 4-6 μ wide, smooth and some with ochraceous content. Pileus with a cuticle of tightly interwoven appressed nongelatinous hyphae with dingy ochraceous content as revived in KOH, 3-5 (18) μ wide, those in the epicuticular zone often with fine granular incrustations both in KOH and Melzer's. Clamp connections absent.

Habit, habitat, and distribution.—Gregarious to scattered under hardwoods, summer and fall after hot weather and heavy showers, rare in Michigan.

Observations.—This species is closest to *B. flammeus* Dick & Snell, but the cap is not viscid and it grows in low, moist, rich hardwoods rather than under spruce, hemlock, and white pine mixtures. We at first tried to identify our material with *B. rhodoxanthus* of Europe, but the latter has a pallid pileus when young and paler colors generally at maturity. The deep purplish red of the pileus and the reticulum of the stipe of *B. rubroflammeus* are constant and distinctive field characters. We have not observed it to be as variable as Dick and Snell indicate for their species.

Material examined.—Washtenaw: Smith 15352, 73004.

187. Boletus luridus Fries

Syst. Myc. 1:391. 1821

Leccinum luridum (Fries) S. F. Gray, Nat. Arr. Brit. Pl. 1:648. 1821.
Tubiporus luridus (Fries) Karsten, Rev. Mycol. 3:16. 1881.
Dictyopus luridus (Fries) Quélet, Enchir. Fung. p. 160. 1881.
Suillus luridus (Fries) Poir. ex O. Kuntze, Rev. Gen. Pl. 3(2):535. 1898.

Illus. Pls. 139-40.

Pileus 5-12 cm broad, convex becoming broadly convex to shallowly depressed at maturity; margin sterile but neither crenate nor in segments, incurved at first; surface dry and unpolished to somewhat shiny, matted-fibrillose or with scattered minute fibrillose scales on the disc, color variable, yellow to olive yellowish with a pinkish brick color, margin in some "pale orange-yellow," disc near "cinnamon-buff" to "clay color" or reddish, soon stained greenish blue where handled. Context thick up to 3 cm, floccose, yellowish ("primrose-yellow") to reddish, with a red line above the tubes when freshly cut, instantly deep blue when injured; odor none, taste mild; dark blue in spots where touched with Melzer's (after the natural bluing has faded).

Tubes free or at least deeply depressed, rarely decurrent, 1-2 cm deep, yellow ("Reed-yellow") but quickly staining greenish blue when injured; pores round small (about 3 per mm), deep red ("Morocco-red" to "Pompeian-red" to "Hays-russet") but soon fading out to orange-red, staining greenish blue when bruised.

Stipe 6-15 cm long, 1-3 cm thick, equal to clavate (base 4 cm thick), often flared at apex, solid, bright yellow within or in the base streaked reddish, soon staining blue to violaceous, surface dry, reticulate at apex or extending nearly to base, pruinose to furfuraceous in addition, yellow ("chrome-yellow") overall or tinged cinnabar to blood-red above and yellow lower down, or yellow above and reddish below.

Spore deposit olive to olive-brown; spores 12-17 X 5.5-7 μ, smooth, narrowly ovoid in face view, somewhat inequilateral in profile, ochraceous tawny in KOH, slightly darker in Melzer's as revived, amyloid when fresh, no apical pore evident.

Basidia 4-spored, 30-36 X 8-10 μ, yellowish in KOH and more so in Melzer's. Pleurocystidia scattered, 33-50 X 8-14 μ, fusoid-ventricose, tips acute, neck often wavy and narrow, hyaline, thin-walled, smooth. Cheilocystidia similar to pleurocystidia or more variable, content red when fresh, yellow in KOH (many clavate cells also present having the same content).

Tube trama divergent and gelatinous from a floccose central strand. Trichodermium of pileus of narrow (3-5 μ wide) hyphae with slightly roughened yellowish walls, becoming grouped into fascicles in age. Context of floccose interwoven hyphae 6-12 μ wide, nonamyloid in any part but mounts violet-fuscous in Melzer's (fleeting-amyloid reaction). Clamp connections absent.

Habit, habitat, and distribution.—Solitary to gregarious in open oak woods, often along edges of dense second-growth stands; common in Michigan but usually not found in any quantity, summer and fall.

Observations.—The wide spores distinguish the species from the previous one, but both are also readily distinguished at sight in the field by the difference in color. Reports can be found in the literature that this is an edible species, but Singer, who certainly knows the fungus as well as anyone, has tried it and found it poisonous. We repeat, again, that the mycophagist should not experiment with any red-pored bolete.

Material examined.—Cheboygan: Smith 25889. Chippewa: Kauffman 1906; Smith 72868. Emmet: N. J. Smith 76A; A. H. Smith 35890, 60788, 63845, 71561, 71800, 71850; Thiers 4419. Mackinac: Smith 25853. Oakland: Smith 64298, 73265, 73299. Schoolcraft: Smith 61048. Washtenaw: Thiers 4559.

188. **Boletus vinaceobasis** Smith & Thiers, sp. nov.

Pileus 4-10 cm latus, siccus, convexus, subspadiceus. Contextus tactu caeruleus. Stipes 7-9 cm longus, 1-2 cm crassus, reticulatus, deorsum vinaceo-ruber; sporae 9-12.5 X 5.5-6 μ. Specimen typicum in Herb. Univ. Mich. conservatum est; prope Pinckney, July, 1967, legit Florence Hoseney 509.

Pileus 4-10 cm broad, convex to nearly plane; surface smooth and matted-fibrillose, dingy ochraceous-brown becoming olive to olivaceous-brown (near "Saccardo's umber"), and about this color as dried, margin even. Context dull yellow but quickly blue when injured, and rather soon fading on stained portions to olivaceous.

Tubes about 1.5 cm deep, depressed around the stipe, yellow with the pores dull red, staining blue when injured; as dried the sides dull olivaceous.

Stipe 7-9 cm long, 1-2 cm thick, equal to pinched off at the base, solid, yellowish within and staining like the pileus, in the base dark vinaceous and this tint still evident as dried; surface with a distinct netlike reticulum over the upper half, yellowish above, vinaceous-red below, the reticulum faintly reddish.

Spores 9-12.5 × 5.5-6 μ, smooth, wall -0.2 μ thick, lacking apical differentiation; shape in profile subelliptic to very obscurely inequilateral in face view narrowly elliptic to slightly ovate, color dull yellow-brown in KOH or Melzer's, a very slight fleeting-amyloid reaction present.

Basidia 4-spored. Pleurocystidia scattered, clavate to fusoid, 28-37 × 9-13 μ, in Melzer's with a dark dingy brown reticulate amorphous content, dull fulvous in KOH. Caulocystidia in patches, 22-46 (60) × 7-12 (15) μ, clavate, clavate-mucronate, or fusoid-ventricose, often with ochraceous to fulvous content as seen revived in KOH.

Pileus cuticle a trichodermium of hyphae 3-6 μ wide, tubular, end-cells tubular or at apex slightly tapered, content often in wads or masses of reddish pigment as revived in Melzer's (ochraceous-brown as revived in KOH), walls smooth and slightly gelatinous as revived in KOH, a yellow pigment diffusing in the mount. Clamp connections absent.

Habit, habitat, and distribution.—Solitary in open oak woods, George Reserve, near Pinckney, July, 1967, Florence Hoseney 509.

Observations.—This species differs from *B. luridus* in distinctly shorter spores and the dark brown pleurocystidia when revived in Melzer's. It has the stature and the same low type of reticulum as *B. rhodoxanthus.* However, it appears most similar to *B. satanoides* (Smotlacha, Česk. Houb. 2:29. 1920). Their spore features appear to distinguish them, however: 11-16 × 5-6.5 μ for *B. satanoides* and 9-12 × 5.5-6 μ for *B. vinaceobasis* as for size, and in shape "schmal mandel-formig mit deutlicher suprahilar-depression" for the former and sub-elliptic to obscurely inequilateral for the latter, which amounts to a significant difference in the appearance of the spore in profile view. Also the stipe in Smotlacha's species appears to be much brighter yellow.

Stirps SUBVELUTIPES

KEY TO SPECIES

1. Many caulocystidia from the middle area of the stipe or above greatly
   elongated to more or less setiform (but thin-walled). 2
1. Caulocystidia clavate to fusoid-ventricose but neck seldom up to 30 μ long . . . 4

   2. Context pallid when fresh; hyphae in cortex of stipe with walls becoming dull brown in KOH as revived . . . . *B. roseobadius*
   2. Context yellow 3

3. Cuticular hyphae of pileus rusty brown to reddish as revived in Melzer's;
   pileus dingy cinnamon at maturity . . . . *B. rufocinnamomeus*
3. Cuticular hyphae red in Melzer's as revived; spores 11-15 ×4-5 μ . . *B. subluridellus*

   4. Stipe typically with pruina arranged in a reticulate pattern near the
   apex; pileus olive-yellow to olive and becoming subsquamulose over
   the disc . . . . *B. pseudo-olivaceus*
   4. Not as above 5

5. Pileus and stipe dark madder-red . . . . see *B. bicolor* var. *borealis*
5. Not as above 6

   6. Context white before changing to blue; pileus dark russet or brown;
   spores 9-12 × 3.5-5 μ (but mostly immature on type) . . . . *B. spraguei*
   6. Context yellow before changing; pileus yellow, orange or red or these
   colors mixed; spores larger than in the above choice 7

7. Spores 14-18 (20) × 5-6.5 (8) μ; pileus drying yellow and base of stipe
   with dark red hairs . . . . *B. subvelutipes* Peck
7. Not with above combination of features . . . . *B. erythropus*

189. Boletus subluridellus Smith & Thiers, sp. nov.

Pileus 5-10 cm latus, convexus, demum late convexus, resinosus,
laete ruber vel badius, tactu obscure violaceus. Contextus laete citrinus
tactu violaceus. Tubuli 6-9 mm longi, lutei tactu violacei; pori carminei.
Stipes 4-9 cm longus, 1.5-2.3 cm crassus, laete citrinus tactu viride
caeruleus. Sporae 11-15 × 4-5.5 μ. Specimen typicum in Herb. Univ.
Mich. conservatum est; prope Ypsilanti, September 3, 1961, legit Smith
6406.

Pileus 5-10 cm broad, convex to irregular in outline, becoming
broadly convex, margin inrolled at first, a very narrow sterile band
present; surface dry and resinous to the touch, velvety to unpolished,
not truly viscid, evenly "vermillion-red" to "bay" or at times orange-
red, instantly dark violet where touched, the surface often pock-marked.
Context bright lemon-yellow but instantly blue where cut, taste slightly
acid-metallic, odor none, $FeSO_4$ blue parts turn yellow and then orange.

Tubes 6-9 mm long, yellow turning blue instantly when cut, depressed around the stipe to nearly free; pores 2-3 per mm, round, carmine or darker red, instantly blue when bruised.

Stipe 4-9 cm long, 1.5-2.3 cm thick, solid, equal, pale lemon-yellow, changing to greenish blue when cut, red in the base especially in the cortex, surface coated over the base with a thin coating of appressed sulphur-colored hyphae, pruinose-furfuraceous overall, not at all reticulate.

Spore deposit olive; spores 11-15 $\times$ 4-5.5 μ, smooth, apical pore minute, color in KOH olive-yellowish (from a deposit) ochraceous-tan from the tube tissue, in Melzer's dingy yellow-brown, shape in face view subfusoid, in profile obscurely to somewhat inequilateral, wall about 0.2 μ thick.

Basidia 4-spored, clavate, 8-12 μ wide. Pleurocystidia clavate-rostrate to fusoid-ventricose, 28-42 $\times$ 6-11 μ, neck about 3 μ wide and often flexuous, apex obtuse, hyaline or with yellow to yellow-brown content in KOH, brownish to hyaline in Melzer's, scarcely projecting beyond sporulating basidia. Cheilocystidia narrowly clavate to fusoid-ventricose, 26-38 $\times$ 4-8 μ, in mass ochraceous-tan in KOH. Caulocystidia versiform: (a) broadly ovate pointed and without a neck, 34-46 $\times$ 12-16 μ, often with adhering tawny debris as revived in KOH; (b) elongate fusoid-ventricose to setiform but with thin walls (26) 40-180+ $\times$ 4-12 μ, hyaline, broader cells with some adhering debris around the ventricose part. Caulobasidioles often with tawny to ochraceous content.

Cuticle of pileus a thick (150+ μ) layer of narrow interwoven (3-5 μ) hyphae with red content in Melzer's and yellow in KOH, at surface arranged in fascicles of parallel hyphae indicating a tendency toward a trichodermial structure, end-cells tubular. Hyphae of subcutis and below them having dark red content in Melzer's, no dextrinoid incrustations noted. Clamp connections absent.

Habit, habitat, and distribution.—Gregarious in grassy oak woods, September, Washtenaw County, Smith 64046.

Observations.—The red cuticular hyphae and a similar pigment in those of the context near the cutis as sections are revived in Melzer's, the presence of long, setiform caulocystidia, the resinous feel of the young pileus when it is touched, the instant color change to dark violet of the cut context, and the numerous small pleurocystidia which scarcely project beyond sporulating basidia are a very distinctive combination of features. When the neck of the pleurocystidium is ochraceous to brown in the mounting medium one is likely to mistake it for a filamentous pseudocystidium.

190. Boletus spraguei Frost

Bull. Buff. Soc. Nat. Sci. 2:102. 1874

"Pileus quite hard, very dark russet or brown, covered with a minute velvety scurf. Tubes very minute, yellow ocher or brownish when cut, around the mouths of a rich dark maroon color, which forms a strong contrast with the light color of the stem, adnate when young. Stem dark brown below, ochraceous at top, smooth above, minutely velvety below, firm, fleshy, slightly constricted in the middle. *Flesh* white, changing to blue, texture firm and fine. The rich color of the pore mouths contrasting with the yellow stem, makes it quite distinct from other species. Spores .0105-.0062 mm."

The original description is given above. The following data were obtained from a Frost specimen at Albany, New York, in the Peck collections.

Spore 9-12 X 3.5-5 μ, smooth, lacking apical differentiation; shape in face view nearly oblong to narrowly elliptic, in profile obscurely inequilateral, color in KOH nearly hyaline (mostly immature), yellowish in Melzer's; wall thin. Hymenium mostly immature but basidia were 4-spored and 7-8 μ broad. Pleurocystidia are present and measure 28-36 X 7-10 μ, most are subfusoid and content not distinctive in KOH or Melzer's; no amyloid reactions seen in or on hyphae or in the wall but some "amyloid" debris present loose in the mounts.

Pileus cuticle a tangled mass of narrow hyphae 3-7 μ wide, with end-cells more or less ascending but not in a true palisade—becoming decumbent as pileus enlarges; the hyphae tubular, cells long, content yellowish in KOH, merely ochraceous in Melzer's and homogeneous, walls thin and smooth, dextrinoid incrustations lacking. Hyphae of sub-cutis with dull yellow to orange-ochraceous content in Melzer's. Clamp connections absent.

The following is a description of Michigan material (Smith 9613) which we believe belongs here:

Pileus 7-13 cm broad, convex to nearly plane, margin even or slightly wavy, surface dry and dull, minutely appressed-fibrillose under a lens when mature, in age the surface layer cracking into minute or easily visible areolae, color often "tawny" over center and yellow on the margin but gradually becoming reddish brown overall, at times finally "Prussian-red." Context yellow but changing to blue quickly when cut; odor and taste not recorded.

Tubes about 1 cm deep, yellowish then quickly blue when bruised, adnate to stipe when young, depressed somewhat in age; pores dark maroon-red ("Dragon's blood-red") but duller by maturity, at edge of pileus sometimes yellowish red.

Stipe 3-6 cm long, 1-2 cm thick solid, yellowish within, but quickly staining bluish; surface evenly furfuraceous, base not hairy, apex pale yellow, toward base reddish to reddish brown, not at all reticulate.

Spores 10-13 (14) X 4-5.5 (6.5) μ when mature, smooth, apical pore lacking, shape in face view subelliptic, to subfusoid, in profile somewhat inequilateral, color in KOH yellow-brown when mature, in Melzer's weakly amyloid at first but soon dextrinoid, wall about 0.3 μ thick.

Basidia 4-spored, 7-10 μ broad, clavate. Pleurocystidia 33-45 X 8-12 μ, fusoid-ventricose with neck more or less elongated and apex subacute, content "empty" in KOH or Melzer's. Cheilocystidia smaller than pleurocystidia and with a red content fresh, as revived dingy ochraceous. Caulocystidia fusoid-ventricose, 30-50 X 8-14 μ, neck elongated and apex obtuse to subacute, with thin hyaline smooth walls in KOH.

Tube trama of the *Boletus* subtype, not amyloid in any way under the microscope but with a distinct fleeting-amyloid reaction when first mounted in Melzer's. Pileus cuticle a tangled layer of interwoven hyphae giving rise to fascicles of 2-3 cells long ascending or forming a loose trichodermium, their content homogeneous and ochraceous in KOH (but fading), in Melzer's more brownish ochraceous to buffy orange-brown, walls thin, smooth (in both KOH and Melzer's) and hyphae tubular, end-cells tubular to weakly cystidioid. Hyphae of subcutis and adjacent context orange-buff in Melzer's but fading. Clamp connections absent.

Observations.—In Smith 9613 the pilei became redder as they aged—a feature of the type—and as dried are dull red to reddish brown, strikingly different from the yellow of *B. subvelutipes.* The tubes are dark maroon as described by Frost. Frost described the context as white, whereas in Smith 9613 it was yellow. This is a serious discrepancy. The immature spores in Smith 9613 are like those of the Frost specimen and the mature spores are larger but still slightly shorter than in *B. subvelutipes.* We believe that these differences are sufficient to justify recognizing Frost's species as distinct from *B. subvelutipes,* at least for the present, and we place Smith 9613 in Frost's species at least until a truly critical study of Subsection *Luridi* can be completed on a broader basis than the present flora. *B. hypocarycinus* Singer is obviously closely related. It is southern in distribution, and according to Singer the terminal cells of the cuticular hyphae are clavate to fusoid and have obtuse tips.

191. Boletus pseudo-olivaceus Smith & Thiers, sp. nov.

Illus. Pls. 141-43.

Pileus 3-10 cm latus, convexus, siccus, adpresse fibrillosus demum fibrillose squamulosus, olivaceus vel olivaceo-brunneus. Contextus ochraceus tactu caeruleus. Tubuli 1 cm longi, lutei tactu caerulei; pori rubidi. Stipes 8-12 cm longus, 1-3 cm crassus, pruinosus, subreticulatus deorsum olivaceo-luteus. Sporae 13-16 X 5-6.5 μ. Specimen typicum in Herb. Univ. Mich. conservatum est; prope Mackinaw City, July 12, 1967, legit Smith 74510.

Pileus 3-10 cm broad, broadly convex; surface dry and unpolished, appearing somewhat fibrillose-scurfy, the margin matted-fibrillose, in age some becoming appressed-squamulose over the disc; color when young "light brownish olive" to "Isabella color" and yellowish to olive-yellow when dried, when fresh soon streaked fuscous where bruised as the aftermath of staining blue; margin even and incurved. Context dingy ochraceous but instantly changing to blue when cut, finally changing to olive-fusocus. Taste mild, odor none.

Tubes up to 1 cm deep, depressed around the stipe, dingy yellow but instantly bluish when cut; pores dark maroon-red (near "Moroccored"), slowly becoming paler and darker in age to almost garnet-brown, evenly colored from pileus margin to stipe.

Stipe 8-12 cm long, 1-3 cm thick at apex, enlarged downward, yellow within, when cut instantly blue, in the base a deep rusty rose; surface pruinose with reddish pruina, purina often arranged in a reticulate pattern over upper half, the flavous ground color showing beneath the particles, lower half with paler yellow ground color and at the base olivaceous to olive-ocher; base lacking any sign of hairs or strigosity.

Spores 13-16 X 5-6.5 μ, smooth, wall -0.5 μ thick, apex furnished with a minute thin spot; shape in face view narrowly elliptic to subfusoid, in profile suboblong with a broad suprahilar depression, color dull tawny-brown in KOH and in Melzer's, the fleeting-amyloid reaction absent to very slight.

Basidia 4-spored, 20-26 X 9-12 μ, with large oil drop when nearly mature. Pleurocystidia rare to scattered, 28-37 X 9-17 μ, ventricose below, neck 2-4 μ wide and tapered to an acute apex, content hyaline in KOH, walls thin and smooth. Cheilocystidia mostly resembling small basidioles but pale rusty brown from color in the wall or content as revived in KOH. Caulocystidia clavate to subelliptic or fusoid-ventricose but the latter rare and resembling pleurocystidia, content typically brownish ochraceous in KOH (cells often revived poorly), 26-38 X 8-14 μ for the clavate to subelliptic cells.

Tube trama typical for the section. Pileus cutis of appressed non-gelatinous hyphae 3-6 μ wide, tubular, the walls smooth to irregular but hyaline or nearly so, terminal cell tubular or apex tapered slightly; an olive-yellow pigment diffusing in mounts of KOH. Clamp connections absent. All hyphae inamyloid.

Habit, habitat, and distribution.—Scattered at edge of a hardwood forest under *Corylus* bushes and near them; Mackinaw City hardwoods, July 12, 1967, Samuel Mazzer, James Bennett, and A. H. Smith 74510, 74511, and 74361.

Observations.—This species differs from *B. subvelutipes* in lacking hairs at the base of the stipe, in having dark maroon-red pores regularly, olive tones in the pileus and as dried the whole basidiocarp is a dingy olive-yellow. The appressed-fibrillose scurfy pileus surface which becomes somewhat fibrillose-squamulose by maturity is a good field character also.

This species is most similar to *B. queletii* as described by Singer (1967), though for a time we regarded it as a variety of *B. subvelutipes*. However, Singer stated that it was: "ohne jede Netzzeichnung" and that the pruina on the stipe was "sehr fein." In *B. pseudo-olivaceus* the stipe is conspicuously pruinose and the pruina is often so arranged as to cause the stipe to appear reticulate, but no veins of tissue are present—just denser aggregations of pruina. The pores are a dark maroon-red overall when young. The pileus cuticle is more of a cutis of appressed filaments than a trichodermium, and in age it breaks up to produce an areolate to fibrillose scaly condition.

Both *B. subvelutipes* and *B. pseudo-olivaceus* grow in the same habitats, and we suspect that there has been gene exchange between them, but the problem needs critical study over a period of years to elucidate it. This whole stirps has always been and will continue to be in the forseeable future, a confusing group because of the "spatter pattern" of the distribution of the characters between the various populations discovered.

192. Boletus roseobadius Smith & Thiers, sp. nov.

Pileus circa 12 cm latus, convexus, siccus, obscure roseus. Contextus pallidus tactu violaceus. Tubuli 12 mm longi, lutei vel crocei; pori coccinei. Stipes circa 10 cm longus, 15 mm crassus, roseo-pruinosus, fractu violaceus, non-reticulatus. Sporae 13-16 X 5-6.5 μ. Specimen typicum in Herb. Univ. Mich. conservatum est; prope Ann Arbor, September 20, 1965, legit Florence Hoseney (Smith 72668).

Pileus 12 cm broad, broadly convex; surface dry and appressed-fibrillose, unpolished, dark rose-red at maturity and dark rose-bay as dried. Context pallid, when cut the surface quickly stained violaceous.

Tubes "ochraceous-buff" to more croceus, staining violet when bruised, 12 mm long, depressed around the stipe; pores minute, flame-scarlet staining bluish olive.

Stipe about 10 cm long, 15 mm thick, equal, within dark red with yellow streaks, staining violet on cut surface; exterior surface lacking any distinct basal mycelium, rose-pruinose above and in upper third with longitudinal lines which darken in drying, not reticulate.

Spores 12-16 X 5-6.5 μ, smooth, with a minute apical hyaline spot, wall about 0.3 μ thick, in profile inequilateral-elongate, in face view somewhat fusiform, dingy ochraceous-brown, near bister in Melzer's (amyloid when fresh but changing on drying).

Basidia 4-spored, clavate, 8-11 μ broad. Pleurocystidia 38-63 (120) X 9-14 μ, fusoid-ventricose with narrow elongated flexuous neck, content often dark yellow-brown in KOH and darker (dark sepia) in Melzer's. Cheilocystidia 21-32 X 3-6 μ, narrowly subfusoid to narrowly clavate, the palisade ochraceous in KOH. Caulocystidia fusoid-ventricose and 28-50 X 9-13 μ with the neck often flexuous and apex subacute, or the cell setiform to narrowly ventricose at base and with a long-tapered neck to over 100 μ long which in some is furnished with one or more short protuberances or branches, some with a secondary septum distal to the ventricose portion, wall smooth and thin, context and wall hyaline to ochraceous in KOH. Subiculum hyphae also hyaline but those of cortex when mounted in KOH quickly very dark brown but slowly fading to pale brown, the walls smooth and thin but colored.

Cuticle of pileus a matted layer of hyphae 2-4 μ wide, with bright red content in KOH when fresh but soon fading to hyaline, in Melzer's as revived the content pinkish red and the color persistent, content of subcuticular and tramal hyphae not distinctive in Melzer's. Clamp connections none.

Habit, habitat, and distribution.—Solitary under low hardwoods, Ann Arbor, September 20, 1965, collected by Florence Hoseney (Smith 72668).

Observations.—The distinguishing features of this species are the greatly elongated pleurocystidia—reminding one of the caulocystidia, the red content of the cuticular hyphae as revived in Melzer's and the dark brown reaction of the hyphae of the cortex of the stipe when revived in KOH. In the field when fresh the longitudinal striations over the upper part of the stipe, the dark red pileus and whitish (pallid) flesh are distinctive. It is close to *B. luridellus,* but the latter has very small pleurocystidia and bright lemon-yellow flesh.

193. Boletus rufocinnamomeus Smith & Thiers, sp. nov.

Pileus 4-12 cm latus, convexus, siccus, subtomentosus, lateritius demum cinnamomeus. Contextus luteus tactu caeruleus. Tubuli adnati, lutei tactu caerulei; pori cinnabarini. Stipes 8-12 cm longus, 12-18 mm crassus, pruinosus, deorsum rubeus, sursum luteus, non-reticulatus. Sporae 14-17 X 4.5-6.5 μ. Caulocystidia 60-200 X 7-15 μ, subaciculata vel fusoide ventricosa. Specimen typicum in Herb. Univ. Mich. conservatum est; prope Burt Lake, Cheboygan County, August 11, 1969, legit Smith 77831.

Pileus 4-12 cm broad, convex becoming plane; surface dry and velvety to subtomentose, becoming unpolished, dull brick-red ("Sanford's brown) when young but with an ochraceous overtone from an evanescent bloom, in age finally dingy cinnamon (near "Sayal-brown"), when dried near "Verona-brown" (not yellow); context yellow but when cut almost instantly blue, taste acidulous, odor none, $FeSO_4$ on blue stained areas causing a change to yellow and finally olive.

Tubes 9-15 mm deep, adnate, becoming depressed at the stipe, yellow; pores 1.5-3 per mm, bright flame-scarlet at maturity, maroon-red when young, quickly blue where injured.

Stipe 8-12 cm long, 12-18 mm at apex, equal, solid, in upper portion yellow within, rusty rose near base, instantly blue when cut; surface yellow above, at or near the base coated with yellow mycelium, pruinose above with the pruina orange-cinnamon to brown, not at all reticulate or pseudoreticulate.

Spores olive-brown in deposit, 14-17 X 4.5-6.5 μ, smooth, apex with a distinct though very minute pore, in profile view ventricose-inequilateral, in face view subfusoid, in KOH pale clay color, in Melzer's dingy yellow-brown, wall less than 0.3 μ thick.

Basidia 4-spored, clavate, 8-11 μ broad. Pleurocystidia rare to scattered, readily collapsing, fusoid-ventricose with obtuse apex (not revived well enough to measure); hymenium readily gelatinizing. Cheilocystidia 26-33 X 7-11 μ, subfusoid, dingy tan to hyaline in KOH. Caulocystidia abundant, many 60-200 X 6-15 μ wide, seta-like to fusoid-ventricose, apex acute to subacute, content hyaline to yellow in KOH, often secondarily septate, walls thin and smooth; some cystidia not proliferated and measuring 36-58 X 10-15 μ, some clavate and 25-40 X 10-16 μ, most cells in the caulohymenium golden ochraceous in KOH.

Tube trama very gelatinous, of the *Boletus* subtype. Pileus cuticle a turf of narrow (2-5 μ) hyphae soon grouped into fascicles, the hyphae 100-300 μ long, septa sparse, cells tubular, end-cells tubular to slightly narrowed at the apex, hyaline or nearly so in KOH and reddish brown in Melzer's (dextrinoid), subcutis tawny in KOH and in Melzer's, walls smooth. Hyphae of adjacent context with ochraceous content in some as revived in Melzer's, the fleeting-amyloid reaction strong on young pilei. Clamp connections absent.

Habit, habitat, and distribution.—Gregarious under brush (*Corylus*) along a woods road in a beech-maple forest, Colonial Point, Burt Lake, Cheboygan County, August 11, 1969, Smith 77831.

Observations.—This is one of the species which has caused untold confusion relative to the recognition of *B. erythropus* and *B. subvelutipes* in this country. Actually, the species belongs in a group with *B. subluridellus* and *B. roseobadius.* The elongated caulocystidia of this group are not cells with a hyphal-like proliferation as found in some species of *Leccinum.* These cystidia are found on the stipe at or above the midportion and are not to be confused with any of the mycelial structures evident around the base. As they are found on young as well as on mature basidiocarps, the age of the latter is not a major factor in their morphological features. *B. rufocinnamomeus* differs from *B. subvelutipes* in that the base of the stipe is not strigose with red hairs even though red pigment is a feature of the base of the stipe of that species, and the pilei as dried retain a dark brown to reddish brown tone.

194. Boletus subvelutipes Peck

Rept. N. Y. State Mus. 2(8):142. 1889

Suillus subvelutipes Kuntze, Rev. Gen. Pl. 3(2):536. 1898.

Illus. Figs. 102, 104-5.

"Pileus convex, firm, subglabrous, yellowish brown to reddish brown, flesh whitish, both it and the tubes changing to blue where wounded; tubes plane or slightly convex, nearly free, yellowish, their mouths small, brownish red; stem equal or slightly tapering upward, firm, even, somewhat pruinose above, velvety with a hairy tomentum toward the base, yellow at the top, reddish-brown below, varied with red and yellow within; spores .00006 to .0007 in. long, .002-.00025 broad.

"Pileus 2-3 in. broad; stem 2-3 in. long, 4-6 lines thick."

Peck compared it to *B. vermiculosus.* We have quoted the original description above and add microscopic data from what appears to be the type: Saratoga, New York, July.

Spores 14-18 (20) X 5-6.5 (8) μ, smooth but with a tendency to be angular, with a minute thin spot at apex, shape in face view fusoid-ventricose, in profile strongly inequilateral and usually with a deep suprahilar depression. Color in KOH pale snuff-brown, in Melzer's reddish tawny (dextrinoid), wall about 0.3-0.5 μ thick.

Basidia 4-spored, 8-11 μ broad, hyaline in KOH and readily gelatinizing. Pleurocystidia rare, 30-42 X 7-12 μ, fusoid-ventricose with obtuse to subacute apex, hyaline in KOH and Melzer's or content weakly ochraceous and homogeneous, thin-walled, walls smooth. Cheilocystidia subventricose to narrowly fusoid-ventricose or mucronate, 16-26 X 4-9 μ, usually yellowish in KOH or Melzer's. Caulocystidia clavate to somewhat cystidioid, 33-50 X 6-14 μ, content ochraceous in KOH, thin-walled, wall smooth.

Tube trama of the *Boletus* subtype, inamyloid under the microscope but with a strong fleeting-amyloid reaction when the trama is crushed in Melzer's. Pileus cutis of long narrow tubular hyphae 2.5-5 μ wide, with thin smooth walls (in either KOH or in Melzer's) and ochraceous content, end-cells elongate with the tips blunt or slightly tapered, no short or inflated cells present, the end-cells not forming a palisade. Hyphae of the subcutis and context mostly ochraceous in KOH. Clamp connections none.

In the type collection insects have destroyed most of the cuticle of the pileus, but there is some evidence to indicate the pileus was yellow as dried. The hairs at the base of the stipe dried very dark red. The diagnostic features of the type are: (1) wide decidedly inequilateral spores, (2) hairs at base of the stipe which are red as dried, (3) white flesh of the pileus before it changes to blue, (4) yellow tones in the pileus when young. The following is a description of Michigan collections which have been placed here.

Pileus 6-13 cm broad, convex to nearly plane to occasionally obscurely umbonate becoming broadly convex to plane, margin even or slightly wavy; surface dry and dull, minutely fibrillose under a lens to velvety-tomentose, with age the cuticle becoming areolate and the areolae tomentose to minutely fibrillose-squamulose, color often "tawny" to "rufous" to "cinnamon-rufous" with a yellow to orange-yellow ("antimony-yellow" to "pale orange-yellow") margin, color soon changing to dark blue to blackish when bruised or on aging. Context moderately thick, soft, bright yellow ("baryta-yellow" to "sulphur-yellow") but changing to blue so quickly when cut that original color is easily missed, odor none, taste mild to slightly acid.

Tubes 1-2 cm deep, adnate sometimes becoming broadly depressed with age, yellowish ("picric-yellow" to "baryta-yellow") but quickly becoming bluish when bruised; pores orange to scarlet or dark red but soon duller and sordid reddish or yellowish red near margin, round, about 2 per mm.

Stipe 3-6 (10.5) cm long, 1-2 cm thick, equal or nearly so, solid, yellow within and soon staining blue, surface evenly furfuraceous to punctate overall and not at all reticulate, base with red strigosity, apex yellow (near "baryta-yellow"), red over basal area, readily staining blue when bruised.

Spore deposit dark olive-brown ("clove-brown"); spores 11-16 X 4.5-5.5 μ, smooth, wall slightly thickened, in face view ventricose-subfusoid, in profile inequilateral, ochraceous in KOH, dingy tan in Melzer's.

Basidia 4-spored, 24-30 X 8-9 μ, yellowish in KOH to nearly hyaline, slightly more yellowish orange in Melzer's. Pleurocystidia abundant, nearly buried in the hymenium to conspicuous, 34-48 X 8-12 μ, fusoid-ventricose, the neck often flexuous, apex acute to subacute, thin-walled, yellowish to hyaline in KOH, lacking any distinctive content in Melzer's. Cheilocystidia similar to pleurocystidia but varying to subfilamentous, content red in water mounts of fresh material, ochraceous in material revived in KOH.

Tube trama of gelatinous hyphae divergent from a central strand. Cuticle of pileus a layer of tangled hyphae 2.5-5 μ wide and at the surface becoming oriented into fascicles of subparallel elements, orange-brown in Melzer's, ochraceous in KOH. Context of interwoven floccose hyphae hyaline in KOH and Melzer's. All hyphae nonamyloid but mounts of hymenophoral tissue in Melzer's dark violet from the bluing reaction (which is evident on material revived in water also). Clamp connections absent.

Habit, habitat, and distribution.—Solitary to scattered under hardwoods, often common in the Mackinac Straits area early in the summer but occurring throughout the state into September. It is one of the first boletes to fruit.

Observations.—The spores of our material are not identical with those of the type. The temporary solution to a disposition of all the Michigan collections referred here is to recognize that the type form still remains to be found in Michigan, and the Michigan variant has yellow flesh and slightly smaller spores than the type. From the present study of the type we may now insist that the type form of the species features preponderantly yellow pilei as dried, that the stipe has red strigosity at the base, that the spores are large (14-18 X 5-6.5 μ), and that setiform caulocystidia are not present. The American taxa of this group simply do not compare well with the most closely related taxa as described from Europe.

195. Boletus erythropus (Fries) Krombholz

Consp. Fung. Esc. p. 24. 1821

Dictyopus erythropus (Fries) Quélet, Enchir. Fung. p. 160. 1886.
Tubiporus erythropus (Fries) Ricken, Vadem. für Pilzfreunde p. 205. 1918.

Illus. Pl. 144.

Pileus 6-12 cm broad, convex, expanding to broadly convex, dry, bright yellow on margin at times and disc yellow-brown ("argus-brown") to reddish ("Sanford's brown") varying to bay-red or rusty rose but then with yellowish overtones from a faint bloom, retaining dark brown to reddish tones in drying or becoming nearly date-brown; context greenish yellow, staining blue quickly when cut, $FeSO_4$ slowly olivaceous, taste mild, odor none.

Tubes 10-12 mm deep, depressed around the stipe, bright yellow, instantly bluish green when cut; pores minute, orange to orange-red, often orange near the stipe and yellow near pileus margin, instantly blue when bruised.

Stipe 6-10 cm long, 10-15 mm thick, equal, solid, yellow within, instantly blue if injured and becoming olive to grayish olive slowly except for red base at least in the cortex; surface bright lemon-yellow and minutely pruinose, pruina orange-cinnamon to red, surface blue where handled, base with yellow mycelium and strigosity at times, rusty red in the base as dried.

Spores 13-16 X 4.5-6 μ, smooth, apical pore minute, color in KOH ochraceous to ochraceous-brown, duller yellow-brown in Melzer's (no amyloid reaction), shape in profile obscurely inequilateral to somewhat inequilateral, subfusoid in face view, wall about 0.2 μ thick.

Basidia 4-spored, clavate, 8-11 μ wide. Pleurocystidia rare to absent, 34-47 X 8-13 μ, fusoid-ventricose, apex subacute, hyaline or yellow in KOH. Cheilocystidia 21-32 X 6-9 μ, subfusoid, ochraceous in KOH. Caulocystidia 31-64 X 7-18 μ, broadly fusoid to fusoid-ventricose, not setiform or proliferated, apex subacute, thin-walled, content hyaline or ochraceous (as revived in KOH), some cells vesiculose and up to 20 μ broad.

Tube trama of the *Boletus* subtype, nonamyloid. Cuticle of pileus of interwoven hyphae 3-5 μ wide, with ends suberect at times (not a true trichodermium), cells tubular, elongate, content ochraceous in KOH, darker yellow-brown in Melzer's but giving to the naked eye a bluish black color to the layer, subcutis not colored distinctively in Melzer's. Clamp connections none.

Habit, habitat, and distribution.—Gregarious in beech-maple and in oak forests, common during warm wet weather in the summer.

Observations.—The Michigan material is not quite typical of the species. The olive $FeSO_4$ reaction and the rare (as far as observed) pleurocystidia are diverging features. Further studies are desirable now that the species with setiform caulocystidia can be readily excluded.

Subsection BOLETI

The tubes are stuffed when young, the stipe is reticulate at least in the upper third, and the flesh is typically white to pallid when fresh. The basidiocarps are robust and fleshy and blue stains are known for only one of our taxa. Some of the best of all edible fungi are found in this group, but for the most part the species are sporadic in appearance in the state or abundant from year to year only in certain localities, such as around Marquette.

KEY TO SPECIES

1. Spores 13-17 × 4-5.5 (17-25 × 6.5-12) μ (the large spores with walls 1-2 μ thick); tubes bright yellow young, brownish yellow mature
   . . . . *B. gertrudiae*
1. Spores and tubes not as above 2
   2. Pileus cuticle more or less a palisade of large end-cells and pileo-cystidia, the latter often with secondarily formed cross walls. 2
   2. Pileus cuticle a trichodermium or of appressed hyphae 4
3. Pores staining bluish when injured . . . . *B. separans* var. *subcaerulescens*
3. Pores not staining blue . . . . *B. separans* var. *separans*
   4. Pileus lemon-yellow splashed ferrugineous to vinaceous-tawny; pores staining vinaceous-cinnamon when bruised . . . . *B. chippewaensis*
   4. Not as above 5
5. Spores 12-15 × 3.8-4.5 μ; content of trichodermial hyphae reddish in Melzer's; clamp connections occasional on hyphae of the basidiocarp. . *B. insuetus*
5. Not as above—spores typically longer and wider 6
   6. Pileus viscid or slightly viscid when fresh and moist . . . .*B. edulis* and variants
   6. Pileus dry and unpolished when fresh 7
7. Trichodermial hyphae roughened as in *Leccinum insigne* (fig. 15). . . *B. atkinsoni*
7. Trichodermial hyphae smooth. 8

8. Pileus dark blackish to dark olive-brown; spores 4-5 μ wide. . . . . . .
   . . . . (but see *B. aereus* of Europe also) *B. variipes* var. *fagicola*
8. Pileus dull tan, smoky brown or crust-brown, typically becoming
   strongly areolate . . . . *B. variipes* var. *variipes*

Stirps GERTRUDIAE

196. Boletus gertrudiae Peck

Bull. N. Y. State Mus. 150. 1911

Pileus 5-12 cm broad, broadly convex, fleshy, glabrous, soft, dry
or nearly so, orange-yellow or brownish yellow, rarely bright yellow;
context white, unchangeable. Tubes long, bright yellow when young,
brownish yellow when mature, adnate or but slightly rounded at the
stipe, pores minute. Stipe rather long, equal or nearly so, solid, glabrous,
yellow above, white below, white within or sometimes more or less
yellow within the upper part.

The above data are from the original description. The following
details are from a study of the type:

Spores 13-17 × 4-5.5 μ (17-25 × 7.5-12 μ), smooth, lacking an
apical pore; color in KOH weakly ochraceous, in Melzer's yellowish to
pale tawny; shape in face view subfusiform, in profile somewhat inequilateral (large spores oblong-angular to broadly fusoid and with the wall
1-2 μ thick), wall in normal spores 0.2-0.3 μ thick, no positive amyloid
reaction of any kind present.

Basidia 4-spored. Pleurocystidia and cheilocystidia not observed
because of the poor preservation of the basidiocarp. Tube trama of
hyphae 4-8 μ wide, not amyloid in any part. Pileus cuticle a trichodermium with the apical cells of the elements clavate to cystidioid and
arranged in a loose palisade, hyaline in KOH, weakly yellow in Melzer's,
gelatinizing in age, no positive amyloid reaction of any kind noted.
Hyphae of subcutis with ochraceous content in Melzer's. No clamp
connections observed (but hyphae poorly preserved).

The tissues of the type are filled with small spores possibly of one
of the Fungi Imperfecti. The spore size given by Peck covers both abnormal and "normal" spores in part. The spore abnormality here is
exactly parallel to that found for *Tylopilus sordidus* in that the wall
thickens greatly, some of the spores are almost fusoid-ventricose, and a
tendency to angularity is present. Since this has been a feature of all
collections of *T. sordidus*, it should be used as a tentative identification

feature in assigning future collections to *B. gertrudiae*. The species needs to be redescribed from such fresh material. We suspect that it will be found to be a readily recognizable species once properly described on a modern basis. To date we have not recognized it in the Michigan flora.

Stirps SEPARANS

197. Boletus separans Peck

Bull. Buff. Soc. Nat. Sci. 1:59. 1873

Boletus edulis subsp. *separans* (Peck) Singer, Amer. Midl. Nat. 37:26. 1947.
Boletus nobilis Peck, Bull. N. Y. State Mus. 94:48. 1904.

var. separans

Illus. Figs. 99-101; Pl. 118 (right).

Pileus 6-15 (20) cm broad, convex with a bent in margin becoming broadly convex or with a low broad umbo; surface dry and unpolished or when wet somewhat tacky, glabrous in age; dark dull red to liver-brown or bay-brown, with an ochraceous undertone near the margin. Context pallid, unchanging, reddish near the cuticle, odor and taste mild in fresh state.

Tubes adnate becoming depressed, whitish, slowly dull yellowish to dingy ochraceous (not olivaceous), unchanging; pores stuffed and white when young, becoming pale yellow, not staining blue, 1-2 per mm, round or nearly so.

Stipe 6-15 cm long, 10-25 mm thick, enlarged downward, finely reticulate as in *B. edulis;* concolorous with or paler than the pileus, glabrous, base whitish at first, white within.

Spores 12.5-16 X 3.5-4.5 μ, narrowly subfusiform in face view, in profile narrowly inequilateral, smooth-walled and wall only slightly thickened, yellow in KOH, yellow to ochraceous-tan in Melzer's (no fleeting-amyloid reaction present), some with an orange-brown content as revived in Melzer's.

Basidia 4-spored. Pleurocystidia none or present only as pseudo-cystidia imbedded in the hymenium. Tube trama gelatinous, more or less divergent from a central strand. Cuticle of pileus a tangled tricho-dermium with end-cells clavate to fusoid-ventricose and 40-80 X 12-18 μ, content yellow to hyaline in KOH, in Melzer's the layer reddish but the content of individual cells ochraceous; laticiferous elements in the context fairly numerous. Clamp connections absent.

Habit, habitat, and distribution.—This species is rare in Michigan. Kauffman collected it near Marquette, which is a region rich in boletes.

Observations.—This is a distinctive species readily recognized by the colors and the almost hymeniform cuticle formed by the enlarged end-cells of the trichodermial elements. The spores are narrower than for the typical form of *B. edulis,* with which species it is classified as a subspecies by Singer (1967).

The following data were obtained from the type: Spores 13-16 × 3.5-4.5 (5) μ, smooth, lacking apical differentiation, shape in face view narrowly subfusoid, the ends obtuse, in profile obscurely inequilateral, suprahilar depression often very shallow; color in KOH weakly ochraceous, in Melzer's bay-red (dextrinoid), wall about 0.3 μ thick.

Basidia 4-spored, 7-10 μ wide, weakly yellowish in KOH, many small yellow globules present along the hymenium in mounts in Melzer's. Pleurocystidia rare, 32-45 × 8-13 μ, fusoid-ventricose, neck often elongated and apex subacute, yellow in KOH or in Melzer's, content homogeneous in Melzer's. Cheilocystidia mostly resembling basidioles or a little wider (6-12 μ), soon gelatinizing in KOH.

Tube trama of the *Boletus* subtype, hyphae lacking any amyloid content or walls and tissue brown as viewed in Melzer's against a white background. Pileus cuticle a hymeniform layer of pear-shaped to clavate end-cells 32-50 × 9-12 μ, walls thin and smooth, content yellow in KOH or Melzer's, no amyloid inclusions present, also in the layer are numerous pileocystidia 40-80 × 9-16 μ or occasionally larger, often multicellular in the neck, smooth, thin-walled and content yellow in Melzer's. Hyphae of subcutis yellow to orange in Melzer's. Clamp connections absent.

Notes: There are more multiseptate pileocystidia in the type than in the Michigan collections, but this hardly justifies recognizing the latter as a distinct taxon. There were faint suggestions of a greenish reaction in KOH on the cuticular elements, but it was not distinctive.

Boletus nobilis Peck is the same as *B. separans* microscopically. The numerous minute oil droplets along the hymenium and throughout the mount are present in Melzer's, the lack of any amyloid reactions in the hyphae or their walls is similar, the spore size and most of all the details of the cuticle of the pileus are alike in both.

197a. **Boletus separans** var. **subcaerulescens** (Dick & Snell)
comb. nov.

Boletus edulis ssp. *subcaerulescens* Dick & Snell, Mycologia 57:455. 1965.

Pileus 5-12 cm broad, obtuse to broadly convex or nearly plane, margin crenate-lobed forming sterile segments of tissue; surface glabrous, uneven, moist but not viscid; color evenly dark vinaceous-brown ("walnut-brown") when young, more purplish brown in age. Context pallid, pale vinaceous-buff near the cuticle; taste mild, odor none, no reaction with $FeSO_4$.

Tubes pallid, becoming dull yellow in age; depressed around the stipe, up to 20 mm deep in largest pilei; pores stuffed when young, becoming pale dull yellow and when bruised staining bluish but the stain soon fading to yellow again.

Stipe 10-15 cm long, 2-3 cm thick at apex, enlarged downward, pallid within, finally yellowish around wormholes, surface concolorous with pileus (dark vinaceous-brown), becoming paler, reticulate overall and reticulum near apex pallid, base with thin white mycelial covering and not discoloring on bruising.

Spores 11-15 X 3.5-4.5 (5) μ, smooth, wall slightly thickened, in face view subfusiform, in profile inequilateral with a broad suprahilar depression, somewhat ventricose in midportion, yellow in KOH, in Melzer's yellow to pale yellow-brown. No fleeting-amyloid reaction present.

Basidia 4-spored, 22-28 X 7-8 μ, clavate, hyaline in KOH and yellowish in Melzer's. Pleurocystidia—none seen. Many plate-like crystals forming in KOH mounts. Tube trama so filled with laticiferous hyphae as to obscure the arrangement of the matrical hyphae.

Pileus cuticle a palisade of clavate to cystidioid end-cells of trichodermial hyphae 2-3 cells long and 4-9 μ wide, the end-cells 25-80 X 10-30 μ, some ventricose-mucronate, some fusoid-ventricose, and some clavate, separating in KOH mounts as if subgelatinous, hyaline in KOH, thin-walled, smooth. Context hyphae 8-15 μ wide and hyaline to yellow in KOH, large (to 15 μ) laticifers very abundant and often lumpy and contorted. Clamp connections absent. All tissues nonamyloid.

Habit, habitat, and distribution.—Scattered under birch, aspen, and maple, near Pellston, July 29, 1961, N. J. Smith, collector (A. H. Smith 63749), and N. J. Smith, September 4, 1965.

Observations.—This variant was assigned to *B. edulis* originally, but the cuticle of the pileus indicates that it is better placed here, and this is correlated with the degree of pigmentation. The blue staining, however, is an aberrant feature for either species.

Stirps INSUETUS

198. Boletus chippewaensis Smith & Thiers, sp. nov.

Pileus 12-16 cm latus, pulvinatus, glaber, subviscidus citrinus demum ferrugineo-maculatus vel vinaceo-fulvus, demum lateritius. Contextus albidus. Tubuli pallide lutei; pori lutei tactu subcaerulei. Stipes 6-10 cm longus, 2.5-3.5 cm crassus, pallide alutaceus, albo-reticulatus. Sporae 11-16 × 5-7 μ. Specimen typicum in Herb. Univ. Mich. conservatum est; prope Sugar Island, July 31, 1965, legit Smith 71914.

Pileus 6-7 cm broad when young, expanding to 12-16 cm when mature, convex, pulvinate or broadly convex; surface glabrous, slightly viscid but soon dry, pale lemon-yellow splashed irregularly with ferruginous to vinaceous-tawny, finally brick-red overall or margin remaining pale yellow. Context thick, firm, white, slowly becoming dingy tan beneath the cuticle after sectioning, odor and taste pleasant but hardly distinctive; olive-gray with $FeSO_4$; with KOH no reaction.

Tubes pale greenish yellow ("primuline-yellow"), free or nearly so, in age olive-ochraceous to olive-brown; pores "primuline-yellow," staining vinaceous-cinnamon when bruised, about 1-2 per mm at maturity.

Stipe bulbous to equal, 6-10 cm long, 2.5-3.5 cm thick at apex, solid and white throughout, scarcely staining when cut; surface dingy pinkish buff to cinnamon-buff beneath a white reticulum, base whitish.

Spores 11-16 × 5-7 μ, smooth, no apical pore evident, in face view mostly broadly fusoid, in profile inequilateral with ·the suprahilar depression rather distinct, in KOH greenish ochraceous, in Melzer's pale dingy yellow-brown (no dextrinoid spores seen), wall about 0.2 μ thick.

Basidia 4-spored, 26-30 × 7-10 μ, clavate, hyaline in KOH, weakly yellowish in Melzer's. Pleurocystidia rare to scattered, narrowly fusoid, 34-46 × 6-10 μ, hyaline in KOH and Melzer's. Cheilocystidia intergrading with basidioles, 18-30 × 7-9 μ, content yellowish in KOH but fading to hyaline.

Pileus cuticle a short trichodermium arising from a layer of gelatinous interwoven hyphae, the turf elements gelatinizing—as revived in KOH, with thickish refractive walls irregular in outline and with some debris adhering, not sharply defined as revived in KOH, in Melzer's the content yellowish and walls apparently soon disintegrating, the apical cell 10-22 × 6-12 μ broad (approximately) and clavate to mitten-shaped, subcuticular hyphae with orange-brown content. Context hyphae hyaline and interwoven, many large laticiferous hyphae up to 15 μ wide and yellow in KOH present. Clamp connections absent.

Habit, habitat, and distribution.—Scattered under mixed conifers and hardwoods, Sugar Island, July 31, 1965, Smith 71914.

Observations.—This is one of the more distinctive species in the *B. edulis* complex. Although the pileus has vinaceous to ferruginous tones fresh, young pilei become yellow in drying ("ocher-yellow"). This color change, the wide spores, pores staining vinaceous when bruised, and the short terminal cells of the trichoderm with their gelatinous halo are outstanding.

199. **Boletus insuetus** Smith & Thiers, sp. nov.

Pileus 3-8 cm latus, convexus, siccus, glaber demum areolatus, pallide fulvus demum ochraceo-tinctus. Contextus albidus, immutabilis; tubuli lutei immutabiles; pori lutei tarde luteo-brunnei. Stipes 6-9 cm longus, 10-16 mm crassus, pallidus, reticulatus, demum luteo-maculatus. Sporae 12-15 X 3.8-4.5 μ. Specimen typicum in Herb. Univ. Mich. conservatum est; prope Colonial Point, Burt Lake, August 8, 1951, legit Smith 37528.

Pileus 3-8 cm broad, convex becoming broadly convex; surface dry, glabrous at first but soon more or less areolate-rimose; color pale tan to pale tawny but developing yellow tones especially in aging ("cinnamon-buff" to "ochraceous-tawny"). Context white, unchanging, taste mild, odor none.

Tubes depressed around the stipe, about 10 mm long, yellow, unchanging; pores about 2 per mm and round, yellow, unchanging or slowly yellowish brown.

Stipe 6-9 cm long, 10-16 mm at apex, clavate becoming equal, pallid within but tending to yellow; surface whitish at base, but with yellow stains, grayish brown over midportion and apex pallid to avellaneous, reticulate over upper half with a fine pallid network.

Spore deposit olive-brown; spores 12-15 X 3.8-4.5 μ, narrowly suboblong in face view to subfusoid, narrowly and obscurely inequilateral in profile, smooth, suprahilar depression broad and usually quite distinct, walls only slightly thickened, dingy ochraceous revived in KOH, many dark reddish brown in Melzer's, at least in part of the spore.

Basidia 4-spored, clavate, 23-30 X 7-9 μ, yellowish in KOH to hyaline. Pleurocystidia—none found. Pileus cutis a collapsed trichodermium of thin-walled smooth hyaline (in KOH) hyphae with reddish content in Melzer's and 6-10 μ broad, end-cells usually cystidioid from being tapered to an obtuse apex. Clamp connections present but rare.

Habit, habitat, and distribution.—Scattered in a mixed forest, Colonial Point, Burt Lake, August 8, 1951, Smith 37528.

Observations.—Peck described the spores of *B. gertrudiae* as 15-20 X 5-6 μ. The stipe was described as yellow above and white below, the tubes as bright yellow when young and brownish yellow when mature. The features of our collection are the occasional clamps on the hyphae of the basidiocarp, the spores 12-15 X 3.8-4.5 μ, the yellow stains on the stipe which is pallid to grayish at the apex, and the generally pale tan pileus. These differences seem sufficient to justify recognizing the species.

Stirps VARIIPES

200. Boletus atkinsoni Peck

Bull. N. Y. State Mus. 94:20. 1905 (Rept. 58)

Pileus 6-10 cm broad, convex, fleshy, expanding to nearly plane, dry, grayish brown to yellowish brown, sometimes minutely rimosely squamulose. Context white, taste mild. Tubes convex, plane or slightly concave in the mass, adnate or slightly depressed around the stipe, 5-8 mm deep; the pores minute, stuffed whitish at first, soon open and yellow or ochraceous. Stipe stout, 5-10 cm long, 1-2.5 cm thick, enlarged either above or below (or both), solid, reticulate above or nearly overall with fine anastomosing brownish lines, pallid.

Spores 10-14 X 3.5-4.5 μ, smooth, lacking an apical spot or pore; shape in face view subfusoid, in profile obscurely inequilateral, the suprahilar depression shallow and broad; yellowish in KOH, only slightly browner in Melzer's, no fleeting-amyloid reaction observed, wall less than 0.2 μ thick.

Basidia 4-spored. Pleurocystidia—none found. Cheilocystidia basidiole-like. Tube trama boletoid. Pileus cuticle a trichodermium of agglutinated hyphae 4-7 μ wide, the cells elongate and tubular, the content yellowish in KOH and Melzer's, the wall with more or less colorless patches of outer wall material resting on a gelatinous layer (much as in *Leccinum insigne*, only the hyphae narrower), the end-cells scarcely inflated, usually tapered to an obtuse apex, no amyloid reactions of any kind observed. Hyphae of subcutis ochraceous in Melzer's. Clamp connections absent.

Observations.—Our description combines Peck's original description with data obtained from the type by Smith. We have as yet not recognized the species in our Michigan flora though the name has been used in the sense of *variipes* on occasion. The roughened hyphae of the fascicles of hyphae forming the tufts on the pileus, and the smaller spores distinguish this species from *B. variipes*. The spores average around 13 μ long, whereas in *B. variipes* they are about 15 μ long.

201. Boletus variipes Peck

Ann. Rept. N. Y. State Mus. 41:76. 1888

Illus. Pl. 145.

Pileus 6-15 (30) cm broad, obtuse to convex, expanding to broadly convex or nearly plane, often deeply rimose on aging; surface soft to the touch, unpolished and dry at first but slightly tacky when rain-soaked, becoming areolate with matted tomentum on the areolae, or merely slightly and irregularly rimose, glabrous in age at times, grayish buff to dingy tan or smoky tan but in drying becoming distinctly ochraceous. Context white, unchanging, odor mild when fresh, resembling *B. edulis* as dried, taste pleasant.

Tubes 1-3 cm deep, becoming depressed around the stipe, white at first, becoming greenish yellow, unchanging when bruised; pores 1-2 per mm, round, stuffed and white at first, ochraceous at maturity, unchanging when bruised.

Stipe 8-15 cm long, 1-3.5 cm thick at apex, equal to clavate when young, over 4-6 cm thick at base in some, solid, white within, unchanging bruised; surface inconspicuously reticulate to near base or only over upper half, concolorous with or paler than the pileus, usually unpolished or dull in appearance.

Spores 12-16 (18) × 3.5-5.5 μ, subfusoid in face view, narrowly inequilateral in profile, with a broad distinct suprahilar depression, smooth wall slightly thickened, yellow in KOH, pale yellowish to tawny in Melzer's, no fleeting-amyloid reaction present.

Basidia 4-spored, 8-10 μ in diameter, clavate. Pleurocystidia—none observed. Cheilocystidia basidiole-like. Pileus cuticle a trichodermium of elements 4-9 μ wide, thin-walled, smooth or with a slight halo (in KOH), cells mostly uninflated and long, the end-cells 30-58 × 5-9 μ, usually tapered to a blunt apex, content yellowish in Melzer's. Hyphae of subcutis with strongly ochraceous content in Melzer's. Clamp connections absent.

Habit, habitat, and distribution.—Gregarious in thin open oak woods or maple-aspen stands on sandy soil, often very abundant in July and August after heavy rains.

Observations.—The microscopic data given above are taken from the type. The species fruits abundantly in hardwood forests containing oak or beech during the summer. The pileus often is coarsely areolate-squamulose as the result of rapid growth and the drying effect of the sun on the cuticle. The color may be gray at first, but soon changes to some shade of cinnamon-buff or yellow-brown. The cuticular hyphae are smooth, and this is the best distinction between this species and *B. atkinsoni,* though, as previously stated, the spores do differ somewhat in length.

201a. **Boletus variipes** var. **fagicola** Smith & Thiers, var. nov.

Illus. Pl. 146.

A typo differt: Pileus atro-brunneus vel obscure olivaceo-brunneus; sporae 13-16 X 4-5 μ, leviter amyloidiis. Specimen typicum in Herb. Univ. Mich. conservatum est; prope Wolverine, August 10, 1968, legit Smith 75914.

Pileus 4-12 cm broad, expanding to plane or nearly so, margin even; surface dry and velvety, matted down in age, color evenly blackish brown ("mummy-brown" to "dark olive-brown") at first, slowly becoming dark yellow-brown ("bister") or dingy cinnamon ("Sayal-brown"), not gray at any stage but in drying some caps dingy pinkish buff. Context white, soon riddled by larvae, taste and odor fungoid, with $FeSO_4$ no reaction, with NH_4OH on cutis bay-brown; not changing to blue when bruised.

Tubes white then olive to olive-yellow, 1 cm or more deep when mature, adnate; pores stuffed, white when young, in age olive-yellow, unchanging when bruised.

Stipe 6-10 cm long, 1-3 cm thick, solid, subequal to obscurely fusiform, pallid within and soon riddled by larvae, cut surface slowly becoming merely brownish; surface with a whitish reticulum on a ground color as dark as the pileus but base typically pallid to white.

Spore deposit olive; spores 13-16 X 4-5 μ, smooth, apex lacking a pore, shape in face view narrowly oblong or nearly so, in profile obscurely to somewhat inequilateral, color in KOH bright ochraceous, in Melzer's with a gray cast at first (weakly amyloid), slowly becoming somewhat dextrinoid; walls 0.2 μ thick.

Basidia 4-spored, about 30 X 10 μ, clavate, hyaline to yellow in KOH and yellowish in Melzer's. Pleurocystidia apparently absent. Pore edges lined by basidia and basidioles.

Tube trama of the *Boletus* subtype, of hyaline, thin-walled, smooth tubular hyphae as revived in KOH; laticiferous elements not conspicuous. Pileus cuticle a tangled trichodermium of elements having thin smooth hyaline walls as revived in KOH and content also hyaline, in Melzer's the walls pale yellow; the terminal cells tubular to flexuous or narrowly clavate to somewhat cystidioid, the former 5-9 μ wide, the latter 8-14 μ wide, length variable, the cells in the trichodermial elements short to long (2, 3, 4, or more times longer than broad but not inflated). Hyphae of subcutis and adjacent context merely with yellowish walls in Melzer's. Clamp connections absent.

Habit, habitat, and distribution.—Gregarious to scattered under beech-maple and some aspen intermixed, Berry Creek, near Wolverine, Cheboygan County, August 10, 1968, Smith 75914.

Observations.—This is a very dark fungus with a velvety dry dull pileus which becomes dingy yellow-brown in age, the stipe is as dark as the pileus but had a fine reticulum of white to pallid lines, the spores are distinctly amyloid at first, and the pileus cuticle is a tangled trichodermium in the cells of which occurs the dark pigment, but when mounted in KOH the color in these cells disappears almost instantly.

Boletus aereus Fries should have a colored reticulum over the stipe. We do not know whether its spores are amyloid or not. Singer considers *B. variipes, B. atkinsoni, B. gertrudiae,* and *B. nobilis* as synonyms of *B. aereus.* A critical study on the basis of hyphal detail and Melzer's reactions in the European members of the group is needed before worthwhile comparisons can be made with the European and American floras in this complex.

Stirps EDULIS

KEY

1. Cuticular hyphae often with diverticulae; pileus ochraceous when young, old, or dried . . . .*B. edulis* var. *ochraceus*
1. Not as above 2
    2. A fleeting-amyloid reaction absent on tissue of the hymenophore; pileus not viscid . . . . *B. edulis* var. *clavipes*
    2. Not as above 3
3. Pileus dull fulvus to crust-brown . . . . *B. edulis* var. *edulis*
3. Pileus ferruginous to rose-red 4
    4. Pileus dark rose-red; pores staining light cinnamon when bruised . . . .*B. edulis* var. *aurantio-ruber* (Ammirati 2004)
    4. Pileus ferruginous to bay-red; pores staining yellowish olive after bruising . . . . *B. edulis* var. *aurantio-ruber*

202. Boletus edulis Bull. ex Fries

Syst. Myc. 1:392. 1821

Leccinum edule (Fries) S. F. Gray, Nat. Arr. Brit. Pls. 1:647. 1821.
Tubiporus edulis (Fries) Karsten, Rev. Mycol. 3:16. 1881.
Dictyopus edulis (Fries) Quélet, Enchir. Fung. p. 159. 1886.

var. edulis

Illus. Pl. 147.

Pileus 8-15 (25) cm broad, convex, becoming broadly convex, rarely nearly plane; surface typically moist but lubricous to subviscid to the touch in wet weather, glabrous, smooth, uneven, wrinkled or somewhat pitted, color variable, pallid to crust-brown or redder brown ("tawny"), most frequently clay color to cinnamon to pale cinnamon-buff. Context firm, 2-4 cm thick, becoming soft in age, white or tinged reddish near the cuticle, unchanging when bruised, odor pleasant, fungoid, taste somewhat nutty.

Tubes adnate but depressed, up to 3 cm deep, stuffed when young, pallid at first but at maturity or in age sordid greenish yellow; pores small and round or nearly so, 2-3 per mm, sometimes staining tawny where bruised, the walls and mouths finally becoming greenish yellow ("Isabella color") or occasionally reddish in age.

Stipe massive, 10-18 cm long, 2-3 (4) cm thick at apex when mature, clavate, 4-6 (10) cm thick at base, solid, white within, unchanging, surface pallid to brownish, paler at apex, finely reticulate over upper third, lower portion usually slightly uneven, occasionally rather brightly colored over midportion ("cinnamon" toward bulb and "light pinkish cinnamon" near the apex).

Spore deposit deep olive-brown; spores (13) 14-17 (19) X 4-6.5 μ, subfusiform in face view, in profile elongate-inequilateral, with a broad suprahilar depression, smooth, wall only slightly thickened, pale tawny in KOH, as revived in Melzer's ochraceous shaded greenish gray but gradually becoming ochraceous to weakly dextrinoid.

Basidia 32-40 X 10-12 μ, hyaline in KOH, 4-spored, clavate. Pleurocystidia scattered, 42-48 (65) X 7-10 μ, narrowly fusoid-ventricose, hyaline in KOH, pale yellowish in Melzer's, smooth, thin-walled.

Tube trama of gelatinous hyphae somewhat divergent from a central strand. Cuticle of pileus a turf (trichodermium) of intertangled (and finally collapsing) hyphae 4-10 μ wide, hyaline and gelatinous in KOH, lacking incrusting pigments; hyphae in hypodermal region floccose,

8-15 μ wide and with ochraceous walls revived in KOH. Clamp connections absent. No amyloid hyphae present, but a fleeting-amyloid reaction does occur on the hymenophoral tissue and the spores.

Habit, habitat, and distribution.—Scattered to gregarious under conifers, late summer and fall, sometimes locally abundant, as around Marquette. Found especially in the pine country after heavy rains in late August and September.

Observations.—This species ranks as one of the best edible fungi in the state. However, no matter what species concept one subscribes to, *B. edulis* is a "collective" species. Singer (1967) recognizes as subspecies many of those we regard as species. The problem of how to classify all the variants will involve much more careful field work than that done in the past and more detailed studies on hyphal characters.

202a. Boletus edulis var. **ochraceus** Smith & Thiers, var. nov.

Pileus circa 12 cm latus, viscidus, luteus; cellulae terminales trichoderma saepe diverticulatae. Specimen typicum in Herb. Univ. Mich. conservatum est; prope Forestville, Marquette County, July 30, 1968, legit J. Ammirati 2013.

Pileus up to 12 cm broad, broadly convex, margin sterile for about 1 mm, glabrous, viscid and when dry sand particles become firmly imbedded in the cuticle, yellowish overall ("pale ochraceous-buff" on margin to near "warm buff" on the disc), pale yellowish as dried. Context white, thick, with reddish tints just beneath cuticle, odor and taste fungoid.

Tubes shallowly depressed, up to 1.5 cm deep, olive-ocher, unchanging when cut, drying dull yellow; pores dingy yellow, staining cinnamon to dull orange-cinnamon.

Stipe up to 14 cm long and 2.5 cm thick at apex, up to 3 cm or more at base, clavate to subequal, solid, whitish to sordid brownish ("pinkish buff" tones in upper portion), finely reticulate above and coarser below, when dried pale buff with reticulum pallid.

Spores 14-18 X 4.5-6 μ, smooth, apex lacking a pore or a thin spot, shape in face view fusiform, in profile elongate-inequilateral; color in KOH olive-ochraceous, in Melzer's giving a fleeting-amyloid reaction then weakly dextrinoid, wall thin (-0.3 μ).

Basidia 4-spored. Hymenophoral trama greenish blue as mounted in Melzer's and viewed with a hand lens against white paper. Pleurocystidia scattered, 36-48 X 8-13 μ, fusoid to fusoid-ventricose, apex subacute, content hyaline in KOH. Pileus cutis of a tangle of gelatinous hyphae 4-10 μ wide, yellowish in KOH and Melzer's under the microscope, terminal cells tubular but in narrow hyphae clavate and with diverticulae near apex (rudimentary dendrophyses). Hyphae of subcutis and adjacent context weakly yellowish in Melzer's under the microscope. Clamp connections absent.

Habit, habitat, and distribution.—Near Forestville in mixed woods, Marquette County, July 30, 1968, Ammirati 2013, at Harlow Creek, Ammirati 2011.

Observations.—The irregular dendrophyses-like branching of some of the terminal cells of the pileus cuticle is a very unusual feature in *Boletus*, and in this taxon is correlated with the very pale colors to make an easily recognized fungus. The pileus margin may become slightly appendiculate-crenate as in section *Leccinum* of *Leccinum.*

202b. Boletus edulis var. clavipes Peck

Rept. N. Y. State Mus. 51:309. 1899

Pileus 5-12 (15) cm broad, convex becoming broadly convex to nearly plane, margin even; surface unpolished at first, merely dull in age (not viscid), glabrous, uneven, color a dull vinaceous-cinnamon becoming flushed with ochraceous first along the margin and then irregularly causing the pilei to be variously colored (some quite yellow in age). Context pallid but vinaceous near cuticle, odor and taste fungoid, no discoloration showing around wormholes, $FeSO_4$ negative.

Tubes 1-1.5 cm deep, depressed around the stipe, stuffed whitish at first, pallid becoming yellow (not greenish until old age); pores minute, yellow, slowly staining brown if injured.

Stipe 6-11 cm long, 1-3 cm thick, clavate becoming equal, stuffed, pallid within and not darkening around larval tunnels, unchanging when cut; surface pallid to brownish overlaid by a fine white reticulum over most of its length, base pallid.

Spores olivaceous in deposit, 16-19 X 4.5-6 μ, smooth, apex lacking a pore or thin spot, in face view fusoid, in profile narrowly inequilateral, color in KOH weakly ochraceous alone or in groups pale clay color, in Melzer's with a faint fleeting-amyloid reaction then weakly dextrinoid, wall about 0.3 μ thick.

Basidia 4-spored, 9-11 μ wide, clavate, yellow fading to hyaline as revived in KOH. Pleurocystidia and cheilocystidia—none observed.

Tube trama of gelatinous divergent hyphae, sections mounted in Melzer's pale brown (no obvious amyloid or fleeting-amyloid reaction), walls and content nonamyloid. Pileus cuticle a tangle of interwoven hyphae 6-12 μ wide with end-cells not significantly differentiated or merely weakly cystidioid, hyphae not in a basically trichodermial pattern, in Melzer's the content ochraceous or orange-ochraceous. Hyphae beneath the cuticle merely yellowish in Melzer's. Clamp connections absent.

Habit, habitat, and distribution.—Scattered under birch and aspen, Wycamp Lake, Emmet County, August 3, 1968, Smith 75838.

Observations.—Mounts of the tissue of the hymenophore and pileus do not show any significant amyloid reaction (they remain yellowish to brownish in the Melzer's as viewed with a hand lens). The pilei were not viscid when fresh, but the cuticle as revived in KOH showed some degree of gelatinization though this does not appear too significant. The field characters are the vinaceous-brown pileus developing yellow tints at least along the margin in age, and the whitish reticulum over the brownish ground color of the stipe.

202c. Boletus edulis var. aurantio-ruber Dick & Snell

Mycologia 57:454. 1965

Pileus 6-10 cm broad, broadly convex with the margin incurved against the tubes, in age broadly convex and with a narrow sterile margin at times breaking into small lobes; surface glabrous, viscid, smooth to rugulose, "ferruginous" (rusty red) overall or only the extreme margin ochraceous, pellicle separable. Context thick, pallid, odor fungoid, taste nutty, slowly staining yellow. around larval tunnels; with KOH pale yellow on context, ochraceous on tubes; with $FeSO_4$ pale olive-gray on flesh, greenish on tubes.

Tubes stuffed white when young, soon yellowish and staining brown where injured, 1-2 cm deep, depressed around the stipe; pores minute and round (2-3 per mm), quickly staining yellowish olive after bruising.

Stipe 8-13 cm long, 2-2.5 cm thick at apex, clavate, solid, pithy, pallid within, surface finely reticulate over upper three-fourths, pallid below, pale vinaceous-cinnamon above, unchanging when bruised.

Spore deposit olive-black fresh; spores 13-17 (19) $\times$ 4.5-6 μ, smooth, lacking an apical pore, pale ochraceous to pale tan in KOH, about the same color in Melzer's but with a grayish shadow, elongate, inequilateral in profile view and with a broad suprahilar depression, subfusoid in face view, wall only slightly thickened.

Basidia 4-spored, 26-35 × 7-9 μ, narrowly clavate, yellowish hyaline in KOH. Pleurocystidia—none found. Pileus cuticle a gelatinized trichodermium matted down to a pellicle, the hyphae 3-7 μ wide.

Habit, habitat, and distribution.—We have one collection by H. V. Smith (Smith 64485) obtained at the Rifle River Recreation area, Lupton, September 29, 1961.

Observations.—The distinctive features of the above collection are the viscid pileus with a separable pellicle and ferruginous-red colors except for the yellow margin, and the pores staining olive yellowish. In Melzer's revived bits of hymenial tissue are greenish to bluish black. When hymenial tissue is revived in KOH much bright yellow pigment diffuses through the mount. According to Dick and Snell the reticulum is brown to black on drying. This could be caused by the specimen standing too long before being dried.

The following description is of a variant from the Huron Mountains: J. Ammirati 2004.

Pileus 10 cm broad, broadly convex, subviscid to viscid, glabrous, vinaceous-red over the disc ("Pompeian-red" to "carmine"), margin paler vinaceous with a distinct yellow undertone and extreme edge yellowish. Context white tinted pinkish, unchanging when cut, odor pungent-fungoid, taste pleasant; when dried pallid-buff.

Tubes lemon-yellow, shallowly depressed, unchanging; pores small, yellow, staining light cinnamon.

Stipe 10 cm long, 2 cm at apex, 3.5 cm at base, clavate, solid, whitish within or near cortex pinkish red; surface with a whitish reticulum over upper portion, base pallid with streaks of cinnamon to pinkish cinnamon, apex cinnamon to pinkish cinnamon beneath the reticulum.

Spores 12-15 × 4-5.5 μ, smooth, apex lacking a pore or thin spot; shape in face view fusoid, in profile elongate-inequilateral, color revived in KOH pale olive-yellow, in Melzer's long retaining a blue-gray shadow (weakly amyloid), wall about 0.2 μ thick.

Basidia 4-spored, 8-11 μ in diameter, clavate, hyaline to yellowish in KOH or Melzer's. Pleurocystidia absent. Cheilocystidia mostly basidiole-like or weakly fusoid-ventricose to an obtuse apex.

Tube trama of the *Boletus* subtype, mounts of hymenophore giving off a yellow pigment in KOH. Pileus cutis a tangled layer of more or less appressed gelatinous hyphae but in places some indications of a trichodermial structure visible, hyphal ends tubular to clavate and up to 7 μ broad at apex, hyphae 4-6 μ broad and gelatinous, with a strong fleeting-amyloid reaction in Melzer's, and in this medium the hyphal tips with yellowish content, the layer beneath the cutis orange-ferruginous in

Melzer's (the cell content homogeneous). Hyphae of subcutis and adjacent context with dingy ochraceous or ochraceous-orange content in Melzer's. Clamp connections none.

Solitary at Harlow Creek, Marquette County, July 30, 1968, Ammirati 2004.

Observations.—On *B. edulis* collected by R. Shaffer (5101) in France, the same fleeting-amyloid reaction is present, and we assume that this reaction is typical of the stirps; its absence in var. *clavipes* may be more significant than we now think. The Ammirati collection may represent a distinct species, but more data is needed. The beautiful red pileus reminds one of *B. speciosus* and *B. regius,* but the color of the flesh and lack of any staining to blue when injured rules them out. For the time being at least, we regard it as an extreme variant of var. *aurantio-ruber.*

Section ALLOSPORI (Singer), comb. nov.

Boletellus section *Allospori* Singer

Farlowia 2:138. 1945

203. Boletus betula Schweinitz

Schr. Naturf. Ges. Leipzig 1:94. 1822

Ceriomyces betula (Schweinitz) Murrill, Mycologia 1:144. 1909.
Boletellus betula (Schweinitz) Gilbert, Bolets p. 108. 1931.

Illus. Pl. 148.

Pileus 3-9 cm broad, convex, becoming broadly convex, glabrous, viscid to glutinous, dingy reddish orange-brown, becoming paler, the red tints receding and yellow becoming dominant, in age "apricot-yellow" (flavous), with a slight sterile margin. Context greenish yellow near the tubes, orange-yellow near cuticle, very soft and readily collapsing, odor and taste not distinctive, no color change to blue or green developing when injured.

Tubes 1-1.5 cm long, depressed around the stipe, dark greenish yellow in section; pores about 1 mm broad, not stuffed when young, round, pale yellow becoming greenish yellow in age but not changing to blue when bruised.

Stipe 10-20 cm long, 7-16 mm thick, equal, solid, reddish within near periphery, central portion pallid or finally pinkish, surface lacerate-reticulate or reticulation less conspicuous, at first yellow above and reddish below or in age reddish throughout, base with white cottony masses of mycelium.

Spore deposit dark olive to olive-brown; spores 15-18 × 6-9 μ, narrowly ellipsoid, ochraceous-brown in KOH and Melzer's or a small number dextrinoid, ornamented with a loose reticulate outer layer and with a distinct apical pore, suprahilar area ornamented, in old spores the exosporium looser and more wrinkled over the apex, the remains near the apiculus more granular.

Basidia 4-spored, 36-45 × 10-15 μ, yellowish in KOH and Melzer's. Pleurocystidia fusoid-ventricose, 36-50 × 9-14 μ, not conspicuous, rare to scattered, apex subacute, content not distinctive.

Tube trama gelatinous and divergent, floccose central strand present, sections when first mounted in Melzer's dark violet and violet pigment diffusing into the mount but hyphal wall not amyloid—the origin of the violaceous pigment not evident.

Pileus context giving the same Melzer's reaction as the tube trama, the hyphae 9-12 μ wide, thin-walled, smooth and yellow in Melzer's. Pileus cuticle a turf of narrow (4-7 μ) hyphae with end-cells subclavate to subcylindric, the cells of the hyphae 60-150 μ long, their walls gelatinous and the cells lacking internal amyloid particles, the content yellow when first revived in KOH but soon fading. Clamp connections absent.

Habit, habitat, and distribution.—Solitary to scattered under mixed pine and oak, southeastern and eastern in distribution but to be expected in the southwestern counties of the state.

Observations.—This is an odd species which deserves more study. It forms a rather isolated type in the family no matter how classified.

PULVEROBOLETUS Murrill

Mycologia 1:9. 1909

The genus is here recognized on the basis of the dry bright yellow veil of the type species. This is the original concept and much narrower than that of Singer (1962), which includes a rather miscellaneous assortment of species.

Type species: *Pulveroboletus ravenelii.*

204. Pulveroboletus ravenelii (Berkeley & Curtis) Murrill

Mycologia 1:9. 1909

Boletus ravenelii Berkeley & Curtis, Ann. Mag. Nat. Hist. II, 12:429. 1853.
Suillus ravenelii (Berkeley & Curtis) Kuntze, Rev. Gen. Pl. 3(2), 536. 1898.

Illus. Pls. 149-50.

Pileus (1) 2-6 (10) cm broad, obtuse to convex to plano-convex to nearly plane; surface dry and pulverulent or at times (when wet) somewhat tacky to viscid but no gelatinizing elements present even when young, bright sulphur-yellow overall and with a pulverulent coating from the remains of a veil, disc becoming orange-red to brownish red ("cinnamon" to "pinkish cinnamon") as pulverulence disappears and the surface then felty-fibrillose; margin incurved, often covered with partial veil material colored "strontium-yellow." Context whitish to yellow, changing slowly to light blue and then drab-brown or dull yellowish where injured, about 1 cm thick, floccose, taste acidulous, odor slight and hardly distinctive.

Tubes 5-8 mm deep, adnate, sometimes becoming slightly depressed around the stipe, soon primuline-yellow to lemon-yellow or more olive tinted ("buffy citrine") changing to greenish blue than dusky brown and finally black when bruised; pores small (2-3 per mm), round or nearly so to angular, bright yellow when undamaged.

Stipe 4-9 (15) cm long, 4-12 (16) mm thick, equal or tapering toward the apex or sometimes variously irregular in shape, solid, whitish within or soon pale yellowish and developing a pink tinge where bruised; covered at least over the lower part with yellow pulverulence from a dry veil, veil rarely leaving a faint annulus; surface dry to moist, colored brilliant yellow, with yellowish to white rhizomorphs around the base.

Spore deposit smoky olive—Coker and Beers; spores $8\text{-}10.5 \times 4\text{-}5\,\mu$, ventricose to more or less ovate in face view, in profile somewhat inequilateral, smooth, hyaline outer sheath inconspicuous, in Melzer's pale rusty brown, dingy ochraceous-tawny in KOH.

Basidia 4-spored, $24\text{-}36 \times 10\text{-}14\,\mu$, pale lemon-yellow in KOH, not as bright in Melzer's. Pleurocystidia scattered, $30\text{-}46 \times 7\text{-}12\,\mu$, sometimes inconspicuous, fusoid-ventricose, hyaline to lemon-yellow in KOH, smooth, thin-walled. Cheilocystidia basidium-like or similar to pleurocystidia.

Tube trama gelatinous and strongly divergent from a distinct central strand, the hyphae nonamyloid. Cuticle of pileus a tangled mass of narrow ($2.5\text{-}6\,\mu$) curved hyphae yellow to hyaline in KOH and reddish brown in Melzer's. Hyphae of the context $5\text{-}10\,\mu$ wide, wormlike in shape, rusty to pale reddish brown in Melzer's. Clamp connections absent.

Habit, habitat, and distribution.—Mostly solitary to scattered in pine woods as well as hardwoods. In Michigan it is most frequent along the south shore of Lake Superior. It fruits during the summer and early fall. It is most abundant in the southeastern United States.

Observations.—Young buttons with the veil intact were sectioned but no sphaerocysts were found, hence the term pulverulence as used for the veil of this species refers to the dry floccose loosely woven texture, not a true pulverulence caused by loose globose cells. The hyphal components are interwoven, smooth-walled, yellow, and branched. The hyphae fade quickly in KOH, but bunches of greenish crystals and amorphous debris often precipitate in KOH mounts. It would, indeed, be difficult to confuse this bolete with any other in the state.

BOLETELLUS Murrill

Mycologia 1:10. 1909

The spores are furnished with more or less longitudinal wings, ridges, or striations. In most other respects the species are quite diverse in some features, as can be seen from the key. However, the Michigan species placed here do seem to form a group of closely related species.

Type species: *Boletellus ananas.*

Singer (1945, 1963) has included species with ornamented spores here as well as some with smooth spores. We find ourselves in strong disagreement with his definition of this genus and the vagueness of his definition of the family Strobilomycetaceae as a whole. For this reason, we adhere to a concept of *Boletellus* with at least one set of features which anyone with a microscope can accurately ascertain and which is consistent with the type of the species.

KEY TO SPECIES

1. Veil present; pileus squamulose . . . . *B. ananas*
1. Veil absent 2

    2. Stipe lacerate-reticulate . . . . *B. russellii*
    2. Stipe surface even or nearly so, or squamulose 3

3. Pileus dark brown to dark vinaceous-brown . . . . *B. chrysenteroides*
3. Pileus rose-red when young or freshly matured 4

    4. Spores 11-14 × 5.5-7 μ . . . . *B. pseudo-chrysenteroides*
    4. Spores 9-12 × 4-5 μ . . . . *B. intermedius*

205. Boletellus ananas (Curtis) Murrill

Mycologia 1:10. 1909

Boletus ananas Curtis, Amer. Journ. Sci. II 6:251. 1848. (Berk. Hook Journ. Bot. 1:101. 1849.)
Strobilomyces pallescens Cooke & Massee, Grevillea 18:5. 1889.
Boletus isabellinus Peck, Bull. Torr. Club. 24:146. 1897.

Pileus 4-10 cm broad, subglobose at first becoming convex-expanded; surface dry and scaly; scales of fascicles of fibers forming large woolly squarrose to subimbricate scales more erect toward the disc; color "deep vinaceous" or "Pompeian-red" and "Isabella color" fibrils, fading in age (vinaceous-red to pale tan with olive-yellow fibrils). Context whitish to yellowish in the pileus, readily staining blue when cut; odor none, taste mild.

Tubes yellow, often tinged with reddish brown, staining blue where injured, eventually becoming yellowish brown or reddish brown all over, depressed around the stipe, 11-16 mm deep; pores yellow, medium to large, irregular, angular, forming a distinctly convex lower surface.

Stipe 6-14 (18.5) cm long, 8-15 mm thick, tapering upward at first, rarely subbulbous or tapered downward, at times nearly equal, solid, white or pallid, often with a red belt near apex ("Pompeian-red" or lighter), becoming sordid pinkish, subfibrillose to fibrous, rarely with an annular remnant from the veil, usually smooth.

Spores 16-20 (23) $\times$ 7.5-9.5 μ, longitudinally ridged by thin wings arranged in a slightly spiraled pattern, apex with a small often indistinct pore.

Basidia 22.5-50 $\times$ 9.5-17 μ, 2- to 4-spored. Cystidia fusoid, mucronate, of ampullaceous to ampullaceous-capitate form, hyaline, brittle, 24-50 $\times$ 5-15 μ. Tube trama bilateral with more colored mediostratum and looser divergent lateral strata, the hyphae lacking clamps; outer layer of veil of hyaline semiopaque, long-cylindric irregularly interwoven, thin-walled hyphae 4-7 μ wide, the terminal cells slightly tapered to an obtuse tip or abruptly broadly rounded or subcapitate.

Chemical reactions: KOH on context yellow changing to brown, pileus surface olive-yellow becoming maroon; NH_4OH on the surface olive-yellow; $FeSO_4$ on context green; with aniline negative.

Habit, habitat, and distribution.—In various kinds of woods, from open pine-oak to dense tropical hammocks frequently on the bases of trees (pine and oak), May to November, southern. As yet not found in Michigan, but to be expected possibly in the southwest corner of the state.

Observations.—Most of the data presented here are taken from Singer (1945) and Coker and Beers (1943). In the dried material we have studied, the spores are dingy yellow-brown in Melzer's (not dextrinoid), the ribs of the spore are as Singer described them, and we found no amyloid reactions of any degree on the spores or hyphae of the basidiocarps.

206. Boletellus russellii (Frost) Gilbert

Bolets p. 107. 1931

Boletus russellii Frost, Bull. Buff. Soc. Nat. Sci. 2:104. 1874.
Ceriomyces russellii (Frost) Murrill, Mycologia 1:144. 1909.
Boletogaster russellii (Frost) Lohwag, in Handel-Mazzetti, Symb. Sin. 2:56. 1937.

Illus. Pl. 151.

Pileus 3-8 (13) cm broad, hemispheric to convex with a strongly incurved margin, expanding to broadly convex but rarely nearly plane or shallowly depressed; surface dry, subtomentose to tomentose, occasionally appearing squamulose, becoming areolate with the tomentum of the areolae aggregated into obscure tufts, colored buffy brown to brownish yellow or olive-gray, at times, however, browner ("Isabella color" to "buffy brown" to "vinaceous-russet" to "tawny-olive"), at times redder ("cinnamon-brown" to "rufous"—Kauffman), yellow context showing between the areolae, sterile margin usually present. Context up to 2 cm thick, yellow to pallid-yellow, often avellaneous around wormholes, unchanging when bruised; taste mild, odor none.

Tubes deeper than thickness of context, adnate becoming depressed around the stipe, often extending for short distances down the stipe as lines, yellowish olive to olive-green ("orange-citrine" to "citrine" to "empire-yellow"), unchanging or brighter yellow when bruised; pores large (1 mm or more wide, angular, dissepiments thin).

Stipe 10-18 cm long, 1-2 cm thick, equal or evenly enlarged downward, often curved at the base, solid, firm, flesh yellow, covered overall with coarse lacerate reticulation, surface including reticulation dull red ("Hays-russet" to "brick-red" to "testaceous"), often viscid near the base but no gelatinous veil present.

Spore deposit dark olive; spores 15-20 X 7-11 μ, in face view elliptic, in profile view obscurely inequilateral, longitudinally striate with deep grooves or wrinkles in outer wall and with a cleft in the wall at the apex, bister in KOH, in Melzer's somewhat dextrinoid.

Basidia 34-40 X 10-15 μ, broadly clavate, hyaline in KOH, yellowish in Melzer's. Pleurocystidia scattered to abundant, 50-70 X 9-15 μ, hyaline in KOH, occasionally with melleous content, yellowish in Melzer's. Cheilocystidia similar to the pleurocystidia, fusoid-ventricose, smooth, thin-walled, apex subacute, inamyloid both as to wall and content.

Tube trama gelatinous and divergent, hyphae hyaline in KOH, yellowish in Melzer's. Pileus trichodermium with the upper cells of the elements short and enlarged and becoming aggregated into fascicles (which are the squamules), content yellowish to hyaline in KOH. Context composed of hyphae interwoven, inamyloid, thin-walled and cells inflated at times to 20 μ wide. Caulohymenium of basidia and cystidia, the cystidia fusoid-ventricose to nearly clavate, mostly hyaline. Clamp connections absent.

Habit, habitat, and distribution.—Solitary to scattered on humus usually under oak, often along openings in woods or scattered in thin woods, late summer. Generally rare, but found in quantity at times in the oak-pine barrens from Roscommon to Topinabee.

Observations.—This is a most easily recognized species in every respect. The conspicuously lacerate-reticulate reddish stipe, the areolate pileus, characteristically long stipe with a viscid base when perfectly fresh, and the longitudinally winged spores are all important features. *Boletus projectellus* has shallower reticulations over the stipe, and the edges of the veins forming them are not lacerate or broken. Under the microscope the spores readily distinguish them, for they are smooth in *B. projectellus*.

Material examined.—Barry: Smith 73157. Cheboygan: Smith 38891, 39269, 62940, 62966, 64855, 8-5-51; Thiers 3748, 3839. Gratiot: Potter 2880, 3802, 7643, 7846, 7974, 11246, 13970. Livingston: Smith 18609. Luce: Smith 38076. Marquette: Kauffman 1580. Oakland: Smith 7112, 7270, 15204. Ogemaw: Shaffer 2711; Smith 67572. Washtenaw: Hoseney 194; Rea 766; Shaffer 2659, 2754; Smith 1713, 1734, 18562, 18666, 64178; Thiers 4324.

207. Boletellus pseudo-chrysenteroides Smith & Thiers, sp. nov.

Illus. Fig. 113; Pl. 152.

Pileus 5-10 cm latus, late convexus, siccus velutinus demum areolatus, obscure roseus, demum olivaceus. Contextus laete luteus tactu caeruleus. Tubuli 8-12 mm longus, lutei tactu caerulei; pori lati (1-1.5 mm), lutei tactu caerulei. Stipes 6-9 cm longus, 9-14 mm crassus, solidus, intus luteo-virgatus tactu caeruleus; extus roseo-pruinosus, sursum luteus, deorsum melleus. Sporae 11-14 × 5.5-7 (8) μ, longitudinaliter striatus. Specimen typicum in Herb. Univ. Mich. conservatum est; prope Manchester, September 15, 1961, legit Smith 64264.

Pileus 5-10 cm broad, broadly convex expanding to plane or nearly so; surface dry and velvety when young, soon becoming areolate as in *B. chrysenteron;* dark rose-red overall ("Pompeian-red") and at first on the tufts of the epicutis adorning the areolae but the tufts slowly becoming somewhat olivaceous; pallid to pinkish in the space between the areolae. Context pale to bright yellow, soon deep blue when injured, odor none, taste acid; with KOH orange, with $FeSO_4$ yellow changing to dingy ochraceous slowly.

Tubes 8-12 mm deep, depressed around the stipe, yellow changing to blue, with decurrent lines down the apex of the stipe; pores wide and angular (1-1.5 mm in diameter), yellow but blue and then brown from injury.

Stipe 6-9 cm long, 9-14 mm thick, equal, solid, within streaked ochraceous and pallid, soon blue where injured; surface red-pruinose to scurfy, about concolorous with pileus or apex yellow, base coated with a dingy honey-yellow mass of soft mycelium.

Spore deposit dark olive-fuscous; spores 11-14 $\times$ 5.5-7 (8) μ; in face view subelliptic, in profile broadly somewhat inequilateral, suprahilar depression broad and often inconspicuous, ornamented with 9-12 sublongitudinal ridges (about as in *B. chrysenteroides*) and with an apical break in the outer wall in the form of a slit, in KOH bright pale amber-brown or more ochraceous, in Melzer's nearly "Brussels-brown" (a dark yellow-brown) or more dingy and many showing a distinct bluish gray shadow when first revived.

Basidia 4-spored, clavate, 20-26 $\times$ 8-9 μ, soon disintegrating in 2 percent KOH, pale yellow in Melzer's and outlines remaining distinct. Pleurocystidia abundant 36-48 $\times$ 9-13 μ, fusoid-ventricose with a long neck and subacute apex, hyaline in KOH, in Melzer's having a dark dingy yellow-brown granular content variously dispersed. Pileus cutis a trichodermium of hyphae 8-12 μ in diameter with dark orange-brown content in Melzer's but no amyloid granules present, walls smooth, thin, usually with a constriction at the septa, end-cells tubular and rounded at the tip or clavate and up to 15 μ or more broad. Clamp connections none.

Habit, habitat, and distribution.—Scattered under beech and other hardwoods on low ground near Manchester, September 15, 1961, Smith 64264 (type), and from near Ann Arbor.

Observations.—The field characters would lead one to place it in the *Boletus fraternus* group. It differs from *Boletellus chrysenteroides* in the copious honey-yellow mycelium around the base of the basidiocarps and in the red of the pileus as well as a quicker and more intense blue reaction when injured.

208. Boletellus chrysenteroides (Snell) Singer

Mycologia 33:422. 1941

Boletus chrysenteroides Snell, Mycologia 28:468. 1936.

Illus. Pl. 153.

Pileus 3-6 cm broad, convex expanding to broadly convex; surface dry, unpolished, velvety to pubescent, "mummy-brown" when young, fading to "Dresden-brown" or "Vandyke-brown" (deep yellow-brown to vinaceous-brown), slowly becoming dull bay, surface becoming areolate-cracked and pale flesh showing at maturity. Context soft-spongy, thinner than the tubes are deep, yellowish white, turning blue when cut.

Tubes "lemon-chrome" to "lemon-yellow" when young and fresh, dingy greenish yellow in age, depressed around the stipe, not stuffed; pores subangular, about 1 per mm, turning blue when bruised.

Stipe 2-4 cm long, 8-12 mm thick, solid, extreme apex yellowish, lower portion dull reddish in age, yellowish within at first and reddish in age, surface becoming sordid where handled and more or less furfuraceous to fibrillose-punctate much as in *Leccinum* but the ornamentation not darkening, sometimes falsely reticulate, sometimes in age the surface fibrils simply matted down.

Spore deposit bister to dark olive-brown; spores (10) 12-16 (22) X 4.6-7.5 (8) μ, very variable in some collections, narrowly ovate to nearly oblong in face view or some more or less ventricose, inequilateral in profile, pale to dark bister in KOH, russet in Melzer's, longitudinally striate, in KOH seen to be longitudinally corrugated and with a pronounced hyaline to yellowish outer sheath.

Basidia 1- to 4-spored, 30-34 X 8-9 μ, hyaline to yellowish in KOH and Melzer's. Pleurocystidia abundant and variable, 36-60 X 8-14 μ, fusoid-ventricose with acute apex or ventricose-capitate and many intermediate in shape, thin-walled, smooth, hyaline to yellowish in KOH and yellow in Melzer's. Cheilocystidia fusoid-ventricose to clavate, or ventricose part in upper third and the apex more or less mucronate.

Tube trama gelatinous and divergent from a central floccose strand. Pileus trichodermium of hyphae with reddish brown content in Melzer's and with short cells near the apex—apical cell cystidioid—and these frequently forming a palisade, subapical cell often inflated (12-20 μ broad and 15-30 μ long), thin-walled, hyaline to yellowish in KOH. Context composed of floccose interwoven hyphae which are non-amyloid, with the content often colloidal (as seen in KOH) and yellowish especially near the trichodermium. Surface of stipe a layer of incrusted interwoven hyphae yellow in KOH. Clamp connections absent.

Habit, habitat, and distribution.—Solitary to gregarious in oak woods, especially around very old decaying logs and stumps, July and August or early September, common at times during warm wet weather in southern Michigan.

Observations.—The spores serve quickly to distinguish this species from the *Boletus chrysenteron-fraternus* complex.

Material examined.—Barry: Mazzer 8-19-66. Cheboygan: Shaffer 1587; Smith 37326, 37556, 63827, 7-23-52; Thiers 3613. Gratiot: Potter 3380, 3804, 7852, 7854, 8154, 8177, 8231, 8386, 8424, 8444, 9830, 10051, 11780. 12663, 12842. 13393. Livingston: Hoseney 508; Smith 7048, 18516, 18615, 64192, 64347. Oakland: Smith 6689, 6959, 7086, 7117, 18745, 73289. Roscommon: Potter 10158b. Washtenaw: Hoseney 195, 543; Kauffman 8-13-15; Smith 1681, 15313, 62661, 64150; Thiers 4521.

209. Boletellus intermedius Smith & Thiers, sp. nov.

Illus. Fig. 114; Pl. 154.

Pileus 4-10 cm latus, convexus demum late convexus, siccus, subtomentosus, roseus demum subolivaceus. Contextus luteus tactu caeruleus. Tubuli 7-12 mm longi, laete lutei, tactu caerulei; pori lati (1 mm) lutei. Stipes 6-12 cm longus, 8-17 mm crassus, intus luteus tactu caeruleus, extus rubro-pruinosus, non-reticulatus. Sporae 9-12 (13) X 4-5 μ. Specimen typicum in Herb. Univ. Mich. conservatum est; prope Pinckney, September 11, 1965, legit R. Homola (Smith 72559).

Pileus 4-10 cm broad, convex with a bent in margin, becoming broadly convex to nearly plane; surface dry and subtomentose (plushlike), when young rose-red ("Pompeian-red") but soon duller from an olive overcast and entire pileus usually olive-gray in age except for rose to pink tints showing where cuticle is cracked, in age the pileus areolate. Context yellow staining blue, taste mild, odor fungoid or lacking, Melzer's on cuticle slowly olive, $FeSO_4$ on context no reaction, KOH yellowish to tan.

Tubes 7-12 mm long, depressed around the stipe, pale bright yellow when young, greenish yellow in age, staining bluish when injured; pores up to 1 mm wide, irregular in outline; breaking downward as in *Boletus.*

Stipe 6-12 cm long, 8-17 mm thick at apex (rarely 3 cm in widest dimension when compressed), solid, interior yellow staining blue, in base of stipe watery greenish yellow, surface longitudinally ribbed to smooth, rhubarb-red from pruina but yellow showing at apex, base pale yellowish to pallid from mycelial coating, nearly white as dried, staining blue where injured, not at all reticulate.

Spore deposit olive. Spores 9-12 (13) X 4-5 μ (some abnormal spores up to 7 μ broad), truncate at apex from a small pore, appearing to be smooth but under a 1.4 NA oil-immersion lens and mounted in Melzer's medium seen to be weakly longitudinally striate on mature spores, amyloid at first but fading to pale bister and some becoming weakly dextrinoid at least in distal part of spore, in profile view bluntly somewhat inequilateral (the suprahilar depression distinct), narrowly ovate to subelliptic to subfusiform (blunt at apex) in face view, wall about -0.2 μ thick.

Basidia 4-spored, 28-33 X 7-10 μ, yellowish in KOH and Melzer's. Pleurocystidia scattered, 36-48 X 7-12 μ, narrowly fusoid, apex sub-acute, hyaline in KOH or Melzer's or in the latter occasionally with amorphous coagulated dark yellow-brown pigment variously dispersed. Cheilocystidia smaller than pleurocystidia and intergrading with basidioles, content ochraceous in KOH or Melzer's. Caulocystidia clavate, yellow in KOH.

Tube trama with subparallel hyphae in the central area, the outer hyphae diverging slightly to the subhymenium; laticiferous hyphae also present. Pileus trichoderm well formed, the apical cell of the filaments cystidiod, 30-120 X 8-15 μ, very heavily but finely incrusted (as mounted in KOH), with fine hyaline to brownish material, in Melzer's the incrustations fewer and more inconspicuous (not tawny as in *B. subtomentosus*), the cells back from the terminal cell elongate and tubular, walls also thin and incrusted. Context hyphae yellowish to hyaline in Melzer's, content not distinctive; laticiferous elements present and with bright yellow homogeneous content in KOH. Clamp connections absent.

Habit, habitat, and distribution.—Collected at the Edwin S. George Reserve, Pinckney, September 11, 1965, by Richard Homola (Smith 72559).

Observations.—This species is easily mistaken for a large *B. chrysenteron,* but the pileus starts out with rose tones and later changes to olivaceous with an undertone of red. The spores are clearly longitudinally striate, but at first glance they appear smooth under low-power oil-immersion lenses. The apex of the spore is blunt and a minute pore is present.

Smith 72458 is interpreted as a second collection of this species. In it small buttons were "snuff-brown," the red developing later. As dried the pilei in both are practically identical in appearance. $FeSO_4$ gave no reaction on the context, and on the cuticle NH_4OH and KOH were both negative. Melzer's gave a bluish spot on the pileus context. In this collection, when the pileus was broken downward, the tubes separated as in *Boletus*. The spores were amyloid as revived from hymenophoral tissue but slowly lost the blue tint.

STROBILOMYCES Berkeley

Decades of Fungi, Hooker's Journ. Bot. 3:77. 1851

Pileus coarsely fibrillose to fibrillose-squamulose and the scales gray to blackish; tubes pallid to cinereous young, blackening in age; spore deposit dark rusty brown to blackish brown and spores both globose and ornamented.

Type species: *Strobilomyces floccopus.*

As found in Michigan this genus and *Boletellus* are amply distinct, but on a worldwide basis it is evident that the 2 genera are closely related. Singer postulated a close relationship with *Porphyrellus,* a genus we include as part of *Tylopilus.*

KEY TO SPECIES

1. Spores reticulate. . . . . *S. floccopus*
1. Spores echinate to verrucose, at times with a broken or partial reticulum . . . . *S. confusus*

210. Strobilomyces floccopus (Fries) Karsten

Bidr. Finl. Nat. Folk 37:16. 1882

Boletus floccopus Fries, Syst. Myc. 1:292. 1821.
Boletus strobilaceus Socopoli. ex Fries, Elench. Fung. 1:127. 1828.

Illus. Pl. 155.

Pileus 4-15 cm broad, pulvinate, finally flattened, dry, between "mummy-brown" and black, or about "sepia" or grayer all over at first or near the margin at maturity; the surface very soon broken up into coarse scales showing the pale context in the cracks, margin typically hung with the remains of the gray to pallid fibrillose to submembranous veil, the pileus often blackish in age. Context white or whitish when injured, discoloring to reddish and finally black or at times blackening directly, odor not distinctive, taste mild, with red stains from KOH and these later becoming brown, NH_4OH ochraceous to lacking a color change, with $FeSO_4$ bluish gray to greenish gray.

Tubes 12-15 mm deep, gray or white at first, becoming dark gray, when injured staining reddish then black, adnate to somewhat decurrent or slightly depressed around the stipe; pores angular, usually sublamellate near the stipe, when mature gray, staining reddish then black when injured; tubes when broken downward splitting rather than separating.

Stipe 5-12 cm long, 12-25 mm thick near apex, solid, subcylindric or enlarged downward, rarely enlarged upward, surface gray to concolorous with the pileus, apex unpolished, striate with longitudinal veins extending down from the walls of the tubes, base tomentose, midportion covered by a thick woolly sheath from the copious soft veil which leaves an annular zone or two to several belts lower down when it breaks, context reddening and finally black when injured.

Spore print black; spores 9.5-15 X 8.5-12 μ, covered entirely by a complete network of lines to form a reticulum, the ridges up to 0.3-1.7 μ high to almost entirely imbedded, the wall rather thick, with a more or less distinct germ pore, buffy brown with deeper brown ornamentation as revived in KOH.

Basidia 30-58 X (7.5) 10-18 μ, 4-spored. Pleurocystidia numerous, 17-90 X 8-26 μ, clavate, mucronate-apiculate to more or less fusoid-ventricose, occasionally containing a brown drop when very young, later brown throughout.

Tube trama with a mediostratum of floccose hyphae with tubular to vesiculose cells, hyphae diverging from the central strand to the subhymenium. Pileus trichodermium of elements with cells 20-88 μ long, cylindric or slightly constricted at the septa, the terminal members often slightly attenuate toward the apex but always rounded, 7-18 μ wide. Clamp connections absent.

Habit, habitat, and distribution.—Solitary to scattered under hardwoods, usually under oak, or in mixed conifer and hardwoods; common but not found in quantity, summer and fall throughout the state.

Observations.—Unfortunately, there is almost unanimous opinion that, though not poisonous, this species has little to recommend it for the table. It is one of the easiest to learn to recognize at sight—seeing it once is usually enough. Because of its shaggy appearance and dull color it has been nicknamed the "Old Man of the Woods."

Material examined.—Barry: Mazzer 4067. Cheboygan: Shaffer 1585; Smith 38205, 63109, 63021; Thiers 3606, 3608, 3760, 3831, 4320. Emmet: Thiers 4157, 4266. Gratiot: Potter 3551, 3534, 7597, 7978, 8394, 12878, 13156, 13352. Livingston: Homola 883, 1924; Hoseney 216, 502; Smith 72722, 72782. Luce: Smith 37487, 41773; Walsh 307. Montcalm: Potter 10095. Ontonagon: Peters 1129.

Washtenaw: Hoseney 523; Kanouse 1927; Kauffman 1905, 1907; Shaffer 2622; Smith 33-639, 1505, 6657, 18392, 18486, 18747, 62626, 62664; Thiers 4511, 4515, 4534, 4558, 4578, 4586, 4597, 4609.

211. Strobilomyces confusus Singer

Farlowia 2:108. 1945

Pileus 3-9.5 cm broad, convex becoming broadly convex, disc flattened in age; surface dry and squamose, color fuscous-black, decorated with very acute erect spines which are denser toward the center and often cristate, fimbriate, or fasciculate, at times compressed, basal diameter of spine not over 2 X 1 mm in dried material, surface appearing areolate-squamose if the spines have been obliterated and then the appearance of the pileus resembles somewhat that of *S. floccopus,* not persistently woolly; veil thin and appendiculate on margin as fragments. Context whitish, quickly reddening in the pileus to carrot color (except for a narrow zone under the cuticle), odor not distinctive.

Tubes whitish gray to gray, becoming blackish even if carefully dried, becoming spotted with "cinnamon" where touched when fresh but the areas finally black, 10-18 mm deep, slightly depressed around the stipe, but also adnate-subdecurrent in small caps; pores moderately wide, about as in *S. floccopus.*

Stipe 4.2-7.8 cm long, 10-20 mm thick, solid, tapering downward, rarely equal, base rarely thickened, white quickly staining red then black, gray, reticulate at apex above the annulus or annular zone, shaggy below it.

Spore deposit black; spores 10.5-12.5 X 9.7-10.2 μ, short-ellipsoid to globose, with very fragmentary dense network or merely echinate, with occasional connected spines or short ridges, the ornamentation 0.3-1.7 μ high; usually partly or almost entirely imbedded in the much paler exosporium, wall rather thick and with an incomplete or complete germ pore.

Basidia (3-) 4-spored, 37-48 X 11-13.5 μ. Pleurocystidia (22) 34-61 X 7.5-26 μ, clavate to clavate-mucronate or fusoid-mucronate, brown as revived, sometimes vesiculose with a narrow neck.

Tube trama with a colored more compactly interwoven mediostratum and exterior to this hyphae diverging to the subhymenium. Cuticle of pileus with tips of the strands of brown parallel hyphae with cells 17-63 X 13-20 μ, some having versiform outgrowths; terminal cells sometimes clavate. Clamp connections none.

Habit, habitat, and distribution.—Solitary to scattered on the ground in woods, often near woodland pools.

Observations.—In Michigan this species has been confused with the more common *S. floccopus* previous to Singer's publication. The more acute erect spinelike scales in conjunction with verrucose rather than reticulate spores distinguishes it, but here in Michigan intermediates appear to be more common than the typical form. Our description is adapted from Singer as our notes were not sufficient for one.

BOLETINELLUS Murrill

Mycologia 1:7. 1909

This genus features ellipsoid to subglobose spores, an hymenophore coarsely boletinoid (varying from merulioid to nearly lamellate), lack of a veil, and the stipe lateral, eccentric or central.

Type species: *Boletinellus merulioides.*

Observations. – In our estimation this species is not at all closely related to *Gyrodon lividus* of Europe. To us the latter seems more closely aligned with *Suillus.* Our opinion is based upon excellent illustrations of the European species, detailed descriptions, and dried specimens. The genus likewise does not appear close to *Phylloporus* of the Paxillaceae. In the latter genus the gills closely resemble those of *Gomphidius* in thickness, attachment, and spacing.

212. Boletinellus merulioides (Schweinitz) Murrill

Mycologia 1:7. 1909

Boletinus merulioides (Schweinitz) Coker & Beers, Bol. of N. Carol. p. 87. 1943.
Gyrodon merulioides (Schweinitz) Singer, Rev. de Mycol. 3:172. 1938.
Daedalea merulioides Schweinitz, Trans. Amer. Phil. Soc. 4(ns)pt. 2:160. 1832.

Illus. Pl. 156.

Pileus 5-12 cm broad, plane to plano-convex with an incurved margin, expanding to plane or shallowly depressed on the disc and with a wavy spreading uplifted margin, margin sterile, entire, incurved to involute when young; surface soft, glabrous to pruinose or with a fine tomentum, dry texture occasionally appearing appressed-fibrillose scaly on the disc, colored various shades of dull yellow-brown when bruised. Context up to 1 cm thick, often thinner, soft but toughish, pale olive-yellow or pinkish next to the cuticle, unchanging or rarely slowly changing slightly to bluish green when cut; taste of raw Irish potatoes, odor not distinctive.

Tubes 3-5 mm deep, strongly decurrent, radiating, light yellow or often with a greenish tint, finally dull ochraceous, when young changing to dark olive and later reddish brown when injured; pores compound, radiating ridges present with crossveins dividing off areas which in turn are further divided into shallower pores.

Stipe 2-4 cm long, 5-10 (25) mm thick, slightly eccentric to eccentric to lateral, expanded upward, base firmly attached, solid, flesh concolorous with that of the pileus, surface concolorous with pileus below and with tubes above, staining reddish brown when injured, base often blackish.

Spore deposit olive-brown ("Isabella color" to "tawny-olive"), spores 7-10 X 6-7.5 μ, in optical section ovate to elliptic varying toward subglobose, smooth, in KOH hyaline to pale olive-ochraceous or pale yellow-brown, in Melzer's yellowish brown to distinctly rusty brown (dextrinoid), smooth, no apical pore present, wall slightly thickened.

Basidia 4-spored, 30-36 X 10-12 μ, hyaline to yellow in KOH, darker yellow in Melzer's. Pleurocystidia 20-35 X 6-9 μ, fusoid-ventricose. Cheilocystidia absent to rare, smooth, thin-walled, hyaline, fusoid-ventricose to rostrate.

Tube trama divergent from a central strand but becoming parallel or nearly so, gelatinous, hyaline in KOH, nonamyloid. Pileus cuticle apparently differentiated from the tramal body as a more compactly interwoven layer, yellow in KOH, pigment diffusing from the hyphae, both cutis and tramal body nonamyloid but in Melzer's incrusting pigment is found on some hyphae of the cuticle and the layer is dark yellow-brown. Clamp connections present.

Habit, habitat, and distribution.—Scattered on moist ground, usually under species of *Fraxinus* in Michigan, late summer and early fall, common in the Lower Peninsula.

Observations.—Sections revived in Melzer's give a "fleeting-amyloid" reaction but none of the hyphae or the spores are truly amyloid. Granular material present in the mounts is violaceous-fuscous both in KOH and Melzer's and hence is not amyloid.

The variant described is var. *opacus* Peck, the type variety being viscid, but we believe this is merely a stage in the development of the basidiocarps much as in many other boletes.

Material examined.—Berrien: Smith 51215, 51227. Clare: Boynton 8-29-49. Chippewa: Povah 8-4-19. Gratiot: Potter 3450, 3784, 3953, 5840, 6218, 6266, 7529, 7609, 7715, 8376, 8737, 9724, 11942, 12479, 12643. Jackson: Mazzer 4124, 4284. Midland: Smith 67790. Oakland: Mazzer 4428; Smith 7621. St. Clair: Boynton 1949, 7-10-49. Washtenaw: Baxter 7-21-20, 7-2120a; Homola 1202; Hoseney 109, 148, 577, 659; Smith 4985, 15245, 62573, 66418, 72830; Thiers 4571.

GASTROBOLETUS Lohwag

Beih. Bot. Centralbl. 422:273. 1926

This is a genus of boletes modified to the extent that the spores are not discharged from the basidia and hence typical spore prints cannot be obtained. The species connect to various genera in the Boletaceae and are to be regarded as modified members of this family and hence derive from it. For further information on the genus see Singer and Smith (1959).

213. Gastroboletus scabrosus Mazzer & Smith

Mich. Bot. 6:60. 1967

Illus. Pl. 157.

Pileus 2.5-6.5 cm broad; convex at first becoming broadly convex at maturity; surface dry and unpolished to subtomentose; dingy yellow-brown with ochraceous undertones, near "snuff-brown" or slightly paler to near "clay color" at the margin. Context 10-14 mm thick, pale buff when first cut, slowly stained dull pinkish buff mottled with pale olive-gray in places; taste mild, odor slight.

Tubes (gleba) 1.5-2 cm deep, curved in a flaring arrangement so that the distal portions of the tubes are oriented away from the stipe in a position approximately horizontal to the ground; pores small, 2-3 per mm soon collapsing and becoming flattened as the pileus expands, dull lemon-yellow at first, becoming dingy olivaceous-yellow (near "old gold") and developing brownish to blackish stains on bruising.

Stipe 7-10 cm long, 9-14 mm thick at the apex, equal with a tapering base; surface faintly striate-subreticulate above, scurfy overall; color yellow above, paler and duller below, scabrose from yellow-brown to orange-brown dots which are larger above and finer and more numerous below; base with a yellowish mycelium. Context pallid-yellow, staining brownish around larval tunnels or where bruised.

Spores 15-18.5 X 4-5 μ, subfusiform, thin-walled, pale brown in KOH. Basidia 22-24 X 9.5-12 μ, 4-spored, hyaline. Pleurocystidia 35-55 X 7-11 μ, fusoid-ventricose, thin-walled, hyaline. Cheilocystidia similar to pleurocystidia. Tube trama bilateral, with a distinct medial layer composed of hyaline subgelatinous hyphae. Pileus trichodermium of loosely interwoven, septate, pale brown hyphae 3-6 μ wide, some with slightly swollen end-cells. Clamp connections absent.

Habit, habitat, and distribution.—Gregarious on sandy soil in a second-growth oak-hickory forest, south of Otis Lake, Barry County, September 1, 1966.

Observations.—This species is obviously closely related to the genus *Leccinum.* However, because of the gastroid hymenophore and the fact that the tubes are arranged so that spores cannot fall out of them, it was described in *Gastroboletus.*

APPENDIX

The following type studies were made in the course of the Michigan project and are included here as a help to other workers since they clarify species concepts often in rather important details. The original descriptions are given for the convenience of the user.

Boletus caespitosus Peck, Torrey Club Bull. 27:17. 1900

Illus. Figs. 115-17.

Pileus broadly convex, to nearly plane, at times slightly concave from the margin becoming elevated, even, brown or blackish brown, the margin often slightly paler or reddish brown. Context slightly tinged red. Tubes adnate or slightly decurrent, yellow, their mouths rather large, angular, concolorous. Stem short, even, solid, glabrous, tapering upward, brown or reddish brown.

Type study. Spores 9-12 X 4-5 μ, smooth, thin-walled, lacking an apical pore, shape in face view narrowly ovate to subelliptic, in profile obscurely inequilateral, yellowish in Melzer's, yellowish in KOH, no sign of any amyloid reaction.

Basidia 4-spored, 24-30 X 7-9 μ, narrowly clavate, yellowish hyaline in KOH or Melzer's. Pleurocystidia scattered, 46-82 X 7-15 μ, narrowly fusoid-ventricose with a long pedicel, tapered to an acute or subacute apex, thin-walled, smooth, content "empty" in KOH or Melzer's.

Tube trama of the *Boletus* subtype, yellowish hyaline throughout in Melzer's. Pileus cutis a tangled layer of hyphae 4-7 μ wide, tubular but curved and intertwining, walls thin, smooth and hyaline, content homogeneous and dull orange in Melzer's for the most part; hyphae of context merely ochraceous in Melzer's; end-cells of cuticular hyphae tubular with blunt apex or slightly tapered to the apex, walls smooth and cell content ochraceous or orange in Melzer's. Clamp connections none.

Notes. *B. caespitosus* lacks the features of *Pulveroboletus* as outlined by Singer (1962). In our estimation it connects to the *B. badius* group, but as boletes go it is not at all closely related to *B. inflexus* or *B. fistulosus*.

398

Boletus firmus Frost, Bull. Buff. Soc. Nat. Sci. 2:103. 1874

Pileus pulvinate, solid, and very firm, gray, slightly tomentose, often lacunose, 2.5-4 in. broad. *Tubes* yellow, mouths tinged with red, unequal, deeply arcuate, adnate. *Stem* solid, hard, 2-4 in. long, yellowish reddish at base, very finely reticulated. Flesh deep yellow or yellowish, changing to blue. A readily distinguished species from its tenacity and generally distorted growth. Spores .0125-.0032 mm. In rich moist woods, July.

We have not examined authentic material of this species, and on the basis of the original description given above, we have not been able to recognize it in Michigan. In the past a species with dull brown pores and a faintly reticulate stipe was placed here, but this, obviously, was incorrect. We suspect the "distorted growth" to indicate the species may have been described from basidiocarps with a systemic infection of some Ascomycete, possibly a *Hypomyces.* Dick (1960) discusses this species and emphasizes that the spores are mostly under 13 μ long.

Boletus fistulosus Peck, Torrey Club Bull. 24:144. 1897

Illus. Figs. 126-27, 130.

Pileus convex, viscid, glabrous, yellow, the margin at first incurved or involute, flesh yellow; tubes plane or subventricose, medium size, round with thin walls, adnate or sometimes depressed around the stem, yellow; stem rather slender, subequal, viscid, glabrous, hollow, yellow, with a white mycelioid tomentum at the base; spores elliptical, .0005 in. long, .00025 in. broad. Pileus about 1 in. broad, stem 2-4 in. long, about 3 lines thick.

Grassy woods, Auburn, Alabama, July, Underwood.

Type study. Spores 11-13 X 4.5-5.5 μ, smooth, wall relatively thin, color in KOH yellow fading to nearly hyaline; in Melzer's pale tawny, in face view broadly subfusiform to elliptic-fusiform, in profile obscurely to somewhat inequilateral.

Basidia 4-spored, 9-11 μ broad, clavate, hyaline in KOH. Pleurocystidia scattered, 32-56 X 5-10 μ, narrowly fusoid to irregular-elongate, with a weakly ochraceous content in KOH, in Melzer's a few seen with dextrinoid, amorphous content.

Tube trama boletoid. Copious masses of a viscous material present which rounds into large globules in KOH making it difficult to see anatomical details. In Melzer's few globules are present, but some slightly dextrinoid debris is present. Pileus cuticle an irregular palisade of clavate to vesiculose cells 8-18 μ wide with pedicels 2-6 μ wide. Hyphae of cuticle proper gelatinized both in KOH and Melzer's. The hyphae of the context and cuticle lacking distinctive content in Melzer's. Clamp connections none.

Notes. We are not inclined to accept that *B. fistulosus* is the same as *B. inflexus* (see Singer, 1947). The most obvious differences are: Sections of *B. fistulosus* when mounted in KOH exhibit so much globular material that anatomical detail is difficult to see. We suspect this species of having a latex. Second, the pseudocystidia have so little content as to be difficult to locate. Finally, there appears to be a cellular epicutis in *B. fistulosus,* whereas in *B. inflexus* such a layer could not be demonstrated though in both the components are reasonably similar anatomically. To us these are significant enough to suggest that further critical studies based on fresh material should be made. Both of these species are relatively unrelated to *Pulveroboletus* in the sense of the type of that genus.

Boletus inflexus Peck, Torrey Club Bull. 22:207. 1895

Illus. Figs. 124-25.

"Pileus convex, glabrous, viscid, yellow, often red or reddish on the disk, the margin thin, inflexed, concealing the marginal tubes, flesh whitish, not changing color where wounded; tubes rather long, adnate, yellowish, becoming dingy-yellow with age, the mouths small, dotted with reddish glandules; stem rather slender, exannulate, solid, viscid, dotted with livid-yellow glandules; spores yellowish, .0004 to .0005 in. long, .00016 to .0002 broad.

"Pileus about 1 in. broad, stem about 2 in. long, 2-4 lines thick.

"Open woods. Trexlertown, September, Herbst."

Type study. Spores 12-15 (16) $\times$ 4-5 μ, smooth, lacking an apical pore, color in KOH bright yellow fading considerably on standing, in Melzer's pale tawny, shape in face view suboblong to narrowly elliptic, in profile obscurely to somewhat inequilateral, wall less than 0.2 μ thick; no amyloid reactions other than weakly dextrinoid evident.

Basidia 4-spored, 5-7 μ wide, yellowish or hyaline in KOH or Melzer's. Pleurocystidia in the form of pseudocystidia scattered, 36-65 $\times$ 5-10 μ, narrowly clavate, in KOH with amorphous ochraceous to orange content which in Melzer's is often dextrinoid.

Pileus cutis a thick layer of tangled narrow (2-7 μ wide) hyphae which gelatinize in Melzer's but not appreciably in KOH, the hyphae tubular and at the surface the hyphal ends clavate to globose and as first revived in KOH with much mucilaginous material adhering and obscuring the outline of the cell, these end-cells not organized into a palisade, in KOH the hyphae of the layer seen to have numerous amorphous refractive yellowish particles scattered through the layer, or the hyphal walls with yellowish to hyaline roughening irregularly distributed; in Melzer's much finely divided, particulate, weakly dextrinoid material present between the hyphae. Hyphae of the context and subcutis lacking distinctive content in Melzer's, but a few yellowish globules scattered in the mount, after 30 minutes, however, some large violet-fuscous globules up to 50 μ in diameter forming in the mounts, their origin not clear. Clamp connections absent.

Notes. This is a most unusual species of bolete. It has pseudocystidia and a unique pileus cuticle. Mounts of the pileus in KOH present a most cluttered appearance caused by the masses of amorphous material, the viscose material surrounding the end-cells, and the refractive particles scattered through the cuticular layer (and some granular material in the cells themselves). A lot of the material is apparently soluble in chloral hydrate. See *B. fistulosus* for a discussion of their similarities. It does not belong in *Suillus*, but may belong in Section *Pseudoleccinum* of *Boletus*.

Boletus modestus Peck, Rept. N. Y. State Mus. 25:81. 1873

"Pileus firm, convex, often irregular, dry, minutely tomentulose, yellowish brown; tubes nearly plane, attached and subdecurrent, pale ochraceous, angular and compound; stem equal, brown, reticulated with darker lines; spores elliptical, .0004' long, .0002' broad; flesh gray to pinkish gray.

"Plants 2' [inches] high, pileus 2" [lines] broad, stem 2"-4" thick. On grassy ground in open woods, Greenbush, August."

Type study. Spores 11-14 × 3.5-4.5 μ, smooth, lacking an apical pore, shape in face view suboblong to subfusiform, in profile obscurely inequilateral, color in KOH dingy cinnamon-buff, in Melzer's pale reddish tan; wall about 0.2 μ thick.

Basidia 4-spored. Pleurocystidia large and conspicuous, 35-44 × 10-15 μ, clavate to clavate-mucronate, filled with dull rusty brown to orange-brown content in KOH and the context as revived in Melzer's much redder (dextrinoid), walls thin, hyaline, smooth.

Pileus cutis a layer of interwoven hyphae 4-8 μ wide, the end-cells tubular to weakly cystidioid, remainder of cells in a filament merely tubular, walls thin and smooth, content brownish ochraceous in KOH but fading on standing, in Melzer's dingy orange brownish and walls poorly defined in some, the content with a tendency to become granular and to aggregate into poorly defined weakly colored masses which are not typical pigment globules. Hyphae of subcutis merely yellowish in Melzer's. Clamp connections none.

Notes. The type is a mixture of 2 species: One basidiocarp has much the appearance of *B. tenax,* including boletinoid pores. These characters are stated in the original description. Two large basidiocarps loose in the box containing the type collection clearly represent a *Tylopilus.* In them the pores obviously were small. Since the name *B. modestus* has been associated with *Tylopilus* in the sense that Murrill thought the species to be identical with *T. felleus,* it is this element for which we have described the microscopic characters. The single basiciocarp may be a *Phylloporus* or a member of the *Subtomentosi* of *Boletus.* The spores are hyaline under the microscope, ellipsoid, 8-10 X 4-5.5 μ, and are weakly amyloid. Since the original description by modern standards clearly includes taxa from 2 different genera and since we do not see any way of deciding accurately which parts of the description apply to which element in the type collection, we believe the species *B. modestus* is best considered a *nomen confusum.*

Boletus multipunctus Peck, Bull. N. Y. State Mus. 54:952. 1902

Illus. Figs. 120, 122-23.

"Pileus fleshy, convex or nearly plane, dry, brownish, ocher, sometimes with a slight reddish tint, the central part adorned with many minute slightly darker areolate spots or dots, flesh whitish, taste mild; tubes small, adnate or depressed about the stem, ventricose in mass, the mouths subrotund, at first whitish becoming greenish yellow, stem equal or tapering upward, pallid, solid, fibrous striate; spores dark olive green, oblong, .00045-.0006 in. long, .00016-.0002 in. broad.

"Pileus 3-5 in. broad; stem 3-5 in. long, 4-8 lines thick. In woods, Bolton, August."

Type study. Spores 9-12.5 X 3-4 μ, smooth, apex lacking a pore, walls thin, shape in face view oblong or nearly so, in profile suboblong to obscurely inequilateral, in KOH ochraceous or paler, in Melzer's very pale tan to ochraceous.

Basidia 4-spored. Pleurocystidia 30-42 X 8-14 μ, narrowly subfusoid to fusoid-ventricose, smooth in KOH, thin-walled, content not distinctive in either KOH or Melzer's. Cheilocystidia 38-27 X 4-8 μ, subfusoid varying to nearly clavate, the layer dull brownish in KOH (edge composed entirely of cheilocystidia).

Pileus cuticle a matted layer of fibrils showing no indications of trichodermial structure, the end-cells of the hyphae narrowly cystidioid, 4-8 μ wide, and walls smooth to roughened lightly by adhering slime (?), the matrical hyphae 3-7 μ broad and not distinctively colored in either KOH or Melzer's. Subcuticular hyphae weakly yellow in Melzer's. Clamp connections absent.

Notes. Murrill thought it might be a synonym of *B. inflexus,* but that is out of the question on the basis of the microscopic characters. It is very close to *B. roxanae,* but the latter has wider cuticular hyphae. However, this character deserves further study from a larger sample of specimens. For the present we recognize *B. multipunctus* as a species close to *B. roxanae* but as yet not recognized from Michigan.

Boletus scabroides Kauffman in Snell, Mycologia 28:472. 1936

Illus. Figs. 90-93.

Pileus convex to plane, 6-10 cm broad; surface slightly viscid when fresh, even, glabrous, minutely rimose in spots; pale tan, ochraceous-cinnamon or brown-ochraceous. Flesh soft, whitish to pale yellowish, changing slightly to bluish or changing not at all; taste and odor mild, except surface bitter. Tubes adnate, becoming depressed and half adnate, as long as the flesh is thick; greenish yellow becoming sordid greenish yellow in age, scarcely changing to blue; mouths angular, some more or less compound, 1-2 mm. Stipe even to somewhat striate, tending toward reticulate, minutely furfuraceous to scurfy or minutely scabrous except at apex; citron-yellow at apex and chrome-yellow at base, in between pallid tinged brownish or perhaps mostly yellow tinged reddish, with scurf sometimes reddish; within yellow, changing very slowly to pale blue; 5-9 cm long, 6-20 mm thick. Spores very pale greenish yellow under the microscope (in old specimens), subfusiform, 11-15 × 3.5-5 μ, some to 18 μ, mostly 12-14 × 3.5-4.5 μ.

Type study. Spores 13-16.5 × 3.5-4.5 μ, smooth, lacking an apical pore, shape in face view elongate-fusiform, in profile elongate-inequilateral, the suprahilar depression broad and pronounced, color in KOH yellowish to yellowish hyaline, in Melzer's pale tan, wall -0.2 μ thick, no sign of any amyloid or fleeting-amyloid reaction seen.

Basidia 4-spored, clavate, 8-9 μ broad, yellowish to hyaline in KOH. Pleurocystidia—none observed. Cheilocystidia not studied (edge of tubes in poor condition). Caulocystidia 30-50 × 8-25 μ, clavate to vesiculose, thin-walled, content not distinctive in KOH.

Tube trama lacking amyloid reaction of any kind, poorly revived but apparently of the *Boletus* type. Pileus cuticle a cellular epithelium, the cells vesiculose to pedicellate-clavate, 18-50 μ broad, subgelatinous in KOH (with a halo around the cell), content yellow in KOH, not distinctive in any way in Melzer's. Context hyphae not distinctive in any way in Melzer's. Clamp connections not found.

Notes. Kauffman had 2 descriptions on separate cards. In one the pileus was described as 6 cm. broad and in the other 10 cm. broad. Snell's description gives 6-10 cm as the range in size of the pileus, and it would seem to follow that these descriptions were used by Snell in compiling the one published. A third card in which no size for the pileus is given is now designated No. 562 (the holotype) in the University of Michigan Herbarium. The microscopic data given above are from this collection. It was collected at Ann Arbor, August 9, 1907, by Kauffman. In it the stipe was described as staining blue. The other 2 collections were from Sault Ste Marie, July 6, 1906, by Pennington, and July 11, 1906, by Kauffman. In the Pennington collection the pileus surface was bitter. Both collections consisted of one basidiocarp each. These were from under hemlock and spruce. It is clear that Snell's description was a composite of Kauffman's data on all 3 collections.

The holotype, in our estimation, is clearly *Boletus subglabripes*— the variant which stains slightly blue. The evidence for this is in the spore size and shape, the cellular cuticle of the pileus, and the scurfy stipe. However, the type is so poorly preserved as to render comparisons with other collections practically useless as far as macroscopic features are concerned. Singer (1947) commented on the blue-staining variant of *subglabripes* to the effect that it had a more subviscid pileus and a redder stipe. Kauffman's description of the type of *B. scabroides* emphasizes that the stipe was stained with red within and without and that the scurfiness was reddish also. Smith's study of the type supports Singer's statement in regard to the slight viscidity since the epithelial cells were clearly subgelatinous (with a "halo") in KOH. In view of these factors we propose that on the basis of the holotype, *B. scabroides* be reduced to synonymy under *B. subglabripes*. It is possible that this variant deserves rank as a variety, but Singer who studied it fresh did not think so, and we have not seen it fresh.

Boletus tennesseensis Snell & Smith, Journ. E. Mitchel Sci. Soc. 56:327. 1940

Illus. Figs. 106-7.

Type study. Spores 6.5-9 × 4.5-5 μ, smooth, thin-walled, lacking an apical pore, ellipsoid, ochraceous to yellowish hyaline in KOH, only weakly tan in Melzer's.

Basidia 4-spored, 20-30 × 8-9 μ, clavate, yellow to hyaline in KOH, yellowish in Melzer's, content not distinctive. Pleurocystidia rare, 28-37 × 9-12 μ, fusoid-ventricose, apex subacute to obtuse, content not distinctive in KOH or Melzer's, mostly nearly buried in the hymenium.

Pileus cutis a matted-down layer of intricately interwoven hyphae 6-9 μ wide, with numerous short cells and relatively few truly elongated cells, walls distinctly roughened or incrusted as revived in KOH (in which medium the hyphae are boldly defined), in Melzer's smooth or nearly so, lacking distinctive content and not boldly defined, the end-cells short, 15-36 × 7-10 μ and obtuse, subcylindric or only very weakly tapered to apex, content not distinctive. Hyphae of subcutis 8-12 μ wide and hyaline to yellowish in KOH or Melzer's, no amyloid material on walls present and no fleeting-amyloid reactions noted. Clamp connections absent.

Notes. The spores, the short, incrusted cells of the pileus cuticle, and lack of any fleeting-amyloid reaction distinguish this species from *B. miniato-olivaceus* either in the sense of Coker and Beers (1943) or the type. Coker and Beers placed it in synonymy with *B. miniato-olivaceus*.

Boletus underwoodii Peck, Torrey Club Bull. 24:145. 1897

Illus. Figs. 118-19.

"Pileus rather thin, convex, becoming nearly plane, slightly velvety, bright brownish red, becoming paler with age, flesh yellow, changing to greenish blue where wounded; tubes adnate or slightly decurrent, greenish yellow, becoming bluish where wounded, their mouths very small, round, cinnabar red, becoming brownish orange; stem equal or slightly tapering upward, somewhat irregular, solid, yellow without and within; spores .0004-.0005 in. long, .0002 in. broad. Pileus 2-3 in. broad; stem 3-4 in. long, 4-6 lines thick.

"Grassy woods, Auburn, Alabama, Underwood."

Type study. Spores 9-12 (3) × 3.5-4.5 μ, smooth, thin-walled, shape in face view narrowly elliptic to subfusiform, in profile obscurely inequilateral, color in KOH weakly yellowish to hyaline, in Melzer's merely weakly yellowish, a very slight fleeting-amyloid reaction present when first revived in Melzer's.

Basidia 4-spored, 8-9 μ broad, hyaline to yellowish in KOH or Melzer's. Pleurocystidia scattered, 25-38 X 5-9 μ, subcylindric to very narrowly fusoid-ventricose, thin-walled, some clavate and 7-12 μ broad and many in KOH with yellow amorphous material adhering to or in the cell (appearing in sections as yellow patches of pigment), masses of amorphous material not distinctively colored in Melzer's.

Tube trama of the *Boletus* subtype, at first with a very fleeting-amyloid reaction; laticiferous elements fairly abundant and bright yellow in KOH. Pileus cuticle a trichodermium of hyphae 2.5-5 μ in diameter, tubular, with yellow content in KOH or Melzer's and content remaining homogeneous in either medium, end-cells not enlarged and apex merely blunt, the hyphae adhering to each other and showing a thin outer halo of gelatinous material. Hyphae of subcutis not distinctive in Melzer's. Clamp connections absent.

Notes. This species is clearly southern, with typical cuticle of the core species of subsection *Luridi*, but is distinguished, as Peck pointed out, by the adnate to somewhat decurrent tubes. We believe that in wet weather the pileus would be at least somewhat viscid—at least this is indicated by the anatomy of the cuticular hyphae. Singer's (1947) synonymy should be disregarded as it now appears that *B. miniato-olivaceus* is not the species he included under that name. *B. underwoodii* appears to be distinct on the basis of the small pale colored spores and features of the cuticle of the pileus.

EDIBILITY

Since the boletes are among the most popular of all edible wild fleshy fungi it seems appropriate to include the following comments and key. The general instructions for cleaning and cooking boletes are as follows: *After you have learned your species,* if possible collect the button or immature basidiocarps. In *Boletus edulis* and its segregates one must be doubly sure of the identification of those variants with red pilei. The stipe is soft and tender at button stages in the *edulis* group, and can be used as well as the cap. Also, at this stage the tubes are not sufficiently developed to make much difference in the way the basidio-carps cook. The rule to follow is that if the stems are tender enough, use them, but if the stem is fibrous and hard, it is better to discard it. At the latter stage insect larvae have usually penetrated to the interior. If only caps with mature or nearly mature tubes are to be used, remove the tubes—they separate readily from the flesh. They are gelatinous at this stage and tend to cook up to a sticky consistency. The surface of the cap should be cleaned with a brush or damp cloth. Washing the caps may cause them to become water-soaked and in this stage they may tend to stew rather that sauté. When collecting, collect as little dirt with the specimens as possible. Do not carry the specimens in plastic bags for periods of an hour or more during hot weather, since mushrooms spoil very readily under conditions of high humidity and temperature. Paper bags are far better than plastic bags for collecting the parts to be eaten.

In the following list of recommended edible species or groups of species attention has been given to those abundant in the region, to ease of identification, and to edible qualities.

1. *Boletus edulis* and related species.

   B. variipes is most common in our oak forests. The variants around *B. edulis,* i.e., var. *clavipes* and *aurantioruber,* are more likely to be found in the Upper Peninsula.

2. *B. ornatipes.* Abundant in the beech-maple areas of the state.

3. *Boletus pallidus.* As there are some reports of a bitter strain in this species, more data on its quality are desirable, but it is very common.

407

4. *Tylopilus indecisus.* Most abundant under oak in hot wet weather; has been confused as a mild-tasting *T. felleus.*

5. *Tylopilus chromapes.* Abundant in June and July and even into August during hot wet weather, and mostly in stands of aspen.

6. *Suillus cavipes.* Fall, under larch in bogs. Beware of poison sumac when collecting this species.

7. *Suillus pictus.* Under white pine, summer and early fall. But we have not found it in plantations of white pine. It appears to favor the trees nearer maturity.

8. *Suillus luteus.* Plantations of Scotch pine, September to November.

9. *Suillus granulatus.* Under white pine more frequently than any other, late summer and fall.

10. *Suillus albidipes.* Under pine, often in plantations of white pine, late summer and fall.

11. *Suillus brevipes.* Late summer and fall under 2 or 3 needle pines.

12. *Gyroporus cyanescens.* Edible and choice (if you can effectively remove the sand), summer and fall, especially along roadsides or beside trails.

FIELD KEY TO MICHIGAN BOLETES TO SPECIES OR GROUPS

(as recognized in the text)

Note that undesirable species are designated with a minus sign (-), and edible species with an *.

1. Cap covered with coarse dry gray to blackish scales; tube layer pallid when young, staining reddish when injured; veil breaking to leave an annulus or zone on the stipe . . . . (-) *Strobilomyces*
1. Not with above combination of features. 2
   2. Stipe having roughness in the form of small to distinct scales, points or fine dots and these either black when young or becoming dark brown to black by maturity or in age, rarely red or reddish brown at first; cap typically dry to velvety and orange reddish to gray-brown to pallid . . . . (*) *Leccinum*
   2. Not as above 3
3. Veil floccose and dry, bright sulphur-yellow . . . . *Pulveroboletus*
3. Veil absent or not as above 4
   4. With any 2 of the following features:
      a) Stipe with glandular dots or smears (the fingers become stained from handling the stipe).

b) Annulus if present gelatinous, at least on the edge.

c) Pores when mature obscurely to distinctly boletinoid (see Pl. 9).

d) Cap viscid to slimy when fresh.

e) Cap dry and squamulose and stipe sheathed with the veil 5

4. Not as above 6

5. Spore deposit as air-dried cinnamon, tan or yellow-brown, rarely olive . . . *Suillus*

5. Spore deposit (air-dried) lilac, purple-brown, vinaceous-brown, or grayish brown with a pinkish tint . . . . *Fuscoboletinus*

    6. Spore deposit pale yellow; tubes white at first; stipe fragile and often hollowed in the base . . . . *Gyroporus*

    6. Stipe solid and becoming toughish in age over lower part: spore deposit olive-brown to rusty brown, vinaceous, vinaceous-brown, chocolate (various shades) or fuscous-brown 7

7. Spore deposit reddish, cinnamon reddish, fuscous-vinaceous to chocolate (various shades); tubes usually white to pallid when young . . . . . . *Tylopilus*

7. Spore deposit amber-brown, dingy yellow-brown, olive-brown, olive or olive-fuscous . . . . *Boletus* and *Boletellus*

(*) *Gyroporus*

As far as known, all the Michigan species are edible.

1. Context of cap instantly blue when injured. . . . .*G. cyanescens*

1. Context not changing to blue or violet readily 2

    2. Cap mineral-red to vinaceous-red . . . .*G. purpurinus*

    2. Cap rusty to chestnut-brown or rich yellow-brown . . . .*G. castaneus*

Fuscoboletinus

1. Cap dry and fibrillose to squamulose, red . . . .*F. paluster*

1. Cap viscid at some stage (especially in age) 2

    2. Tubes white to pallid or grayish when young 3

    2. Tubes yellow or soon becoming so 6

3. Cap covered with a chocolate colored slime when young . . . . *F. serotinus*

3. Cap not as above 4

    4. Cap context pallid staining blue if injured . . . .*F. aeruginascens* (Pl. 42)

    4. Cap context as above but not staining blue, though tubes may stain slightly . . . .*F. grisellus* (Pl. 43)

5. Cap with conspicuous dry floccose squamules over a viscid layer . . .*F. spectabilis*

5. Cap glabrous when young . . . . *F. glandulosus*

Suillus

As far as is known, the Michigan species are all edible but some are not very good. It is important to know these as well as the best ones. The key is to major groups and to the more important species. To identify *Suilli* accurately use the technical key in the text. By weight, more pounds of *Suillus* are collected in the state for food than of any other group of boletes.

1. Veil thick, tough, gelatinous; growing under oak or (rarely) other hard-woods . . . . (-) *Suillus sphaerosporus*
1. Growing under conifers; veil (if gelatinous) shrinking markedly on breaking but veil absent to marginal in some 2

   2. Annulus present on the stipe, or an annular zone visibly marking the line where the veil broke 3
   2. Veil if present marginal (not leaving remains on the stipe) 6

3. Stipe hollow in base . . . . (*) *S. cavipes*
3. Stipe typically solid 4

   4. Cap dull rose-red from a fibrillose covering, stipe sheathed with same material; growing with white pine . . . . (*) *S. pictus*
   4. Not as above 5

5. Annulus with a violaceous-brown zone on outer (under) side and this material soon gelatinous . . . . (**) *S. luteus*
5. Not as above. The remainder of the annulate group are difficult to identify and not very good for the table because of thin flesh or somewhat inferior flavor and cooking quality . . . .
. . . . see (-) *S. proximus*, (-) *S. grevillei*, (-) *S. subluteus*, (-) *S. acidus*

   6. A marginal veil visible on young fruit bodies 7
   6. Lacking a marginal veil 8

7. Stipe 3-8 (11) mm thick, relatively long and slender, conspicuously glandular, dotted; cap usually spotted near the margin with brown patches of veil material . . . . (-) *S. americanus*
7. Stipe (8) 10-30 mm thick, typically short . . . .
. . . . see (*) *S. albidipes* and *S. glandulosipes*

   8. Stipe white and lacking glandular dots at first . . . . (*) *S. brevipes*
   8. Stipe very soon developing distinct colored glandular dots 9

9. Cap white; growing under white pine . . . . (-) *S. placidus*
9. Cap if white at first soon becoming colored 10

   10. Injured flesh and/or tubes staining blue . . . .(-) *S. tomentosus*
   10. Not staining as above . . . . see descriptions
     of (-) *S. punctipes*, *S. subaureus*, and *S. hirtellus* also (*) *S. granulatus*

Tylopilus

Many species in this genus are inedible simply because of the bitter taste, a feature more widely prevalent among the species of this genus than in any other. Those without a bitter taste are likely to be quite desirable because of their thick, tender flesh. This and the large size of the basidiocarps of many species stimulates the collector's interest. Our best species is *B. indecisus*.

1. Stipe base bright yellow, surface above this furfuraceous to punctate with pinkish ornamentation at first, pink often showing prominently on the cap. . . . . (*) *T. chromapes*
1. Not as above. 2
 2. Cap dark olive-brown to gray-brown; spore deposit color in the chocolate-colored series (and the tubes at maturity dark dingy grayish brown to reddish brown) 3
 2. Cap dark vinaceous-brown to various shades of dingy cinnamon to dull orange-brown; spore deposit vinaceous to vinaceous-cinnamon (the tubes and pores showing this at maturity) 6
3. Pore edges cinnamon to dark coffee-brown when immature . . . *T. pseudoscaber*
3. Pore edges not differently colored from the sides 4
 4. Tubes when young (before being colored by spores) weakly yellow . . . . *T. cyaneotinctus*
 4. Tubes white to pallid or grayish at first 5
5. Stipe with a short pseudorhiza (root) . . . . *T. umbrosus*
5. Stipe not rooting . . . . . see *T. sordidus, T. fumosipes,* and *T. porphyrosporus*
 6. Taste bitter, stipe reticulate at least at apex 7
 6. Not as above 9
7. Cap white, surface wrinkled like parchment paper. . . . . (-) *T. intermedius*
7. Not as above. 8
 8. Cap and stipe violaceous when young; stipe reticulate only near the apex . . . . (-) *T. plumbeoviolaceus*
 8. Not as above . . . . (-) *T. felleus* and variants
9. Taste bitter, stipe not reticulate, young cap dark vinaceous-brown . . . . (-) *T. rubrobrunneus*
9. Taste mild, stipe at apex scarcely reticulate or reticulum extending down the stipe a short distance. . . . . (*) *T. indecisus*

(*) *Leccinum*

Since all the known species in the state which have been tested have been found edible, we recommend them on the basis that if poisonous species occur, we should have learned about it by this time

The quality of the species is generally good, though not as good as that of *B. edulis*. It should be remembered, however, that insect larvae soon riddle the context of these species, and one should try to collect them when they are developing rapidly and can be collected in quantity. Considerable sorting is almost always necessary even when collecting immature specimens. In some species the context darkens naturally and is dark when cooked. This in no way impairs edibility. People living in the areas where aspen and birch are the most numerous trees in the forest will collect this genus frequently, since it forms mycorrhiza with these trees. *Leccinum* fruits throughout the season from late May to late September or October, and the Michigan flora is one of the most extensive known for the genus.

Boletus

Keyed here under the above generic name are the edible species in *Boletus* and *Boletellus* (both as groups and single species insofar as these are pertinent). One should not experiment indiscriminantly on the edibility of species in these two genera as some poisonous species occur here and there are many for which we have no data on edibility.

1. Stipe lacerate-reticulate and typically long for the width of the cap
. . . . .Sect. *Laceripedes*
1. Not as above 2

  2. Taste of raw flesh distinctly acrid or peppery . . . . Sect. *Piperati*
  2. Taste mild to bitter but not sharp 3

3. Pores red, red-brown, or dark brown in immature stages . . . (-) Subsect. *Luridi*
3. Pores not differently colored than the sides of the tubes 4

  4. Stipe in some degree (at times only over apical region) having a fine
    to medium-coarse network or reticulum . . . . Sect. *Boletus*
  4. Stipe naked, pruinose to scurfy or with coarse raised lines at times
    forming a wide-meshed reticulum 5

5. Cap distinctly velvety to subtomentose at first and usually conspicuously
  areolate-cracked in age (fig. 113) . . . .(-) Sect. *Subtomentosi*
5. Cap merely unpolished at first or glabrous or viscid 6

  6. Stipe furfuraceous to punctate when young . . . . . . Sect. *Pseudoleccinum*
    Boletus subglabripes is our common species in this group and we
    have had reports that it is a good edible species. We can add
    nothing from personal experience.
  6. Stipe pruinose to naked . . . . (-) Sect. *Pseudoboleti*

Section *Boletus,* Subsection *Luridi*

For safety's sake the members of this subsection should not be used for food in our region. Some are known to be poisonous, and a number of Michigan species have not been tested. The species are to a large extent identified by microscopic characters (see p.361).

Section *Boletus*
(excluding subsect. *Luridi*)

1. Tubes white at first and the pores usually overgrown with a white mantle of hyphae (i.e., they are "stuffed" when young), becoming greenish yellow to olive-yellow or olive brownish; stipe very finely reticulate; spore deposit olive to olive-brown . . . . (*) Subsect. *Boleti*
1. Tubes yellow from early stages on 2
    2. Taste of context typically bitter; stipe finely reticulate over the upper part (but see *B. ornatipes* also) 3
    2. Stipe conspicuously reticulate and yellow; tubes bright yellow from youth to old age . . . .(*) *B. ornatipes*
3. Cap whitish to pale olive-buff; stipe bright red and usually rather long and slender. . . . . (-) *B. inedulis*
3. Not as above, taste mild to bitter. 4
    4. Cap red at first, soon fading to buffy brown . . . . . . (-) *B. pseudopeckii*
    4. Cap persistently bright rose-red . . . .*B. speciosus*
      This species is reported as good by some authors, but we refuse to recommend it since the red-capped boleti are very difficult to identify and some are poisonous.

Section *Subtomentosi*

This section encompasses a large and confusing group of species some of which are edible and choice and some very probably poisonous. Anyone interested in the group should use the key running through the text and emphasizing microscopic features. Some progress may be made by consulting the photographs and reading the appropriate descriptions: for instance *B. mirabilis,* which unfortunately is rare in Michigan, and *B. projectellus* are good edible species. Those particularly to be avoided are in the *Fraterni* and feature in addition to cap color a color change to blue on the tubes if the latter are injured, and yellow tubes throughout the life of the hymenophore. When the taxonomy of this section becomes stabilized and we have a more complete inventory of the recognizable variants occurring in the state, we shall very likely be able to recommend another dozen species to the collector.

BIBLIOGRAPHY

Bartelli, Ingrid. 1967. 1966–The Year of the Boletes. Mich. Bot. 6:51-52.

Bartelli, Ingrid, and Alexander H. Smith. 1964. Notes on Interesting Mushrooms from the Upper Peninsula. Mich. Bot. 3:83-86.

Coker, William C., and Alma H. Beers. 1943. The Boletaceae of North Carolina. Chapel Hill, N. C.: Univ. N. Carolina Press, pp. 96. Pls. 1-65.

Dick, E. A. 1960. Notes on Boletes XII. Mycologia 52:130-36.

Fries, E. M. 1874. Hymenomycetes Europaei. 755 pp. Upsalla.

Frost, C. C. 1877. Catalogue of Boleti of New England. Bull. Buff. Soc. Nat. Sci. II. 100 pp.

Groves, J. Walton, and Sheila C. Thomson. 1955. Notes on Fungi from Northern Canada. II. Boletaceae. Canadian Field-Nat. 69:44-51.

Mazzer, Samuel J., and Alexander H. Smith. 1967. New and Interesting Boletes from Michigan. Mich. Bot. 6:57-67.

Miller, O. K., and Roy Watling. 1968. The Status of *Boletus calopus* Fr. in North America. Notes from the Royal Bot. Garden Edinb. 28(3):317-26.

Moser, M. 1967. Die Röhrlinge und Blätterpilze. Basidiomyceten II. 443 pp. Gustav Fischer. Stuttgart.

Murrill, W. A. 1940. Additions to Florida Fungi–II. Torrey Club Bull. 67:57-66.

Perreau-Bertrand, Jacqueline. 1961. Recherches sur les ornementations sporales et la sporogenèses chez quelques espèces des generes *Boletellus* et *Strobilomyces* (Basidiomycètes). Ann. des Sc. Nat. Bot. 12e ser.:400-88.

Secretan, L. 1833. Mycographie Suisse. Vol. II. 576 pp. Geneva.

Singer, Rolf. 1945. The Boletineae of Florida with Notes on Extralimital Species I. The Strobilomycetaceae. Farlowia 2(1):97-141.

————— 1945. The Boletineae of Florida with Notes on Extralimital Species II. The Boletaceae (Gyroporideae). Farlowia 2(2):223-303.

————— 1946. The Boletineae of Florida with Notes on Extralimital Species IV. The Lamellate Families (Gomphidiaceae, Paxillaceae, and Jugasporaceae). Farlowia 2(4):527-67.

————— 1947. The Boletoideae of Florida. The Boletineae of Florida with Notes on Extralimital Species III. Amer. Midl. Nat. 37:1-135.

————— 1962. The Agaricales in Modern Taxonomy. Ed. II. 915 pp. J. Cramer. Weinheim.

415

Singer, R., W. H. Snell, and E. A. Dick. 1963. The Genus Fuscoboletinus. Mycologia 55:352-57.

————. 1965. Die Röhrlinge Teil—I:131 pp., 21 pls. Jules Klinkhardt, Bad Heilbrunn.

————. 1967. Die Röhrlinge Teil—II: 151 pp., 26 pls. Jules Klinkhardt, Bad Heilbrunn.

Smith, Alexander H. Sept. 1965. New and Unusual Basidiomycetes with Comments on Hyphal and Spore Wall Reactions with Melzer's Solution. Mycopathol. et Mycol. Applicata 26(4):385-402.

————. 1966. A Note on *Psiloboletinus.* Mycologia 58:332-36.

Smith, Alexander H., and Rolf Singer. 1959. Studies on Secotiaceous Fungi—IV. Gastroboletus, Truncocolumella, and Chamonixia. Brittonia 11:205-23.

Smith, Alexander H., and Harry D. Thiers. June 1964. A Contribution toward a Monograph of the North American Species of *Suillus.* Ann Arbor. Pp. 1-116, 46 pls.

————. 1966. Further Notes on *Suillus:* the *S. granulatus* Problem. Mycologia 58:469-74.

————. 1967. Comments on *Suillus amabilis* and *Suillus lakei.* Mycologia 59:361-67.

————. 1968. Notes on the Genus *Suillus* (Boletaceae). Mich. Bot. 7:14-18.

————. 1968. The Generic Position of *Boletus subglabripes* and *Boletus chromapes.* Mycologia 60:943-49.

————. 1968. A Note on Four Species of *Tylopilus.* Mycologia 60:949-54.

Smith, Alexander H., Harry D. Thiers, and Orson K. Miller. June 1965. The Species of *Suillus* and *Fuscoboletinus* of the Priest River Experimental Forest and Vicinity, Priest River, Idaho. Lloydia 28:120-38.

Smith, Alexander H., Harry D. Thiers, and Roy Watling. May 1966. A Preliminary Account of the North American Species of *Leccinum* Section *Leccinum.* Mich. Bot. 5(3A):131-78.

————. May 1967. A Preliminary Account of the North American Species of *Leccinum,* Sections *Luteoscabra* and *Scabra.* Mich. Bot. 6(3A):107-54.

————. In press. Notes on Species of *Leccinum,* I. Additions to Section *Leccinum.*

Snell, W. H. 1936. Notes on Boletes V. Mycologia 28:463-75.

Thiers, H. D. 1966. California Boletes II. Mycologia 58:815-26.

Watling, Roy. 1961. Notes on British Boleti. Trans. Bot. Soc. Edinb. (1961) 39, pt. 2:196-205.

————. 1964. Notes on British Boleti: III. Trans. Bot. Soc. Edinb. 39:475-88.

————. 1965. Notes on British Boleti: IV. Trans. Bot. Soc. Edinb. (1965) 40(1):100-120.

————. 1968. Records of Boleti and Notes on Their Taxonomic Position. Notes from the Royal Bot. Garden Edinb. 28(3):301-15.

ILLUSTRATIONS

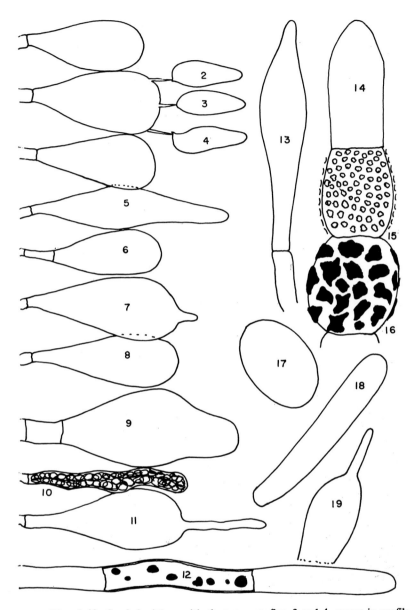

Figs. 1-19: fig. 1, basidium with three spores; figs. 2 and 4, spores in profile view; fig. 3, a spore in face view; fig. 5, a fusoid-ventricose cystidium; fig. 6, a basidiole; fig. 7, a clavate-mucronate cystidium; fig. 8, a basidiole; fig. 9, broadly fusoid-ventricose cystidium; fig. 10, a pseudocystidium; fig. 11, a rostrate cystidium; fig. 12, a cuticular hypha showing pigment globules; fig. 13, an end-cell cystidioid in shape; fig. 14, a bullet-shaped end-cell as in *Leccinum insigne*; fig. 15, cuticular cell of *Leccinum insigne;* fig. 16, a trichodermial cell showing pigment incrustations as in *Boletus chrysenteron;* figs. 17-18, disarticulated cells from hyphae of a trichodermium; fig. 19, rostrate terminal trichodermial cell.

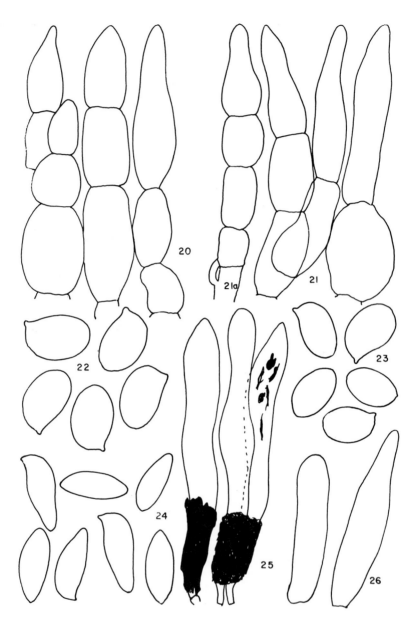

Figs. 20-26: fig. 20, trichodermial elements of *Gyroporus purpurinus*; fig. 21-21a, trichodermial elements of *G. castaneus*; fig. 22, spores of *G. purpurinus*; fig. 23, spores of *G. castaneus*; fig. 24, spores of *Suillus pictus*; fig. 25, pleurocystidia of same; fig. 26, cells from trichodermial elements of pileus of same.

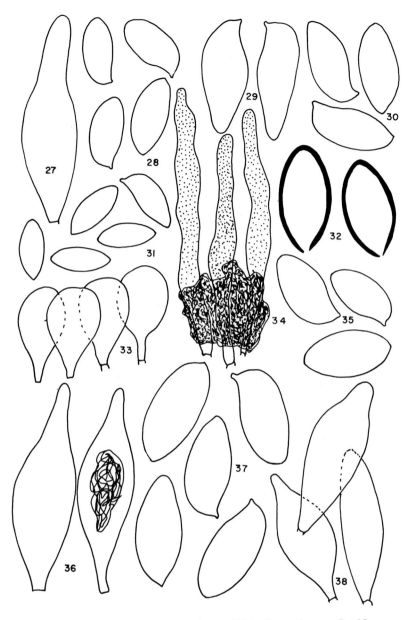

Figs. 27-38: fig. 27, pleurocystidium of *Tylopilus umbrosus;* fig. 28, spores of same; figs. 29-30, spores of *T. nebulosus*; fig. 31, spores of *Suillus americanus*; figs. 32 and 35, spores of *T. sordidus*; fig. 33, cheilocystidia of same; fig. 34, pleurocystidia of *Suillus americanus*; fig. 36, pleurocystidia of *T. sordidus*; fig. 37, spores of *T. pseudoscaber*; fig. 38, cheilocystidia of *T. pseudoscaber*.

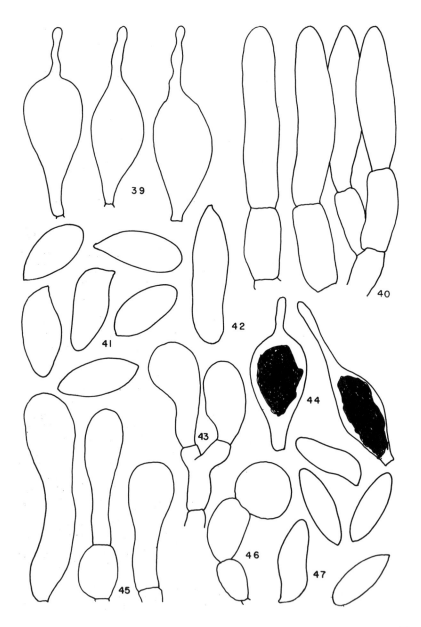

Figs. 39-47: fig. 39, pleurocystidia of *Tylopilus pseudoscaber*; fig. 40, trichodermial element of *T. fumosipes*; figs. 41-42, normal spores and 1 abnormal elongated spore of same; fig. 43, cheilocystidia of same; figs. 45-46, caulocystidia of same; fig. 44, pleurocystidia of *T. indecisus*; fig. 47, spores of same.

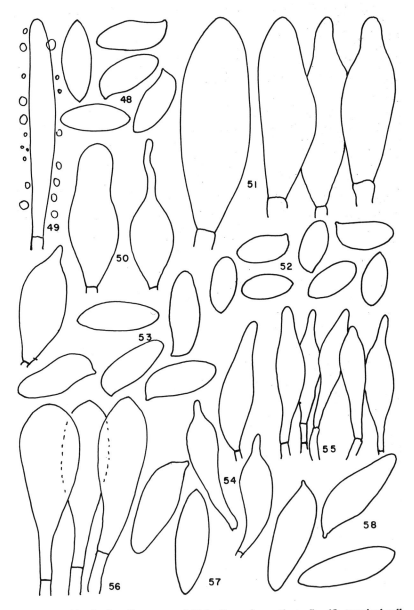

Figs. 48-58: fig. 48, spores of *Tylopilus subpunctipes*; fig. 49, terminal cell from cuticle of same; fig. 50, pleurocystidia of *Boletus chrysenteron* var. *sphagnorum*; fig. 51, end-cells of cuticular elements of same; fig. 52, spores of *Tylopilus badiceps*; fig. 53, spores of *B. chrysenteron* var. *sphagnorum*; fig. 54, pleurocystidia of *T. badiceps*; fig. 55, cheilocystidia of same; fig. 56, caulocystidia of *Leccinum atrostipitatum*; figs. 57-58, spores of same.

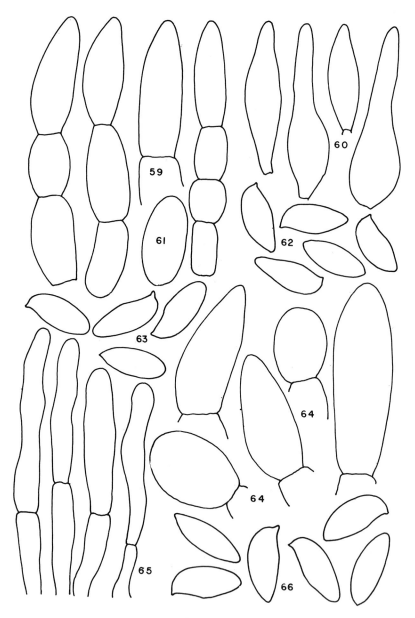

Figs. 59-66: fig. 59, trichodermial elements of *Boletus roxanae*; fig. 60, pleurocystidia of same; fig. 61, disarticulated cell from cuticle hypha of same; fig. 62, spores of same; fig. 63, spores of *B. peckii*; fig. 64, end-cell of cuticle hypha of *B. illudens*; fig. 65, cuticular hyphae of *B. peckii*; fig. 66, spores of *B. illudens*.

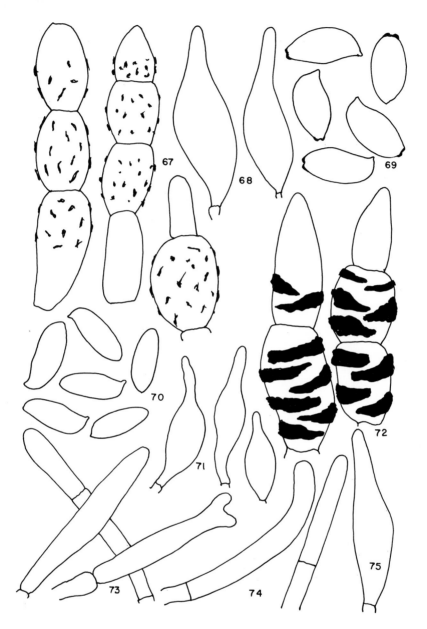

Figs. 67-75: fig. 67, trichodermial elements of *Boletus truncatus*; fig. 68, pleurocystidia of same; fig. 69, spores of same; fig. 70, spores of *B. bicolor*; fig. 71, pleurocystidia of *B. illudens*; fig. 72, trichodermial cells from *B. chrysenteron*; figs. 73-75, terminal cells of cuticular hyphae of *B. bicolor*.

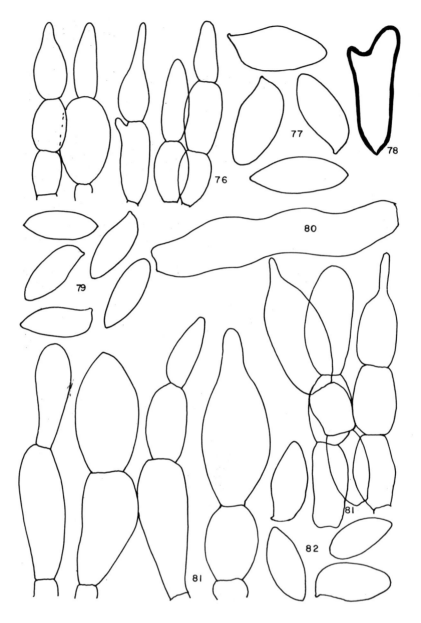

Figs. 76-82: fig. 76, trichodermial hyphae from *Boletus fraternus*; figs. 77-78, normal and 1 abnormal spore of same; fig. 79, spores of *B. rubeus*; fig. 80, cell from tramal hypha; fig. 81, trichodermial elements from pileus of *B. rubellus*; fig. 82, spores from *B. rubellus* var. *flammeus*.

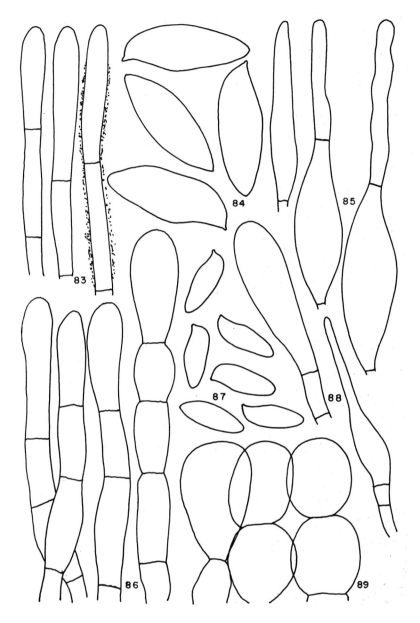

Figs. 83-89: fig. 83, cuticular hyphae of *Leccinum atrostipitatum*; fig. 84, spores of *L. snellii*; fig. 85, caulocystidia of same; fig. 86, trichodermial hyphae of same; fig. 87, spores of *Boletus affinis* var. *affinis*; fig. 88, caulocystidia of *Leccinum snellii*; fig. 89, epithelial cells from pileus of *Boletus affinis* var. *affinis*.

Figs. 90-97: fig. 90, cells from epithelium of *Boletus scabroides*; fig. 91, spores of *B. scabroides*; figs. 92-93, caulocystidia of same; fig. 94, spores of *B. longicurvipes*; fig. 95, spores of *B. rubropunctus*; fig. 96, spores of *B. affinis* var. *affinis*; fig. 97, caulocystidia of *B. rubropunctus*.

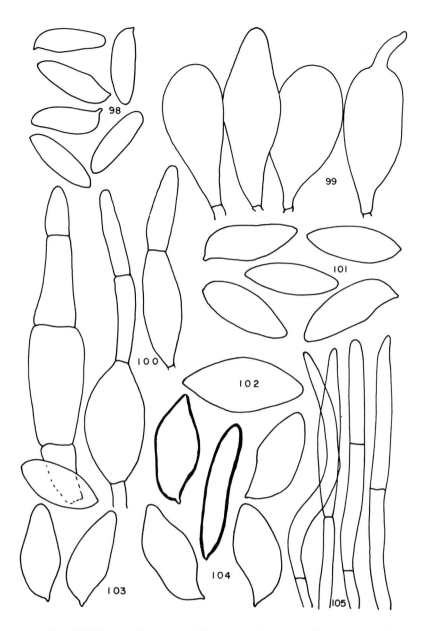

Figs. 98-105: fig. 98, spores of *Boletus speciosus*; fig. 99, cuticular cells of pileus of *B. separans*; fig. 100, pileocystidia of *B. separans*; fig. 101, spores of same; figs. 102 and 104, spores of *B. subvelutipes;* fig. 103, spores of *B. vermiculosus*; fig. 105, cuticular hyphae of *B. subvelutipes*.

Figs. 106-14: fig. 106, cuticular hyphae of *Boletus tennessensis*; fig. 107, spores of same; fig. 108, spores of *B. affinis* var. *maculosus*; figs. 109-10, spores of *B. holoroseus*, normal and abnormal respectively; fig. 111, epithelium of *B. affinis* var. *maculosus*; fig. 112, thick-walled abnormal spores of *B. holoroseus*; fig. 113, spores of *Boletellus pseudo-chrysenteroides*; fig. 114, spores of *Boletellus inter-medius*.

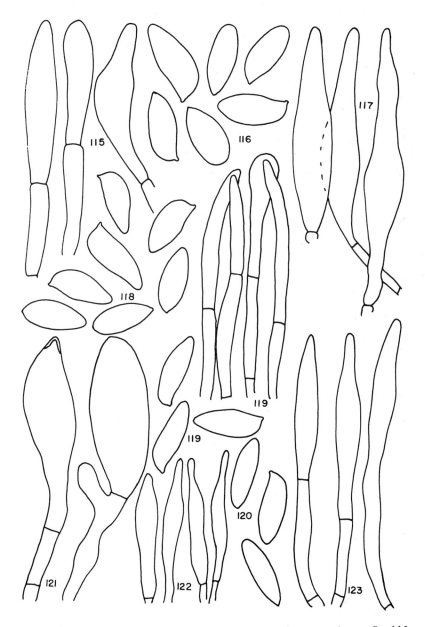

Figs. 115-23: fig. 115, cuticular elements of *Boletus caespitosus*; fig. 116, spores of same; fig. 117, pleurocystidia of same; fig. 118, spores of *B. underwoodii*; fig. 119, fascicle of trichodermial hyphae of same; fig. 120, spores of *B. multipunctus*; fig. 121, caulocystidia of *Leccinum areolatum*; fig. 122, cheilocystidia of *Boletus multipunctus*; fig. 123, cuticular elements of same.

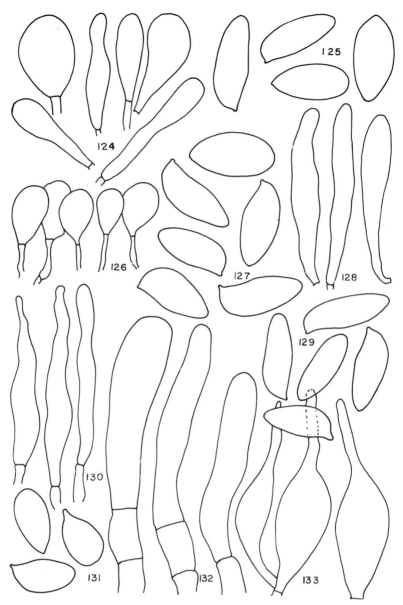

Figs. 124-33: fig. 124, end-cells of cuticular hyphae of *Boletus inflexus*; fig. 125, spores of same; fig. 126, cuticle of *B. fistulosus*; fig. 127, spores of same; fig. 128, pleurocystidia of *B. inflexus*; fig. 129, spores of *B. glabellus*; fig. 130, pleurocystidia of *B. fistulosus*; fig. 131, spores of *B. glabellus*; fig. 132, cuticular hyphae of same; fig. 133, pleurocystidia of same.

PLATES

PLATE 1

Smith 62899

Gyroporus cyanescens   × 1

PLATE 2

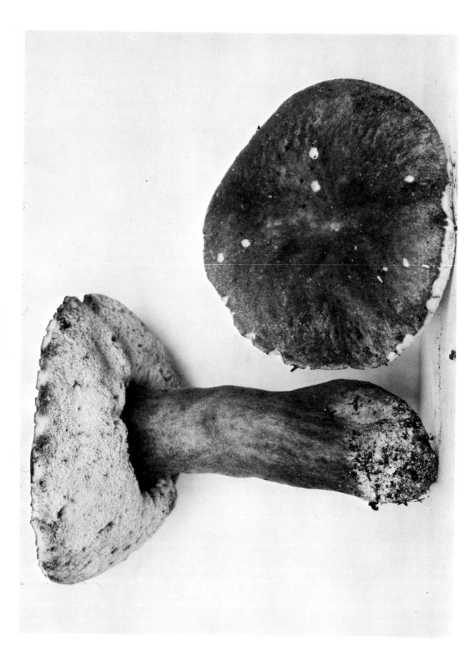

Gyroporus castaneus   × 1   Smith 63867

PLATE 3

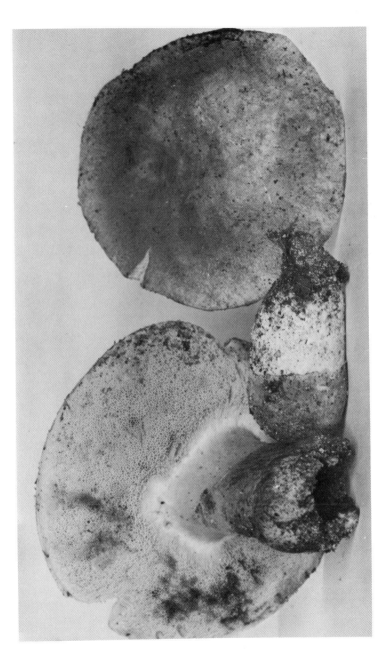

Gyroporus castaneus   × 1

Smith 57398

PLATE 4

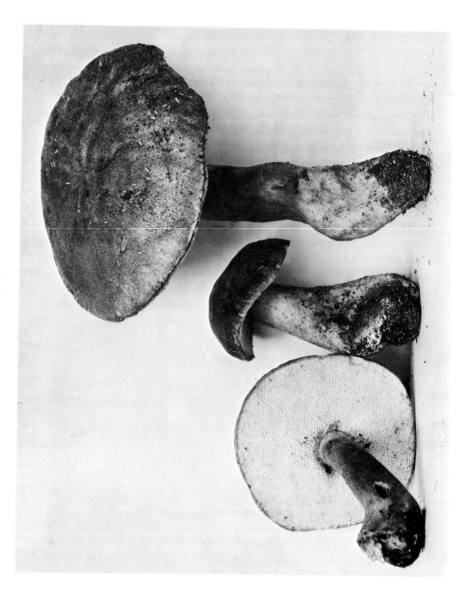

PLATE 5

Suillus sphaerosporus   × 1 (young)

PLATE 6

Smith 72471

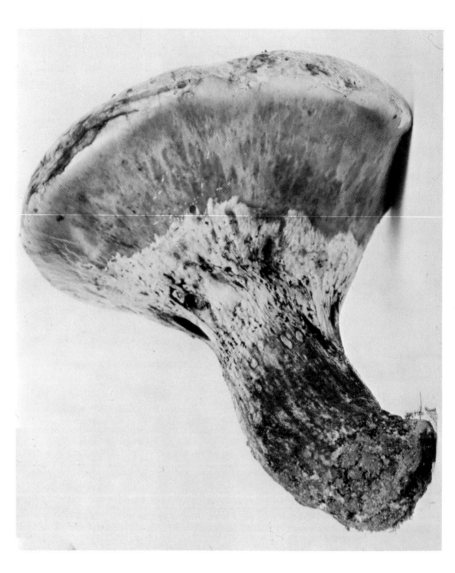

Suillus sphaerosporus   × 1 (showing double veil)

PLATE 7

Suillus sphaerosporus   × 1 (near maturity)

Smith 72471

PLATE 8

Suillus cavipes   x 1

PLATE 9

Smith 11634

Suillus cavipes   x 1

PLATE 10

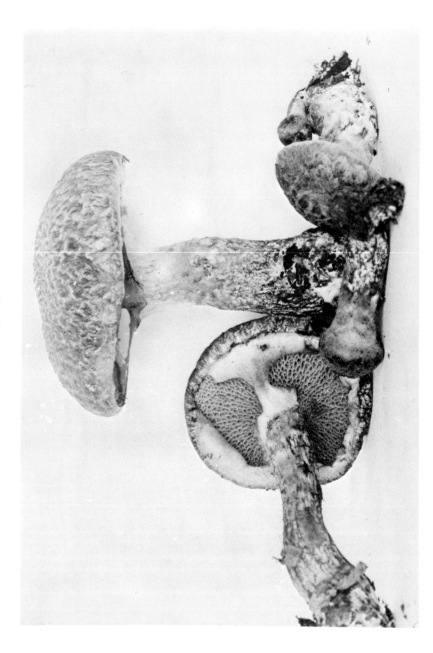

Suillus pictus × 1

PLATE 11

Suillus proximus   x 1   Smith 64545

PLATE 12

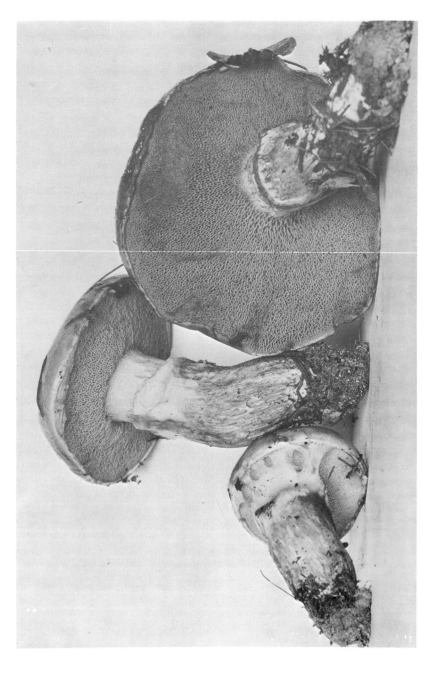

Suillus grevillei   × 1   Smith 54702

PLATE 13

Suillus tomentosus × 1

Smith 53554

PLATE 14

Suillus hirtellus   × 1   Smith 72396

Boletus edulis   × 1   Photo Smith

PLATE 15

Suillus subaureus   × 1

Smith 75692

PLATE 16

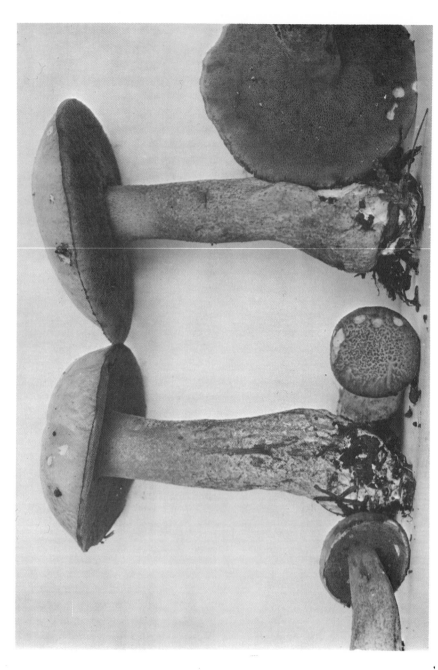

Suillus punctipes   × 1

Suillus luteus   x 1

PLATE 18

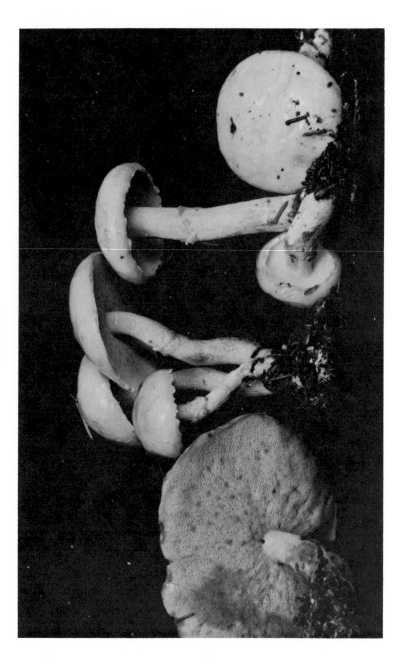

Smith 73307

Suillus acidus var. *acidus*   × 1

PLATE 19

Suillus intermedius   × 1

Smith 72414

PLATE 20

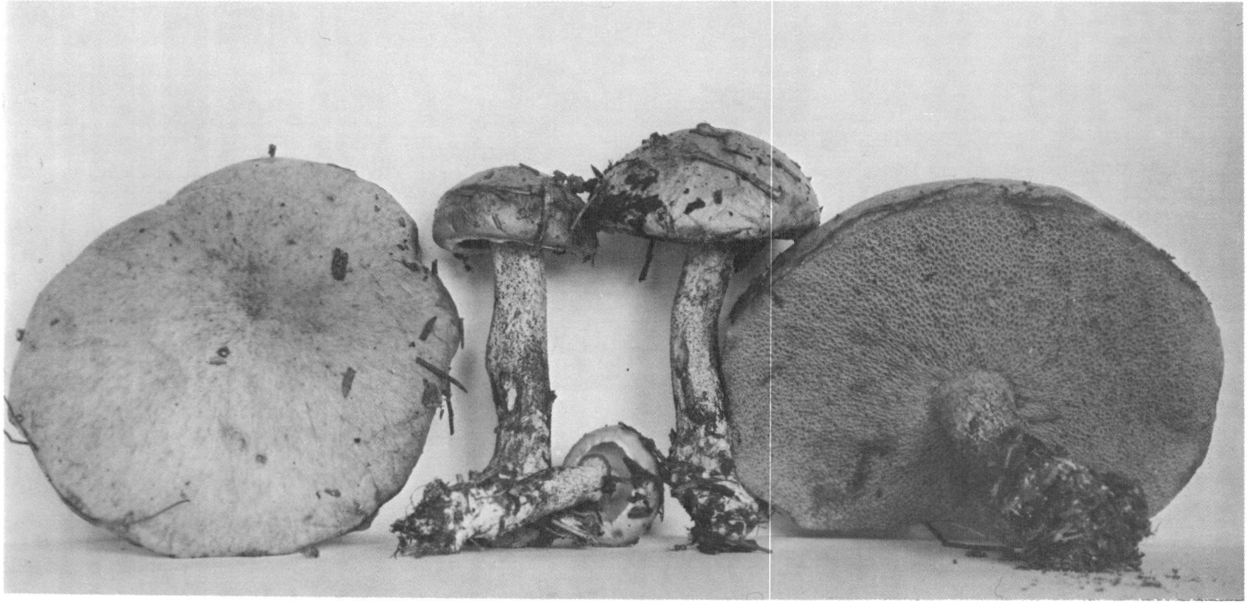

Suillus subalutaceus   × 1

PLATE 21

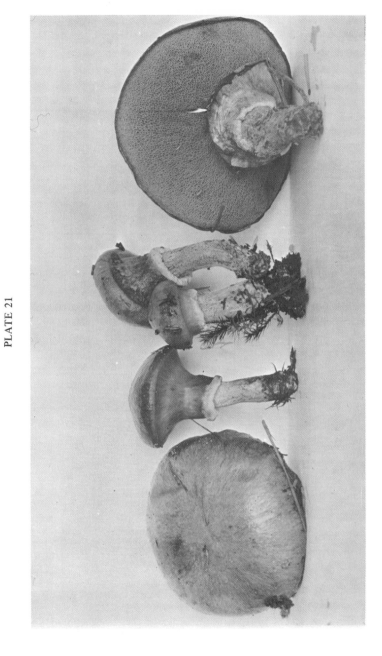

Suillus subluteus   × 1

Smith 58109

PLATE 22

Suillus albidipes   × 1   Smith 64747

PLATE 23

Suillus albidipes   × 1   Photo Smith

Suillus albidipes   × 1   Photo Smith

PLATE 24

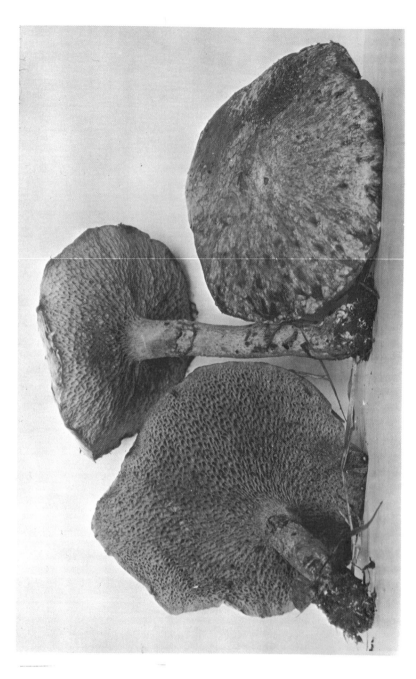

Suillus sibiricus   x 1

Smith 54266

PLATE 25

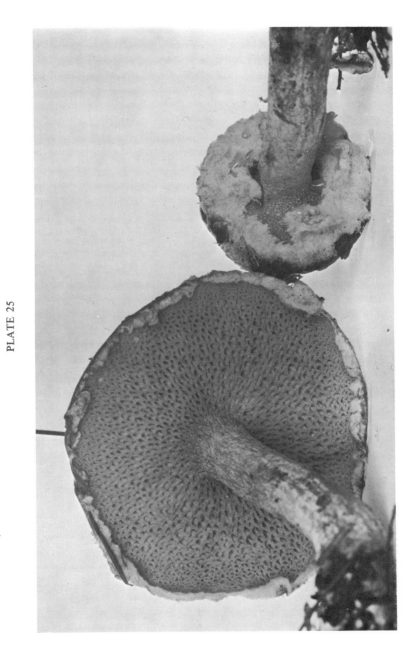

Suillus americanus   × 1-1/2

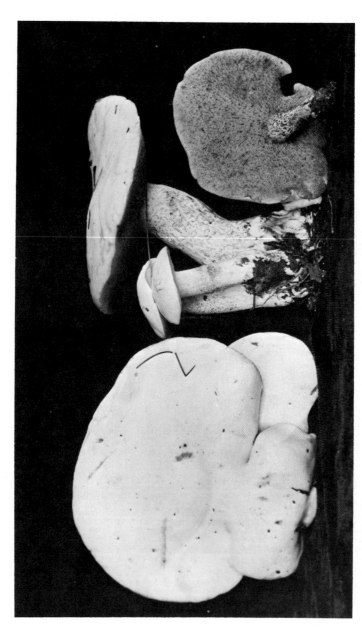

PLATE 26

Smith 58177

Suillus placidus　　　× 1

PLATE 27

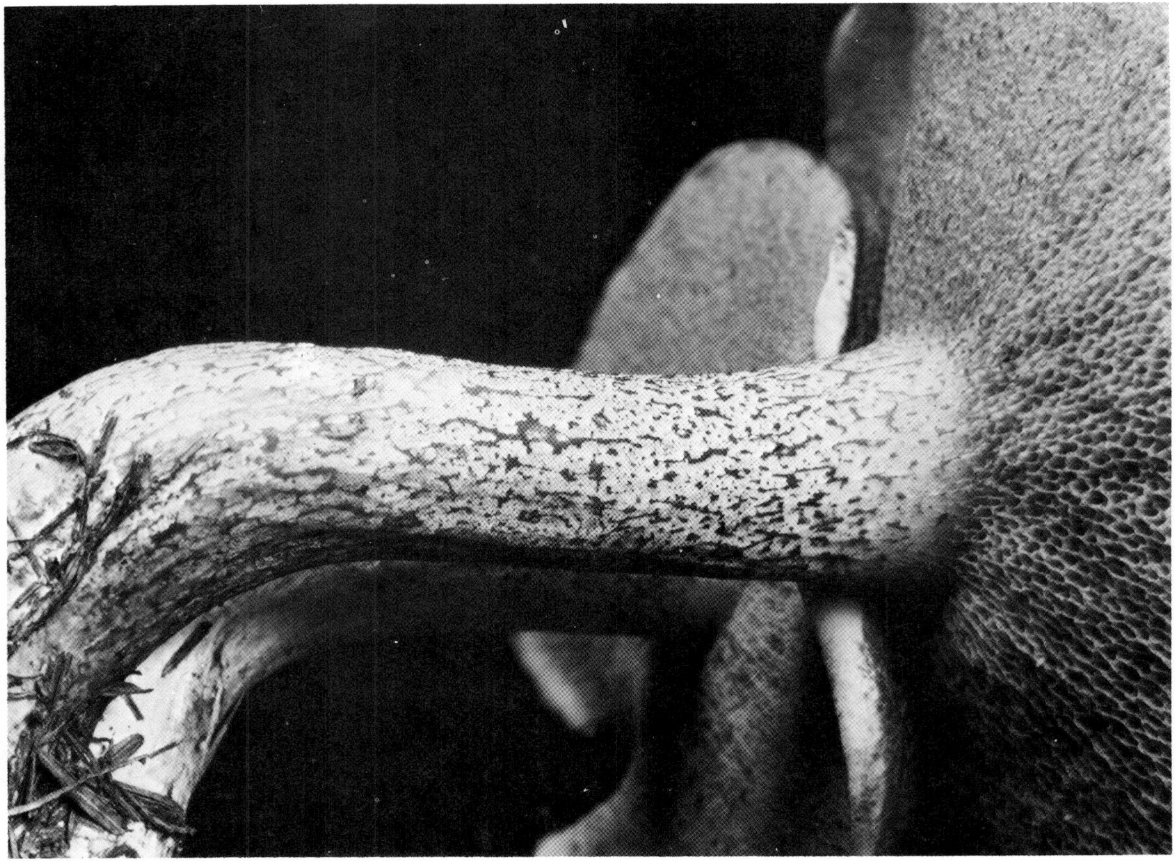

Suillus placidus   × 2

PLATE 28

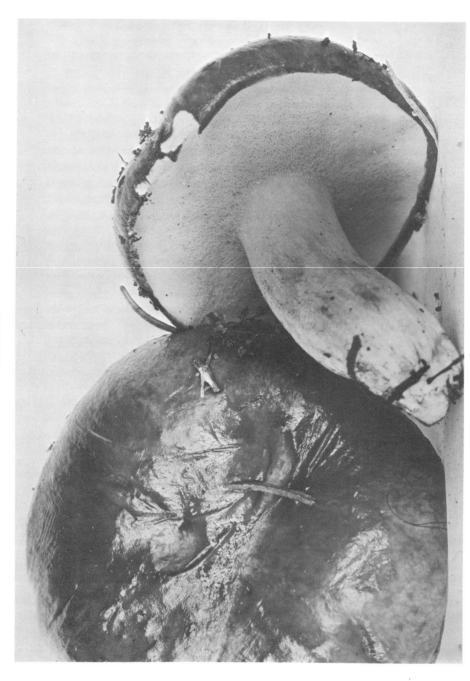

Suillus brevipes × 1

PLATE 29

Suillus brevipes var. *subgracilis*   × 1

Smith 75376

PLATE 30

Suillus lactifluus　　× 1　　　　　　　　　　Photo Smith

Suillus lactifluus　　× 1　　　　　　　　　　**Photo Smith**

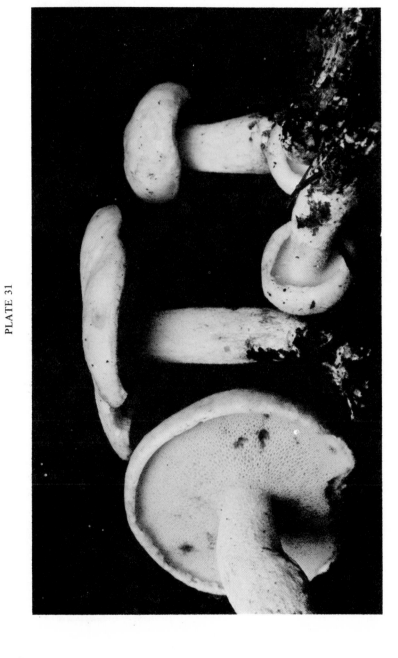

PLATE 31

Suillus unicolor   × 1

Smith 74648

PLATE 32

Smith 54075

Suillus granulatus   x 1

PLATE 33

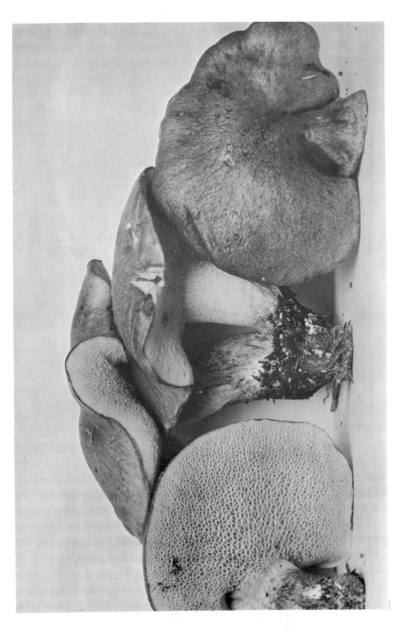

Suillus punctatipes   x 1/2

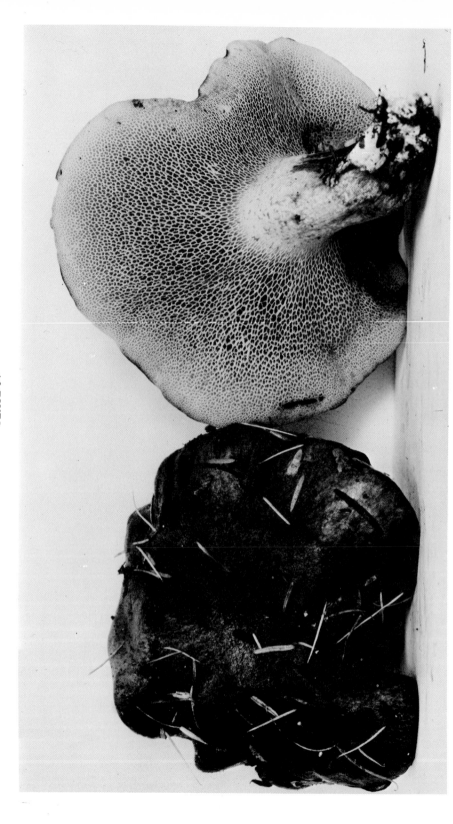

PLATE 34

Smith 70833

Suillus punctatipes   x 1/2

PLATE 35

Fuscoboletinus paluster   x 1

PLATE 36

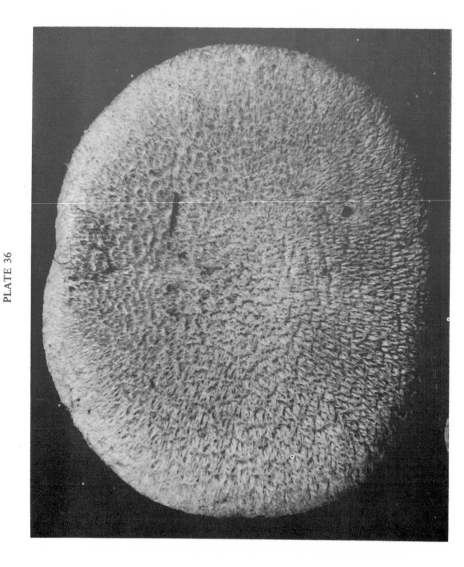

Fuscoboletinus ochraceoroseus   × 1

PLATE 37

Fuscoboletinus ochraceoroseus   × 1

Smith 44938

PLATE 38

Fuscoboletinus glandulosus   × 1

Smith 36874

PLATE 39

Fuscoboletinus glandulosus   x 1

K. A. Harrison 8452

PLATE 40

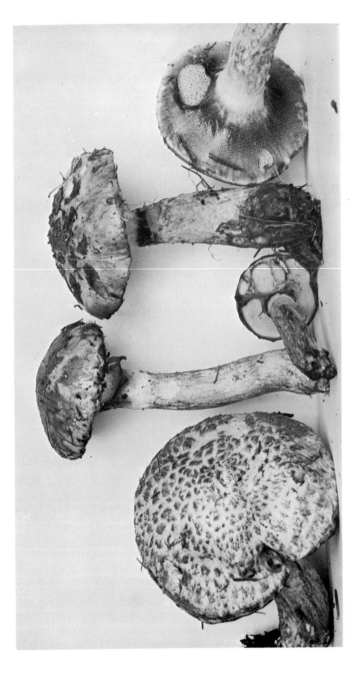

Smith 11016

Fuscoboletinus spectabilis   × 1

PLATE 41

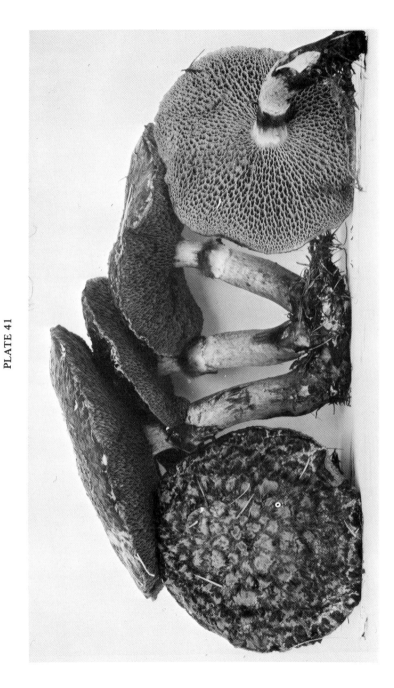

Fuscoboletinus spectabilis × 1

PLATE 42

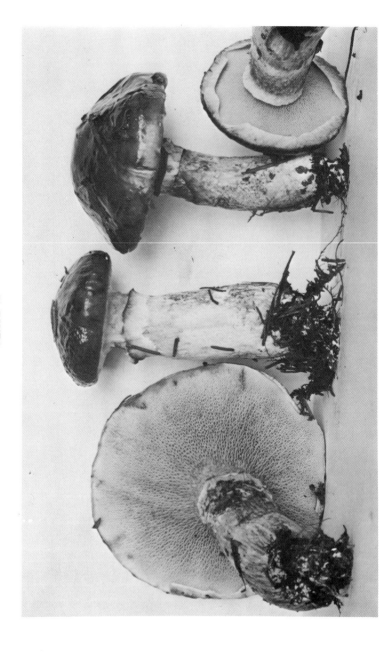

Fuscoboletinus aeruginascens   × 1

Smith 34178

PLATE 43

Fuscoboletinus grisellus   × 1   Smith 64652

...scoboletinus grisellus   × 1   Smith 34250

PLATE 44

Tylopilus chromapes   x 1

Smith 72375

PLATE 45

Smith 63009

Tylopilus chromapes × 1

PLATE 46

Tylopilus gracilis   × 1

Smith 62970

Tylopilus pseudoscaber   × 1

Smith 77126

PLATE 48

Tylopilus porphyrosporus   x 1/2   Smith 14185

PLATE 49

Tylopilus porphyrosporus   x 1

PLATE 50

Tylopilus sordidus × 1

Smith 64188

PLATE 51

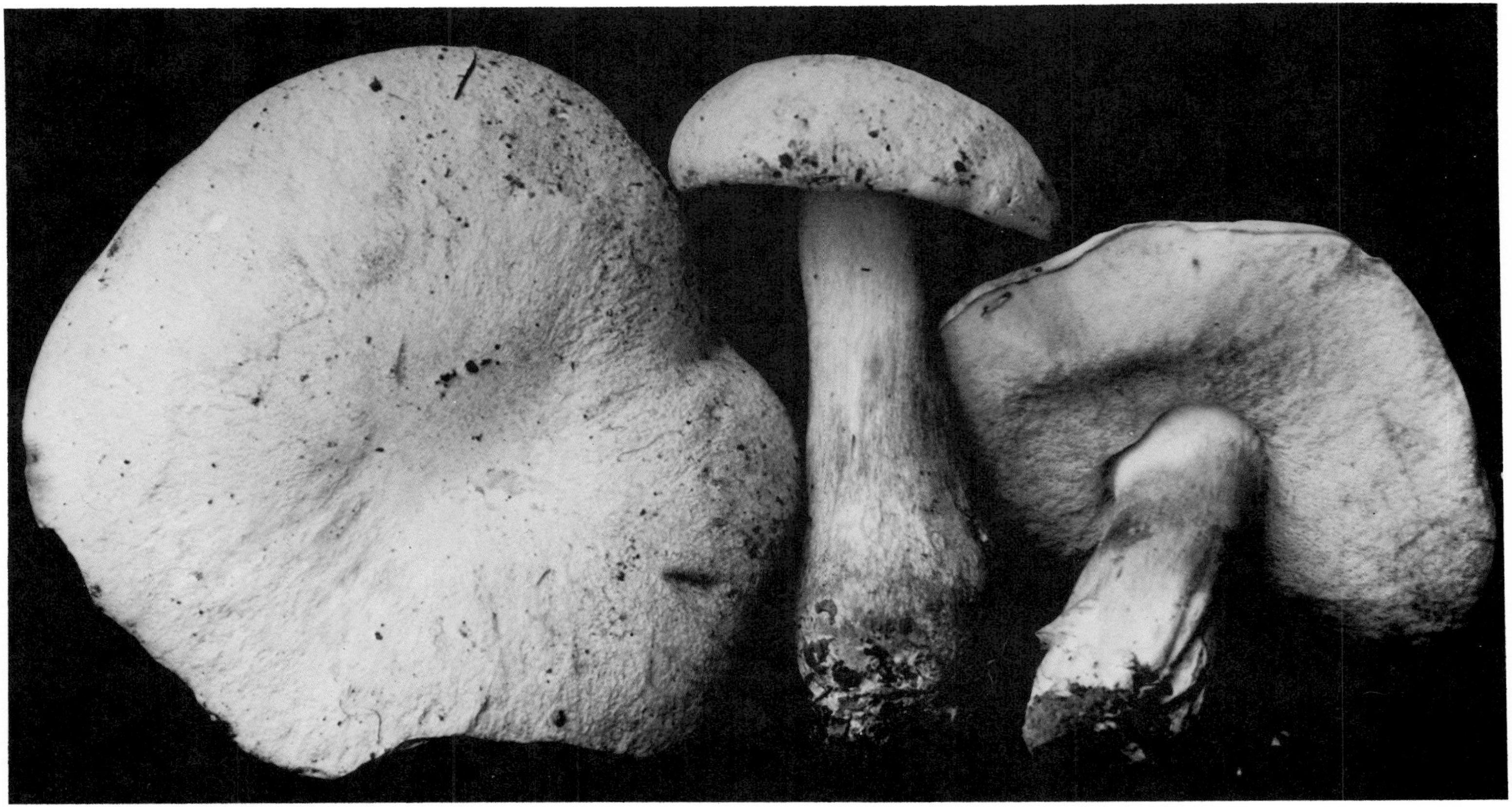

Tylopilus intermedius × 1

PLATE 52

Tylopilus plumbeo-violaceus   × 1   Smith 73296

Tylopilus plumbeo-violaceus   × 1   Smith 73211

PLATE 53

Tylopilus felleus   × 1

Smith 71892

PLATE 54

Tylopilus felleus   × 1

Smith 25983

PLATE 55

Tylopilus rubrobrunneus   × 1

Smith 62543

PLATE 56

Photo Smith

Tylopilus rubrobrunneus   × 1/2

PLATE 57

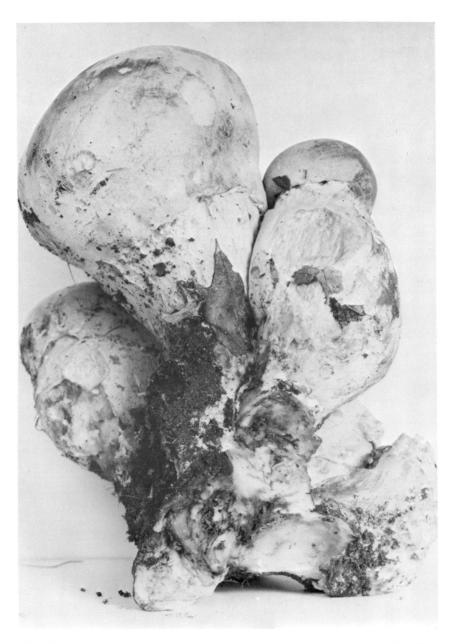

Tylopilus rubrobrunneus   × 1   Parasitized

PLATE 58

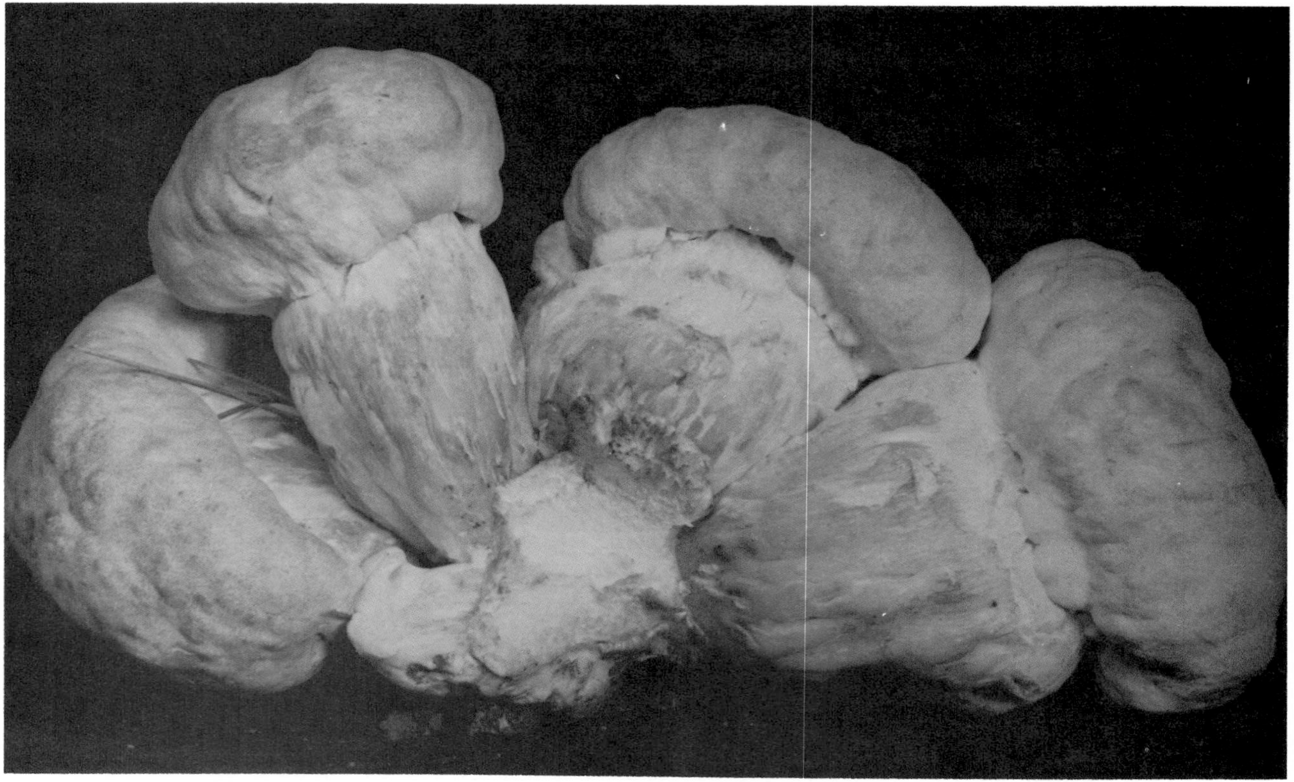

Tylopilus rubrobrunneus   × 1   Parasitized

PLATE 59

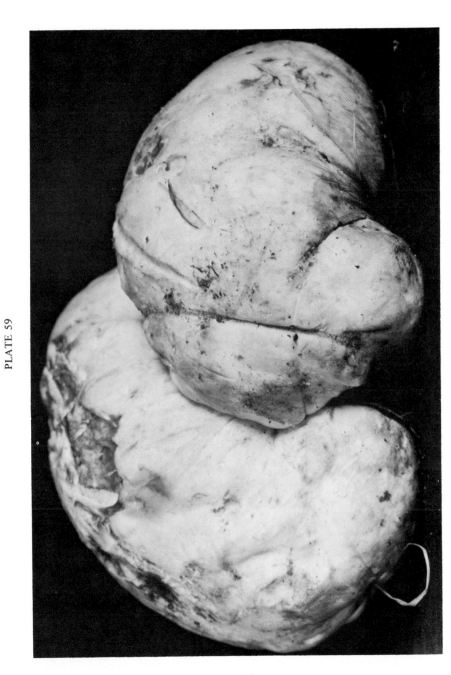

Tylopilus rubrobrunneus   × 1

Parasitized

PLATE 60

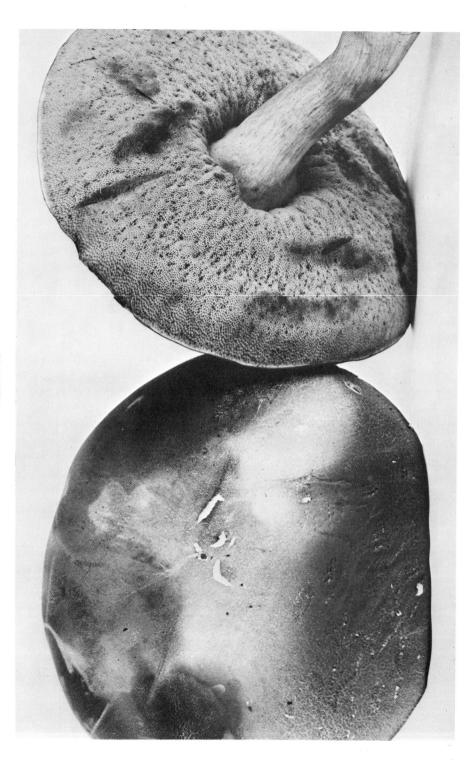

Tylopilus indecisus   × 1

PLATE 61

Smith 73247

Tylopilus subpunctipes   × 1

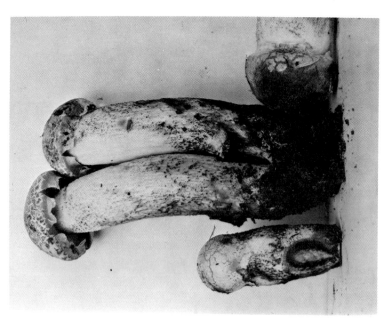

PLATE 62

Smith 73255   (young)   Smith 64492

Leccinum potteri   × 1   *Leccinum potteri*   × 1

PLATE 63

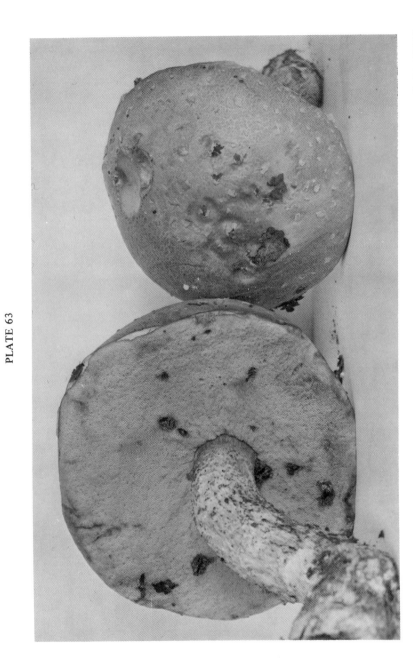

Leccinum potteri   x 1

Smith 72529

PLATE 64

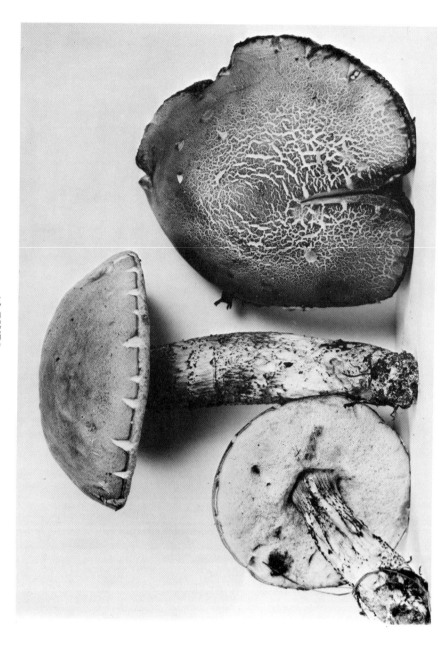

Smith 75837

Leccinum uliginosum   × 1

PLATE 65

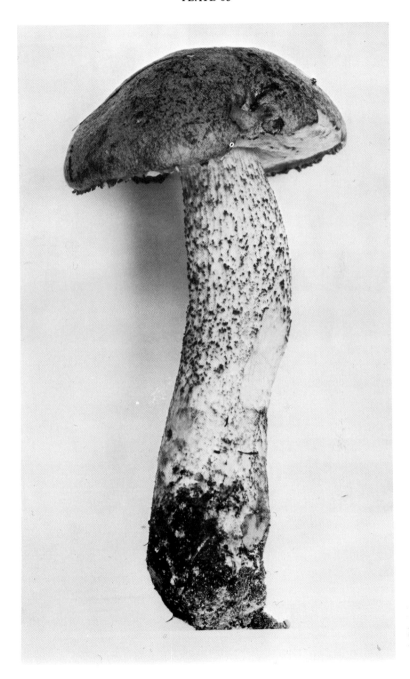

Leccinum obscurum   × 1   Smith 72566

PLATE 66

Smith 74341

Leccinum subspadiceum   × 1

PLATE 67

Leccinum ambiguum × 1

Smith 73371

PLATE 68

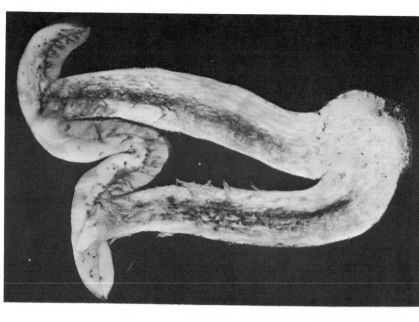

Leccinum atrostipitatum   × 1/2+   N. J. Smith 003

Leccinum atrostipitatum   × 1-   Photo A. H. Smith

PLATE 69

Leccinum atrostipitatum   x 1/2

Smith 71855

PLATE 70

Leccinum aurantiacum var. *aurantiacum*　　× 1

PLATE 71

Leccinum cinnamomeum var. *cinnamomeum*   x 1

PLATE 72

Leccinum ochraceum   x 1   Smith 72411

PLATE 73

Leccinum subtestaceum var. *subtestaceum*   × 1

Smith 58018

PLATE 74

Leccinum subtestaceum var. *subtestaceum* × 1 Photo Smith

PLATE 75

Leccinum insolens var. *insolens*   × 1   Smith 74546

PLATE 76

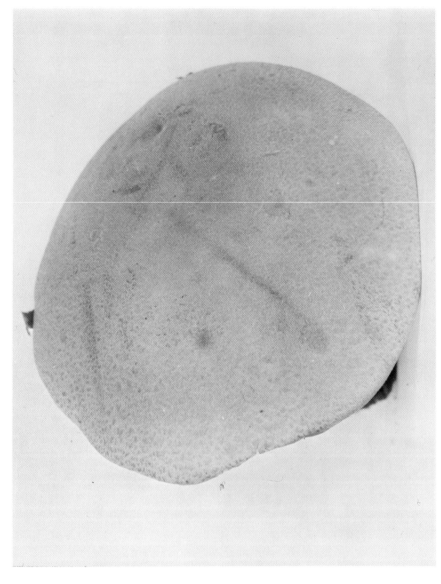

Leccinum insigne var. *insigne*   × 1

PLATE 77

Leccinum insigne   × 1

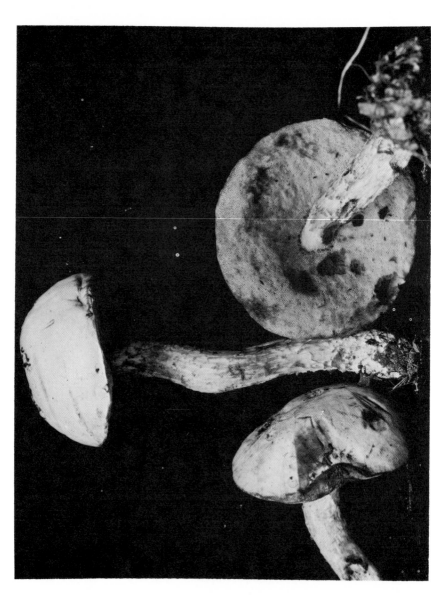

PLATE 78

Leccinum olivaceopallidum　　×　1

PLATE 79

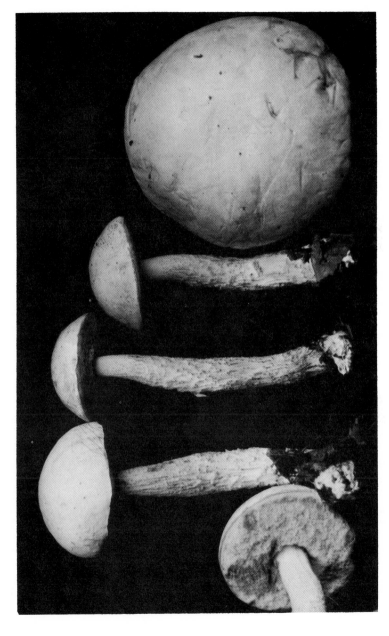

Leccinum holopus var. *holopus*   × 1

Smith 67636

PLATE 80

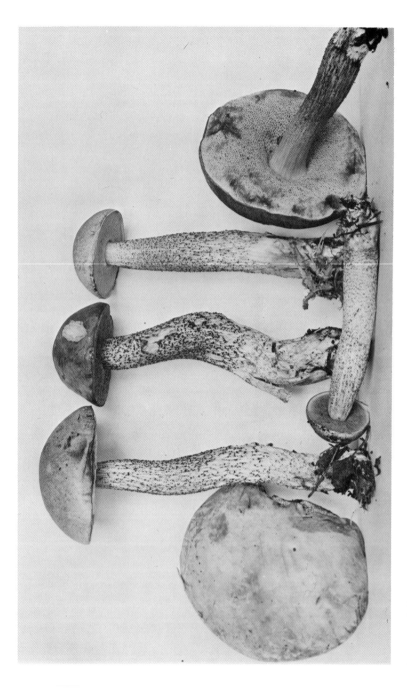

Smith 57894

Leccinum holopus var. *americanum*   x 1

PLATE 81

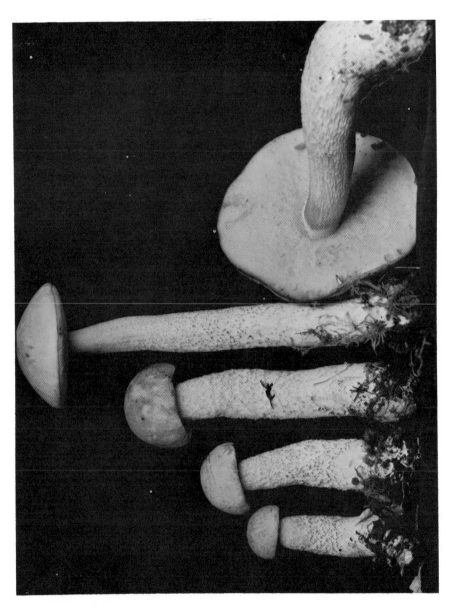

Leccinum variabile × 1

PLATE 82

Leccinum olivaceoglutinosum   x 1

PLATE 83

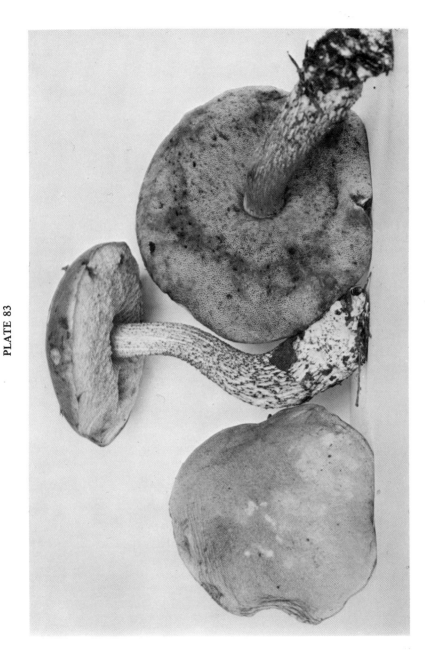

Leccinum subleucophaeum   × 1

PLATE 84

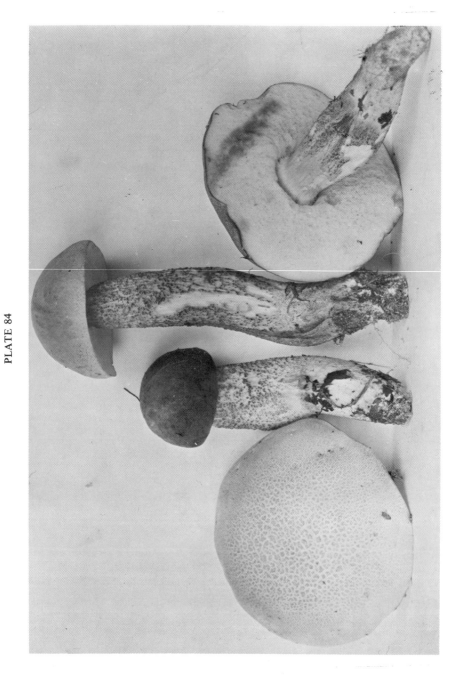

Leccinum griseonigrum × 1

PLATE 85

Leccinum disarticulatum × 1

PLATE 86

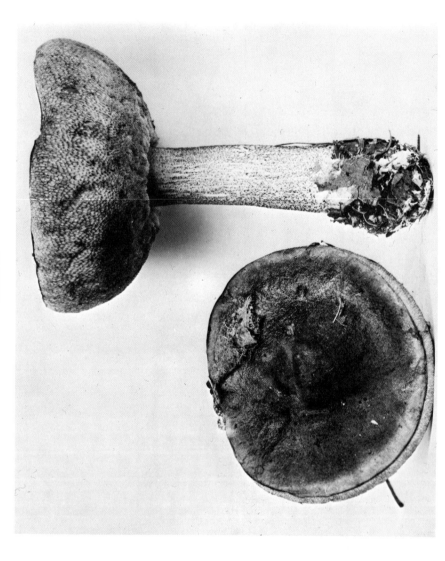

Smith 75336

Leccinum coffeatum   × 1

PLATE 87

Leccinum rimulosum   x 1

Smith 75513

PLATE 88

Leccinum scabrum var. *scabrum*   × 1

Smith 66882

PLATE 89

Leccinum snellii   × 1   Smith 72103

Leccinum snellii   × 1   Smith 72103

PLATE 90

Leccinum griseum   × 1

PLATE 91

Leccinum luteum   × 1

Smith 73252

PLATE 92

Boletus sphaerocephalus   × 1   Smith 67090

PLATE 93

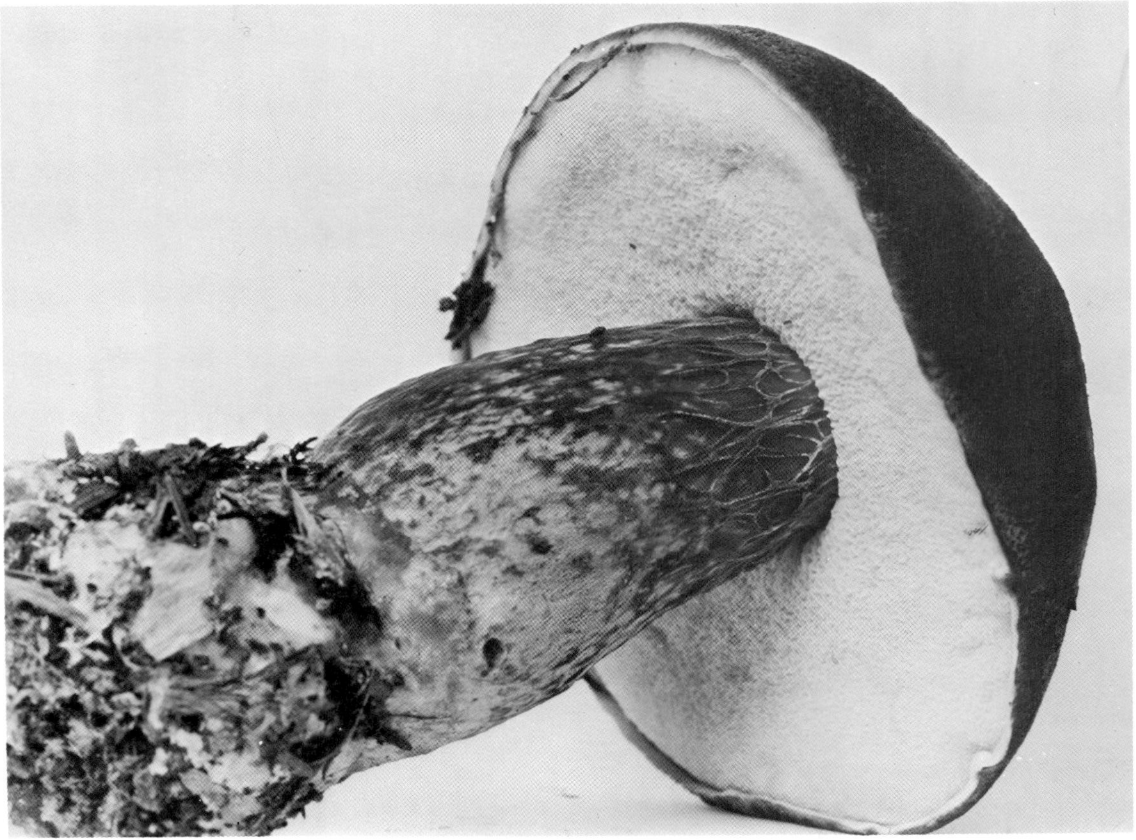

Boletus mirabilis   x 1

PLATE 94

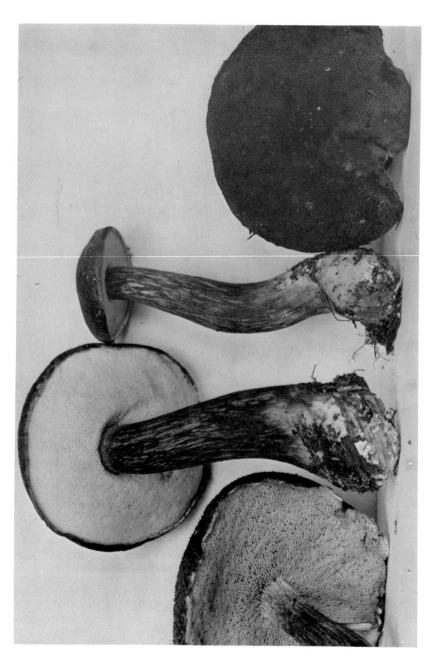

Smith 70582

Boletus mirabilis   × 1

PLATE 95

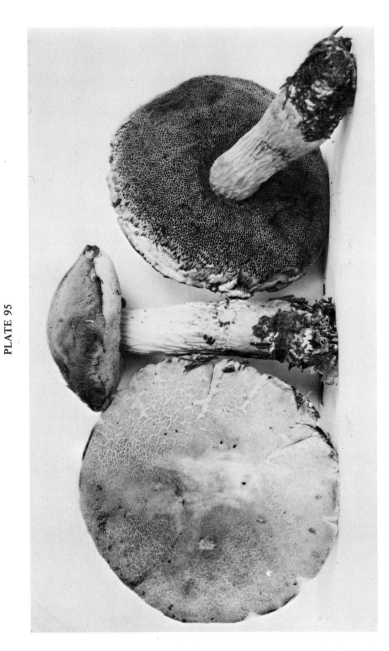

Boletus projectellus × 1

Smith 67677

PLATE 96

Boletus parasiticus　×　1　　　　　Smith 64208

Boletus parasiticus　×　1　　　　　Photo Smith

PLATE 97

Boletus affinis var. *affinis*   x 1

PLATE 98

Boletus roxanae   × 1   Smith 64140

PLATE 99

Smith 75298

Boletus subpalustris   x 1+

PLATE 100

Boletus illudens   × 1

PLATE 101

Boletus affinus x 1 Smith 75735

Boletus minutiporus x 1 Smith 75467

PLATE 102

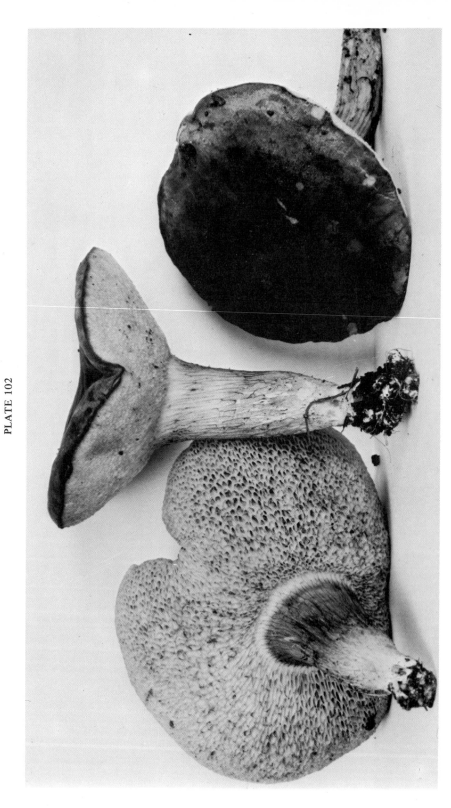

Smith 72556

Boletus tenax   x 1

PLATE 103

Boletus tenax   × 1/2   Smith 72556

Boletus subtomentosus var. *perplexus*   × 1   Photo Smith

PLATE 104

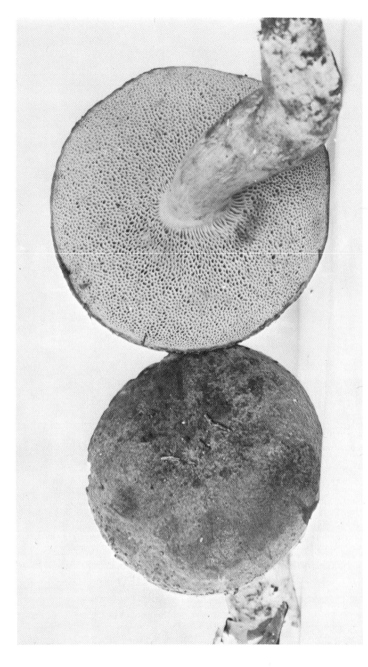

Boletus subtomentosus var. *subtomentosus*   x 1-

Smith 67195a

PLATE 105

Boletus chrysenteron   x 1   Smith 63008

PLATE 106

Boletus patriciae   x 1   Smith 73236

Boletus campestris   x 1   Smith 72961

PLATE 107

Boletus subdepauperatus   × 1   Smith 75592

Boletus rubellus var. *flammeus*   × 1   Smith 73026

PLATE 108

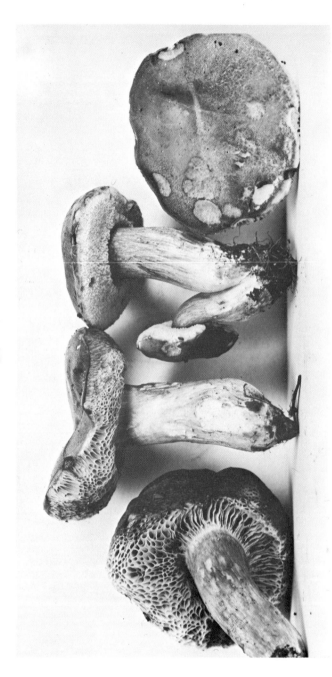

Photo Smith

Boletus subfraternus   x 1

PLATE 109

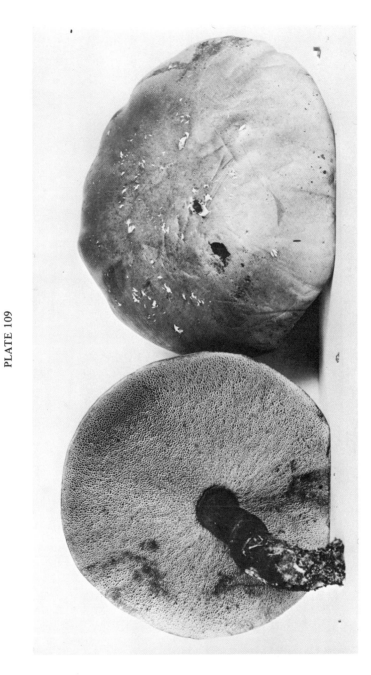

Smith 75969

Boletus bicolor   x 1

PLATE 110

Boletus pseudosensibilis   × 1

PLATE 111

PLATE 112

Smith 75744

Boletus sensibilis   × 1

PLATE 113

Boletus truncatus   × 1

Smith 67089

PLATE 114

Smith 38254

Boletus piperatus   x 1

Boletus amarellus   × 1   Photo Smith

Boletus piperatus   × 1   Smith 72164

PLATE 116

Boletus pallidus   × 1

PLATE 117

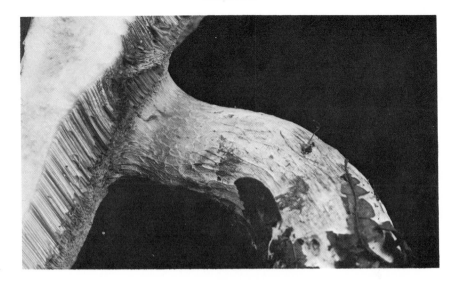

Boletus pallidus   x 1   Smith 75912

Boletus huronensis   x 1   Smith 76005

PLATE 118

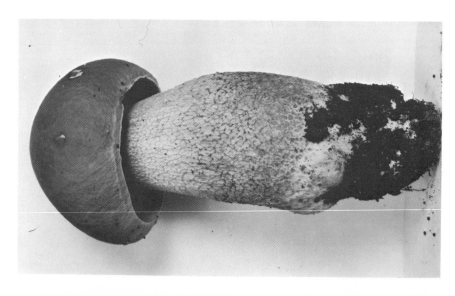

Boletus huronensis   × 1   Smith 76005   Boletus separans   × 1   Smith 75783

PLATE 119

Smith 63817

Boletus badius   x 1

PLATE 120

Photo Smith

Boletus badius   x 1

PLATE 121

Boletus pseudosulphureus   × 1

PLATE 122

Boletus pulverulentus   × 1

PLATE 123

Boletus pulverulentus   x 1

Smith 73303

PLATE 124

Boletus subglabripes   × 1

PLATE 125

Boletus flavorubellus   × 1   Smith 62872

Boletus longicurvipes   × 1   Smith 10320

PLATE 126

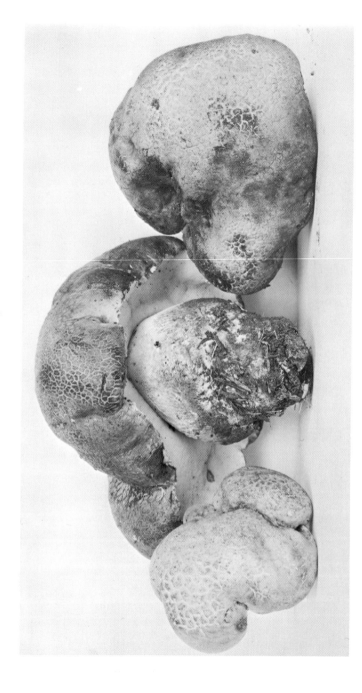

Boletus calopus   x 1/2

PLATE 127

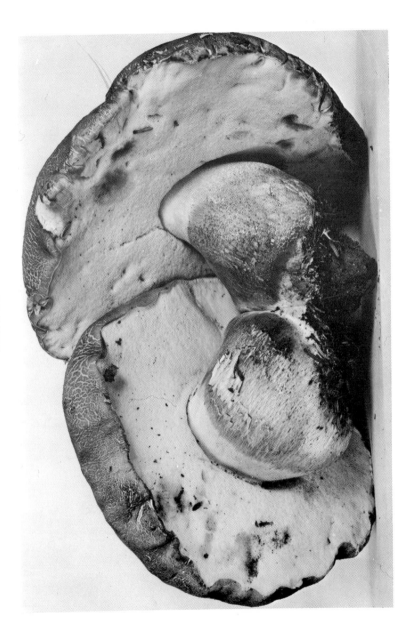

Photo Smith

Boletus calopus   x 1

PLATE 128

Smith 73003

Boletus inedulis   × 1

PLATE 129

Boletus inedulis   × 1   Smith 64141

PLATE 130

Smith 64160

Boletus speciosus   x 1

PLATE 131

Boletus ornatipes   × 1

PLATE 132

Boletus griseus   × 1

PLATE 133

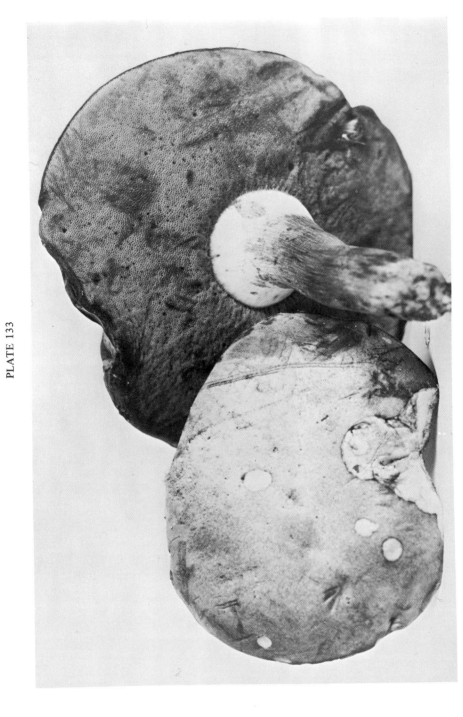

Boletus subgraveolens   × 1

PLATE 134

Smith 64046

Boletus vermiculosoides   × 1

PLATE 135

Boletus vermiculosoides × 1

PLATE 136

Boletus frostii   × 1

PLATE 137

Boletus frostii   x 1

PLATE 138

Smith 15352

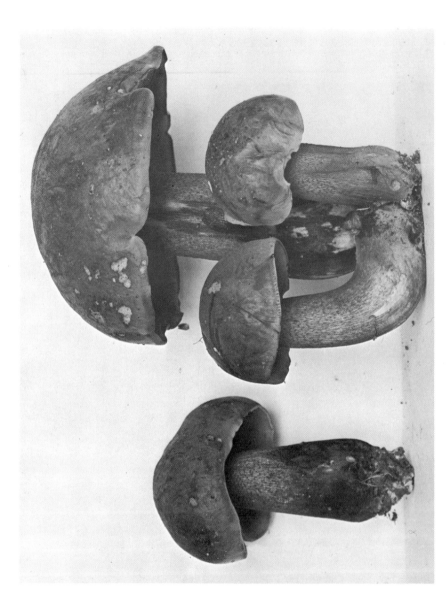

Boletus rubroflammeus   x 1

PLATE 139

Boletus luridus   x 1

PLATE 140

Boletus luridus × 1

PLATE 141

Boletus pseudo-olivaceus × 1

Smith 74510

PLATE 142

Boletus pseudo-olivaceus   × 1   Smith 74510

PLATE 143

Boletus pseudo-olivaceus   x 1

Smith 75318

PLATE 144

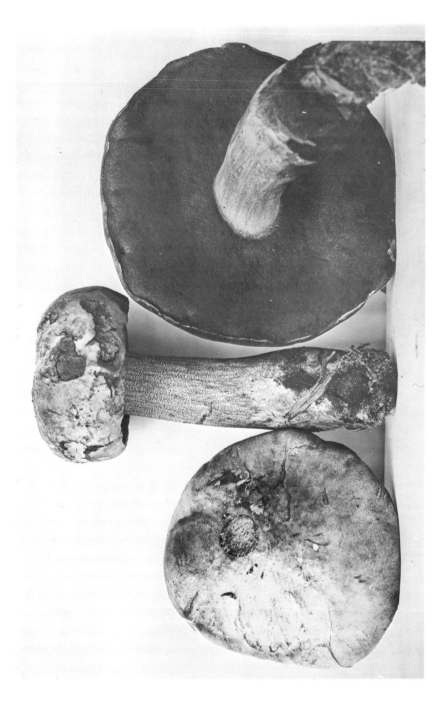

Boletus erythropus f. *michiganensis*   × 1

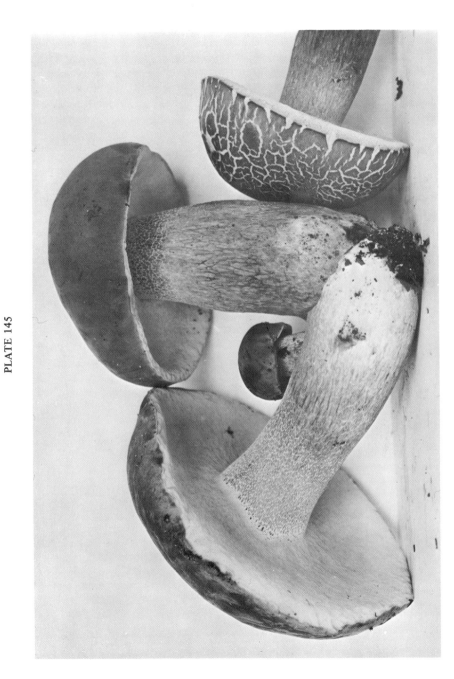

PLATE 145

Smith 63758

Boletus varipes   × 1

PLATE 146

Boletus variipes var. *fagicola*   × 1   Smith 75743

PLATE 147

Boletus edulis   × 1

Photo Smith

PLATE 148

Boletus betulae   × 1   Smith 10314

PLATE 149

Pulveroboletus ravenelii   × 1

PLATE 150

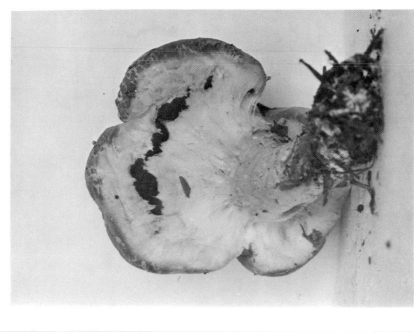

Smith 67856 *Pulveroboletus ravenelii* × 1

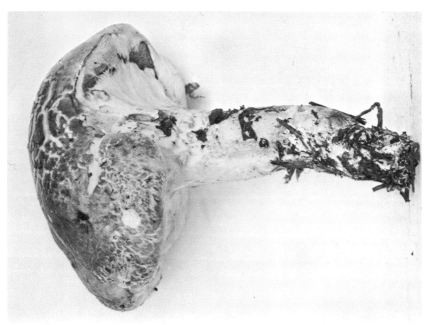

Pulveroboletus ravenelii × 1 Smith 63856

PLATE 151

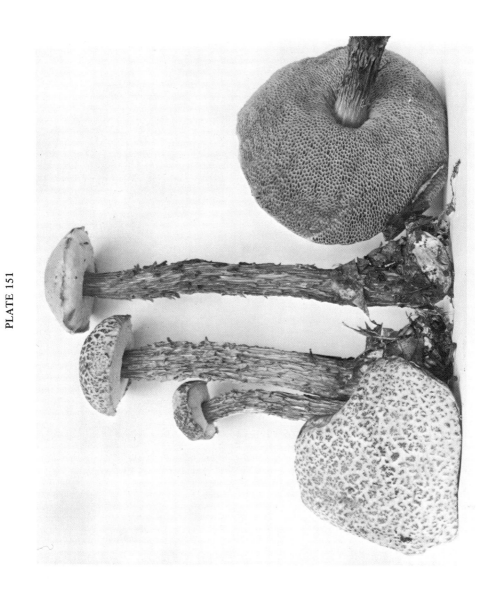

Boletellus russellii   x 1/2

PLATE 152

Smith 64264

Boletellus pseudo-chrysenteroides   x 1

PLATE 153

Boletellus chrysenteroides   × 1

PLATE 154

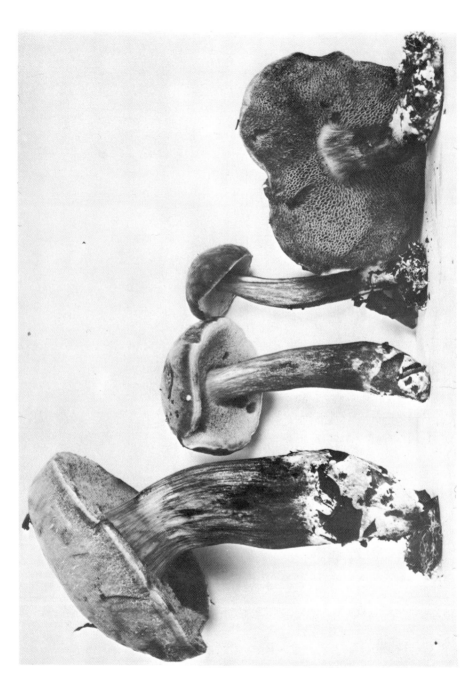

Boletellus intermedius   × 1

Smith 72559

PLATE 155

Strobilomyces floccopus   × 1

Smith 63021

PLATE 156

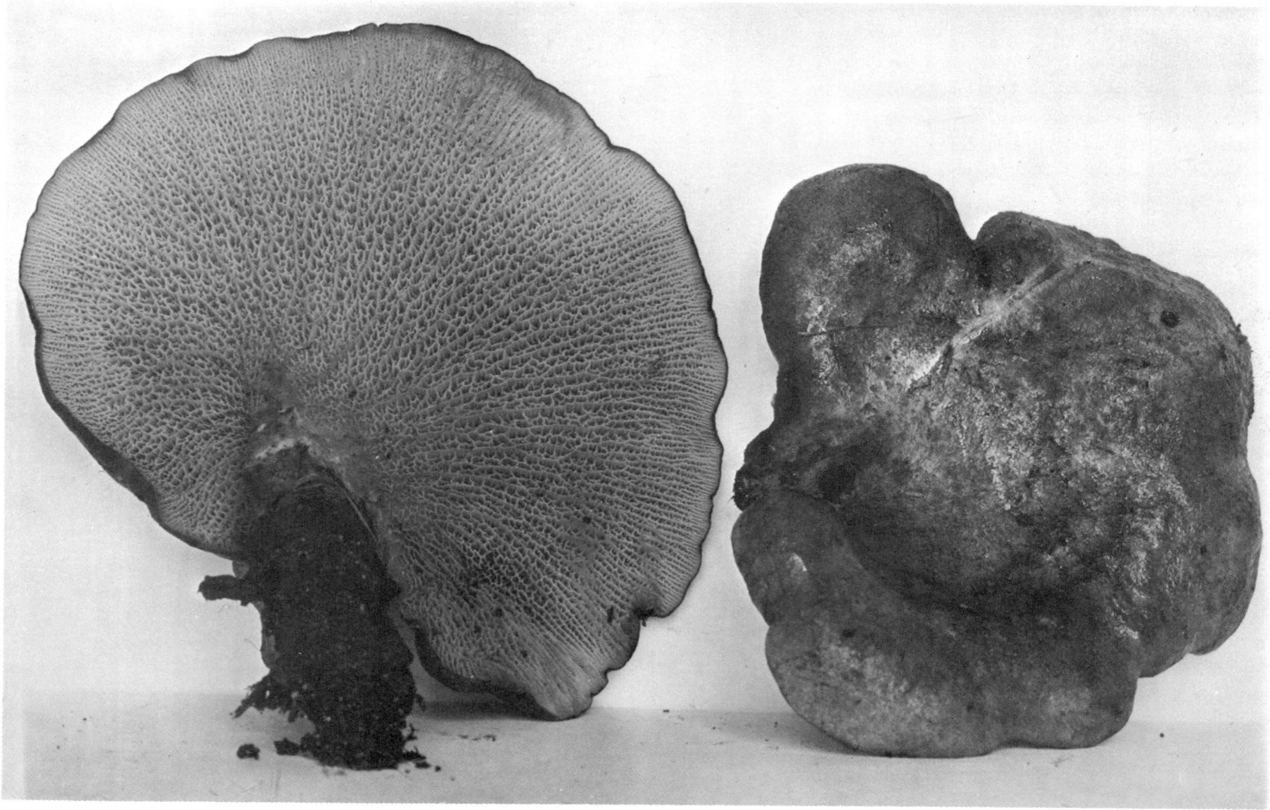

Boletinellus merulioides   x 1

PLATE 157

Gastroboletus scabrosus   × 1   Photo Smith

INDEX

Page numbers in boldface refer to page on which description occurs.